# MOLECULAR STRUCTURE AND ENERGETICS

## Volume 4

### Biophysical Aspects

# MOLECULAR STRUCTURE AND ENERGETICS

## Volume 4

### Series Editors
## Joel F. Liebman
*University of Maryland Baltimore County*

## Arthur Greenberg
*New Jersey Institute of Technology*

### Advisory Board

© 1987 VCH Publishers, Inc. New York, New York

Distribution: VCH Verlagsgesellschaft mbH, P.O. Box 1260/1280, D-6940 Weinheim,
    Federal Republic of Germany

USA and Canada: VCH Publishers, Inc., 303 N.W. 12th Avenue, Deerfield Beach, FL 33442-1705, USA

# MOLECULAR STRUCTURE AND ENERGETICS

## Volume 4

### Biophysical Aspects

Edited by

**Joel F. Liebman**
**and**
**Arthur Greenberg**

Joel F. Liebman
Department of Chemistry
University of Maryland Baltimore County
Catonsville, Maryland 21228

Arthur Greenberg
Division of Chemistry
New Jersey Institute of Technology
Newark, New Jersey 07102

**Library of Congress Cataloging-in-Publication Data**

Molecular structure and energetics

  Includes bibliographies and indexes
  Contents: v. 1 Chemical bonding models --
  v. 2 Physical measurements --
  [etc]. --
  v. 4 Biophysical aspects
  1. Molecular structure—Collected works.
2. Molecular dynamics—Collected works
I. Liebman, Joel F.   II. Greenberg, Arthur
QD461.M629   1987      541.2′2      86-19250
ISBN 0-89573-336-6

Printed in the United States of America.

ISBN-0-89573-336-6   VCH Publishers
ISBN 3-527-26478-7   VCH Verlagsgesellschaft

*To Ellis H. Rosenberg (1936–1986)*
*publisher and friend*

# SERIES FOREWORD

Molecular structure and energetics are two of the most ubiquitous, fundamental and, therefore, important concepts in chemistry. The concept of molecular structure arises as soon as even two atoms are said to be bound together since one naturally thinks of the binding in terms of bond length and interatomic separation. The addition of a third atom introduces the concept of bond angles. These concepts of bond length and bond angle remain useful in describing molecular phenomena in more complex species, whether it be the degree of pyramidality of a nitrogen in a hydrazine, the twisting of an olefin, the planarity of a benzene ring, or the orientation of a bioactive substance when binding to an enzyme. The concept of energetics arises as soon as one considers nuclei and electrons and their assemblages, atoms and molecules. Indeed, knowledge of some of the simplest processes, eg, the loss of an electron or the gain of a proton, has proven useful for the understanding of atomic and molecular hydrogen, of amino acids in solution, and of the activation of aromatic hydrocarbons on airborne particulates.

Molecular structure and energetics have been studied by a variety of methods ranging from rigorous theory to precise experiment, from intuitive models to casual observation. Some theorists and experimentalists will talk about bond distances measured to an accuracy of 0.001 Å, bond angles to 0.1°, and energies to 0.1 kcal/mol and will emphasize the necessity of such precision for their understanding. Yet other theorists and experimentalists will make equally active and valid use of such seemingly ill-defined sources of information as relative yields of products, vapor pressures, and toxicity. The various chapters in this book series use as their theme "Molecular Structure and Energetics", and it has been the individual authors' choice as to the mix of theory and of experiment, of rigor and of intuition that they have wished to combine.

As editors, we have asked the authors to explain not only "what" they know but "how" they know it and explicitly encouraged a thorough blending of data and of concepts in each chapter. Many of the authors have told us that writing their chapters have provided them with a useful and enjoyable (re)education. The chapters have had much the same effect on us and we trust readers will share our enthusiasm. Each chapter stands autonomously as a combined review and tutorial of a major research area. Yet clearly there are interrelations between them and to emphasize this coherence we have tried to have a single theme in each volume. Indeed the first four volumes of this series were written in parallel, and so for these there is an even higher degree of unity. It is this underlying unity of molecular structure and energetics with all of chemistry that marks the series and our efforts.

Another underlying unity we wish to emphasize is that of the emotions and of the intellect. We thus enthusiastically thank Alan Marchand for the opportunity to write a volume for his book series, which grew first to multiple volumes, and then became the current, autonomous series for which this essay is the foreword. We likewise thank Marie Stilkind of VCH Publishers for her versatility, cooperation and humor. We also wish to emphasize the support, the counsel, the tolerance and the encouragement we have long received from our respective parents, Murray and Lucille, Murray and Bella; spouses, Deborah and Susan; parents-in-law, Jo and Van, Wilbert and Rena; and children, David and Rachel. Indeed, it is this latter unity, that of the intellect and of emotions, that provides the motivation for the dedication for this series:

*"To Life, to Love, and to Learning".*

Joel F. Liebman
Baltimore, Maryland

Arthur Greenberg
Newark, New Jersey

# PREFACE

This volume relates selected aspects of physical and organic chemistry to biological activity at a molecular level. The emphasis, of course, is on those properties related to molecular structure and energetics.

Chapter 1, by Greenberg and Darack, summarizes the known chemistry of polycyclic aromatic hydrocarbons (PAH) under ambient atmospheric conditions. It relates basic molecular properties, such as shape, ionization potential and proton affinity, to results of atmospheric simulation studies and both approaches to real-life studies. While numerous PAH are carcinogenic, many of their derivatives are even more bioactive and the origin and fate of these compounds has obvious health implications.

Chapter 2, by Liebman, introduces the reader to some "back-of-the-envelope" techniques for estimation of gas-phase proton affinities for small molecules. Although the methods are not highly accurate, they provide good qualitative predictions, models which themselves shed light on the protonated species and allow one to understand why one protonation site is preferred over another in a polyfunctional molecule.

Chapter 3, by Meot-Ner (Mautner), and Chapter 4, by Deakyne, compare respectively experimental and theoretical studies of ionic hydrogen bonds. Aside from providing an understanding of the relative strengths of general hydrogen bonds, such studies represent the beginnings of quantitative knowledge of solution-phase chemistry since solvent clusters built around ions are the first steps to their solvation.

The fifth chapter, by Katz, applies the principles of ionic hydrogen bonding to studies of interactions modeled for serine protease and a substrate. The relationships between enthalpy of binding and rate of catalysis are explored in order to better understand the true catalytic role of an enzyme.

Chapter 6, by George, Bock and Trachtman, continues the modeling of biologically-interesting molecules as they define and quantitate empirical resonance energies for amides, esters, anhydrides and other conjugated species. The subsequent chapter, by Anteunis and Sleeckx, delves into the conformational analysis of peptides with emphasis on proline-containing molecules. Proline is well known to introduce considerable rigidity and other interesting features. The chapter concludes with the symbols, nomenclature and conformation notation currently employed in protein chemistry.

Linear free energy relationships (LFER) have long been employed to understand organic reaction mechanisms while their application to deducing enzyme mechanisms and understanding drug action is more recent. Chapter 8, by Topsom, describes his group's approach toward understanding the individual electronic components of substituent effects. The methodology rests upon ab initio molecular orbital calculations and provides qualitative insight as well as quantitative predictions of molecular properties. In chapter 9, Charton defines a new set of substituent constants useful for describing the behavior of groups attached to cationic carbon. Photo-electron and mass spectroscopic data have been employed to derive these numbers.

Chapter 10, by T. Gund and P. Gund, outline the basic aspects of computer modeling of molecules. This technique has found particularly extensive application in the design of pharmaceuticals. Their chapter also lists, without specific endorsements or claim of exhaustive coverage, hardware and software available in the 1985 marketplace. The volume concludes with a chapter by Hansch in which he unifies correlation analysis, LFER, and molecular modeling with the study of enzyme structure and activity.

When a picture of the active site drawn by a computer merges with one deduced by application of correlation analysis, when experimental or theoretical results on gas-phase species are used to aid the understanding of solution effects and macromolecules, the unified outlook is esthetically pleasing and scientifically sound.

# CONTRIBUTORS

Marc J. O. Anteunis
Laboratory for Organic Chemistry
State University of Ghent
Krijgslaan 281 (S-4bis)
B-9000 Ghent
Belgium

Charles W. Bock
Chemistry Department
Philadelphia College of Textiles and Science
Philadelphia, Pennsylvania 19144

Marvin Charton
Chemistry Department
School of Liberal Arts and Sciences
Pratt Institute
Brooklyn, New York 11205

Faye B. Darack
Chemistry Division
New Jersey Institute of Technology
Newark, New Jersey 07102

Carol A. Deakyne
Department of Chemistry
Air Force Geophysics Laboratory
Hanscom Air Force Base, Massachusetts 01731

Philip George
Biology Department
University of Pennsylvania
Philadelphia, Pennsylvania 19104

Arthur Greenberg
Department of Chemistry
New Jersey Institute of Technology
Newark, New Jersey 07102

Peter Gund
Chemical and Biological Systems
Merck Sharp & Dohme Research Laboratories
Rahway, New Jersey

Tamara Gund
Department of Chemistry
New Jersey Institute of Technology
Newark, New Jersey 07102

Corwin Hansch
Department of Chemistry
Pomona College
Claremont, California 91711

Howard Edan Katz
AT&T Bell Laboratories
Murray Hill, New Jersey 07974

Joel F. Liebman
Department of Chemistry
University of Maryland Baltimore County
Catonsville, Maryland 21228

Michael Meot-Ner (Mautner)
Center for Chemical Physics
National Bureau of Standards
Gaithersburg, Maryland 20899

Jef J. M. Sleeckx
Laboratory for Organic Chemistry
State University of Ghent
Krijgslaan 281 (S-4bis)
B-9000 Ghent
Belgium

Ronald D. Topsom
Department of Chemistry
La Trobe University
Bundoora
Melbourne, Victoria 3083
Australia

Mendel Trachtman
Chemistry Department
Philadelphia College of Textiles and Science
Philadelphia, Pennsylvania 19144

# CONTENTS

# Atmospheric Reactions and Reactivity Indices of Polycyclic Aromatic Hydrocarbons

## Arthur Greenberg and Faye B. Darack

**New Jersey Institute of Technology, Newark, New Jersey**

## CONTENTS

## 1. INTRODUCTION

Polycyclic aromatic hydrocarbons (PAH) are ubiquitous air pollutants arising from incomplete combustion of organic matter.[1-7] They are often included in a larger class termed polycyclic organic matter (POM), which contains PAH, their derivatives (nitro-PAH, quinones, etc.), as well as het-

MOLECULAR STRUCTURE AND ENERGETICS, Vol. 4

TABLE 1-1. Benzo[a]pyrene Emission Rates for Combustion Sources[10]

| Fuel type | User | Emission rate (ng/Btu) |
|---|---|---|
| Solid | | |
| Coal | Utilities | $5.6-7.0 \times 10^{-2}$ |
| Coal | Residential | 0.12–61 |
| Wood | Residential | 27–6300 |
| Oil | | |
| Heating | Residential | $< 4-5 \times 10^{-2}$ |
| Miscellaneous distillates | Commercial/Industrial | $< 2-4.7 \times 10^{-2}$ |
| Residual | Utilities/Commercial/Industrial | $4.3 \times 10^{-4}$ |
| Natural Gas | | |
| Heating | Residential | $< 2 \times 10^{-2}$ |
| Motor fuels | | |
| Gasoline | Autos/trucks | 0.6 |
| Diesel | Trucks/buses | 2.3 |

eroanalogues having nitrogen and sulfur as part of the ring system. Recently the term "PICs" has been employed to describe products of incomplete combustion, which include the PAH. PICs are considered to make the largest contribution to carcinogenicity attributable to air pollution.[8] The first association of cancer with an environmental pollutant was made in 1776 when Percival Pott attributed the abnormally high incidence of scrotal cancers in chimney sweeps to their exposure to chimney soot coupled with poor hygiene. We know now that the PAH are responsible for a large amount of the carcinogenic activity of this material. Considerable attention has been devoted to one particular PAH, benzo[a]pyrene (1, termed B[a]P), which is a potent carcinogen. Frequently, this single pollutant is measured as a generic index of the carcinogenicity of airborne particulate matter.[9] Table 1-1 lists B[a]P emission levels from a variety of environmentally significant sources.[10] Particularly noteworthy are the high levels from domestic wood and coal combustion, despite the large uncertainties in the numbers.

During the 1970s investigators realized that B[a]P, and presumably all other PAH, cannot directly react with DNA but do so via metabolites, such as the anti-7,8-diol-9,10-epoxide 2, termed "ultimate carcinogens."[11,12] Compound 2 is a potent alkylating agent, as are most chemicals capable of directly attacking DNA and proteins. The most widely held view today is that PAH capable of forming particularly stable (usually "bay-region") carbonium ions from their metabolites are most carcinogenic.[11]

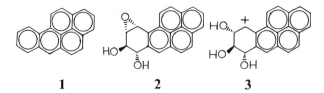

1                    2                    3

The importance of metabolic activation is emphasized by the observation that B[a]P and other PAH are not mutagenic in the Ames assay unless mammalian liver enzyme extract is added to the *Salmonella* culture.[13] In contrast, many other carcinogens, including some PAH derivatives, are direct mutagens requiring no metabolic activation. While the majority of carcinogens are mutagens,[13] the reverse statement is not necessarily true, and thus mutagenicity is not equivalent to carcinogenicity.

This chapter examines the reactivity of PAH under atmospheric and simulated atmospheric conditions. This issue is of concern to environmental scientists for many reasons, but for the present, the following rationale is presented for study of ambient PAH reactivity:

1. Study of lifetimes of airborne PAH will indicate whether chemical loss mechanisms or physical loss mechanisms (eg, Stokes sedimentation, wet removal) are limiting; knowledge of lifetimes of airborne PAH will lead to understanding of the efficacy of their long-range transport.

2. Many investigators have found "extra mutagenicity" in POM fractions beyond what is expected for PAH; apparently PAH derivatives are important contributors to this effect, most notably in direct (unactivated) mutagenicity.

3. Numerous PAH derivatives, including nitro compounds, quinones, and ketones, have been identified in airborne particulates. How does their formation relate to PAH reactivity?

4. Are the foregoing PAH derivatives formed in ambient air or upon generation of the parent in the presence of exhaust gases? Or, are they merely sampling artifacts? That is, is conversion catalyzed on the collection filter?

5. In studying long-range transport, the question of "aging" of the aerosol enters. Will PAH reactivity significantly alter the identities of the compounds in the aerosol? Are "aged" airborne particulates more or less carcinogenic than newly generated particulates?

6. Do different PAH sources (eg, automobiles, coal-fired furnaces) have different "PAH profiles" (ie, PAH ratios)? If so, can one monitor airborne particulate PAH and factor out different contributors, or does the atmosphere affect an equilibrium that tends to "lose the memory" of its sources?

7. Winter/summer PAH ratios are fairly large. Is this simply a result of greater atmospheric input during the winter heating season or is it primarily due to atmospheric decomposition, which should peak during the summer?

This chapter outlines the problems of PAH chemical reactivity and summarizes results of simulated environmental reactions. In addition, a number of indices of PAH reactivity are assembled and summarized. Usually, discussions of atmospheric reactivity of PAH do not include a summary of reactivity indices of PAH. However, by presenting both discussions in one chapter, we are able to beg a simple question: that is, should one examine the reactivities of many PAH under a few different reaction conditions or should one measure the reactivities of a selected few PAH under many conditions? Finally, we briefly examine some environmental data with reference to the questions raised above, to see what light such phenomenological data can shed on these questions.

## 2. PHYSICAL AND CHEMICAL PATHWAYS FOR LOSS OF POLYCYCLIC AROMATIC HYDROCARBONS

The stability of PAH under ambient conditions is a very important topic for environmental scientists. There are many reasons for this significance. First, the lifetime of airborne PAH governs the extent of their atmospheric transport, and this property is important in understanding, for example, the effect of a heavily industrialized area on the carcinogenic burden of a distant rural region. That there exists long-range PAH transport is indisputable. For example, Blumer had demonstrated the presence of background levels of B[a]P of 7 ppm in remote areas of the Swiss Alps,[14,15] while LaFlamme and Hites[16] found a decrease in PAH levels in sediments as a function of distance from urban areas, and Bjorseth and co-workers showed long-range PAH transport into Norway from industrial regions in Northern Europe.[17] The critical question in this regard is whether the dominant loss mechanism is physical or chemical. Most PAH are associated with fine particulate matter ($<1$ $\mu$m in aerodynamic diameter) as a result of their intimate association with particulates immediately after combustion.[18] Airborne particulates smaller than about 50 $\mu$m are classified as suspended particulate matter [reported as total suspended particulates (TSP)]; particulates smaller than about 10–15 $\mu$m are classified as inhalable particulate matter (IPM), which largely escapes entrapment by the mucous membranes of the nasal passages and throat; and fine (submicrometer) particulates can penetrate the alveoli and damage the lungs. Typically, particle settling velocities are calculated using Stokes' law. For a 1-$\mu$m particle ($d = 2$ g/cm$^3$) in air at 20°C a settling (Stokes' sedimentation) velocity of about $6 \times 10^{-5}$ m/s is calculated. If suspended in air at 20 m, the particle's lifetime will be about 4 days. (Submicrometer particles could have lifetimes, in principle, on the order of weeks.) If a constant velocity/direction wind of 4 m/s is assumed, such particles can travel as far as 1400 km. Moreover, if one envisions transport of a packet of air, it is likely that high concentrations of particulates can be transported long distances.[17] Obviously, other physical removal mechanisms, including impaction and wet (precipitation) removal, are also operative. The wet cycle in Middle Europe, for example, is about 7 days; thus wet removal is the limiting mechanism for removal of the smallest particulates. The other extreme PAH loss mechanism would be chemical losses having lifetimes shorter than about 4 days. A recent study indicates that the more reactive PAH such as B[a]P have their lifetimes determined by chemistry.[20] However the conclusions are derived on the basis of model chemical studies not necessarily relevant for environmental conditions, and we feel the question remains open.

Aside from studies of PAH transport, there are many other reasons for the interest in PAH reactivity under ambient conditions. It is known that PAH exposed to air on glass fiber filters as well as to oxides of nitrogen (usually accompanied by some nitric acid) will be converted to direct mutagens requiring no metabolic activation.[21,22] Nitro-PAH are responsible for an important part of the direct mutagenicity.[23] Thus, benzo[a]pyrene is converted to its 1-, 3-, and 6-nitro derivatives (4–6) as well as dinitro compounds. These observataions sparked an interesting debate,[24,25] since the major by-product, 6-nitrobenzo[a]pyrene (6) is a much weaker carcinogen than the parent hydrocarbon. However, the other two derivatives as well as some dinitro compounds are potent carcinogens. Furthermore, the noncarcinogenic PAH pyrene is converted to the carcinogenic derivative 1-nitropyrene (7) and 1,8-dinitropyrene (8), one of the most potent direct mutagens known.[26] Nitro-PAH have been reported in ambient air particulates and die-

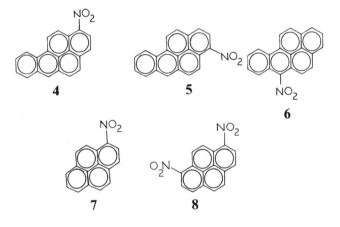

sel exhaust particulates.[27–32]

PAH are also known to react with active oxygen species including ozone, singlet oxygen, and hydroxyl radical. Thus, benzo[a]pyrene exposed on glass fiber filters to ozone formed as major products the mild direct mutagenic quinones 9–11.[33] Such quinones have been found in environmental sam-

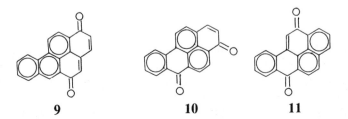

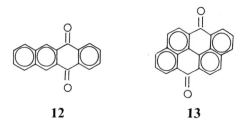

**12**              **13**

ples.[34−36] Naphthacene-5,12-dione (**12**) and anthanthrone (**13**), derivatives of two particularly reactive PAH, have been identified in environmental samples.[34,36]

Polycyclic aromatic ketones (PAK), including **14–18**, have been identified in environmental samples.[36,37] The mutagenic impact of the PAK fraction is not yet known.[37] The ketones are presumably derived from highly reactive

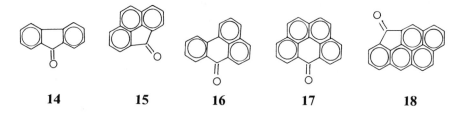

**14**          **15**          **16**          **17**          **18**

methylene-bridged PAH which, in turn, appear to be produced from methyl-PAH.[38] Compounds **14**, **15**, and **18** are derivatives of fluorene, benzofluorenes, and related compounds, some of which (eg, **19**, **20**) have been identified in air.[39] Some of these hydrocarbons, including 1-methyl-4*H*-cyclopenta[*def*]phenanthrene and 4*H*-cyclopenta[*def*]chrysene, are potent mutagens.[38a] The latter is a potent tumorigen, although not as potent as 5-methylchrysene.[38a] Compounds **16** and **17** are derived from benzophenalenes, none of which have been reported in ambient air, presumably because of their high reactivity.[37,40]

**19**              **20**

A particularly striking observation is the analysis of the metabolite benzo[*a*]pyrene-4,5-oxide (**21**) as a product of the reaction of ozone and B[*a*]P on a glass fiber filter.[33] This compound is a potent direct mutagen. It has not yet been reported in environmental samples, possibly because of the reactivity endowed by the epoxide ring. An interesting oxygenated compound reported in diesel exhaust particulates is pyrene-3,4-dicarboxylic anhydride (22).[41] This weak, direct-acting mutagen is considered to be an example of a stable class possibly derived from a cyclopenteno-PAH such as cyclopenta[cd]pyrene (23). It has been duly noted that while compounds such as 23 are found at high levels in newly formed emissions, aged aerosols have but small amounts of these reactive compounds.[42–44] Thus, although

**21**          **22**          **23**

levels of 23 roughly 13 times that of B[a]P have been reported in automobile emissions and tunnels,[45] relatively low levels are found in the environment.[46] K-Region lactones closely related to 15 have also been reported in diesel and ambient particulate matter.[7c]

# 3. STABILITY OF THE POLYCYCLIC AROMATIC HYDROCARBONS

The reactivity of polycyclic aromatic hydrocarbons follows striking topological trends that continue even to large polybenzenoid structures en route to graphite. Thus, in Chapter 2 of Volume 2 in this series, Stein and Brown show how topology influences the edge reactivity of these polybenzenoid species. In his classical works on PAH,[1a,b] Clar notes the increased reactivity in linear acenes, which is attributed to the constraint that only a single aromatic sextet can be drawn for this class (eg, anthracene, 24). As the acene grows, it becomes increasingly polyolefinic and reactive. Thus, the orange-

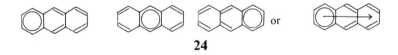

or

**24**

yellow compound naphthacene (25), which has been reported in coal tar,[1a] rapidly transforms to the 5,12-quinone 12 when shaken in xylene solution with air in the presence of UV light.[3] Thus, it is not surprising that the parent hydrocarbon is not reported in air while the quinone is.[34,36] Although this could be an artifact of reactivity during laboratory manipulation, it is more likely that oxidation occurred shortly after formation and during atmospheric residence. The reactivity of the dark green PAH hexacene (26) is so

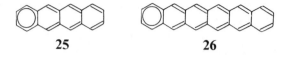

high that the compound has never been isolated in pure form.[1a,b]

In contrast to anthracene, phenanthrene (27), with one angular benzo ring, is a much less reactive molecule. This is readily rationalized using the Clar approach in terms of two aromatic sextets. This also explains why phenanthrene reacts primarily at $C_{9,10}$ (K region) and can even add $Br_2$ across this olefinlike bond without catalysis. While this approach anticipates the decreased reactivities of benz[a]anthracene (28) and chrysene (29) relative ot tetracene, it makes no obvious differentiation between the latter isomers. Triphenylene (30) is an example of a "fully benzenoid" PAH, and its extreme stability (eg, it will not be protonated or sulfonated in concentrated sulfuric acid[1a,b]), is consistent with this view. Another fully benzenoid PAH is dibenzo[fg,op]naphthacene (31). A shortcoming of Clar's sextet is that it

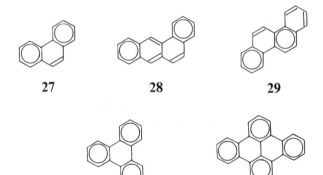

makes no special provision for molecules such as naphtho[1,2,3,4-def]chrysene (32), a less stable isomer of 31, which can be represented by three degenerate structures having three sextets.

32

Another useful way of rationalizing PAH stability employs the resonance concepts of Pauling[47] and Wheland.[48] Thus, the more equivalent "Kékulé-type" resonance structures a PAH has, the more stable it is. Various ways of obtaining structure count (SC) for PAH and related reaction intermediates have been summarized by Herndon.[49] Thus, Scheme I shows that while

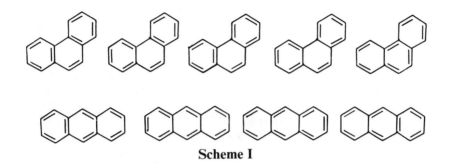

**Scheme I**

five such structures can be drawn for phenanthrene, only four can be written for the more reactive anthracene. Herndon[49] notes a simple technique for obtaining SC for neutral hydrocarbons using the methodology of perturbational molecular orbital theory.[50] Thus, one obtains the maximum value possible by deleting one carbon from the even-alternant PAH in order to generate an odd-alternant PAH. One then obtains the lowest whole-number *unnormalized* coefficients of the nonbonding molecular orbital (NBMO) in the usual manner[50] and sums the coefficients of the carbons attached to the deleted carbon. The examples in Scheme II illustrate this technique, which

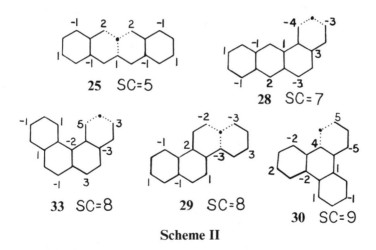

**Scheme II**

further indicates that chrysene (**29**) is indeed more stable than benz[*a*]anthracene (**28**). The technique under discussion, however, falsely predicts equivalent stability for benzo[c]phenanthrene and chrysene; there-fore, this discrepancy will be discussed shortly. A closely related parameter termed the corrected structure count (CSC) is identical to the structure count for fully benzenoid PAH but lower than SC for a molecule such as benzocyclobutadiene. In this molecule, SC is 3 but CSC is 1, since the two resonance structures having cyclobutadiene structures are not counted for CSC.[51a] The relationship between SC and CSC for nonalternant PAH has also been discussed.[51a] The resonance energy of a PAH is linearly related to ln(CSC) [or ln(SC) for fully benzenoid PAH].[49,51a]

Hückel molecular orbital (HMO) theory has enjoyed excellent success in its application to aromatic molecules.[52] Resonance energies (eg, difference between total $\pi$ energy of a PAH and the total $\pi$ energy of an equivalent number of ethylenes) should be, in principle, a measure of stability. Further refinements include comparison with different standards (eg, 1,3-butadiene) or new parameterizations [eg, resonance energy per $\pi$ electron (REPE)]. It is abundantly clear that resonance energies are strongly topology dependent, and Zander[53] has shown a simple linear relationship between resonance energy and "angularity" (the sum of 120° angles in point graphs of each mol-ecule) of isomers. He also shows clearly different relationships of molecular size versus REPE for different angularity series of PAH.[53] It is well to keep in mind that part of this different behavior for topologically distinct PAH classes arises from the fact that the HMO technique is a one-electron esti-mate. That is, it "loads" each electron into an orbital without correction for the repulsive effects of other electrons in the molecule. It is worthwhile not-ing that, as found by Ham and Ruedenberg,[51b] the CSC can be obtained as the square root of the product of the eigenvalue coefficients of the Hückel molecular orbitals.[51a]

The total $\pi$ energies and gas phase enthalpies of formation[54] of anthracene (19.31$\beta$; 54.4 kcal/mol) and phenanthrene (19.45$\beta$; 50.0 kcal/mol) are in the proper direction. A difference of 2.5 kcal/mol would have been predicted using $\beta$ = 18 kcal/mol. The total $\pi$ energies of the five isomeric $C_{18}H_{12}$ mol-ecules are shown in Scheme III, along with experimental standard gas phase enthalpies of formation $(\Delta H_f^\circ)$[54] and the "angularity."[53] Except for benzo[c]phenanthrene, the resonance energy order and the order of ther-

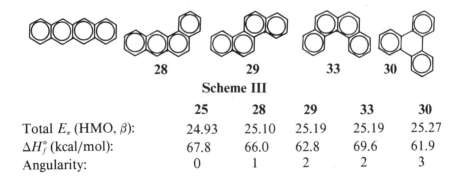

|  | 28 |  | 29 |  | 33 |  | 30 |
|---|---|---|---|---|---|---|---|

**Scheme III**

|  | 25 | 28 | 29 | 33 | 30 |
|---|---|---|---|---|---|
| Total $E_\pi$ (HMO, $\beta$): | 24.93 | 25.10 | 25.19 | 25.19 | 25.27 |
| $\Delta H_f^\circ$ (kcal/mol): | 67.8 | 66.0 | 62.8 | 69.6 | 61.9 |
| Angularity: | 0 | 1 | 2 | 2 | 3 |

modynamic stabilities are in good agreement. If benzo[c]phenanthrene were planar, intolerably large steric repulsions would occur between its terminal rings. Thus it distorts from planarity, incurring an increment of strain energy absent in its isomers. It has been termed "tetrahelicene" to emphasize the deviation from planarity and, with its larger homologues, exhibits other anomalous behavior. The total $\pi$ energies of pyrene and fluoranthene are almost equal, but the $\Delta H_f^\circ(g)$ of fluoranthene is almost 20 kcal/mol higher. This emphasizes the dominant influence of the $\sigma$ framework, differences in strain, and the inadequacies of the HMO approach in treating isomers that differ so greatly in structural type. While one might anticipate that ab initio molecular orbital calculations should give very good $\Delta H_f^\circ(g)$ and MO properties, they would be too time-consuming to perform at this level.

Numerous techniques have been employed for quantitating or predicting reactivity of polycyclic aromatic hydrocarbons. We have employed some of these in this chapter and will attempt to see what insights they provide. Table 1-2 lists both experimental and calculational parameters employed in the present work for PAH of 24 carbons or less. Most PAH studies emphasize compounds in this range, although larger hydrocarbons have been studied.[55]

We begin by briefly examining the experimental gas phase standard enthalpy of combustion $\Delta H_f^\circ(g)$ and note that relatively few data exist for most PAH.[54] However, estimation schemes have been employed to predict values,[56-59] and we employed the methodology of Stein and co-workers[57-59] as well as the macroincrementation approach described in Chapter 6 of Volume 3, by Liebman. The Stein approach[57-59] involves group increment additivity as illustrated for benzo[a]pyrene in Scheme IV. To illustrate the Lieb-

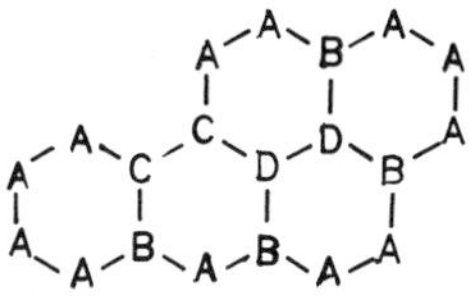

$$\Delta H_f \text{(kcal/mol)}$$

A = 3.30
B = 4.80
C = 3.70
D = 1.45
$\Delta H_f^\circ(g) = 12A + 4B + 2C + 2D = 69.1$ kcal/mol

**Scheme IV**

man macroincrementation approach, we estimate the value for indeno[1,2,3-cd] fluoranthene (**54**):

$$\Delta H_f^\circ(\mathbf{54}) = \Delta H_f(\mathbf{37}) + \Delta H_f(\mathbf{37}) - \Delta H_f(\mathbf{34})$$

explicit correction for differences in resonance energies between PAH, the great topology dependence of resonance energies makes this correction implicit in the Stein group increments.

From the point of view of reactivity, one should not expect to gain much insight from comparison of $\Delta H_f^\circ(g)$ values. However, it is interesting to note that fluoranthene is much higher in energy than pyrene and that the ben-

**Table 1-2.** Thermochemical and Reactivity Indices of Polycyclic Aromatic Hydrocarbons Shown in Figure 1-1

| PAH | $\Delta H_f(g)$ (kcal/mol)[a] | Ionization potentials (eV)[b] | | REPE[c] | SC[d] | Localization energies[e] | | | $\ln[SC_R/SC]$[g] | PA (kcal/mol)[h] | $(\log f \times 10^7)$[i] | $k_{NO_2}$[j] | $6 + \log k_{DA}$[k] |
|---|---|---|---|---|---|---|---|---|---|---|---|---|---|
| | | IP$_1$ | IP$_2$ | | | $L_r^+$ | $L_r$ | $N_u^f$ | | | | | |
| Naphthalene (34) | 35.8* | 8.15 | 8.88 | 0.065 | 3 | 2.30 | 2.30 | 1.81 | 0.85 | 194.7 | 3.06 | 1.00 | NR |
| Acenaphthylene (35) | 61.6* | 8.22 | 8.39 | 0.039 | 3 | — | — | — | 1.20 | 203.5 | — | — | — |
| Anthracene (24) | 54.4 | 7.41 | 8.54 | 0.047 | 4 | 2.01 | 2.01 | 1.26 | 1.39 | 207.0 | 7.10 | — | 3.36 |
| Phenanthrene (27) | 50.0* | 7.86 | 8.15 | 0.055 | 5 | 2.30 | 2.30 | 1.79 | 0.96 | 198.7 | 3.21 | 1.04 | NR |
| Pyrene (36) | 49.9* | 7.41 | 8.26 | 0.051 | 6 | 2.19 | 2.19 | 1.51 | 1.25 | 206.1 | 5.91 | 36 | — |
| Fluoranthene (37) | 69.2* | 7.95 | 8.1 | 0.048 | 6 | 2.34 | 2.21 | — | 1.20 | 199.3 | 3.96 | 2.9 | — |
| Aceanthrylene (38) | 80.2 | (7.38) | (7.95) | 0.038 | 4 | 2.04 | 1.94 | — | | | | | |
| Acephenanthrylene (39) | 75.8 | (7.58) | (8.10) | 0.043 | 5 | 2.12 | 2.12 | — | | | | | |
| Cyclopenta[cd]pyrene (40) | 76.0 | (7.45) | (7.76) | 0.041 | 6 | 2.08 | | | | | | | |
| Benzo[ghi]fluoranthene (41) | 74.5 | (7.68) | (8.09) | 0.048 | 8 | 2.32 | 2.22 | | | | | | |
| Naphthacene (25) | 67.8* | 6.97 | 8.41 | 0.042 | 5 | 1.93 | 1.93 | 1.02 | 1.69 | 217.8 | — | — | 4.97 |
| Benz[a]anthracene (28) | 66.0* | 7.41 | 8.04 | 0.050 | 7 | 2.05 | 2.05 | 1.35 | 1.42 | — | — | — | 2.13 |
| Chrysene (29) | 62.8* | 7.59 | 8.10 | 0.053 | 8 | 2.25 | 2.25 | 1.67 | 1.18 | 201.6 | 4.09 | 7.4 | NR |
| Benzo[c]phenanthrene (33) | 69.6* | 7.61 | 8.03 | 0.053 | 8 | 2.30 | 2.30 | 1.79 | 1.10 | — | — | — | NR |
| Triphenylene (30) | 61.9* | 7.87 | 7.87 | 0.056 | 9 | 2.37 | 2.37 | 2.00 | 0.94 | 198.5 | 2.79 | 1.3 | NR |
| Benzo[a]pyrene (1) | 69.1 | 7.12 | 8.00 | 0.049 | 9 | 1.96 | 1.96 | 1.15 | 1.61 | | | | |
| Benzo[e]pyrene (42) | 67.0 | 7.43 | 8.04 | 0.052 | 11 | 2.23 | 2.23 | 1.63 | 1.29 | | | | |
| Perylene (43) | 73.7* | 7.00 | 8.55 | 0.048 | 9 | 2.14 | 2.14 | 1.33 | 1.49 | 211.4 | (6.2) | | 0.047 |
| Benzo[j]fluoranthene | 87.4 | (7.41) | (7.83) | 0.047 | 9 | 2.21 | 2.18 | — | 1.36 | | | | |
| Benzo[k]fluoranthene (45) | 85.4 | (7.22) | (8.15) | 0.047 | 9 | 2.13 | 2.13 | — | | | | | |
| Benz[e]acephenanthrylene (46) | 85.4 | (7.63) | (7.88) | 0.049 | 10 | 2.28 | 2.11 | — | 1.33 | | | | |
| Benz[a]aceanthrylene (47) | 83.2 | (7.19) | (7.96) | 0.048 | 8 | 2.08 | 1.94 | — | | | | | |
| Cyclopenta[de]naphthacene (48) | 94.0 | (7.00) | (7.85) | 0.036 | 5 | 2.00 | 1.88 | — | | | | | |
| Benz[k]acephenanthrylene (49) | 92.1 | (7.22) | (7.86) | 0.041 | 7 | 2.10 | 1.99 | — | | | | | |
| Dibenzo[bh]biphenylene (50) | 136.7 | 7.61 | 8.90 | 0.038 | 10 | 2.20 | 2.20 | 1.63 | 1.25 | | | | |

| Compound | | | | | | | | | | | |
|---|---|---|---|---|---|---|---|---|---|---|---|
| Benzo[ghi]perylene (51) | 72.2 | 7.19 | 7.86 | 0.051 | 14 | 2.22 | 2.22 | 1.55 | 1.39 | 208.5 | |
| Dibenzo[def,mno]chrysene (52) | 74.2 | 6.92 | 8.08 | 0.045 | 10 | 1.93 | 1.93 | 1.03 | 1.76 | | |
| Indeno[1,2,3-cd]pyrene (53) | 83.2 | (7.26) | (7.95) | 0.047 | 12 | 2.22 | 2.22 | — | 1.41 | | |
| Indeno[1,2,3-cd]fluoranthene (54) | 102.6 | (7.57) | (7.68) | 0.044 | 13 | | | — | | | |
| Pentacene (55) | 84.6 | 6.61 | 8.32 | 0.038 | 6 | 1.84 | 1.84 | 0.80 | 2.01 | | 4.23 |
| Benzo[a]naphthacene (56) | 82.4 | 6.96 | 7.92 | 0.045 | 9 | 1.95 | 1.95 | 1.09 | 1.70 | | NR |
| Picene (57) | 78.0 | 7.50 | 7.74 | 0.053 | 13 | 2.25 | 2.25 | 1.67 | 1.24 | 203.4 | |
| Benzo[b]chrysene (58) | 80.2 | 7.20 | 8.24 | 0.049 | 11 | 2.03 | 2.03 | 1.29 | 1.49 | | 1.82 |
| Pentaphene (59) | 82.4 | 7.27 | 7.39 | 0.047 | 10 | 2.05 | 2.05 | 1.36 | 1.57 | | |
| Benzo[c]chrysene (60) | 79.1 | (7.48) | (7.63) | 0.053 | 13 | | | 1.71 | 1.20 | | |
| Benzo[g]chrysene (61) | 76.9 | (7.43) | (7.94) | 0.054 | 14 | | | 1.70 | 1.23 | | |
| Dibenz[ah]anthracene (62) | 80.2 | 7.38 | 7.80 | 0.051 | 12 | 2.13 | 2.13 | 1.51 | 1.39 | | 1.02 |
| Dibenz[a]anthracene (63) | 80.2 | 7.40 | 7.79 | 0.051 | 12 | 2.08 | 2.08 | 1.44 | 1.45 | | 1.02 |
| Benzo[b]triphenylene (64) | 78.0 | 7.39 | 7.91 | 0.053 | 13 | 2.12 | 2.12 | 1.50 | 1.37 | | 1.83 |
| Coronene (65) | 77.1 | 7.36 | 7.36 | 0.053 | 28 | 2.31 | 2.31 | 1.80 | 1.22 | 205.0 | 3.87 |
| Dibenzo[fg,op]naphthacene (66) | 78.7 | 7.40 | 7.83 | 0.055 | 20 | | | 1.79 | 1.25 | | |
| Benzo[rst]pentaphene (67) | 83.2 | 7.07 | 7.81 | 0.048 | 14 | 1.95 | 1.95 | 1.11 | 1.74 | | |
| Dibenzo[b,def]chrysene (68) | 83.2 | 6.82 | (8.18) | 0.047 | 13 | 2.21 | 2.21 | 1.01 | 1.85 | | |
| Dibenzo[def,p]chrysene (69) | 85.2 | 7.07 | 7.72 | 0.050 | 16 | 1.99 | 1.99 | 1.22 | 1.68 | | |
| Naphtho[1,2,3,4-def]chrysene (70) | 80.9 | 7.11 | 7.87 | 0.051 | 17 | | | 1.36 | 1.59 | | |

[a]Values marked with asterisks are experimental and from Reference 54; other values are estimated using methods in References 57–59.
[b]From References 51 and 68–70; values in parentheses are estimated using Equation 1-1. Sources of HMO theory parameters include: (a) Streitwieser, A., Jr.; Brauman, J.I.; Coulson, C.A. "Supplemental Tables of Molecular Orbital Calculations". Pergamon Press: Oxford, 1965. (b) Coulson, C.A.; Streitwieser, Jr., A.; Poole, M.D.; Brauman, J.I. "Dictionary of Pi-Electron Calculations". W.H. Freeman: San Francisco, 1965. (c) Heilbronner, E.; Straub, P.A. "Hückel Molecular Orbitals, HMO". Springer-Verlag: Berlin, 1966. (d) Zahradnik, R.; Pancir, J. "HMO Energy Characteristics". IFI/Plenum Publishers: New York, 1970.
[c]Resonance energy per electron (see text).
[d]Resonance structure count (see text).
[e]Localization energies (see text).
[f]Dewar number (see text).
[g]Logarithmic ratio of the structure count of a reactive intermediate (eg, protonated PAH) to that of the PAH; after Herndon (see text).
[h]Gas phase proton affinities (see Reference 76).
[i]Rate constants of protodetritiation in trifluoroacetic acid (see References 77–81).
[j]Rate constants for nitration (see References 81, 83, 84, 88, and 89).
[k]Rate constants for Diels–Alder reactions with maleic anhydride (see References 70 and 97).

zofluoranthenes **44** and **45** and the closely related compounds **46** and **47** are also much higher in energy than the benzopyrenes and perylene. Our later discussion will indicate that **44–46** are present in the atmosphere in concentrations comparable to the benzopyrenes. Furthermore, perylene, although thermodynamically much more stable than the benzofluoranthenes, is present in much smaller quantities. Thus, thermodynamic stability is not a useful predictor and kinetic considerations must enter. These could center on preferred formation pathways,[57–59] which are dependent on temperature (optimum PAH formation temperatures for the benzopyrenes and benzofluoranthenes are 710–720°C) and fuel.[60] Thus, pyrolysis of butylbenzene at 700°C gives a fluoranthene/pyrene ratio of 0.54, while pyrolysis of propylbenzene at this temperature yields a corresponding ratio of 3.[60] An extreme example of fuel dependency is provided by the finding of a relatively high concentration of picene (**57**) in brown coal emissions.[61] Picene is not found in such high concentrations in the usual coal, oil, gasoline, or diesel emissions, and its presence in brown coal emissions is attributed to the presence of structurally related triterpenoids in the brown coal.[61]

Furthermore, one must consider reactivity under the conditions of formation and in the ambient environment. The relatively small amount of perylene formed[62] appears to reflect small emissions rather than high ambient reactivity. Although the $C_{22}H_{12}$ forms most commonly found in air particulates are benzo[*ghi*]perylene (**51**) and indeno[1,2,3-*cd*]pyrene (**53**), small amounts of dibenzo[*def,mno*]chrysene (anthanthrene, **52**) are found, and the quinone has been observed in air samples. The much less thermodynamically stable indeno[1,2,3-*cd*]fluoranthene (**54**) has also been reported in tar samples[60] and emissions from brown coal[61] and peat.[62] Interestingly, both dibenzo[*b,def*]chrysene-7,14-dione (**71**) and benzo[*rst*]pentaphene-5,8-dione (**72**) yield **54** on vacuum flow pyrolysis.[63] This could represent a case

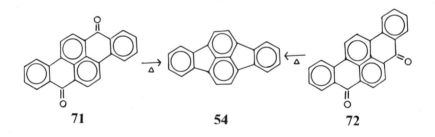

**71**                     **54**                     **72**

of generation of hydrocarbons by the oxidation products of more common PAH.

Resonance energies would at first appear to provide a direct handle on reactivity while also bearing some relationship to thermodynamic stability. While one might directly compare, for example, Hückel $\pi$-electronic energies, these increase for larger systems. It is tempting simply to divide the electronic energy by the number of carbon atoms or to define resonance energies relative to ethylenes and divide by the number of carbon atoms, but Hess and Schaad[64] have provided a more useful definition. They note that if reference is made to acyclic polyenes for calibration of resonance energies, topology must be taken into account. They have developed a set of eight structural parameters (Scheme V), in which two are arbitrarily assigned a value $2.0000\beta$. This allows one to calculate a model $\pi$ energy, compare it with the Hückel $\pi$ energy, and divide the difference by the number of $\pi$ electrons to obtain the REPE.

| Type of bond | Designation | Increment ($\beta$) |
|---|---|---|
| $H_2C=CH$ | $E_{23}$ | 2.0000 |
| $HC=CH$ | $E_{22}$ | 2.0699 |
| $H_2C=C$ | $E_{22'}$ | 2.0000 |
| $HC=C$ | $E_{21}$ | 2.1083 |
| $C=C$ | $E_{20}$ | 2.1716 |
| $HC-CH$ | $E_{12}$ | 0.4660 |
| $HC-C$ | $E_{11}$ | 0.4362 |
| $C-C$ | $E_{10}$ | 0.4358 |

$$E_{add} = 4E_{22} + 4E_{21} + E_{20} + 3E_{12} + 6E_{11} + 3E_{10} = 24.21\beta$$

$$E_{Hückel} = 25.10\beta$$

$$REPE = \frac{(25.10 - 24.21)}{18} = 0.050\beta$$

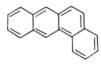

**Scheme V**

An example of this calculation is provided in the scheme for benz [a]anthracene. Comparison of REPE values offers some interesting insights. Thus, in the $C_{18}H_{12}$ series the order of the thermodynamic stabilities (**30** > **29** > **28** > **25**) is the same as the REPE trend except for benzo[c]phenanthrene (**33**), which distorts from planarity. It must be noted that different Kékulé-type structures of the same PAH may have different REPE values, but the differences are insignificant; thus an evaluation of any good Kékulé structure can be employed.[64] It is obvious from Table 1-2 that within an isomeric series (eg, the $C_{22}H_{14}$, **55–64**), higher REPE values are associated with higher SC values. While the REPE parameter is a very powerful indicator of aromaticity, antiaromaticity, and nonaromaticity,[64] its utility in predicting reactivity differences for PAH, where the differences are smaller, is limited. REPE values for the benzofluoranthenes (and other non-alternant PAHs) are very low and suggest that these compounds are more reactive than benzo[*a*]pyrene, contrary to fact. Similarly, with this parameter, cyclopenteno compounds such as cyclopenta[*cd*]pyrene (**40**) appear to be even more reactive than they actually are. The differences in REPE between alternant PAH are also often very small.

Dibenzo[*bh*]biphenylene (**50**) introduces two additional problems, also obviously understood by Hess and Schaad. First, a square, four-membered ring introduces some of the antiaromaticity of the cyclobutadiene system. Second, ring strain decreases the thermal stability of this compound. Dibenzo[*bh*]biphenylene is an interesting possibility as an environmental by-product. It could arise via pyrolysis of 2,2'-binaphthyl, a well-known combustion by-product,[60] and is stable to 400°C[65] (linear cyclobutanoid PAH are more stable than nonlinear isomers, in contrast to acenes).

# 4. CHEMICAL REACTIVITY

## A. Thermal Stability

Perhaps the most simple reaction to relate to hydrocarbon stability is thermolysis. An extensive investigation of 84 PAH using differential thermal analysis (DTA) under an argon atmosphere was undertaken by Lewis and Edstrom.[66] Unreactive hydrocarbons were taken as those that were unchanged at 750°C. Of the series of compounds in Table 1-2 investigated in the Lewis–Edstrom study, only acenaphthylene (**35**), napthacene (**25**), pentacene (**55**), dibenzo[*def,mno*]chrysene (**52**), and benzo[*rst*]pentaphene (**67**) exhibited such reactivity.

Acenaphthylene's behavior was particularly noteworthy, since it polymerized at temperatures below 300°C in contrast to the others that reacted above 400°C. This reflects the olefinic character of the cyclopenteno ring and actually neatly relates to the very low REPE value (0.039) for this molecule. It is also noteworthy because other reactivity parameters do not indicate such striking reactivity.

Herndon has analyzed the thermal stabilities above in terms of the reactivity of PAHs with free radicals. In addition to the free valence parameter ($F_r$) derived from Hückel theory, he employs parameters known as the structure count ratio (SC ratio) and the Dewar reactivity number ($N_u$). The Dewar reactivity number,[50] which is based on the perturbational treatment of Hückel molecular orbital calculations, is illustrated in Scheme VI for calculating the relative reactivities at the 7- and 14- positions of dibenz[*aj*]anthracene (**63**). The trial reactive position is "saturated" (corresponding to attack by a free radical, cation, or anion) and *normalized* coef-

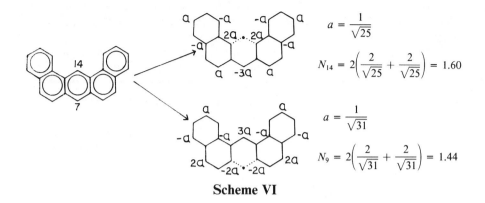

**Scheme VI**

ficients of the previously attacked atoms in the NBMO of the remaining odd-alternant hydrocarbon are summed and multiplied by 2. Low Dewar numbers connote high reactivity. Thus, the 7-position of dibenz[$aj$]anthracene is more reactive than the 14-position. Clearly naphthacene ($N_5$ = 1.02) is much more reactive than its isomer benz[$a$]anthracene ($N_7$ = 1.35) toward reactions forming Wheland-type intermediates and considerably less reactive than pentacene ($N_u$ = 0.80). Dewar reactivity numbers cannot, however, be calculated for such nonalternant PAH as fluoranthene, which contain odd-membered rings.

Herndon's structure count technique involves calculation of the ratio of the structure count of the reactive intermediate divided by the structure count of the PAH itself (ie, $SC_{ratio}$ = ($SC_I/SC_R$). In the present case, values of ln $SC_{ratio}$ are listed. The method for obtaining $SC_I$ is illustrated in Scheme VII for the 9-position of anthracene. It involves "saturating" the reaction site and adding the absolute values of the lowest whole number ratios of coefficients for the *entire* odd-electron system. (Recall that $SC_R$ is obtained by summing the absolute values of only coefficients attached to the "saturation" site, which itself can be at any position in the molecule.) A high $SC_{ratio}$ means that there is increased resonance stabilization in the interme-

$$SC_R = 4 \qquad SC_{ratio} = 4.000$$
$$SC_I = 16 \qquad \ln SC_{eratio} = 1.386$$

**Scheme VII**

diate relative to the PAH and thus high ratios imply high reactivities. This technique can be applied to nonalternant hydrocarbons but, unlike the situation for alternant PAH, the structure count of their radicals does not correspond to the number of canonical Kékulé structures. Herndon describes a method for obtaining the CSC for these systems.[67]

Both the Dewar reactivity number and the Herndon structure count techniques can be done using pencil and paper. An older parameter is the Hückel molecular orbital localization energy $L_r$. The $L_r$ value is the same for radical attack ($L_r^.$) and cation attack ($L_r^+$) on alternant PAH. It corresponds to the difference in the resonance energies in saturating a position with a radical or cation (or in general, an electron pair acceptor). While $L_r^.$ = $L_r^+$ for alternant hydrocarbons, generally $L_r^. \neq L_r^+$ for nonalternant

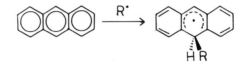

$$E_\pi = 14a + 19.31\beta$$

$$E_\pi = 13a + 17.30\beta$$

$$L_r^\circ = (19.31 - 17.30) = 2.01$$

hydrocarbons such as fluoranthenes. Low $L_r$ values correspond to high reactivity. Herndon correlated experimental reactivities and found better correlations with his structure–resonance theory ($SC_{ratio}$) technique than with $L_r^.$, $N_u$, and $F_r$.[67]

## B. One-Electron Oxidation

When one considers attack of free radicals, and electron pair acceptors on PAH, it would appear that the ability of these compounds to lose electrons may furnish a measure of reactivity. Thus, the ionization potentials of PAH are of obvious interest, and useful compendia including discussions are available.[51,68–70] Photoelectron spectroscopy has been extremely useful in obtaining such data and indeed in assigning correct structures to PAH whose structures had been incorrectly assigned.[71a] It was noted that NMR and mass spectrometry have very limited utility in PAH analyses, while UV, which is an extremely powerful technique, is limited by the broadness of bands. Photoelectron spectroscopy, which like UV, depends on electronic levels, produces sharp bands of great utility in structural assignments. These bands are readily attributable to $\pi$ molecular orbitals, since the onset of the $\sigma$ levels usually occurs above 10.6 eV.[71a] It must, of course, be noted that for certain highly symmetric PAH (eg, triphenylene and coronene), the highest occupied molecular orbitals (HOMOs) are doubly degenerate. Another useful technique for differentiating PAH isomers is charge exchange chemical ionization mass spectrometry (CECIMS).[71b] Equation 1-1 was derived for the data-condensed PAH series relating the ionization of $\pi$ orbital $i$ with its energy in $\beta$ units ($x_i$). If Hückel calculations are made self-consistent with respect to bond orders, another correlation becomes evident (Equation 1-2). A later version (Equation 1-3)[71a] also corrected for molecular size ($N$ = number of C atoms).

$$IP_i(eV) = (5.893 \pm 0.041) + (2.884 \pm 0.038)x_i; \, SE = 0.142 \text{ eV} \quad (1\text{-}1)$$
$$IP_i(eV) = (5.652 \pm 0.034) + (3.214 \pm 0.032)x_i'; \, SE = 0.109 \text{ eV} \quad (1\text{-}2)$$

where

$$x_i' = x_i = F \sum_{uv} 2C_{ui}C_{vi}(\% - P_{uv})$$

Further correlations were obtained between the UV and visible bands and between the first and

$$IP_i(eV) = 5.391 + \frac{5.549}{N} + 3.215 \left[ x_i^\circ + 0.788 \sum_{uv} 2C_{\mu i}C_{vi}(P_{\mu v} - \%) \right] \quad (1\text{-}3)$$

second ionization potentials of alternant PAH (Equations 1-4 to 4-6)[71a]:

$$E_a = -3.750 + 0.924(0.582IP_1 + 0.418IP_2) \quad (1\text{-}4)$$
$$E_p = -5.015 + 1.190(1.172IP_1 - 0.172IP_2) \quad (1\text{-}5)$$
$$E_B = -6.232 + 1.386(0.495IP_1 + 0.505IP_2) \quad (1\text{-}6)$$

Resonance theory has also been employed for estimation of $IP_1$ data.[51] Thus, equation 1-7 was obtained for 198 alternant and nonalternant PAH. The first term in the equation represents the valence state ionization potential of an $sp^2$ carbon atom, while the remaining terms represent the resonance energies of the neutral PAH and its radical cation, respectively.[51] Calculation of CSC was discussed earlier and is trivial for benzenoid alternant PAH. Calculation of CSC(PAH$^+$) is much more complex, since it involves Dewar-type as well as Kékulé-type resonance contributors and "separation" of + and · assignments. Algorithms for calculating this parameter have been published.[51]

$$IP_1(eV) = 11.55 + 1.35 \ln[CSC(PAH)] - 1.15 \ln[CSC(PAH^+)] \quad (1\text{-}7)$$

Since one school of thought holds that radical cations of PAH may be important contributors to their carcinogenic behavior,[73] IPs would appear to be a significant property. The low IPs of methyl-substituted PAH appear to be consistent with their high benzylic reactivities (Scheme VIII).[73] Radical cations of unsubstituted PAH may dimerize in the absence of nucleophiles or react with nucleophiles to form quinones or other substituted species (Scheme IX). Furthermore, one might imagine compounds in polluted

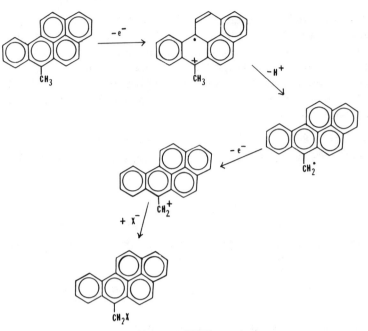

**Scheme VIII**

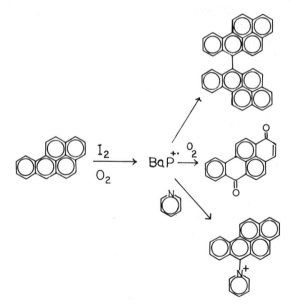

**Scheme IX**

air that have high electron affinities, which might allow them to induce formation of PAH$^+$. An example is peroxyacetyl nitrate (PAN), an important component of photochemical smog. It is also important to note that despite its high IP (8.22 eV), acenaphthylene is reactive thermally (see earlier discussion) due to the presence of the olefinlike cyclopenteno ring. Other cyclopenteno compounds seem to exhibit reactivities having no relation to their ionization potential. Thus, if Equation 1-1 is employed to predict the IP of cyclopenta[cd]pyrene (40): Hückel HOMO, 0.5393$\beta$; IP$_1$ (calc) = 7.45 eV, it would appear that this PAH should not have been very reactive. Equation 1-1 tends to underestimate IP of analogous compounds, predicting values of only 7.73 eV for acenaphthylene and 7.58 eV for acephenanthrylene. Generally, the cyclopenteno-PAH compounds have IPs very similar to the PAH to which the five-membered rings are fused. However, their high environmental reactivities cause them to disappear rapidly in the environment.[43] Thus, IP is a very poor measure of environmental stability for these compounds. Since these compounds react through the olefinlike cyclopenteno ring, aceanthrylene (38), which has only recently been synthesized and characterized,[74,75] readily polymerizes in the presence of acid and is easily hydrogenated. This is consistent with the absence or very low levels of these compounds in aged aerosols.[6]

## C. Reaction with Nitrogen Dioxide and Nitric Acid

Attack of electrophilic reagents such as sulfur trioxide, nitrogen dioxide, and nitric acid on PAH is of potential importance in determining atmospheric reactivites. Furthermore, it is of considerable concern in health because nitro and dinitro PAH are frequently more carcinogenic than the parent compounds. The $L_r^+$ parameter, provided in Table 1-2, as well as the ln SC$_{ratio}$ and $N_u$ parameters, should indicate reactivity. A compendium[76] of gas phase proton affinities has been published recently and values for PAHs are listed in Table 1-2. The correlation coefficient with $L_r^+$ is only 0.82 for the 11 alternant PAH in this table, while the correlation with $N_u$ is 0.88. However, the correlation with ln SC$_{ratio}$ is 0.98 (with SC$_{ratio}$, $R$ = 0.99). The predicted value for the nonalternant PAH fluoranthene is about 4.7 kcal/mol greater than the experimental value.

Experimental data exist on the protodetritiation of PAH now normalized to 100% trifluoroacetic acid at 70°C.[77-81] In all cases studied, the most active detritiation site is the one predicted using the $L_r^+$, $N_u$, or SC$_{ratio}$ techniques and shown in Figure 1-1. The data also are listed in Table 1-2 in the form of log($f \times 10^7$), where $f$ is the specific rate factor. Herndon has shown that his resonance structure technique is a much better predictor of protodetritiation rates than Hückel parameters even when helicenes are examined.[80,82]

It is important to note that the first systematic measurements of reactivity toward substitution were made by Dewar, who examined rates of nitration.[83,84] The rates and positions of substitution were essentially those pre-

34

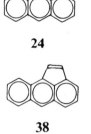
35

24

27

36

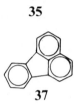
37

38

39

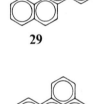
40

41

25

28

29          33          30          1

42          43          44          45

46          47          48          49

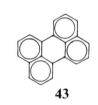

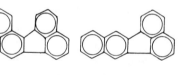

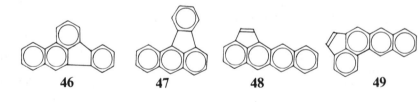

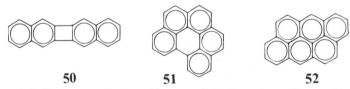
50          51          52

**Figure 1-1.** Structures of polycyclic aromatic hydrocarbons discussed in this chapter and listed in Table 1-2.

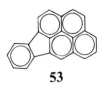

**53**

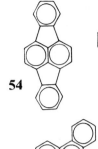

**54**

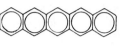

**55**

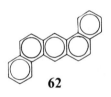

**56**

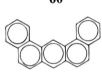

**57**

**58**

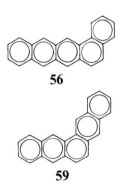

**59**

**60**

**61**

**62**

**63**

**64**

**65**

**66**

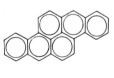

**67**

**68**

**69**

**70**

dicted by simple Hückel techniques. Triphenylene is somewhat anomalous in that the 1- and 2-positions exhibit equal rates of nitration despite the HMO theory prediction that the 1-position is more reactive. The predicted order is observed in the protodetritiation reaction, which is much less subject to steric hindrance at the crowded 1-position than is nitration.[77,81] The 7-position of fluoranthene, which is expectedly less reactive than the 3-position is anomalously slow in the nitration reaction due to steric hindrance, while the 2-position of indeno[1,2,3-cd]fluoranthene nitrates slowly for other more subtle reasons.[85] It is useful to note in passing the effects of alkylation and methylene bridging on the rates of protodetriatiation of PAH (Scheme X).[86,87] Note that the proton affinities of 1-methylnaphthalene

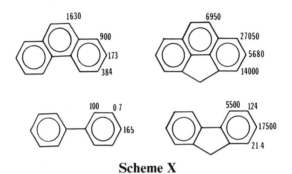

**Scheme X**

(200.7 kcal/mol) and 2-methylnaphthalene (200.0 kcal/mol) increase relative to naphthalene (194.7 kcal/mol). Similarly, the proton affinities of 2-methylanthracene (210.3 kcal/mol) and 9-methylanthracene (213.9 kcal/mol) increase relative to anthracene (207.0 kcal/mol).[76]

We also briefly note a parallelism between electrophilic aromatic substitution and solvolysis of the carbinyl derivative substituted at the positon of substitution.[81] Thus, as the 9-position of anthracene is most rapidly substituted by electrophilic reagents, so also is the 9-carbinyl tosylate most readily solvolyzed. A series of (Brown and Okamoto) substituent constants, $\sigma^+$,

were derived from acetolysis studies using $\rho = -5.71 \pm 0.09$, which was obtained for activated benzyl tosylates.[88] Although protodetritiation rates have been employed to derive many new $\sigma^+$ values,[77-81] potential problems with this approach have been noted.[88,89] The experimental $\sigma^+$ values showed a fairly poor correlation with the Hückel parameter $[E_{\pi ArCh2+} - E_{\pi ArH}]$. This again points up inadequacies in the Hückel technique. It is interesting that the bay region carcinogenicity theory makes use of at least semiquantitative correlation with $[E_{\pi ArCh2+} - E_{\pi ArH}]$. Again, the resonance structure theory provides very good correlations for these series yielding a correlation coefficient of 0.993 between $\sigma^+$ and $\ln(SC_{ratio})$.[90]

## D. Photooxidation and Ozonolysis

Photooxidation and reaction with ozone are probably two of the most important chemical atmospheric disappearance pathways for PAH.[2,6] Photooxidation involves initial formation of the excited singlet PAH (which may dimerize), followed by intersystem crossing to yield the triplet, which can form singlet oxygen from the normal atmospheric triplet. Singlet oxygen reacts with a PAH, such as benz[a]anthracene in a manner analogous to the Diels–Alder reaction of molecules such as maleic anhydride, to form an endoperoxide, which subsequently yields the quinone (Scheme XI).[91] Recent work has shown specifically that anthracene, phenanthrene, 9,10-diphenylanthracene, fluoranthene, pyrene, benz[a]anthracene, chrysene, benzo[a]pyrene, and 9-fluorenone are all efficient singlet oxygen sensitizers. The $^1O_2$ thus produced rapidly oxidizes proximate PAH (model study on fluidized bed of PAH-coated Chromosorb 102).[92] Note that methyl-sub-

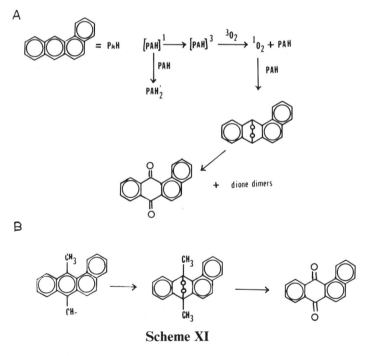

**Scheme XI**

stituted compounds, such as the powerful carcinogen 7,12-dimethylbenz[a]anthracene (DMBA), can be demethylated by photooxidation. Benzo[a]pyrene produces quinones **9–11** even though endoperoxides are sterically impossible. Such compounds may come from initially formed dioxetanes.[7b] Photooxidation studies on alumina have indicated formation of a variety of oxygenated compounds in addition to quinones.[93] These have tentatively been assigned as dialdehydes, hydroxy derivatives, ketones, coumarins, and xanthones. Prominent dialdehydes are formed from the 9,10-phenanthrene-like (K-region) bonds of benz[a]anthracene and dibenz[ac]anthracene (benzo[b]triphenylene) (see Scheme XII), although dialdehydes from oxidation of the terminal aromatic rings are also found.[93]

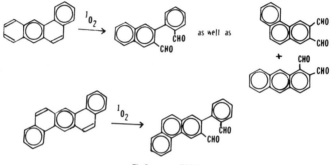

**Scheme XII**

Ozonolysis of PAH produces a variety of derivatives that have been rationalized on the basis of 1,2-addition, 1,4-addition, and simple electrophilic attack. Scheme XIII is a partial description of some of the products and

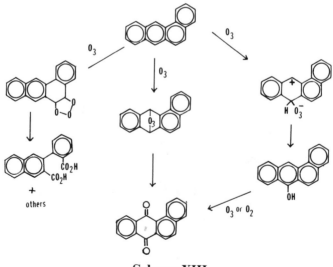

**Scheme XIII**

postulated reaction pathways for benz[*a*]anthracene. Benzo[*a*]pyrene similarly forms quinones **9–11**, epoxide **21**, phenols, and a variety of ring-opened oxygenated products.[94]

In assessing reactivity, it is useful to recall additional theoretical parameters. For example, Brown[95,96] developed the concept of paralocalization energy (*P*), which is the difference in $\pi$ energies of the starting compound and the 1,4-adduct. For example, Scheme XIV shows that the *P* value for naphthacene is less negative than that for benz[*a*]anthracene, indicating that the former is more reactive. Dewar's perturbation molecular orbital (PMO) approach has also been used to approximate paralocalization energies,

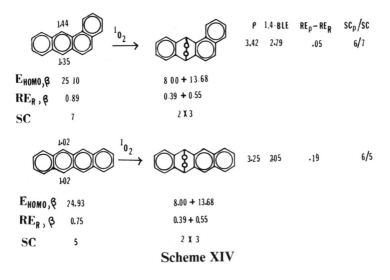

$$E_{HOMO},\beta \quad 25.10 \qquad 8.00 + 13.68$$
$$RE_R,\beta \quad 0.89 \qquad 0.39 + 0.55$$
$$SC \quad 7 \qquad 2 \times 3$$

**Scheme XIV**

although Dewar prefers the more general term 1,4-bislocalization energy (1,4-BLE). 1,4-BLE is the simple sum of the Dewar numbers ($N_u$, see previous discussion). It is worth commenting here that this explicitly reveals the interdependency between reactivity parameters, since $N_u$ is a measure of reaction with $Y^+$, while a sum of $N_u$ is a measure of Diels–Alder reactivity. A more refined approach employs the difference in the Hess–Schaad resonance energies ($\Delta R$; see earlier discussion) between the para-adduct ($RE_p$) and the PAH ($RE_R$).[97] Finally, the attractively simple resonance theory structure count technique of Herndon, discussed earlier, can also be employed to assess the relative changes in resonance energies.[98]

Compendia of second-order rate constants of PAH with maleic anhydride under uniform conditions have been published[70,97] and these data are listed in Table 1-2 in the convenient form (6 + log $k_{DA}$).[70] Correlation of Diels–Alder reactivities with Brown paralocalization (*P*), Dewar PMO, Hess–Schaad ($\Delta RE$), and Herndon structure count [ln(SC/SC)] parameters shows roughly comparable agreement, with slightly better correlation for the Hess–

Schaad parameter.[97] However, the Herndon approach has the powerful virtue of "back-of-the-envelope" simplicity.

One of the most interesting observations associated with Diels–Alder reactivities is that the first ionization potential is a rather poor indicator of reactivity. Biermann and Schmidt noted, for example, that angular annelation has relatively little effect on $IP_1$, yet produces dramatic effects on the rate constants.[70] However, they found that the greater the value $\Delta IP$ ($= IP_2 - IP_1$), the greater the reactivity ($6 + \log k_2 = 0.029 + 3.983\ \Delta IP$), although a weighted equation gave significantly better correlation.[97] These trends are illustrated in Scheme XV and the accompanying table, which gives half-lives for photochemical reactivity of anthracenes and their annelated homologues.[93] These reactivities nicely parallel the Diels–Alder reactivities with maleic anhydride, thus implying that both ozone and singlet oxygen may react in the same way with these compounds.[93]

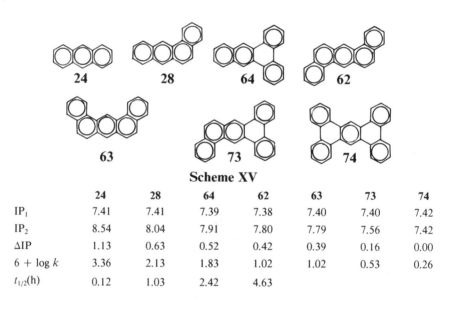

**Scheme XV**

|  | 24 | 28 | 64 | 62 | 63 | 73 | 74 |
|---|---|---|---|---|---|---|---|
| $IP_1$ | 7.41 | 7.41 | 7.39 | 7.38 | 7.40 | 7.40 | 7.42 |
| $IP_2$ | 8.54 | 8.04 | 7.91 | 7.80 | 7.79 | 7.56 | 7.42 |
| $\Delta IP$ | 1.13 | 0.63 | 0.52 | 0.42 | 0.39 | 0.16 | 0.00 |
| $6 + \log k$ | 3.36 | 2.13 | 1.83 | 1.02 | 1.02 | 0.53 | 0.26 |
| $t_{1/2}(h)$ | 0.12 | 1.03 | 2.42 | 4.63 |  |  |  |

Definitions analogous to paralocalization (ortholocalization) and 1,4-bislocalization energy (1,4-BLE) (1,2-bislocalization energy, 1,2-BLE) exist and have been applied toward understanding olefinlike reactivity including vicinal bromination and epoxidation. Thus, the most reactive double bond in phenanthrene is the 9,10-bond, as expected on the basis of structure **27**. Similarly, the formation of benzo[a]pyrene-4,5-epoxide (**21**) is consistent with the relatively low 1,2-BLE ($2.02\beta$) for the 4,5-bond in B[a]P, which could be compared with the high value for the 1,2-bond in triphenylene ($2.99\beta$).

# 5. SIMULATIONS OF ENVIRONMENTAL CONDITIONS

In 1956 Falk, Markul, and Kotin,[98] attempting to square their observations of PAH profiles in ambient air with source emissions, conducted studies of decomposition in the absence or presence of light and in the absence or presence of synthetic smog (ozone plus hydrocarbons). The results are somewhat clouded by volatilization losses (eg, phenanthrene). However, their most striking findings include the report that PAH adsorbed on soot exhibited negligible disappearance in the presence of air and light. This finding was later confirmed and extended by Korfmacher and co-workers,[100] who indicated that PAH, including B[a]P and anthracene, adsorbed on coal fly ash are extremely resistant to photochemical decomposition. This property has raised doubts[100] with respect to the frequently invoked supposition that photochemical decomposition plays an important role in atmospheric PAH losses, a view first seriously posed by Tebbens and associates.[101] These workers indicated a half-life with respect to photodecompositon of B[a]P on soot of less than 40 min, a value *enhanced* relative to glass beads, and also showed that PAH adsorbed to alumina are highly susceptible to photodecomposition.

Photodecomposition has been demonstrated on silica gel[91,102a] and alumina plates,[102a] while it is significantly reduced on cellulose or acetylated cellulose plates.[102a] Inscoe[102a] examined a total of 15 PAH and found that of this group, only phenanthrene, chrysene, triphenylene, and picene did not suffer reaction. Fatiadi[102b] followed up this study by showing that the photochemical decomposition of pyrene in soil produced 1,1'-bipyrenyl, pyrenediones, and pyrenediols. The finding of pyrenediones is consistent with photooxidation as the decomposition pathway, which in turn is consistent with later results.[91] Furthermore, the four unreactive PAH do not undergo Diels–Alder reaction with maleic anhydride (see Table 1-2),[70] consistent with the implicit similarity of the addition to the singlet oxygen cycloaddition. Photochemical decomposition on silica gel has been employed to estimate an atmospheric lifetime of 10 h for benzo[a]pyrene under summer conditions in Northern Europe,[103] which would correspond to a maximum transport distance of about 140 km.[103] Although airborne particulate matter has considerable silicaceous content, it would appear that soot is a more relevant medium and here, as noted, there appears to be a consensus supporting little photoreactivity under ambient conditions. Blau and Güsten[103] note that in winter, the PAH decompositon rate can be lower by a factor of up to 10 due to the lower zenith angle of the sun.

Since our later discussion will show that the ratio between winter and summer levels is of this order of magnitude, it might appear that Blau and Güsten are correct and that photodecompositon might be important. However, we feel that there are other explanations for the seasonal difference and that the argument concerning the relevancy of soot is compelling. Never-

TABLE 1-3. PAH Photochemical Lifetimes on Silica Gel and Their
Translation into Transport Ranges[a]

| PAH | Lifetime (h) | R (km) | $t_{1/2}$ (h) |
|---|---|---|---|
| Anthracene | 0.02 | 0.3 | 0.2 |
| Phenanthrene | 1100 | 15000 | — |
| Fluoranthene | 110 | 1600 | — |
| Pyrene | — | — | 4.2 |
| Benz[a]anthracene | 2.5 | 35 | 4.2 |
| Benzo[a]pyrene | 10 | 140 | 5.3 |
| Benzo[e]pyrene | — | — | 21.1 |
| Benzo[j]fluoranthene | 14 | 200 | — |
| Benzo[k]fluoranthene | 3 | 40 | 14.1 |
| Benz[e]acephenanthrylene | 80 | 1100 | 8.7 |
| Benz[a]aceanthrylene | 0.25 | 3.3 | — |
| Dibenz[ah]anthracene | — | — | 9.6 |
| Benzo[b]triphenylene | — | — | 9.2 |

[a]Results obtained using summertime northern European meteorological conditions[103] as well as photochemical half-lives on TLC plates.[104,105] (See references for specific reaction conditions and ozone concentrations that differ.)

theless, for the sake of comparison we list in Table 1-3 the observed lifetimes of PAH on silica gel.[103] In addition, we list photodecomposition half-lives on thin-layer chromatography (TLC) plates determined a few years earlier.[104,105] A study of photochemical reactivity of PAH in solution employed correlation with free valence ($F_r$), $L_r$, and $\sigma_r$, and the authors concluded that photooxidation involved PAH radicals.[106] The view that radical cations play a primary role in photooxidation has been mentioned elsewhere.[7b]

Model studies employing TLC are, of course, two-dimensional and less realistic than the three-dimensional and controlled study achievable using a fluidized bed.[107] These studies reemphasize the importance of the substrate in determining photochemical activity. Thus, the half-life of pyrene on Carbosieve 5 is reported to be about 20–30 h in air in the presence or absence of UV light.[107] The half-life of pyrene on coal fly ash is reported to be about the same, while curiously, it is reported to be reduced to 6–8 h in the presence of UV light.[107] A very interesting and important study of eight different coal fly ashes as well as model substrates showed significant differences in PAH reactivity. Thus, dark coal fly ashes, which tend to be relatively rich in carbon and iron, render adsorbed PAH unreactive because of their "inner filter" absorption of incident light.[108] In contrast, fly ashes relatively low in these substances are lighter in shade and their adsorbed PAH are photoreactive, as they are on silica and alumina.[108]

In contrast to the marked stabilization to photooxidation imparted by adsorption to (dark) fly ash, certain adsorbed PAH seem to be much more susceptible to nonphotochemical oxidation.[109] The compounds most susceptible, which have benzylic hydrogens, include fluorene, 3-methylcholanthrene, and 9,10-dimethylbenz[a]anthracene.[109] The latter two compounds are very potent carcinogens but appear to occur in extremely minute quantities if they are indeed detectable in airborne particulates.[109] Fluorenone (14) is present in a much greater quantity than fluorene and, in general, methylated PAH are present in very low quantities.[110] Curiously, the potent carcinogen 7,12-dimethylbenz[a]anthracene is unreactive in the dark when adsorbed to fly ash.[109] Other studies also clearly show the sensitivity of alkylated PAH to heat[66,111] and oxidation.[112] Oxidation of the benzylic carbon of 1,10-methanophenanthrene yields the corresponding lactone.[7c]

The role of the adsorbent is very crucial in determining realistic conditions of PAH reactivity. Thus, only recently has the rate constant for reaction of hydroxyl radical with a gas phase PAH (naphthalene) been determined.[113] Reaction with hydroxyl radical represents the most important loss process for most atmospheric organics. The experimental result with naphthalene indicates that under normal daytime atmospheric OH radical concentrations ($1 \times 10^6$ molecule/cm$^3$), the lifetime of (gas phase) naphthalene will be about 12 h. Subsequent studies of gas phase phenanthrene and anthracene found lifetimes of 9 h and 2 h, respectively.[114] In fact, the rate constants for reaction of these three small PAH as well as benzene and biphenyl show a good correlation with $IP_1$, consistent with the view that radical attack involves addition to the ring; this is also consistent with the correlations found with Nielsen's nitration studies, as reported in Reference 114. However, while naphthalene, anthracene, and phenanthrene have considerable volatility (they are lost during high-volume sampling), volatility losses are negligible for benzo[a]pyrene and comparably sized and larger PAH.[115-118] Thus, gas phase reactions for these species are not directly relevant to the particulate phase. Furthermore, the question arises whether hydroxyl could penetrate the surface of the particle. Hydroxyl-PAH are prone to react with $NO_3$ to yield aryloxy radicals, which rapidly react with oxygen to form quinones and with $NO_2$ to yield nitrophenolics.[7b] Reaction with hydroxyl radical is the major atmospheric removal mechanism for volatile aromatics such as benzene, toluene, and xylenes. The latter compounds are much more reactive than the unsubstituted parent benzene. Ratios of toluene to benzene and xylenes to benzene have thus been employed to assess aerosol transport.[119] It would be interesting to see whether reaction with hydroxyl radical in the gas phase could be the dominant removal mechanism for semivolatile substituted PAH such as methylnaphthalenes, methylated anthracenes and phenanthrenes, or even tetracyclic PAH.

Ozone levels in summer are much higher than in winter. For example, Table 1-4 lists data for $O_3$, $NO_2$, and $SO_2$ at a sampling site in Bayonne, New Jersey, during 1983. While no strong seasonal trends are found for $NO_2$ and $SO_2$, the summer levels of $O_3$ are more than five times those of the winter levels. This might strongly suggest that reaction with ozone represents an important PAH loss mechanism in the summer. It has also been noted that

TABLE 1-4. Monthly Averages of Three Gaseous Pollutants at the Bayonne, New Jersey, Veterans' Park Trailer During 1983[a]

| Month | Pollutant (ppm) | | |
|---|---|---|---|
| | $NO_2$ | $O_3$ | $SO_2$ |
| January | 0.035 | 0.006 | 0.020 |
| February | 0.033 | 0.012 | 0.018 |
| March | 0.020 | 0.018 | 0.013 |
| April | 0.036 | 0.022 | 0.012 |
| May | 0.029 | 0.032 | 0.010 |
| June | 0.045 | 0.034 | 0.015 |
| July | 0.033 | 0.038 | 0.012 |
| August | 0.037 | 0.036 | 0.012 |
| September | 0.040 | 0.026 | 0.014 |
| October | 0.027 | 0.010 | 0.016 |
| November | 0.032 | 0.008 | 0.019 |
| December | 0.024 | 0.007 | 0.019 |

[a]Data courtesy of Bureau of Air Pollution, New Jersey Department of Environmental Protection.

carbonaceous particulates such as soot are strong adsorbers of ozone, thus providing an effective concentration mechanism placing the reactive gas in proximity to surface-bound PAH.[120] This might be the cause of the accelerated reactivity of soot-bound PAH with ozone and other gases. As noted earlier, decreased photochemical reactivity appears to result from physical shielding of the PAH in particulate crevices and/or absorption of incident light by the particulate material. Reactions of PAH on model surfaces are quite rapid. For example, the half-life of B[a]P exposed in a petri dish to 190 ppb $O_3$ is 0.62 h,[105] and the half-life of B[a]P on a glass fiber filter exposed to 100–200 ppb $O_3$ was found to be about 1 h, with epoxide 21 as a major product.[33] Losses of B[a]P on glass fiber filters exposed to ambient air have been correlated with daily ozone levels.[121] Table 1-5 lists half-lives of some PAH as exposed to 0.2 ppm $O_3$ in the dark and also in light.[104,105] It is clear that benzo[e]pyrene is much less reactive to ozone than benzo[a]pyrene. The two nonalterant PAH benz[e]acephenanthrylene and benzo[k]fluoranthene, are much less reactive than even B[e]P. Light did not appreciably change the half-life of B[a]P, B[e]P, or dibenz[ah]anthracene and its isomer benzo[b]triphenylene.[104,105] However, dramatic effects were seen for other PAH. Van Vaeck and Van Cauwenberghe[120] found that the half-lives of most of the PAH they examined, adsorbed to diesel particulates ($< 0.5$ $\mu$m) and exposed to 1.5-ppm levels of $O_3$, were in the range of 0.5–1 h. They also found the benzofluoranthenes, which include benz[e]acephenanthrylene, to be more stable than the other PAH examined. Perylene, normally generated at very low levels (see earlier discussion), disappears very rapidly, and its isomer B[a]P reacts significantly more rapidly

TABLE 1-5. Heterogeneous Photooxidation and Ozonolysis Half-lives of PAH on TLC Plates[a]

| PAH | Half-life (h) | | |
|---|---|---|---|
| | Ozonolysis in dark (0.2 ppm) | Photooxidation in air | Photooxidation and ozonolysis |
| Anthracene | 1.23 | 0.2 | 0.15 |
| Benz[a]anthracene | 2.88 | 4.2 | 1.35 |
| Dibenz[ah]anthracene | 2.71 | 9.6 | 4.8 |
| Benzo[b]triphenylene | 3.82 | 9.2 | 4.6 |
| Pyrene | 15.72 | 4.2 | 2.75 |
| Benzo[a]pyrene | 0.62 | 5.3 | 0.58 |
| Benzo[e]pyrene | 7.6 | 21.1 | 5.38 |
| Benz[e]acephenanthrylene | 52.7 | 8.7 | 4.2 |
| Benzo[k]fluoranthene | 34.9 | 14.1 | 3.9 |

[a]Adapted from Reference 6 (Table 3-5, based on the work of Katz et al[104] and Lane and Katz.[105])

than B[e]P. Cyclopenta[cd]pyrene and benz[a]anthracene also disappear rapidly, but a significant part of this is a result of their volatilization.[120] The authors of this work note that the size fraction of diesel particulate (<0.5 μm), which corresponds to the fraction having the highest PAH concentrations, has a large specific surface area, thus putting the PAH in contact with carbon rather than silica and rendering 90% of the PAH available for reaction. They thus emphasize the inapplicability of model studies on silica particles. A recent study employed adsorption of B[a]P and perylene on fused silica plates and monitoring of PAH fluorescence upon exposure to ozone.[122] Among the noteworthy results were the finding of a half-life of only 2–6 min for B[a]P in 0.2 ppm O₃ and a much longer half-life (132 min) for perylene under the same conditions. The half-life obtained from B[a]P is the shortest yet reported under ozone. The authors also noted that unaggregated PAH are far more reactive than aggregated PAH, thus emphasizing another problem in simulation experiments.

As noted above, reactions with singlet oxygen as well as ozone produce PAH quinones. It is important to ascertain these PAH disappearance pathways by observing ambient concentrations of the products. In a very interesting study, Pierce and Katz[123] separated the quinone fraction from the PAH and then reduced it to the parent PAH with apparent recoveries of 70–85%. They found that the ratio percent of B[a]P-quinones to B[a]P increased from 3–4% in the winter months to almost 9% in August. It must be kept in mind that B[a]P is one of the more reactive PAHs and the quinones may include those formed during or shortly after combustion and during atmospheric residence, as well as on the filter during collection. Benz[a]anthracene (B[a]A) yields quinones under exposure to ozone and UV light considerably more slowly than benzo[a]pyrene. Thus, it does not appear too surprising that the maximum ratio percent of B[a]A-quinones to B[a]A—namely, 0.9%, also in August—is about a factor of 10 lower than that for B[a]P. The winter value for the ratio percent of B[a]A-quinones to B[a]A was only 0.1%.[123] The summer ratios are probably higher than they could be in reality because some volatility losses of B[a]A occur. Pierce and Katz also noted the presence of relatively large amounts of benzanthrone in their samples but were not able to propose a reasonable mechanism for its formation, especially in light of their simulations.[123]

Nitration of PAH is generally considered to be a very important atmospheric reaction. This is not, however, because nitration is a removal pathway for a significant fraction of the airborne PAH. In fact reported ambient levels of nitro- and dinitro-PAH are very much lower than those of the corresponding PAH,[124-126] although it might be argued that this is due to the high reactivity of the nitro-PAH. It is important to remember that reactions with nitrating species may yield hydroxy compounds and quinones in addition to nitro-PAH. However, many of these compounds, especially the dinitro-PAH, are extremely potent direct mutagens. A recent study of diesel particulate extract ($CH_2Cl_2$) in which 65 fractions were tested by unactivated Ames assay, found particularly significant activity in 8 fractions.[127] Within these fractions, 30–40% of the total direct mutagenic activity was attributed to only six compounds: 1-nitropyrene, 3- and 8-nitrofluoranthene, and 1,3-, 1,6-, and 1,8-dinitropyrene, with the last compound alone contributing between 10 and 19% of the direct mutagenic activity.[127]

In Table 1-6 we list the results of a solution phase study of PAH lifetimes and corresponding reactivity classes derived from these results and other correlations.[128,129] The solution phase involved 0.1–3 ppm concentrations of PAH in a water–methanol–dioxane mixture having 0.16 M $HNO_3$ and 0.016 M $NaNO_2$. These results are essentially in agreement with those of Dewar and co-workers.[83-84] However, Nielsen[128,129] finds for anthanthrene and perylene reactivities that he notes are anomalously large. Additionally, the high rate of reaction reported for picene, which Nielsen indicates did not give such clear-cut results, is clearly out of line with all expectations based on data in Table 1-2, which indicate that it should be very unreactive. Nielsen[128,129] found good correlations of ln $t_{1/2}$ against the position of the first p band, against the first IP and against a term to be defined as ($a \times 100/n_C n_H$) where $a$ is the maximum number of Clar aromatic sextets for a given PAH (see earlier discussion) and $n_C$ and $n_H$ are the number of carbons and hydrogens, respectively. Thus, for anthracene ($a = 1$), the value is 0.71 and for phenanthrene, ($a = 2$), the value is 1.43.

In placing compounds into different reactivity groups (I = most reactive; VI = least reactive), Nielsen employed extrapolations of reaction conditions and judicious interpretation of his own results. Class I includes compounds such as naphthacene, pentacene, and benzo[a]naphthacene, which are not found in ambient air samples. Dibenzo[ah]pyrene (ie, dibenzo[b,def]chrysene, (68) is placed in class I, but the basis for this assignment is not obvious, nor does it appear to fit the calculated reactivity parameters (Table 1-2) for this compound. There are other assignments that are

TABLE 1-6. PAH Half-lives in Solution and on Soot in the Presence of
Oxides of Nitrogen as well as Nielsen's Classification[a]

| PAH | Data from Ref. 128 | | Data from Ref. 24 |
|---|---|---|---|
| | $t_{1/2}$ (sol) | Class[b] | $t_{1/2}$ (soot) |
| Anthanthrene | 2.2 min | II | 3.7 days |
| Perylene | ~ 3 min | II | — |
| 9,10-Dimethylanthracene | 11 min | (I–II) | — |
| 9-Methylanthracene | 34 min | (I–II) | — |
| Picene | 100 min | III | — |
| 2-Methylanthracene | 140 min | (II) | — |
| Anthracene | 190 min | II | — |
| Benzo[a]pyrene | 200 min | II | 7 days |
| 1-Methylanthracene | 260 min | (II) | — |
| Benzo[ghi]perylene | 5.5 days | III | 8 days |
| Pyrene | 5.7 days | III | 14 days |
| Coronene | 23 days | IV | 29 days |
| Chrysene | 24 days | IV | 26 days |
| Dibenz[ah]anthracene | 29 days | IV | — |
| Benz[a]anthracene | 34 days | III | 11 days |
| Benzo[e]pyrene | > 100 days | IV | 24 days |
| Benzo[b]fluorene | > 200 days | V | — |
| Benzo[ghi]fluoranthene | >200 days | V | — |
| Benz[e]acephenanthrylene | > 300 days | V | — |
| Benzo[k]fluoranthene | > 300 days | V | — |
| Triphenylene | > 300 days | V | — |
| Phenanthrene | > 70 days | V | 30 days |
| Fluoranthene | > 400 days | V | 27 days |
| Benzo[j]fluoranthene | > 800 days | V | — |
| Indeno[1,2,3-cd]pyrene | > 800 days | V | — |

[a]Based on reactivity with related systems, where class I is most reactive and class VI is
least reactive.
[b]Other PAH in classes: I, benzo[a]naphthacene, dibenzo[b,def]chrysene, pentacene,
naphthacene; II, cyclopenta[cd]pyrene, dibenzo[def,p]chrysene, benzo[rst]pentaphene,
dibenzo[ac]naphthacene; III, benzo[g]chrysene, naphtho[1,2,3,4-def]chrysene; IV,
benzo[c]chrysene, benzo[c]phenanthrene, dibenzo[fg,op]naphthacene, dibenz[aj]an-
thracene, benzo[c]triphenylene; V, acenaphthylene.

not easily derived from data in Table 1-2. Nielsen assumes that the reactivities of the dibenzanthracenes (62–64) should be virtually identical because
the values of IP$_1$ are virtually equal. We earlier noted the observations of
Biermann and Schmidt[70] concerning the inappropriateness of simply equating reactivity and IP data. For this series, $L_r^+$ is also similar, with the slightly
lower value for dibenz[aj]anthracene predicting greater reactivity, but probably not enough to place it into a different class. Nielsen notes that the nonalternant PAH including fluoranthene, benzofluoranthenes, and
indeno[1,2,3-cd]pyrene are less reactive than predicted on the basis of the
p-band and aromaticity index criteria.

The third column of Table 1-6 lists data from exposure of freshly generated soot particles to air containing about 10 ppm oxides of nitrogen ($NO_x$: considered to contain 1-2 ppm $NO_2$ and 8-9 ppm of relatively unreactive NO).[24] In typical urban atmospheres in close proximity to traffic, $NO_x$ levels are usually only about 1 ppm.[24] Thus, the values in the third column of Table 1-6 would seem to be considerably shorter than realistic lifetimes (ie, in considerable excess of 7 days for B[a]P). Other studies have shown very rapid but low yield conversion of PAH under $NO_2$. For example, dinitro-B[a]P was formed from B[a]P on fly ash under 1.33 ppm $NO_2$ in 4 h.[130] On the other hand, Pitts and associates find 40% conversion of B[a]P adsorbed on glass fiber filters exposed to 1 ppm $NO_2$ for 8 h.[21] This might be due to a catalytic effect of glass fiber filters, which is why some groups of investigators employ Teflon-coated filters.[131] An important observation is that the reactivity of PAH toward $NO_2$ increases significantly with the acidity of the substrate.[132] The lowest reactivity is observed on soot. This is also somewhat consistent with the work of Brorstrom and colleagues,[133] in which PAH on "side-by-side" air particulates suffer the greatest losses on exposure to about 120 ppb airborne $HNO_3$. These workers found significant degradation under $NO_2$ and estimated that 0.2 ppm $NO_2$ could decompose 40% of BaP during collection,[133] possibly implying that $HNO_3$ is the actual nitrating agent. This is also consistent with the view that the reactivity of $NO_2$ with PAH depends strongly on the moisture of the air.[134] The major atmospheric fate of the nitro-PAH appears to be conversion to quinones.[21]

At realistic atmospheric concentrations of $SO_2$ ($< 5$ ppm), there is reported to be no significant reaction of this gas with PAH.[24,132] However, it is possible that $SO_2$ and sulfuric acid catalyze nitration. Furthermore, one might anticipate possible reactions with $SO_2$ in stacks, where high concentrations are accompanied by high temperature. [Similarly, the half-lives of PAH injected into hot diesel exhaust (300°C) are on the order of 0.01-0.4 s.[135]] On the other hand, $SO_x$ and $NO_x$ become more oxidized and acidic as they reside in air. Reactivity increases even as concentration decreases. Low levels of $SO_3$ are known to form PAH-sulfonic acids, and not much is known concerning their toxicity.[7b]

The previously cited results appear to strongly imply consensus on the reactivity of PAH under ambient conditions, during atmospheric residence, and/or during collection. It is clear that the nature of the adsorption medium is crucial and that the concentrations of airborne reactants should be realistic. However, in the face of this apparent consensus comes a report by Grosjean, Fung, and Harrison[136] of a study of B[a]P, perylene, and 1-nitropyrene adsorbed on fly ash, diesel exhaust, and ambient air particulates loaded onto glass fiber and Teflon filters. Each experimental trial employed six such loaded filters (three glass, three Teflon) plus a glass blank and a Teflon blank. These researchers found no detectable decomposition in the dark using pure filtered air, pure air containing about 100 ppb $NO_2$ (from which nitric acid was removed using nylon filters), about 100 ppb $O_3$, about 100 ppb $SO_2$, and ambient air. They found no evidence for formation of new chemical products from the three compounds and have attempted to demonstrate the consistency of their results with those of earlier investigations. They note that the concentrations of gases employed were much more relevant to ambient conditions than most earlier studies, which had very high gas levels often accompanied by unrealistically high PAH levels. However, even when (HNO$_3$-free) NO$_2$ levels were increased by a factor of 7 and B[a]P concentration was increased by a thousand-fold, no evidence of nitration was observed. On the other hand, passive exposure to nitric acid vapor in pure air yielded virtually 100% conversion of B[a]P after 20 h and the products appeared to be nitro-B[a]P, possibly a dinitro-B[a]P, and perhaps B[a]P-quinones. Previously, it was noted that nitro-PAH may be photooxidized to yield quinones. The implications of the study by Grosjean and associates are as follows:

1. Previous studies of PAH reactivity with NO$_2$ did not exclude nitric acid, which was present at artificially high levels.
2. The principle reactant in such studies appears to be nitric acid.
3. Nitric acid was undoubtedly present during the experiments in which particulates were exposed to particle-free ambient air; however, realistically low NO$_x$ levels require exceedingly low HNO$_3$ levels, which do not appear to cause significant reaction.
4. It would appear that sampling losses and losses during atmospheric residence should be negligible even for reactive PAH such as B[a]P and perylene. It is unfortunate that analogous data have not yet been obtained for more reactive carcinogens such as cyclopenta[cd]pyrene.

# 6. PHENOMENOLOGICAL ESTIMATIONS OF THE ATMOSPHERIC REACTIVITY OF POLYCYCLIC AROMATIC HYDROCARBONS

The atmospheric simulation experiments discussed in the preceding sections have given rise to conflicting estimates of PAH lifetimes. These discrepancies are due to the extreme complexity in simulating realistic environmental conditions, particulate matter, and modes of adsorption.

Another approach to understanding ambient PAH reactivities is to carefully analyze environmental data to see whether they incorporate such information. We term this a phenomenological approach, since one may discern patterns of chemical reactivity without knowing what the reactions are. While one may obtain circumstantial evidence (eg, inverse correlation between PAH concentrations and ozone levels), laboratory simulations and mechanistic investigations must eventually be performed. Still, since the issue of chemical versus physical disappearance mechanisms has yet to be resolved, this is an important area for investigation.

One of the most striking findings concerning PAH is the extreme seasonal variation in their concentrations. Thus, our study of B[a]P levels at 27 New

TABLE 1-7. Comparison of Monthly Means for
Benzo[a]pyrene and Total Suspended Particulates (TSP) for
27 Sites in New Jersey Throughout 1982.[a]

| Month | Benzo[a]pyrene (ng/m$^3$) | TSP ($\mu$g/m$^3$) |
|---|---|---|
| January | 0.79 | 39.6 |
| February | 0.67 | 45.7 |
| March | 0.54 | 49.5 |
| April | 0.28 | 44.3 |
| May | 0.24 | 51.3 |
| June | 0.17 | 62.8 |
| July | 0.10 | 50.1 |
| August | 0.09 | 44.3 |
| September | 0.13 | 53.0 |
| October | 0.37 | 50.1 |
| November | 0.57 | 38.8 |
| December | 1.22 | 39.6 |

[a]From Reference 10.

Jersey sites during 1982 found levels in the highest month (December) that were about 13 times those in the lowest month (August) (see Table 1-7).[10] This is not inconsistent with other studies,[46,137] in which we found the average concentrations during 6-week periods in January and February to be about five times those in July and August, since the 6-week campaigns did not focus on minimum and maximum months. The levels of TSP show less seasonal variation and tend to be higher during the summer months (Table 1-7). Another particle-bound pollutant, lead, which is not reactive, tends to have winter concentrations about twice as high as summer concentrations.[138] Since in New Jersey gasoline consumption is constant throughout the year,[10] this result is probably due to lower average inversion heights in the winter and seasonal variability in wind speed.

Is the great difference in seasonal PAH levels a consequence of greater chemical reactivity during the summer or greater input during the winter? A glance at Table 1-4 might immediately suggest an obvious inverse correlation with ozone, thus implying a specific disappearance pathway. But this, of course, would assume that a mere correlation proves cause and effect. Furthermore, high ozone levels accompany increased sunlight, which may increase photoreactivity. If motor vehicle traffic were the major source of PAH in New Jersey (not immediately unreasonable, given the state's high traffic density and low use of coal and wood for heating), one would expect fairly constant input throughout the year. The observed seasonal trends would thus suggest the importance of chemical disappearance pathways.

What are the major PAH emission sources in New Jersey? A survey of emission rates of various fuels (Table 1-1) and their annual usage in New Jersey led to the results portrayed for B[a]P in Table 1-8.[10] One must note that the greatest uncertainties are associated with wood combustion, involv-

TABLE 1-8. Estimated Annual, Heating Season, and Nonheating Season Benzo[a]pyrene Emission Rates for New Jersey[10]

| Fuel type | User | Estimated emission rate (ng/Btu) | Estimated emissions (kg) | | |
|---|---|---|---|---|---|
| | | | Annual | Heating season[a] | Nonheating season |
| **Solid** | | | | | |
| Coal | Utilities | $6.1 \times 10^{-2}$ | 4.5 | 1.9 | 2.6 |
| Coal | Residential | 37.7 | 3.8 | 3.8 | — |
| Wood | Residential | 227 | 6,129 | 6,129 | — |
| | | | 6,137 | 6,135 | 2.6 |
| **Oil** | | | | | |
| Heating | Residential | $2.6 \times 10^{-3}$ | 0.4 | 0.4 | — |
| Miscellaneous distillates | Commercial/industrial | $2 \times 10^{-4}$ | 0.2 | 0.1 | 0.1 |
| Residual | Utilities/commercial/ industrial | $4.3 \times 10^{-4}$ | <0.1 | — | — |
| | | | 0.6 | 0.5 | 0.1 |
| **Natural gas** | | | | | |
| Heating | Residential | $2.0 \times 10^{-4}$ | 0.1 | 0.1 | — |
| **Motor fuels** | | | | | |
| Gasoline | Autos/trucks | 0.6 | 228 | 95 | 133 |
| Diesel | Trucks/buses | 2.3 | 85 | 35 | 50 |
| | | | 313 | 130 | 183 |
| | | Total | 6451 kg | 6266 | 186 |

[a]November–March.

ing as much as an order of magnitude in residential use and two orders of magnitude in emission rates. We arrived at the intuitively reasonable conclusion that about 98% of the summer PAH comes from motor vehicle emissions. However, we were surprised to be obliged to attribute 98% of statewide B[a]P emissions to wood combustion. Even within the confines of the uncertainties in wood B[a]P emissions, the small percentage of households (1–2% in 1982) that burn wood clearly seem to make a major impact. The reason is that B[a]P emissions from residential wood combustion yield about 100,000 times more B[a]P per Btu than heating oil combustion. The total winter levels are estimated to be about 35 times those of the summer levels, in surprisingly decent agreement with the factor of 13 between December and August. Moreover, winter B[a]P levels appear to reflect the population density of wood burning.[10] At first glance then, it would seem that differences in seasonal output account for the relative B[a]P concentrations. However, the analysis is very approximate and phenomenological to the extreme. Moreover, even if conversion to PAH derivatives is small, this investigation is still vital, since some of the derivatives are extremely potent carcinogens and direct mutagens.

That the most reactive PAH are rapidly transformed in the environment is well established. Thus, as noted earlier, one finds cyclopenteno-PAH such as acephenanthrylene (39) in recent emissions but not in ambient particulate matter.[6,43] Cyclopenta[cd]pyrene is found at levels more than 10 times those of B[a]P in the emissions of automobiles and in traffic tunnels, but only very low levels are found in ambient air.[45] Naphthacene is not found in ambient air samples but its quinone is. Methyl-PAH are produced in abundance in new emissions but found at low levels in ambient samples.

What of the remaining PAH that are not linear acenes, alkyl-PAH, or cyclopenteno-PAH? The reader is reminded of Grosjean's[136] work on the relatively reactive B[a]P and perylene, which showed negligible decomposition except in the presence of unnaturally high concentrations of nitric acid. We also remind the reader of the low levels of B[a]P-quinones, possible end products of reactions with singlet oxygen, ozone, and $NO_2$ found during the summer, as well as the even lower levels of nitro-PAH.

An interesting approach to assessing ambient PAH reactivity was proposed by Duval and Friedlander.[139] In advancing a receptor model based on chemical mass balance, they employed Equation 1-8 to incorporate the effects of decomposition for a reactive pollutant. These workers derived the following decay factors ($a_{ij}$): B[a]P (0.48 ± 0.21); B[e]P (1.04 ± 0.29), which is assumed to be a stable species; benzofluoranthenes (0.98 ± 0.26); benzo[ghi]perylene (0.83 ± 0.19); and anthanthrene (0.21 ± 0.11).

$$p_i = \sum_{j=1}^{p} a_{ij} x_{ij} v_{ij} \tag{1-8}$$

where $p_i$ = mass concentration of species $i$ measured at receptor site

$a_{ij}$ = dimensionless decay factor (fraction of species $i$ emitted from source $j$ remaining at receptor site)

$x_{ij}$ = dimensionless ratio of mass of species $i$ to the reference species in the emission from source $j$

$v_{ij}$ = mass concentration of reference species (unreactive) from source $j$ at the point of measurement

These factors were derived by observing PAH concentrations in California near a freeway and comparing them with data from automobile emissions. Unfortunately, the automobile data were derived from European studies and also different analytical procedures. The decay factors certainly reflect our ideas concerning the relative reactivities of these PAH.

Another approach was employed by Nikolaou and co-workers,[140] who observed that the daily high PAH concentration ($C_M$) in Paris occurs at about 9 A.M. and the daily minimum concentration ($C_m$) occurs at about 4 P.M. regardless of season. They defined a ratio of these levels ($\rho = C_M/C_m$) for each PAH, and to correct for daily and seasonal variations in total PAH levels, they normalized values with respect to the unreactive PAH fluoranthene:

$$R = \frac{\rho_{PAH}}{\rho_{fluoranthene}} \tag{1-9}$$

Values of this relative stability index are reproduced in Table 1-9, where high $R$ connotes high reactivity. The observed $R$ values have some corre-

TABLE 1-9. Values of Empirical Relative Stability
Parameters, $(R)^a$

| PAH | R (Winter) | (Summer) |
|---|---|---|
| Benzo[a]pyrene | 3.5 | 7 |
| Coronene | 3 | 3 |
| Benz[a]anthracene | 3 | 3 |
| Dibenz[ah]anthracene | 2.3 | 4.7 |
| Benzo[ghi]perylene | 1.7 | 4.3 |
| Indeno[1,2,3-cd]pyrene | 1.7 | 4.0 |
| Benzo[k]fluoranthene | 1.7 | 3.0 |
| Benz[e]acephenanthrylene | 1.7 | 3.0 |
| Anthracene | 1.7 | 3.0 |
| Chrysene | 1.3 | 3.0 |
| Benzo[e]pyrene | 1.3 | 2.7 |
| Pyrene | 1.7 | 1.7 |
| Fluoranthene | (1.0) | (1.0) |

[a]From reference 140, with fluoranthene taken as the standard.

lation with reactivities derived elsewhere. Thus, B[a]P is seen to be much more reactive than B[e]P and the benzofluoranthenes. It also appears that reactivity increases during the summer. However, anthracene appears to be anomalously stable, and we are concerned about the lack of correction for volatilization losses. Furthermore, normalization to another unreactive and less volatile PAH (eg, B[e]P or a benzofluoranthene) would yield a much smaller R scale and smaller apparent seasonal changes. A systematic error or other artifact in fluoranthene analyses would drastically effect the study's conclusions.

Our observations[46] have included little significant change in PAH profiles between seasons for PAH of significantly different reactivities. We would expect the more reactive PAH to have concentrations lower, relative to less reactive PAH, in the summer than in the winter. Our findings indicate this to be true for volatile PAH such as chrysene, benz[a]anthracene, pyrene, and cyclopenta[cd]pyrene. However, we find the largest winter/summer difference for the last-named PAH and feel that it also reflects the high reactivity of this (and similar) cyclopenteno-PAH.[46]

In summary, it appears that only extremely reactive PAH such as cyclopenteno-PAH, alkyl-PAH, and linear acenes, as well as certain highly reactive condensed PAH (eg, anthanthrene) suffer extensive atmospheric decomposition. It would seem that B[a]P and other moderately as well as less reactive PAH disappear largely by physical mechanisms. The greatest losses almost certainly occur during the summer, when insolation as well as concentrations of $O_3$, $^1O_2$, and OH are relatively high. Also, summer particulates tend to be more lightly shaded than winter particulates and thus less effective as "inner filters." However, even minor pathways producing carcinogenic derivatives are important from the health perspective. The high winter/summer PAH ratio in New Jersey is largely the result of high winter input rather than extensive summer decomposition.

We conclude this chapter with one last, if somewhat simplistic, idea for gaining evidence concerning the nature of atmospheric species that attack PAH. Table 1-2 demonstrated that the $L_r^+$ and $L_r^{\cdot}$ values for fully benzenoid (even-alternant) PAH are equal, while these values are frequently not equal for nonalternant PAH. This well-known principle[52] is related to the fact that the most reactive positions to cations and radicals are often different in the nonalternant PAH. Scheme XVI shows that while the same position in

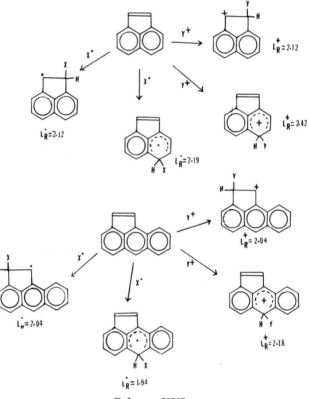

**Scheme XVI**

acenaphthylene is attacked by cations and radicals ($L_r^+ = L_r^{\cdot} = 2.12$), different positions are susceptible in aceanthrylene where $L_r^+ = 2.04$ and $L_r^{\cdot} = 1.94$ for the most reactive positions. A glance at Table 1-2 indicates that cations and radicals attack the same position on benzo[$k$]fluoranthene but different positions on, for example, benz[$e$]acephenanthrylene. Thus, one might gain insight into the nature of the attacking species through careful analysis of the isomeric distribution of monosubstituted derivatives of some of these nonalternant PAH.

## ACKNOWLEDGMENTS

We gratefully acknowledge collaborative studies with Drs. Ronald Harkov and Judy Louis and Ms. Leslie McGeorge of the Office of Science and Research of the New Jersey Department of Environmental Protection (NJDEP), which largely funded this research. The Office of Air Quality of NJDEP has been extremely helpful and supportive and collected all samples for the 1982 B[a]P study.

We have also benefited from discussions and collaboration with Drs. Joan Daisey and Paul Lioy of the New York University College of Medicine, Dr. Edward J. Lavoie of the Naylor-Dana Institute, Drs. Steven Stein and Robert Brown of the National Bureau of Standards, and Prof. Joel F. Liebman of the University of Maryland Baltimore County. Studies at the New Jersey Institute of Technology were ably assisted by Dina Natsiashvili, Nina and Alexander Kovalyov, Clint Brockway, and Janis Racz.

## REFERENCES

1. (a) Clar, E. "Polycyclic Hydrocarbons". Academic Press: London, 1964. (b) Clar, E. "The Aromatic Sextet". Wiley: London, 1972.
2. National Academy of Sciences. "Particulate Organic Matter". NAS: Washington, D.C., 1972.
3. Lee, M.L.; Novotny, M.V.; Bartle, K.D. "Analytical Chemistry of Polycyclic Aromatic Compounds". Academic Press: New York, 1982.
4. Grimmer, G. "Environmental Carcinogens: Polycyclic Aromatic Hydrocarbons". CRC Press: Boca Raton, Fla., 1983.
5. Bjorseth. A., Ed. "Handbook of Polycyclic Aromatic Hydrocarbons". Dekker: New York, 1983.
6. National Academy of Sciences. "Polycyclic Aromatic Hydrocarbons: Evaluations of Sources and Effects". NAS: Washington, D.C., 1983.
7. (a) Nikolaou, K.; Masclet, P.; Mouvier, G. Sci. Total Environ. 1984, 32, 103–132. (b) Nielsen, T.; Ramdahl, T.; Bjorseth, A. Environ. Health Perspect. 1983, 47, 103–114. (c) Pitts, J.N., Jr. ibid, 1983, 47, 115–140. (d) Van Cauwenberghe, K.; Van Vaeck, L. Mutat. Res. 1983, 116, 1–20. (e) Edwards, N.J. J. Environ. Qual. 1983, 12, 427.
8. Haemisegger, E.; Jones, A.; Steigerwald, B.; Thomsen, V. "The Air Toxics Problem in the United States: An Analysis of Cancer Risks for Selected Pollutants", EPA-450/1-85-001, U.S. Environmental Protection Agency; Washington, D.C., May, 1985.
9. Faoro, R.B.; Manning, J.A. J. Air Pollut. Control Assoc. 1981, 31, 62–64.
10. Harkov, R.; Greenberg, A. J. Air Pollut. Control Assoc. 1985, 35, 238–243.
11. Jerina, D.M.; Yagi, H.; Lehr, R.E.; Thakker, D.R.; Schaeffer-Ridder, M.; Karle, J.M.; Levin, W.; Wood, A.W.; Chang, R.L.; Conney, A.H. In "Polycyclic Hydrocarbons and Cancer", Vol. 1; Gelboin, H.V.; and Ts'O, P.O.P., Eds.; Academic Press: New York, 1981, pp. 173–188.
12. (a) Harvey, R.G. Acc. Chem. Res. 1981, 14, 218–226. (b) Harvey, R.G. Am. Sci. 1982, 70, 386–393.
13. Ames, B.N.; McCann, J.; Yamasaki, E. Mutat. Res. 1975, 31, 347–363.
14. Blumer, M.; Blumer, W.; Reich, T. Environ. Sci. Technol. 1977, 11, 1082–1084.

15. Blumer, M.; Youngblood, W.W. *Science* **1975**, *188*, 53–55.
16. LaFlamme, R.E.; Hites, R.A. *Geochim. Cosmochim. Acta* **1978**, *42*, 289–303.
17. (a) Lunde, G.; Bjorseth, A. *Nature* **1977**, *268*, 518–519. (b) Bjorseth, A.; Lunde, G.; Lindskog, A. *Atmos. Environ.* **1977**, *13*, 45–53. (c) Bjorseth, A.; Sortland Olufsen, B. In Reference 5, Chapter 12.
18. Miguel, A.H.; Friedlander, S.K. *Atmos. Environ.* **1978**, *12*, 2407–2413.
19. See Reference 6, Chapter 3.
20. Muller, J. *Sci. Total Environ.* **1984**, *36*, 339–346.
21. Pitts, J.N., Jr.; Van Cauwenberghe, K.A.; Grosjean, D.; Schmid, J.P.; Fitz, D.R.; Belser, W.L., Jr.; Knudson, G.B.; Hynds, P.M. *Science*, **1978**, *202*, 515–518.
22. Pitts, J.N., Jr. *Phil. Trans. R. Soc. London Ser. A.*, **1979**, *290*, 551–576.
23. Rosenkranz, H.S. *Mutat. Res.* **1982**, *101*, 1–10.
24. Butler, J.D.; Crossley, P. *Atmos. Environ.* **1981**, *15*, 91–94.
25. Andrews, P.A.; Bryant, D.; Vitakunas, S.; Gouin, M.; Anderson, G.; McCarry, B.E.; Quilliam, M.A.; McCalla, D.R. In "Polynuclear Aromatic Hydrocarbons: Formation, Metabolism and Measurement", Cooke, M.; and Dennis, A.J., Eds.; Battelle Press: Columbus, Ohio: 1983, pp. 89–98.
26. Pitts, J.N., Jr.; Grosjean, J.; Mischke, T.M.; Simmon, V.F.; Poole, D. *Toxicol. Lett.* **1977**, *1*, 65–70.
27. Wang, Y.Y.; Rappaport, S.M.; Sawyer, R.F.; Talcott, D.; Wei, E.T. *Cancer Lett.* **1978**, *5*, 39–47.
28. Jager, J. *J. Chromatogr.* **1978**, *152*, 575–578.
29. Tokiwa, H.; Nagawa, R.; Ohnishi, Y. *Mutat. Res. 1981*, 91, 321–325.
30. Schuetzle, D.; Riley, T.L.; Prater, T.J.; Harvey, T.M.; Hunt, D.F. *Anal. Chem.* **1982**, *54*, 265–271.
31. Gibson, T.L. *Atmos. Environ.* **1982**, *16*, 2037–2040.
32. D'Agostino, P.A.; Narine, D.R.; McCarry, B.E.; Quilliam, M.A. In "Polynuclear Aromatic Hydrocarbons: Formation, Metabolism, and Measurement", Cooke, M.; and Dennis, A.J., Eds.; Battelle Press: Columbus, Ohio, 1983, pp. 365–377.
33. Pitts, J.N., Jr.; Lokensgard, D.M.; Ripley, P.S.; Van Cauwenberghe, K.A.; Van Vaeck, L.; Shaffer, S.D.; Thill, A.J.; Belser, W.L., Jr. *Science* **1980**, *210*, 1347–1349.
34. Cautreels, W.; Van Cauwenberghe, K. *Atmos. Environ.* **1976**, *10*, 447–457.
35. Pierce, R.C.; Katz, M. *Environ. Sci. Technol.* **1976**, *10*, 45–51.
36. Konig, J.; Balfanz, E.; Funcke, W.; Romanowski, T. *Anal. Chem.* **1983**, *55*, 599–603.
37. Ramdahl, T. *Environ. Sci. Technol.* **1983**, *17*, 666–670.
38. (a) Lavoie, E.; Rice, J. Unpublished observations. (b) Adams, J.D.; Lavoie, E.J.; Hoffmann, D. *J. Chromatogr. Sci.* **1982**, *20*, 274–277.
39. Grimmer, G.; Bohnke, H.; Glaser, A. *Erd-Kohle Erdgas Petrochem.* **1977**, *30*, 411–417.
40. Streitwieser, A., Jr.; Word, J.M.; Guibe, F.; Wright, J.S. *J. Org. Chem.* **1981**, *46*, 2588–2589.
41. Rappaport, S.M.; Wang, Y.Y.; Wei, E.T.; Sawyer, R.; Watkins, B.E.; Rapoport, H. *Environ. Sci. Technol.* **1980**, *14*, 1505–1509.
42. See Reference 6, Chapter 3.
43. Lee, M.L.; Prado, G.P.; Howard, J.B.; Hites, R.A. *Biomed. Mass Spectrom.* **1977**, *4*, 182–186.
44. Krishman, S.; Hites, R.A. *Anal. Chem.* **1981**, *53*, 342–343.
45. Grimmer, G.; Naujack, K.W.; Schneiden, D. In "Polynuclear Aromatic Hydrocarbons", Bjorseth, A.; and Dennis, A.J., Eds.; Battelle Press: Columbus, Ohio, 1981, pp. 107–125.
46. Greenberg, A.; Darack, F.; Harkov, R.; Lioy, P.; Daisey, J. *Atmos. Environ.* **1985**, *19*, 1325–1339.
47. Pauling, L. "The Nature of the Chemical Bond". Third ed., Cornell University Press: Ithaca, N.Y., 1961.
48. Wheland, G.W. "The Theory of Resonance and Its Application to Organic Chemistry". Wiley: New York, 1944.
49. Herndon, W.C. *J. Chem. Educ.* **1974**, *51*, 10–17.
50. Dewar, M.J.S.; Dougherty, R.C. "The PMO Theory of Organic Chemistry". Plenum Press: New York, 1975.
51. (a) Eilfeld, P.; Schmidt, W. *J. Electron. Spectrosc. Related Phenom.* **1981**, *24*, 101–120. (b) Ham, N.S.; Ruedenberg, K., *J. Chem. Phys.*, **1958**, *29*, 1215–1223.

52. Streitwieser, A., Jr. "Molecular Orbital Theory for Organic Chemists". Wiley: New York, 1961.
53. Zander, M. In "Handbook of Polycyclic Aromatic Hydrocarbons". Bjorseth, A., Ed.; Dekker: New York, 1983, pp. 1–25.
54. Pedley, J.B.; Rylance, J. "Sussex-N.P.L. Computer-Analysed Thermochemical Data: Organic and Organometallic Compounds". University of Sussex, Brighton, U.K., 1977.
55. Novotny, M.; Hirose, A.; Wiesler, D. *Anal. Chem.* **1984,** *56,* 1243–1248.
56. Somayajulu, G.R.; Zwolinski, B.J. *J. Chem. Soc. Faraday Trans. 2* **1975,** 1928–1941.
57. (a) Stein, S.E.; Golden, D.M.; Benson, S.W. *J. Phys. Chem.* **1980,** *81,* 314–317. (b) Stein, S.E.; Barton, B.D. *Thermochim. Acta* **1981,** *44,* 265–281.
58. Stein, S.E. *J. Phys. Chem.* **1978,** *82,* 566–571.
59. Greenberg, A.; Stein, S.E.; Brown, R. *Sci. Total Environ.* **1984,** *40,* 219–230.
60. Badger, G.M., *Prog. Phys. Org. Chem.* **1965,** *3,* 1–40.
61. Grimmer, G.M.; Jacob, J.; Naujack, K.-W.; Dettbarn, G. *Anal. Chem.* **1983,** *55,* 892–900.
62. Alsberg, T.; Stenberg, W. *Chemosphere* **1979,** *8,* 487–496.
63. Schaaden, G. *Z. Naturforsch. B: Anorg. Org. Chem.* **1980,** *35,* 1328.
64. Hess, B.A., Jr.; Schaad, L.J. *J. Am. Chem. Soc.* **1971,** *93,* 305–310.
65. Friedman, L.; Lindow, D.F. *J. Am. Chem. Soc.* **1968,** *90,* 2324–2328.
66. Lewis, I.C.; Edstrom, T. *J. Org. Chem.* **1963,** *28,* 2050–2057.
67. Herndon, W.C. *Tetrahedron* **1982,** *38,* 1389–1396.
68. Levin, R.D.; Lias, S.G. "Ionization and Appearance Potential Measurements, 1971–1981", NSRDS-NBS 71. Department of Commerce, National Bureau of Standards: Washington, D.C., 1982.
69. Boschi, R.; Clar, E.; Schmidt, W. *J. Chem. Phys.* **1974,** *60,* 4406–4418.
70. Biermann, D.; Schmidt, W. *J. Am. Chem. Soc.* **1980,** *102,* 3163–3173, 3173–3181.
71. (a) Clar, E.; Robertson, J.M.; Schlogl, R.; Schmidt, W. *J. Am. Chem. Soc.* **1981,** *103,* 1320–1328. (b) Simonsick, W.J., Jr.; Hites, R.A. In "Polynuclear Aromatic Hydrocarbons: Mechanisms, Methods and Metabolism", Cooke, M.; and Dennis, A.J., Eds.; Battelle Press: Columbus, Ohio, 1985, pp. 1227–1237.
72. Schmidt, W. *J. Chem. Phys.* **1977,** *66,* 828–845.
73. (a) Cavalieri, E.; Rogan, E. In "Polynuclear Aromatic Hydrocarbons: Formation, Metabolism and Measurement", Cooke, M.; and Dennis, A.J., Eds.; Battelle Press: Columbus, Ohio, 1983, pp. 1–26. (b) Cavalieri, E.; Rogan, E.; Warner, C.; Bobst, A. In "Polynuclear Aromatic Hydrocarbons: Mechanisms, Methods and Metabolism", Cooke, M.; and Dennis, A.J., Eds.; Battelle Press: Columbus, Ohio, 1985, pp. 227–236.
74. Plummer, B.F.; Al-Saigh, Z.Y.; Arfan, M. *J. Org. Chem.* **1984,** *49,* 2069–2070.
75. Becker, H.-D.; Hansen, L.; Andersson, K. *J. Org. Chem.* **1985,** *50,* 277–279.
76. Lias, S.G.; Liebman, J.F.; Levin, R.D. *J. Phys. Chem. Ref. Data* **1984,** *13,* 695–808.
77. Baker, R.; Eaborn, C.; Taylor, R. *J. Chem. Soc. Perkin Trans. 2* **1972,** 97–101.
78. Ansell, H.V.; Hirschler, M.M.; Taylor, R. *J. Chem. Soc. Perkin Trans. 2* **1977,** 353–355.
79. Le Guen, J.; Taylor, R. *J. Chem. Soc. Trans. 2* **1974,** 1274–1277.
80. Archer, W.J.; Taylor, R.; Gore, P.H.; Kamounah, F.S. *J. Chem. Soc. Perkin Trans. 2* **1980,** 1828–1831.
81. Streitwieser, A., Jr.; Lewis, A.; Schwager, I.; Fish, R.W.; Labana, S. *J. Am. Chem. Soc.* **1970,** *92,* 6525–6529.
82. Shawali, A.S.; Hassaneen, H.M.; Parkanyi, C.; Herndon, W.C. *J. Org. Chem.* **1983,** *48,* 4800–4803.
83. Dewar, M.J.S.; Warford, E.W.T. *J. Chem. Soc.* **1956,** 3570–3572.
84. Dewar, M.J.S.; Mole, T.; Warford, E.W.T. *J. Chem. Soc.* **1956,** 3576–3580, 3581–3586.
85. Davies, A.; Warren, K.D. *J. Chem. Soc. B* **1968,** 1337–1339.
86. Bancroft, K.C.C.; Bott, R.W.; Eaborn, C. *J. Chem. Soc.* **1964,** 4806–4811.
87. Archer, W.J.; Taylor, R. *J. Chem. Soc. Trans. 2* **1981,** 1153–1156.
88. Streitwieser, A., Jr.; Hammond, H.A.; Jagow, R.H.; Williams, R.M.; Jesaitis, R.G.; Chang, C.J.; Wolf, R. *J. Am. Chem. Soc.* **1970,** *92,* 5141–5150.
89. Streitwieser, A., Jr.; Mowery, P.C.; Jesaitis, R.G.; Lewis, A. *J. Am. Chem. Soc.* **1970,** *92,* 6529–6533.
90. Herndon, W.C. *J. Org. Chem.* **1975,** *40,* 3583–3586.
91. Fox, M.A.; Olive, S. *Science* **1979,** *205,* 582–583.

92. Eisenberg, W.C.; Taylor, K.; Cunningham, D.L.B.; Murray, R.W. In "Polynuclear Aromatic Hydrocarbons: Mechanisms, Methods and Metabolism", Cooke, M.; and Dennis, A.J., Eds.; Battelle Press: Columbus, Ohio, 1985, pp. 395–410.
93. Konig, J.; Balfanz, E.; Funcke, W.; Romanowski, T. In "Polynuclear Aromatic Hydrocarbons: Mechanisms, Methods and Metabolism", Cooke, M.; and Dennis, A.J., Eds.; Battelle Press: Columbus, Ohio, 1985, pp. 739–748.
94. Van Vaeck, L.; Broddin, G.; Van Cauwenberghe, K. *Biomed. Mass Spectrom.* **1980**, *7*, 473–483.
95. Brown, R.D. *J. Chem. Soc.* **1950**, 691–697, 3249–3254.
96. Brown, R.D. *J. Chem. Soc.* **1951**, 1612–1614, 3129–3131.
97. Hess, B.A., Jr.; Schaad, L.J.; Herndon, W.C.; Biermann, D.; Schmidt, W. *Tetrahedron* **1981**, *37*, 2983–2987.
98. Swinbourne-Sheldrake, R.; Herndon, W.C.; Gutman, I. *Tetrahedron Lett.* **1975**, 755–758.
99. Falk, H.L.; Markul, I.; Kotin, P. *A.M.A. Arch. Ind. Health* **1956**, *13*, 13–17.
100. Korfmacher, W.A.; Wehry, E.L.; Mamantov, G.; Natusch, D.F.S. *Environ. Sci. Technol.* **1980**, *14*, 1094–1099.
101. (a) Tebbens, B.D.; Thomas, J.F.; Mukai, M. *Am. Ind. Hyg. Assoc. J.* **1966**, *27*, 415–422. (b) Thomas, J.F.; Mukai, M.; Tebbens, B.D. *Environ. Sci. Technol.* **1968**, *2*, 33–39.
102 (a) Inscoe, M.N. *Anal. Chem.* **1964**, *36*, 2505–2506. (b) Fatiadi, A. *Environ. Sci. Technol.* **1967**, *1*, 570–572.
103. Blau, L.H.; Güsten, H. In "Polycyclic Aromatic Hydrocarbons: Physical and Biological Chemistry", Cooke, M.; Dennis, A.J.; and Fisher, A.J., Eds.; Battelle Press: Columbus, Ohio, 1982, pp. 133–144.
104. Katz, M.; Chen, C.; Tosine, H.; Sakuma, T. In "Polynuclear Aromatic Hydrocarbons: Third International Symposium in Chemistry and Biology—Carcinogens and Mutagens", Jones, P.W.; and Leber, P., eds.; Ann Arbor Science Publishers, Ann Arbor, Mich.: 1979, pp. 171–189.
105. Lane, D.A.; Katz, M. In "Fate of Pollutants in the Air and Water Environments", Part 2, Suffet, I.H., Ed.; Wiley: New York, 1977, pp. 137–154.
106. Paalme, L.; Uibopuu, H.; Rohtla, I.; Pahapill, J.; Goubergrits, M.; Jacquignon, P.C. In "Polynuclear Aromatic Hydrocarbons: Formation, Metabolism", Cooke, M.; and Dennis, A.J., Eds.; Battelle Press: Columbus, Ohio, 1983, pp. 999–1008.
107. Daisey, J.M.; Lewandowski, C.G.; Zorz, M. *Environ. Sci. Technol.* **1982**, *16*, 857–861.
108. (a) Yokley, R.A.; Garrison, A.A.; Wehry, E.L.; Mamantov, G., *Environ. Sci. Technol.* **1986**,*20*, 86–90.(b) Wehry, E.L.; Mamantov, G.; Garrison, A.A.; Yokley, R.A.; Engelbach, R.J. Presentation of Ninth International Polynuclear Aromatic Hydrocarbon Symposium, Columbus, Ohio, October, 1984; personal communication from G. Mamantov.
109. Korfmacher, W.A.; Natusch, D.F.S.; Mamantov, G.; Wehry, E.L. *Science* **1980**, *207*, 763–765.
110. Lao, R.C.; Thomas, R.S.; Oja, H.; Dubois, E. *Anal. Chem.* **1973**, *45*, 908–915.
111. Adams, J.D.; Lavoie, E.J.; Hoffman, D. *J. Chromatogr. Sci.* **1982**, *20*, 247–277.
112. Jensen, T.E.; Hites, R.A. *Anal. Chem.* **1983**, *55*, 594–599.
113. Atkinson, R.; Aschman, S.M.; Pitts, J.N., Jr. *Environ. Sci. Technol.* **1984**, *18*, 110–113.
114. Biermann, H.W.; MacCleod, H.; Atkinson, R.; Winer, A.M.; Pitts, J.N., Jr. *Environ. Sci. Technol.* **1985**, *19*, 244–248.
115. Commins, B.T.; Lawther, P.J. *Br. J. Cancer* **1958**, *12*, 351–358.
116. Galasyn, J.F.; Hornig, J.F.; Soderberg, R.H. *J. Air Pollut. Control Assoc.* **1984**, *34*, 57–59.
117. You, F.; Bidleman, T.F. *Environ. Sci. Technol.* **1984**, *18*, 330–333.
118. Yamasaki, H.; Kuwata, K.; Miyamoto, H., *Environ. Sci. Technol.* **1982**, *16*, 189–194.
119. Roberts, J.M.; Fehsenfeld, F.C.; Liu, S.C.; Bollinger, M.J.; Hahn, C.; Albritton, D.L.; Sievers, R.E. *Atmos. Environ.* **1984**, *18*, 2421–2432.
120. Van Vaeck, L.; Van Cauwenberghe, K. *Atmos. Environ.* **1984**, *18*, 323–328.
121. Peters, J.; Siefert, B. *Atmos. Environ.* **1980**, *14*, 117–119.
122. Wu, C.-H.; Salmeen, I.; Niki, H. *Environ. Sci. Technol* **1984**, *18*, 603–607.
123. Pierce, R.C.; Katz, M. *Environ. Sci. Technol.* **1976**, *10*, 45–51.
124. Jager, J. *J. Chromatogr.* **1978**, *152*, 575–578.
125. (a) Gibson, T.L.; Tironi, G. In "Polynuclear Aromatic Hydrocarbons: Mechanisms, Methods and Metabolism", Cooke, M.; and Dennis, A.J., Eds.; Battelle Press: Columbus, Ohio, 1985, pp. 463–474. (b) Siak, J.; Chan, T.L.; Gibson, T.L.; Wolff, G.T., *Atmos. Environ.* **1985**, *19*, 369–376.

126. Tokiwa, H.; Nakagawa, R.; Ohnishi, Y. *Mutat. Res.* **1981,** *91,* 321–325.
127. Salmeen, I.T.; Pero, A.M.; Zator, R.; Schuetzle, D.; Riley, T.L. *Environ. Sci. Technol.* **1984,** *18,* 375–382.
128. Nielsen, T. Nordic PAH—Project Report No. 10, May, 1981. A Study of the Reactivity of Polycyclic Aromatic Hydrocarbons. Central Institute for Industrial Research, Oslo, Norway.
129. Nielsen, T. *Environ. Sci. Technol.* **1984,** *18,* 157–163.
130. Jager, J.; Hanus, V. *J. Hyg. Epidemiol. Microbiol. Immunol.* **1980,** *24,* 1–15.
131. Lee, F.S.-C.; Pierson, W.R.; Ezike, J. In "Polynuclear Aromatic Hydrocarbons: Chemistry and Biological Effects", Bjorseth, A.; and Dennis, A.J., eds.; Battelle Press: Columbus, Ohio, 1980, pp. 543–563.
132. Hughes, M.M., Natusch, D.F.S.; Taylor, D.R.; Zeller, M.V. In "Polynuclear Aromatic Hydrocarbons: Chemistry and Biological Effects", Bjorseth, A.; and Dennis, A.J., Eds.; Battelle Press: Columbus, Ohio, 1980, pp. 1–8.
133. (a) Brorstrom, E.; Grennfelt, P.; Lindskog, A.; Sjodin, A.; Nielson, T. In "Polynuclear Aromatic Hydrocarbons: Formation, Metabolism, and Measurements", Cooke, M.; and Dennis, A.J., Eds.; Battelle Press: Columbus, Ohio, 1983, pp. 201–210. (b) Brorstrom, E.; Greenfelt, P.; Lindskog, A. *Atmos. Environ.* **1983,** *17,* 601–605.
134. Lindskog, A. *Environ. Health Perspect.* **1983,** *47,* 81–84.
135. Schuetzle, D. *Environ. Health Perspect.* **1983,** *47,* 65–80.
136. Grosjean, D.; Fung, K.; Harrison, J. *Environ. Sci. Technol.* **1983,** *17,* 673–679.
137. Harkov, R.; Greenberg, A.; Darack, F.; Daisey, J.M.; Lioy, P.J. *Environ. Sci. Technol* **1984,** *18,* 287–291.
138. Lioy, P.J.; Daisey, J.M.; Atherholt, T.; Bozzelli, J.; Darack, F.; Fisher, R.; Greenberg, A.; Harkov, R.; Kebbekus, B.; Kneip, T.J.; Louis, J.; McGarrity, G.; McGeorge, L.; Reiss, N.M. *J. Air Pollut. Control Assoc.* **1983,** *33,* 649–657.
139. Duval, M.M.; Friedlander, S.K. "Source Resolution of Polycyclic Aromatic Hydrocarbons in the Los Angeles Atmosphere: Application of a Chemical Species Balance Method with First-Order Chemical Decay". Environmental Protection Agency Report EPA-600/S2-81-161; EPA: Washington, D.C., January 1982.
140. Nikolaou, K.; Masclet, P.; Mouvier, P. *Sci. Total Environ.* **1984,** *36,* 383–388.

CHAPTER 2

# Are the Numerical Magnitudes of Proton Affinities Intuitively "Plausible"?

## Joel F. Liebman

University of Maryland Baltimore County,
Catonsville, Maryland

## CONTENTS

© 1987 VCH Publishers, Inc.
MOLECULAR STRUCTURE AND ENERGETICS, Vol. 4

*Statements in angular brackets such as this, ≪ ... ≫, are intended to supplement the text by offering details of arguments, additional examples, and extensions of the ideas presented. They may be omitted in an initial reading without loss of continuity.*

# 1. INTRODUCTION: PHILOSOPHY AND DEFINITIONS

Using a collection of well-defined techniques, measurements of the heat of protonation of gaseous neutral molecules have been made[1-3] on about 800 species. These have been supplemented by quantum chemical calculations on numerous others.[1-3] The protonation process may be recognized as an integral component of counterion-free, solvent-free, isolated molecule, acid–base chemistry. As such, the associated energetics is often of seminal interest to chemists interested in gas phase phenomena as well as to those interested in condensed media, to physical chemists as well as to organic chemists, to scientists interested in ions and in neutral species, and to experimentalists as well as theoreticians. The energy of reaction of a molecule with a proton (or more properly the negative of the enthalpy of reaction) is known as the "proton affinity" (PA).

Yet knowledge of the data alone does not allow one to estimate what would be expected for some unknown base. This chapter presents several methods for producing estimates in simple and general fashion. These methods are not designed to compete with the data derived from the various high-accuracy experimental techniques or the rigorous quantum chemical calculational methods. The methods described here are far too coarse for that. Rather, they complement the approaches arising from the physical chemist's rigor and the organic chemist's intuition. While proton affinities of whole classes of organic bases have been discussed in the framework of substituent effects, those for simple species either are used generally as input numbers or have been totally ignored.

To use the approaches described in this chapter, no instrumentation more sophisticated than the proverbial "back of the envelope" and no mathematics except addition and subtraction will be needed, in contrast to the more rigorous, demanding, and expensive literature techniques as well as the approaches that call for more knowledge of molecular structure and energetics. There is no requirement for a suitably pure, volatile, and available sample; there is no computer-based molecular size or other financial restriction on what may be studied. In that the methods in this section are truly general, one can talk about any compound one wishes, independent of whether it has ever been synthesized.

This chapter also discusses the extent to which the numerical values of the proton affinities are intuitively "plausible." Of course we recognize that to the extent that a numerical value of any physical property is accurately measured and/or computed, "it is what it is." Our perception of the "plausibleness" of a given value does not, and cannot, affect its intrinsic accuracy. However, it can affect its perceived utility: a plausible value is one that we use with confidence because it is safely in accord with our experience as well as our biases.

## 2. PROTON AFFINITIES AND HOMOLYTIC BOND STRENGTHS TO HYDROGEN

In the current case, a quick perusal of the recent compilation[3] of evaluated proton affinity data shows that numerous values lie between 180 and 230 kcal/mol. By contrast, the homolytic bond strength or dissociation energy of a typical single bond to hydrogen in a neutral species, $D(Z-H)$, is rarely more than 100 kcal/mol. Why are the energies associated with these two different types of bond to hydrogen so different? We assert that this large numerical difference presumably arises from the fundamental distinction between *the* general $Z-H$ single bond found in most neutral compounds, which almost always arises from the interaction of two atoms with a single electron apiece, and the $B-H^+$ bond formed with most bases, which almost always arises from two electrons in the base interacting with the zero-electron proton. We need to know how to compare these types of bond to permit an a priori estimate of proton affinities.

In writing this section, we both state and extensively (if not excessively) illustrate estimation rules as they apply to a collection of bases. The specific examples clearly represent subjective choices. However, rather than choosing a truly arbitrary set of species that might have been of interest solely to the author, we presented estimates for the proton affinities of the 11 "special" compounds that are the reference species in the recently evaluated compilation (Reference 3). We remind the reader that the following PAs are given in kilocalories per mole, in decreasing order of numerical value: $NH_3$, 204.0; $CH_2=C=O$, 198.0; $(CH_3)_2C=CH_2$, 195.9; $CH_3CH=CH_2$, 179.5; $CH_2O$, 171.7;

≪This is the value for protonation on the oxygen site to form the more stable cation, $CH_2OH^+$. Formation of $CH_3O^+$ by C-protonation also is discussed in this chapter.≫

$H_2O$, 166.5; $C_2H_4$, 157.4;

≪This corresponds to formation of the "classical" ethyl ion, that is, the ion with the $CH_3CH_2^+$ structure as opposed to the more stable, bridged, "nonclassical" structure. We obtained 157.4 kcal/mol by combining the experimental value of the PA of ethylene, 162.6 kcal/mol, with the theoretical energy difference of the two isomers derived from the best (highest level) quantum chemical calculation,[4] 5.2 kcal/mol. The choice to emphasize the classical isomer was made on two grounds: (a) since few "simple" carbocations are nonclassical, the bridged structure is of less "pedagogical" use, and (b) it is not altogether apparent how much to trust *any* estimate involving such a feature because the bridging (ie, three center–two electron bonding) is rather unique.≫

CO, 141.7;

≪This is the value for protonation on carbon as required to form the more stable cation, $HCO^+$. Formation of $HOC^+$ by O-protonation also is discussed in this chapter.≫

$CO_2$, 130.9; O, 116.3; and $O_2$, 100.9.

This chapter presents nonexperimentally, non-quantum-chemical calculationally determined upper and lower bounds to the desired proton affinities. By such "theoretical bracketing," we hope to show that the experimental results are reasonable and consistent, not merely correct. Furthermore, we wish to encourage readers to be more confident in their estimation of proton affinities of species that may be of more immediate interest.

## 3. BOUNDS AND INTERRELATIONSHIPS: STANTON'S THEOREMS

To provide upper and/or lower bounds for *any* property, it is generally necessary to recast the problem in terms of new equalities, inequalities, and interrelationships involving known quantities. In the particular, the proton affinity of an arbitrary species Z should be identified not only with the dissociation energy of $ZH^+$ corresponding to deprotonation to form Z and $H^+$, but also with the homolytic bond energy to form $Z^+ + H$:

$$PA(Z) = D(Z-H^+) = D(Z^+-H) + IP(H) - IP(Z) \qquad (2\text{-}1)$$

This juxtaposition of the two bond energies, $D(Z-H^+)$ and $D(Z^+-H)$ reminds us of Stanton's theorem,[5] the first interrelationship to be used here:

1. If the ionization potential of an element with nuclear charge z, M(z), is less than that of hydrogen, 13.6 eV,

≪We adopt here the standard, but admittedly inconsistent, convention wherein bond energies and proton affinities are given in kilocalories per mole while ionization potentials (IP) are given in electron-volts. Note that 1 eV = 23.0609 kcal/mol.≫

then the bond strength in the ion, $D(M(z + 1)^+ - H)$, is greater than that for the isoelectronic neutral,[6] $D(M(z) - H))$. However, if the IP of the element $M(z)$ is "significantly greater" than that of H, then the bond strength in the ion, $D(M(z + 1) - H^+)$, is less than that for the neutral. When the IPs of $M(z + 1)$ and H are close, say within 1 eV, this inequality suggests that $M(z + 1)H^+$ also has a higher bond strength than in the neutral. For example, since the IP of Cl is 13.0 eV, below that of H, the theorem predicts that the bond strength in $HCl^+$ is higher than that of SH. This is confirmed experimentally[7]: the values are 104 and 85 kcal/mol, respectively. A contrasting case is found in the second-row analogue to Cl. Because the IP of F is 17.4 eV, a quantity significantly higher than that of H, it is predicted that the dissociation energy of $HF^+$ is less than that of OH. Experiment confirms this: numerically, we have 78 and 104 kcal/mol, respectively.

Because only six elements (He, N, F, Ne, Ar, and Kr) have ionization potentials larger than that of H, the value of interrelationship 1 may be questioned. Furthermore, as written, it is limited to diatomic hydrides. For polyhydrides, Stanton used the same derivation to prove interrelationship 2:

2. If the ionization potential of an element $M(z)$ is less than that of hydrogen, 13.6 eV, then the ion $M(z + 1)H_n^+$ has a greater total bond strength than the isoelectronic neutral, $M(z)H_n$. However, if the IP of $M(z)$ is significantly higher than that of hydrogen (say by more than 1 eV), then the total bond strength in the ion is less than that for the neutral. In the latter form, comparisons were made between polyhydrides with one "heavy" (nonhydrogenic) atom such as $CH_4$ and $NH_4^+$. To compare the total bond strengths of these two species (ie, the energies needed to dissociate $CH_4$ into C + 4H and $NH_4^+$ into N + 3H + $H^+$), Stanton used a rather primitive value of the proton affinity of $NH_3$ to obtain the heat of formation of $NH_4^+$. In that IP($N$) = 14.5 eV, and so is not significantly greater than that of H, the ion dissociation energy was expected to be larger than that of the neutral. This was verified. With more modern numbers, one still confirms this: for $CH_4$ and $NH_4^+$, the values of the dissociation energies are 397 and 483 kcal/mol, respectively.

Must one always consider complete dissociation of a polyhydride? No. There is a much more recent interrelation[8] between the PAs of bases containing a given constant central atom M.

3. For any element M, $n$ is greater than $m$, and with $MH_n$ and $MH_m$ both Lewis bases (electron pair donors), $PA(MH_n) > PA(MH_m)$.

In principle, at least, interrelationship 3 may be used to compare all the individual bond strengths of a cationic polyhydride with each other and thus with neutral polyhydrides via interrelationships 1 and 2. For example, interrelationship 3 allows one to compare $NH_4^+$, $NH_3^+$, $NH_2^+$, and $NH^+$, and so by use of interrelationships 1 and 2, one may also compare $CH_4$, $CH_3$, $CH_2$, and CH. This analysis suggests that a cation $(M(z + 1)H_n)^+$ may be recast as $(M(z + 1)H_{n-1}-H)^+$, an alternative formulation that will prove immediately suggestive.

Can the foregoing inequalities be used in reverse? That is, can one use them to derive proton affinities for arbitrary molecules? Interrelationship 3 and the associated rewriting of $(M(z + 1)H_n)^+$ as $(M(z + 1)H_{n-1}-H)^+$ implicitly suggest lumping together all but one of the hydrogens on the central atom. Moreover, if such lumping is legitimate, why must one be limited to the case of all the attached atoms being hydrogens? Although we have only heuristically derived relationship 4 from Stanton's logic, there appears to be nothing invalid in its use:

4. For arbitrary M(z) and its associated attached groups (ligands) Rs, if the ionization potential of the fragment RsM(z), is less than that of H, 13.6 eV, then the bond strength in the ion, $D(RsM(z + 1)^+-H)$, is greater than that for the isoelectronic neutral, $D(RsM(z)-H))$. However, if the IP of RsM(z) is significantly higher than that of H, then the bond strength in the ion, $D(RsM(z + 1)-H^+)$, is less than that for the neutral. When the IP of RsM(z + 1) is higher than but adequately close to that of H, this inequality is modified to suggest that these also have higher bond strengths in the cation than in the neutral. For example, since the IP of $NH_3$ is 10.2 eV, considerably less than that of H, it is correctly predicted that $D(NH_3^+-H)$ will be larger than $D(CH_3-H)$; the values are 124 and 105 kcal/mol, respectively.

Much as there are few elements with ionization potentials larger than that of H, there are very few compounds with such high ionization potentials. For example, of all the chemically bound diatomic molecules, only $H_2$, HF, CN, CO, $N_2$, and $F_2$ have such high IPs, while $CO_2$ is likewise alone among measured triatomics. As such, the reader may inquire why we bother with the formal statement of Stanton's interrelationships if the sign of the inequality is almost always in one direction. Why not simply say that the bond to hydrogen in a cation is generally stronger than that in the isoelectronic neutral? The answer is twofold:

i. As more completely written out, these interrelationships provide a more accurate inequality than if abridged.
ii. They facilitate the use of, and even suggest, new interrelationships.

## 4. BOUNDS AND INEQUALITIES: METHYL SUBSTITUENT EFFECTS

As an example of such a new interrelationship, we start with the observation that for most species of the type RsM(z)-H, there are more than one hydrogen or even general groups attached to M but not to each other that are lumped together as R. As each hydrogen is sequentially replaced by some substituent, is the effect the same? Are the substituent effects linearly additive, synergistic, or of lessening effect with each new substituent? Most often, the observed phenomenon is the last. Equivalently, we may invoke a long-known general "folklore" principle of the energetics of ions and neutrals alike, the leveling of substituent effects, here labeled interrelationship 5:

5. The sequential substitution or replacement of hydrogens by other groups generally results in changes that decrease in magnitude as the number of substituents increases. For example, consider the four "methylated" amines: $(CH_3)_3N$, $(CH_3)_2NH$, $CH_3NH_2$, and $NH_3$, having three, two, one, and zero methyl groups with associated proton affinities 225.1, 220.6, 214.1, and 204.0 kcal/mol. Sequential methylation of ammonia is seen to increase the proton affinities by 10.1, 6.5, and 4.5 kcal/mol.

The last set of interrelationships to be presented here is derived from the observation that methyl groups stabilize positive charges more than the hydrogen they can replace. This suggests interrelationship 6 for all X.

6. Methylation increases proton affinities; that is,

$$PA(CH_3X) > PA(HX) \qquad (2\text{-}2)$$

This finding is confirmed by the proton affinities for the three methylated amines and ammonia. Interrelationship 6 may be augmented by 6′, a finding that is consistent with the parallel behavior[9] of proton affinities and ionization potentials.

≪This parallel behavior has been shown to be general and to be derivable by recasting the proton affinity of a molecule Z in terms of the homolytic bond strength to hydrogen by its ion, the hydrogen affinity, $D(Z^+ - H)$. Making use of the definitions of the terms contained within, for all Z, one finds the identity $D(Z^+ - H) = PA(Z) + IP(Z) - IP(H)$. Since $D(Z^+ - H)$ is found to be nearly a constant for members of a related series of compounds, (eg, see Reference 9), a decrease in IP is matched by a comparable increase in PA. "Related series" is usually taken in a rather narrow sense (eg, the compounds to be studied are *only* the homologous, primary amines). However, it remains a useful guide when used with lessened restrictions (eg, the compounds to be studied are *all* amines).≫

6′. Methylation decreases ionization potentials, that is,

$$IP(CH_3X) < IP(HX) \qquad (2\text{-}3)$$

Finally, recall that methyl groups are generally the most innocuous of substituents when both electronic and steric effects are considered.

≪This may appear to contradict the "methyl versus ethyl dichotomy" as discussed in Volume 2 of this series by Liebman and Van Vechten, and by

Liebman alone in Chapters 6 of Volume 3. What is said here is that for any X, the thermochemistry of $CH_3X$ generally resembles HX more than RX does for any R. The chapters cited said that for any X and R chosen from the set of $n$-alkyl substituents, the thermochemistry of a general RX is better mimicked by R $= C_2H_5$ than $CH_3$. For example, consider the thermochemical property of proton affinity with X $= NH_2$; that is, consider primary amines and their protonation enthalpies. Numerically, the following evaluated values (kcal/mol) are found R $=$ H, 204.0; R $= CH_3$, 214.1; R $= C_2H_5$, 217.0; R $= n-C_3H_7$, 217.9; R $= n-C_4H_9$, 218.6. It is seen that the difference in PAs of methylamine and ethylamine is larger than between ethylamine and $n$-propylamine, or between $n$-propylamine and $n$-butylamine. Methyl is an anomalous alkyl substituent compared to its higher homologues. However, $CH_3$ substitution on $NH_3$ results in a smaller change than with any of its higher homologues; that is, the difference in the PAs of ammonia and methylamine is less than that between ammonia and any other alkylamine.

This suggests the final interrelationship derived from heuristic modification of Stanton's results:

7. For arbitrary M(z) and its associated attached groups (ligands) Rs, if the IP of the fragment, RsM(z), is less than that of $CH_3$, 9.8 eV, then the bond strength in the ion, $D(RsM(z + 1)^+ - CH_3)$, is greater than that for the isoelectronic neutral, $D(RsM(z) - CH_3))$. However, if the IP of the fragment, RsM(z), is significantly higher than that of $CH_3$, then the bond strength in the ion, $D(RsM(z + 1) - CH_3^+)$, is less than that for the neutral. When the IP of RsM(z + 1) is higher than, but adequately close to, that of $CH_3$, this inequality is modified to suggest that these also have higher bond strengths in the cation than in the neutral.

In accord with the above-mentioned IPs for $CH_3Cl$ and $CH_3F$, the carbon–halogen bonds in $CH_3F^+$ and $CH_3Cl^+$ (48 and 52 kcal/mol) are weaker than the carbon–chalcogen bonds in $CH_3O$ and $CH_3S$ (90 and 65 kcal/mol), respectively (derived from the data in References 6 and 9). Interrelationship 7 has another, more powerful use: it facilitates the tying together of proton and methyl cation affinities, $D(RsM(z + 1) - CH_3^+)$. For example, one may compare $CH_3CO^+$ with $CH_3CN$ in two ways: (a) knowing the relative IPs of CO and $CH_3$ allows a comparison of the bond energies $D(CH_3^+ - CO)$ and $D(CH_3 - CN)$, and (b) knowing the relative IPs of H and $CH_2CO$ allows a comparison of the bond energies $D(H - CH_2CO^+)$ and $D(H - CH_2CN)$. It would appear that the energetics of two of the "special" benchmark compounds, CO and $CH_2C=O$, are inseparable, and additionally related to the simple, but important species $CH_3CN$. Both comparisons are made below.

Let us now proceed—carefully, thoroughly, and dare we admit, laboriously—through the 11 benchmark compounds and the enumerated interrelationships. The purpose of this exercise is to illustrate how to use these interrelationships so that the reader may apply them to cases of PAs that are less exactly known.

## 5. AMMONIA

While this section deals explicitly with the specific question of the proton affinity of $NH_3$, the more general question of the PAs for the classically important organic bases, the amines, is also addressed inferentially. Applications of the various interrelationships are now discussed in turn. Interrelationship 1 clearly is inapplicable because it applies only to diatomic species. Interrelationship 2 asserts that $NH_4^+$ cannot be *less* bound than $CH_4$. Assuming they are equally bound results in a lower bound for $PA(NH_3)$ of 118 kcal/mol. This value is low enough to encourage us to recommend disregarding inequality 2 except in a historical sense. Nonetheless, it immediately, and correctly, suggests that proton affinities will be rather large numbers. Interrelationship 3, when used with the experimental value of $PA(NH_2)$, presents a better lower bound, 187 kcal/mol.

≪This value of $PA(NH_2)$ is somewhat different from that used in the original paper (Reference 8). This is because the ionization potentials and bond strengths used are close to but not identical with those currently preferred. However, no qualitative difference in the set of numbers employed there and those now in use is found for any species of interest.≫

For interrelationship 4, let z = 6 and take RsM(z) as $-CH_3$ and RsM(z + 1)$^+$ as $-NH_3^+$. This gives 185 kcal/mol as a result, comparable to the lower bound from interrelationship 3. By reexpressing proton affinities, $D(RsM(z + 1)-H^+)$, in terms of bond energies, $D(RsM(z + 1)^+-H)$, and recalling that the ionization potentials of most organic species are less than that of H, a new interrelationship arises. Interrelationship 8 is almost always true:

8.

$$PA(RsM(z + 1)) = D(RsM(z + 1)-H^+)$$
$$= D(RsM(z + 1)^+-H) + IP(H) - IP(RsM(z + 1))$$
$$> D(RsM(z)-H) + IP(H) - IP(RsM(z + 1))$$
$$> D(RsM(z)-H) + IP(H) - IP(RsM(z + 1))$$

While interrelationship 8 rarely contains any new numerical information not deducible from earlier interrelationships, it is nonetheless conceptually useful. For example, making use of the observation that bond strengths in neutral molecules are seemingly less affected by substituents than ionization potentials, one confirms the earlier assertion that for a "homologous" series of bases, the ionization potential and proton affinity have long been known to parallel.[9]

Returning to the proton affinity of ammonia, interrelationship 5 and the proton affinity values for mono-, di-, and trimethylamine can be used to present upper bounds. More precisely, one may derive:

$$PA((CH_3)_2NH) - PA(CH_3NH_2) < PA(CH_3NH_2) - PA(NH_3) \qquad (2\text{-}4)$$

and

$$PA((CH_3)_3N) - PA((CH_3)_2NH) < PA(CH_3NH_2) - PA(NH_3) \qquad (2\text{-}5)$$

or equivalently by simple algebraic manipulations:

$$PA(NH_3) < 2PA(CH_3NH_2) - PA((CH_3)_2NH) \qquad (2\text{-}6)$$

and

$$PA(NH_3) < PA(CH_3NH_2) - PA((CH_3)_2NH) + PA((CH_3)_3N) \qquad (2\text{-}7)$$

These "predictions" for $PA(NH_3)$ are also confirmed, numerically: $204.6 <$ 208 and $204.6 < 219$ kcal/mol, respectively.

In the absence of numerical expressions comparing general proton and methyl cation affinities, interrelationship 6 is without use in the study of the proton affinity of $NH_3$. Likewise, there are interrelationships between the proton affinities of parent and methylated species. Many interrelationships of both types exist, but in the name of brevity all but one are ignored in this section. From the numbers in Reference 3, one finds that the difference in proton affinities between HX and $CH_3X$ is nearly the same for a second-row element and its third-row congener:

| X | Difference in PA (kcal/mol) |
|---|---|
| OH | 15 |
| SH | 17 |
| F | 33 |
| Cl | 33 |

For $X = PH_2$, the difference is 15 kcal/mol. Making use of the experimental proton affinity of 214 kcal/mol and assuming the same difference for $X = NH_2$ as for $PH_2$ results in a predicted proton affinity of $NH_3$ of about 200 kcal/mol. Combining all these results in the theoretical bracketing value of $187 < PA(NH_3) < 208$ kcal/mol, with a preference for, or predilection toward, the upper bound. The uncertainty or spread in values is large, but the work and expense needed to acquire the values is minimal. The resulting "prediction" is valid—the experimental value is 204 kcal/mol.

## 6. KETENE

Now consider the protonation of $CH_2=C=O$. Interrelationships 1, 2, and 3 do not apply because the composition of the compound is inappropriate. To apply interrelation 4, the proton affinity is again recast in terms of $C-H$ bond strengths and ionization potentials. Since the ionization potential of ketene, 9.6 eV, is less than that of H, the first inequality suggests $D(H-CH_2CO^+) > D(H-CH_2CN) = 93$ kcal/mol. The PA of ketene is thus at least 185 kcal/mol.

To use interrelationship 5 requires us to know the proton affinities of both $CH_3CH=C=O$ and $(CH_3)_2C=C=O$. We do not know the latter. Nonetheless, we may use it in the simpler form of interrelationship 6, which asserts that because of the greater stabilizing power for positive charges of $CH_3$ than of H, $CH_3CH=C=O$ will have a higher proton affinity than $CH_2=C=O$ itself. An upper bound to $PA(CH_2=C=O)$ of 199 kcal/mol is thus suggested. This can be refined somewhat by noting that the protonated forms of $CH_2=C=O$ and $CH_3-C\equiv N$ are isoelectronic, as are the protonated forms of $CH_3CH=C=O$ and $CH_3CH_2-C\equiv N$. Consequently, the increase of proton affinity associated with methylation of $CH_3-CN$ to form $CH_3CH_2-CN$ should be comparable to that of going from ketene to methylketene. The difference of the proton affinities for the two nitriles is 5 kcal/mol, suggesting that $PA(CH_2=C=O)$ should be about 194 kcal/mol. $CH_3CO^+$ and $CH_3CN$ are also isoelectronic, and so letting $RsM(z) = CN$ and $RsM(z + 1) = CO$, inequality 7 also gives an upper bound. Since the ionization potential of CO, 14.0 eV, far exceeds that of $CH_3$, there is the immediate prediction that $D(CH_3^+-CO) < D(CH_3-CN)$. The heat of formation of acetyl cation is thus more than 113 kcal/mol, and so the proton affinity of ketene is less than 241 kcal/mol. This last inequality says little. However, combining the earlier, tighter and thus more useful inequalities, one correctly deduces $185 < PA(CH_2=C=O) < 199$ kcal/mol. Indeed, given the foregoing analysis of the proton affinities of nitriles, there is a preference toward the higher value. This conclusion is confirmed by the experimental value of 198 kcal/mol.

## 7. ISOBUTENE

Now consider protonated $(CH_3)_2C=CH_2$, that is, *tert*-butyl cation. None of the first three interrelationships is a priori applicable. In principle, interrelationship 4 may be useful. While $RsM(z + 1)$ may naturally be $(CH_3)_2CCH_2$, no direct experimental thermochemical data are available on the either of the choices for $RsM(z)$: $(CH_3)_2BCH_2$ or $(CH_3)_2CBH_2$. It has been suggested[10] that the $\alpha$-C$-$H bond strength in $CH_3CH_2B(CH_2CH_3)_2$ is about 80 kcal/mol. Let us assume this value to be the same as the C$-$H bond strength in $(CH_3)_3B$. Interrelationship 4 suggests that this is a lower bound to the $C-H$ bond strength in *tert*-butyl cation. Combining the experimental heats of formation of H and $[(CH_3)_2C=CH_2]^{\ddagger}$ results in an upper bound of 180 kcal/mol for the heat of formation of *tert*-butyl cation and a lower bound of 179 kcal/mol for the proton affinity of isobutene.

In the spirit of interrelationship 5, one may view *tert*-butyl cation as "trimethyl methyl cation." One could thus note what sequentially affixed methyl groups do to methyl cation and compare the effect to those when the groups are affixed to methyl radical. Intuitively, we expect a larger effect for the former. Done in steps, this would involve a discussion of the mono- and dimethylated species, more commonly known as "classical" ethyl and *sec*-propyl cations, as discussed later.

Most appropriate here, however, is a discussion of the cumulative effect of three methyl groups. Indeed our expectations are dramatically confirmed: the ionization potential of $CH_3$ is more than 3 eV higher than of *tert*-butyl radical. A useful comparison is with trimethylamine and ammonia and their ionization to their radical cations. In that carbenium ions have a vacant p orbital while in aminium ions this orbital is half-filled, hyperconjugation-derived logic suggests that removal of an electron from either trimethylamine or ammonia yields more localized cations than result from the removal of an electron from *tert*-butyl or methyl radical. As such, substituent effects are presumably smaller in the former than in the latter case. The following inequality is thus expected:

$$IP(CH_3) - IP((CH_3)_3C) > IP(NH_3) - IP((CH_3)_3N) \qquad (2\text{-}8)$$

This is confirmed using the most recent ionization potentials: $9.8 - 6.7 > 10.2 - 7.8$ eV. Taking the heat of formation of *tert*-butyl radical as 12 kcal/mol, the interrelationship then results in an upper bound for the heat of formation of *tert*-butyl ion of 192 kcal/mol and thus a lower bound for the proton affinity of isobutene is determined to be 176 kcal/mol.

Interrelationship 6, which asserts that methyl groups are rather innocuous, can also be used to methylate isobutene and so compare isobutene with 2-methyl-1-butene, wherein the saturated methyl group of isobutene has been methylated. The proton affinity of the latter olefin has not been measured. However, the proton affinity of the isomeric 2-methyl-2-butene is known to be 196 kcal/mol, and so the PA of isobutene is presumably comparable to that of this latter isomer. This value can be refined. Since both methylbutenes protonate to the same carbocation, the difference in heats of formation will be identical to the difference in proton affinities. A proton affinity of 2-methyl-1-butene of 198 kcal/mol is found, and is suggested as an upper bound for that of isobutene. This logic can be refined still more by relating olefins to amines: again, in particular isobutene with trimethylamine and 2-methyl-1-butene with $N,N$-dimethylethylamine. Dimethylethylamine has a proton affinity 3 kcal/mol more than trimethylamine and an IP some 2 kcal/mol less. This suggests that methylation of a methyl group on trimethylamine stabilizes positive nitrogen by about 3 kcal/mol. The stabilization of positive nitrogen is presumably comparable to that expected for the stabilization of positive carbon, and so we conclude that the proton affinity of 2-methyl-1-butene should be about 3 kcal/mol higher than isobutene. The PA of isobutene is thus expected to be 195 kcal/mol. The experimental number is 195 kcal/mol, in excellent agreement.

Interrelationship 7 could also be used. Neopentane ionizes to form a radical cation that can formally be considered to be a weak complex of $(CH_3)_3C^+$ and $CH_3^{\cdot}$. The binding energy of this complex is bounded from below by the binding energy of the Lewis acid $(CH_3)_3B$ with the one-electron donor $CH_3$. By assuming that the last energy is zero and using the 10.2-eV ionization potential of neopentane and the well-established heats of formation of $C(CH_3)_4$ and $CH_3$, one finds that the heat of formation of *tert*-butyl cation is at least 160 kcal/mol. Equivalently, the highest possible proton affinity for isobutene is 202 kcal/mol. In summary, interrelationships 5 and 7 predict $179 < PA((CH_3)_2C=CH_2) < 202$ kcal/mol, while interrelationship 6 allows a more definite prediction of about 195 kcal/mol in complete agreement with experiment.

## 8. PROPENE

Now consider protonated $CH_3CH=CH_2$, which may also be recognized as isopropyl or *sec*-propyl ion $(CH_3)_2CH^+$, or "dimethyl methyl cation." The considerations for this cation mimic that for *tert*-butyl ion. That is, from application of substituent leveling effects and interrelationship 5, the following inequality is expected:

$$IP(CH_3) - IP(NH_3) + IP((CH_3)_2NH) > IP((CH_3)_2CH) \qquad (2-9)$$

Using experimental values for the heats of formation of all four neutrals results in a lower bound of 157 kcal/mol for $PA(CH_3CH=CH_2)$. Equivalently, from substituent leveling effects and interrelationship 5 again, one derives:

$$IP((CH_3)_2CH) - IP((CH_3)_3C) > IP(CH_3)_2NH) - IP((CH_3)_3N) \qquad (2-10)$$

resulting in an upper bound of 180 kcal/mol for $PA(CH_3CH=CH_2)$.

Interrelationship 6 allows comparison of the proton affinity of propene with those of its methylated derivatives 1-methylpropene (2-butene), 2-methylpropene (isobutene), and 3-methypropene (1-butene). Proceeding with either 1- or 3-methylpropene, as was done for the methylated isobutenes, results in a prediction of about 180 kcal/mol, while with 2-methypropene, the prediction is 192 kcal/mol. We argue that the latter conflict with the first derived upper bound relates to the fact that in the 2-methyl derivative the new methyl group is closer to the cationic center than in the 1- and 3-methylated cases. Thus we expect the predictions of the 1- and 3-methylpropenes to be more reliable. The proton affinity of propene is thus determined to lie between 157 and 180 kcal/mol, with a preference for the higher values. It is encouraging, and surprising, that the upper bound is so tight: the experimental value is 179.5 kcal/mol.

## 9. FORMALDEHYDE

Let us now consider $CH_2O$ and commence with the O-protonated ion. Both $CH_2O$ and $(CH_2OH)^+$ are rather structurally "decorated" species compared to the binary hydrides ammonia, isobutene, and propene, and so new, more imaginative interrelationships may be suggested. For example, note that the positive charge is not localized and that this ion may be alternatively viewed as a oxonium ion, $CH_2{=}O^+H$, or as a carbonium ion, $^+CH_2{-}OH$. To the extent that the former description is valid, the effect of C-methylation on the PA should mimic that of another oxonium ion, protonated methanol. In accord with interrelationship 6, sequential C-methylation of formaldehyde and methanol increases the PA in both cases: $CH_2O$, 171.4; $CH_3CHO$, 186.6; $(CH_3)_2CO$, 196.7 and $CH_3OH$, 181.9; $CH_3CH_2OH$, 188.3; $(CH_3)_2CHOH$, 190.8 kcal/mol. However, the sensitivity of the proton affinities to C-methylation differs dramatically in the two cases. In the alcohol case there is nearly total leveling (cf interrelationship 5) after one methyl has been introduced, that is, $PA(CH_3CH_2OH) \simeq PA((CH_3)_2CHOH$. However, introduction of a second methyl in the formaldehyde case produces a change smaller than that accompanying the introduction of the first, but still significant. Combining both statements results in the inequality:

$$(PA(CH_3)_2CO) - PA(CH_3CHO)) - (PA(CH_3CHO) - PA(CH_2O)) >$$
$$(PA(CH_3)_2CHOH)) - PA(CH_3CH_2OH))$$
$$- (PA(CH_3CH_2OH) - PA(CH_3OH)) \quad (2\text{-}11)$$

This is confirmed numerically, $25 > 9$ kcal/mol, suggesting that there is considerable charge on carbon. This is confirmed by related trends on the PA of ethylene and its C-methylated derivatives in which there is even more charge on carbon: $C_2H_4$, 157.4; $CH_3CHCH_2$, 179.5; $(CH_3)_2CCH_2$, 195.5 kcal/mol. Equivalently, the following PA inequality appears plausible:

$$PA(CH_3CHCH_2) - PA(CH_2CH_2)$$
$$> PA(CH_3CHO) - PA(CH_2O)$$
$$> PA(CH_3CH_2OH) - PA(CH_3OH) \quad (2\text{-}12)$$

The indicated net numbers 22, 15, and 6 kcal/mol are in accord with the interrelationship above and may be used to provide upper and lower bounds on the proton affinity of formaldehyde of 180 and 165 kcal/mol, respectively.

If the description of positive carbon is to be taken literally, then protonated formaldehyde is synonymous with hydroxymethyl cation. Hydroxyl groups unequivocally stabilize carbonium ions. For example, the reaction:

$$CH_3OH + CH_3^+ \rightarrow CH_2OH^+ + CH_4 \qquad (2\text{-}13)$$

is exothermic by 64 kcal/mol. Intuitively, a hydroxyl group should stabilize a carbonium ion more than it does a keto group, since the latter has a lower positive charge on the carbon. Thus it is not surprising that the following reaction is endothermic by 29 kcal/mol:

$$CH_2O + CH_2OH^+ \rightarrow HCOOH + CH_3^+ \qquad (2\text{-}14)$$

It is well established that it is stabilizing to have both a keto and a hydroxyl group on the same carbon (eg, carboxylic acids have considerable resonance energy); see, for example, Chapter 6 in this volume and Reference 11. For example, from the awareness that the following reaction is highly exothermic (35 kcal/mol):

$$CH_2O + CH_3OH \rightarrow HCOOH + CH_4 \qquad (2\text{-}15)$$

it is unequivocally demonstrated that hydroxyl groups highly stabilize carbonium ions, a hardly astonishing conclusion.

≪Two much longer and more thorough discussions of the effects of substituents in stabilizing (or destabilizing) carbonium ions are Chapters 8 and 9, by Topsom and Charton, respectively.≫

That is, there is considerable electron donation from the oxygen to the carbon and so protonated formaldehyde has considerable oxonium ion behavior.

It is well established that formaldehyde is protonated on the oxygen. Nonetheless, a brief discussion giving an estimate of the relative stabilities of C- and O-protonated formaldehyde may still be of interest. A lower bound to the heat of formation of the former ion, $CH_3O^+$, may be derived using interrelationship 7. Since the ionization potential of O, 13.6 eV, significantly exceeds that of $CH_3$, it is anticipated that $D(CH_3^+ - O)$ will be less than $D(CH_3 - N)$. The latter may be found from knowledge of the energy difference of $CH_3N$ and $CH_2=NH$ and the heats of formation of $CH_3$, N, and $CH_2=NH$. The difference has been accurately calculated quantum chemically[12] to be 46 kcal/mol, while the heats are known from experiment. Combining everything, a lower bound to the heat of formation of $CH_3O^+$ is 237 kcal/mol, resulting in an upper bound to the C-proton affinity of formaldehyde of 94 kcal/mol. Since the heat of formation of the O-protonated species is 168 kcal/mol, there is a difference of at least 74 kcal/mol, a value comparable to the 92 kcal/mol determined by another high-accuracy quantum chemical calculation.[13]

The value of 74 kcal/mol for the difference of the PAs for C- versus O- of formaldehyde was derived by assuming that the $CH_3^+ - O$ bond strength is the same as that found in $CH_3 - N$. Should the cation be less bound, the 74 kcal/mol difference of the two PAs will grow. The members of the isoelectronic pair of $CH_3OH_2^+$ and $CH_3NH_2$ are related to each other the same way as $CH_3O^+$ and $CH_3N$ are. From solely experimental numbers, the $CH_3^+ - OH_2$ bond strength in protonated methanol is found to be 68 kcal/mol, while for $CH_3NH_2$, the corresponding $C - N$ bond strength is 84 kcal/mol. In this all-experimental case the $C - N$ bond is 16 kcal/mol stronger than the isoelectronic $C^+ - O$ bond. Using 16 kcal/mol as the difference in the more theoretical case of $CH_3O^+$ and $CH_3N$ results in a difference of C- and O-PAs of 90 kcal/mol. This is almost identical to that directly, and much more rigorously computed.

## 10. WATER

The proton affinity of $H_2O$ may be estimated from the more conventional interrelationships. Interrelationship 2 is valid, but as with the proton affinity of $NH_3$, it gives little information—the total dissociation energy of $H_3O^+$ of 388 kcal/mol is far greater than the 280 kcal/mol for $NH_3$. Interrelationship 3 and the proton affinity of OH suggest a lower bound of 139 kcal/mol for $PA(H_2O)$. Since $IP(H_2O) = 12.6$ eV, less than that of hydrogen, interrelationship 4 suggests:

$$D(H_2O^+ - H) > D(H_2N - H) \qquad (2\text{-}16)$$

This gives as a lower bound for $PA(H_2O)$ of 130 kcal. Interrelationship 5 gives the upper bound:

$$2PA(ROH) - PA(R_2O) > PA(H_2O) \qquad (2\text{-}17)$$

for arbitrary R. This is confirmed for R = $CH_3$, $C_2H_5$, n- and i-$C_3H_7$: the suggested upper bounds are the nearly constant 175.7, 176.4, 175.3, and 176.5 kcal/mol. This suggests that analogues of inequality 7 for substituents other than methyl are reasonable.

Comparison of the proton affinities for second- and third-row elements as was done in the case of $NH_3$ is also useful here. The PAs of $H_2S$, $CH_3SH$, and $CH_3OH$ suggest a proton affinity of $H_2O$ of about 165 kcal/mol. The proton affinity of water is thus bounded: $139 < PA(H_2O) < 176$ kcal/mol, with a preference for the higher value, a result confirmed by the experimental value of 166.5 kcal/mol.

## 11. ETHYLENE

Turning now to the proton affinity of $C_2H_4$, we admitted in Section 1 that only the protonation process that forms the classical isomer of ethyl cation will be discussed. Moreover we note that, paralleling the comment in the earlier discussion of $PA((CH_3)_2C=CH_2)$, there are no experimental thermochemical data on the appropriate boron compounds such as the isoelectronic $CH_3BH_2$, no less on $CH_2BH_2$, $CH_3BH_2 \cdot CH_3$, or $CH_3CH_2 \cdot BH_3$. These values would be needed to employ most of the earlier inequalities. However, before despairing, note that upper bounds for the proton affinity to form the classical isomer are easily derived from the leveling interrelationship 5. That is, for any R, it is expected that:

$$2[PA(RCHCH_2)] - PA(R_2CCH_2) > PA(C_2H_4) \qquad (2-18)$$

For R = $CH_3$, F, and $C_6H_5$, the proposed upper bounds are 163, 174, and 192 kcal/mol. Since all the cations implicit on the left-hand side are "classical," this inequality is for the appropriate $C_2H_5^+$ ion. That the least upper bound is found for R = $CH_3$ exemplifies the optimism earlier expressed about using methyl substitution. Analogously, inequality 5 would suggest that:

$$\Delta H_f(CH_3^+) + \Delta H_f((CH_3)_2CH^+) > 2\,\Delta H_f(CH_3CH_2^+) \qquad (2-19)$$

This results in an upper bound of 226 kcal/mol for the heat of formation of "classical" ethyl cation. A lower bound of 150 kcal/mol is thus suggested for $PA(C_2H_4)$. An encouragingly narrow range 150 kcal/mol $< PA(C_2H_4) < 163$ kcal/mol is thus derived, and is seen to be in accord with the accepted value, 157.4 kcal/mol.

## 12. CARBON MONOXIDE

In investigating the C-proton affinity of CO, it may be tempting to try to employ inequality 3 and compare this quantity to the C-proton affinity of $CH_2O$. So doing results in the erroneous prediction that $PA(CO) < 94$ kcal/mol. This effort failed because of a bond order change on protonation: while in a classical description of the C-protonation of CO to form $HCO^+$, there need not be any loss of $C-O$ multiple bonding, C-protonation of $CH_2O$ results in total loss of $C-O$ double bond character. Interrelationship 4 does not encounter this problem, and so a more reasonable expression to use. Since the $IP(CO) = 14.4$ eV, and so is "rather" close to that of hydrogen, $PA(CO)$ is expected to be greater than $D(H-CN)$, 125 kcal/mol. With but one replaceable hydrogen, there is no way to have leveling effects needed to employ interrelationship 5.

However, to the extent that there is any charge on the carbon in the acyl cation, it is unambiguous that methyl will stabilize $HCO^+$ more than it does HCN. That is, inequality 6 suggests:

$$\Delta H_f(CH_3CO^+) - \Delta H_f(HCO^+) < \Delta H_f(CH_3CN) - \Delta H_f(HCN) \qquad (2-20)$$

This is confirmed experimentally giving a lower bound to the heat of formation of $HCO^+$ of 171 kcal/mol. Equivalently, there is the upper bound of 158 kcal/mol for $PA(CO)$. From the numerous, long-known similarities of CO and $N_2$, the following near equality is suggested:

$$\Delta H_f(CH_3CO^+) - \Delta H_f(HCO^+) \approx \Delta H_f(CH_3NN^+) - \Delta H_f(HNN^+) \quad (2\text{-}21)$$

This suggests that the heat of formation of $HCO^+$ is 189 kcal/mol and that $PA(CO)$ is 150 kcal/mol. This value, although neither a lower nor an upper bound, may safely be assumed to be rather reliable. The experimental value is 141.7 kcal/mol.

It is well established that C-protonation of CO is considerably favored over O-protonation. Narrow bounds and reliable estimates are unlikely for the O-protonation of CO—not only are there no handles for substituents, but almost no experimental data exist on carbocations of the general type $RC^+$ for *any* R except R = H. Nonetheless, let us try. A simple estimate for the stability of the O-protonated species that is without any doubt a lower bound assumes equality of $-OH$ stabilization on $CH_3^+$ and on $CH^+$. That is, assume thermoneutrality for the reaction:

$$CH_2OH^+ + CH^+ = CH_3^+ + COH^+ \quad (2\text{-}22)$$

This results in a predicted heat of formation of 293 kcal/mol for $COH^+$, and so a lower bound of 66 kcal/mol is suggested for the O-proton affinity of CO.

Alternatively, one may assume an equality:

$$D(H-CO^+) - D(H-CN) = D(H-OC^+) - D(H-NC) \quad (2\text{-}23)$$

The difference of the C- and O-proton affinities of CO is thus comparable the difference of heats of formation of HCN and HNC.

$$\begin{aligned} PA_C(CO) - PA_O(CO) &= D(H-CO^+) - D(H-OC^+) \\ &= D(H-CN) - D(H-NC) \\ &= \Delta H_f(HCN) - \Delta H_f(HNC) \quad (2\text{-}24) \end{aligned}$$

This difference for HCN and HNC equals 19 kcal/mol, suggesting that the O-proton affinity of CO is about 120 kcal/mol. To see that this is no doubt too high, consider the appropriate states for CN and $CO^+$ that would bond with hydrogen atoms to form HCN, HNC, $HCO^+$, and $HOC^+$. Both CN and $CO^+$ have ground state $^2\Sigma$ electron configurations. Intuitively, most of the unpaired electron density is on the carbon, as shown by the resonance structures $\cdot C\equiv N:$ and $\cdot C\equiv O:^+$. These have precisely the singly occupied orbitals needed to form HCN and $HCO^+$ upon "reaction" with H. However, to form HNC and $HOC^+$, we need an excited $^2\Sigma$ state of CN and $CO^+$ with most of the unpaired electron density *not* on the carbon as in, for example, $:C\equiv N\cdot$ and $:C\equiv O\cdot^+$. These two states lie 74 and 131 kcal/mol (3.2 and 5.7 eV) above the ground states.[14] Equivalently, there is an effective ionization potential of CO of 14.4 + 5.7 eV. This "IP" is high enough to ensure that unambiguously $D(H-NC)$ will be greater than $D(H^+-OC)$: the O-proton affinity of CO is no doubt less than 120 kcal/mol. This result is compatible with the value determined by the best quantum chemical calculation, 102 kcal/mol.

## 13. CARBON DIOXIDE

There are no hydrogens on $CO_2$ to allow for methylation or introduction of any other substituents. This provides complications because few of our interrelationships are usable. However, the study of the proton affinity of $CO_2$ is simplified because the IP of $CO_2$ is 13.8 eV, a quantity nearly identical to that of H. As such, there is no ambiguity as to the predictions from interrelationship 4.

$$PA(CO_2) = D(OCO^+ - H)$$
$$> D(H-NCO), D(H-OBO), \text{ and } D(H-OCN) \qquad (2\text{-}25)$$

The first two neutral bond strengths[15] on the right are experimentally 113 and 114 kcal/mol, resulting in the two lower bounds for the proton affinity of $CO_2$ of 108 and 109 kcal/mol. No experimental data are available for the $H-OCN$ bond strength. However, because the homolytic dissociation products of HOCN and HNCO are the same, and HOCN is quantum chemically calculated to be (21 kcal/mol) less stable than HNCO,[16] the inequality above is automatically confirmed for HOCN as well.

Note that the proton affinity of $CO_2$ is remarkably low. Moreover, recall that the proton affinities of two of the reference species, $CH_2=C(CH_3)_2$ and $CH_2=C=O$, are nearly equal, a finding that would suggest that $CH_3-$ stabilizes $-C^+(CH_3)_2$ and $-CO^+$ by comparable amounts. That does not mean, of course, that another stabilizing group, say $-OH$, will affect the two types of carbocation equally. Nonetheless, the proton affinities of $O=C(CH_3)_2$ and $O=C=O$ would have been expected to be more comparable than they are—197 and 130 kcal/mol. Why is $CO_2$ *so* unbasic?

We suggest the simple answer that $CO_2$, not its protonated derivative, enjoys considerable resonance stabilization.[17] The neutral $CO_2$ has the following resonance structures:

$$O=C=O \leftrightarrow {}^-O-C\equiv O^+ \leftrightarrow {}^+O\equiv C-O^- \qquad (2\text{-}26)$$

The resultant stabilization energy is quantitated by comparison of the earlier reaction (2-15) involving formaldehyde and formic acid:

$$CH_2O + CH_3OH \rightarrow HCOOH + CH_4 \qquad (2\text{-}15)$$

noted to be exothermic by 35 kcal/mol, with a related reaction for $CO_2$:

$$2CH_2O \rightarrow CO_2 + CH_4 \qquad (2\text{-}27)$$

Reaction 2-27 is exothermic by 60 kcal/mol. If 35 kcal/mol can be said to be the resonance energy of formic acid, as it often is, then 60 kcal/mol can likewise be said to be the resonance energy of $CO_2$. To discuss relative basicities, one must also consider the resonance energies associated with the protonated species. Protonation of the "keto" oxygen in HCOOH transforms the two major, but inequivalent resonance structures of the neutral:

$$O=C(H)-OH \leftrightarrow {}^-O-C(H)=O^+H \qquad (2\text{-}28)$$

into two equivalent structures in the cation:

$$H^+O=C(H)-OH \leftrightarrow HO-C(H)=O^+H \qquad (2\text{-}29)$$

Species that have equivalent resonance structures usually are considerably stabilized. It is thus not surprising that experimental proton affinity measurements show HCOOH to be more basic than $CH_2O$. The corresponding new reaction involving protonated formic acid:

$$CH_2OH^+ + CH_3OH \rightarrow HC(OH)_2^+ + CH_4 \qquad (2\text{-}30)$$

is exothermic by 42 kcal/mol. By contrast, protonated $CO_2$ lacks the equivalent resonance structures corresponding to the two dipolar forms of the neutral. As such, the corresponding reaction involving protonated $CO_2$:

$$CH_2OH^+ + CH_2O \rightarrow HOCO^+ + CH_4 \qquad (2\text{-}31)$$

is expected to be less exothermic than for the parent neutral, and in fact is exothermic by *only* 10 kcal/mol. This confirms that a large part of the weakness of the basicity of $CO_2$ is due to preferential stabilization of the neutral.

## 14. ATOMIC OXYGEN

The IP of O is fortuitously nearly identical to that of atomic hydrogen. Thus, there is no doubt that interrelationship 1 is applicable, and so:

$$D(H-O^+) = D(H^+-O) = PA(O) > D(H-N) \qquad (2\text{-}32)$$

Equivalently, a lower bound of 80 kcal/mol is suggested for $PA(O)$. Likewise, from the interrelationship 3, one may deduce:

$$PA(HO) > PA(O) \qquad (2\text{-}33)$$

or equivalently, there is an upper bound of 139 kcal/mol for the proton affinity of O.

In some of the preceding sections, ancillary information was used to suggest values of proton affinities and so provide guidance on effectively narrowing the range of the upper to lower bounds. Can this be done here? From the rigorous identities:

$$\Delta H_f(OH^+) = \Delta H_f(OH) + IP(OH) \qquad (2\text{-}34)$$

and

$$PA(O) = \Delta H_f(O) + \Delta H_f(H^+) - \Delta H_f(OH^+) \qquad (2\text{-}35)$$

and the approximate equality of the ionization potentials of OH and OF,

≪This is derived from application of the "$\pi$-fluoro effect," the principle that asserts that the IP corresponding to removal of a $\pi$ electron from a planar molecule is essentially unchanged if some (or all) of the hydrogens are replaced by fluorine. (See Reference 18.)≫

the value of 117 kcal/mol is suggested for $PA(O)$. This new result does not particularly help us to decide whether the upper or lower bound earlier presented is more reliable. Rather, it suggests a value for the proton affinity of atomic oxygen more toward the middle of the allowed range. This conclusion, and the suggested number, are almost exactly correct: the experimental number is 116.3 kcal/mol.

## 15. MOLECULAR OXYGEN

The final species to be discussed is $O_2$. Since its IP is 12.2 eV, it is unambiguously predicted by inequality 3 that:

$$D(O_2^+ - H) > D(ON - H) \tag{2-36}$$

From inequality 2-36, the rather low $H - N$ bond strength of HNO (50 kcal/mol) and the ionization potentials of $O_2$ and H, a lower bound of 82 kcal/mol is thus suggested for $PA(O_2)$. This value refers to formation of the singlet state of $HO_2^+$, $H - O^+ = O$, since it isoelectronically corresponds to the ground state singlet for HNO, with its Lewis structure, $H - N = O$. As such, 82 kcal/mol is even more definitively too small because $HO_2^+$ may well be a ground state triplet, $H - O - O^+$. The naive assumption that the same singlet–triplet energy difference applies to $HO_2^+$ as to $O_2$ ($\simeq 23$ kcal/mol) results in a predicted proton affinity of $O_2$ of at least 105 kcal/mol. There is little doubt that this prediction is too high because the singlet–triplet gap used is too large.[19] High-accuracy quantum chemical calculations[20] give a singlet–triplet gap of about 8 kcal/mol. The proposed bounds, 82 and 105 kcal/mol, are indeed correct—the experimental value is 100.9 kcal/mol. An experimental determination of the singlet–triplet energy difference for $HO_2^+$ is suggested.

## ACKNOWLEDGMENTS

The author would like to thank Drs. J.E. Del Bene, D.A. Dixon, S.G. Lias, M. Meot-Ner, and D. Van Vechten for numerous discussions, incisive questions, and clarifying remarks.

## REFERENCES

1. As documentation of the great importance of proton affinities in chemistry, we need merely note the following contributions to these volumes in which this quantity plays a seminal

role: Charton (Vol. 4, Chapter 9), Deakyne (Vol. 4, Chapter 4), Del Bene (Vol. 1, Chapter 9), Dixon and Lias (Vol. 2, Chapter 7), Liebman and Simons (Vol. 1, Chapter 3), Liebman (Vol. 3, Chapter 6), Meot-Ner (Mautner) (Vol. 4, Chapter 3), and Topsom (Vol. 4, Chapter 8.)

2. Taft, R.W. *Prog. Phys. Org. Chem.* **1983**, *14*, 248.
3. Lias, S.G.; Liebman, J.F.; Levin, R.D. *J. Phys. Chem. Ref. Data* **1984**, *13*, 695.
4. (a) Raghavachari, K.; Whiteside, R.A.; Pople, J.A.; Schleyer, P.v.R. *J. Am. Chem. Soc.* **1981**, *103*, 5649. (b) Del Bene, J.E.; Frisch, M.J.; Raghavachari, K.; Pople, J.A. *J. Phys. Chem.* **1982**, *86*, 1529.
5. Stanton, R.E. *J. Chem. Phys.* **1963**, *39*, 2368.
6. For a thorough discussion of isoelectronic reasoning, see Chapter 2 of Volume 1 by Bent. Numerous chapters in this series make implicit or explicit application of this conceptual approach to molecular structure and energetics.
7. Unless explicitly said, all experimental energetics quantities in this chapter are taken from the forthcoming evaluated compendium on gas phase ion energetics by Lias, S.G.; Liebman, J.F.; Holmes, J.L.; Bartmess, J.E.; Levin, R.D.; (manuscript in preparation for *J. Phys. Chem. Ref. Data*). In this compendium, ionization potentials and heats of formation were often derived from a collection of measurements from different sources. For example, the energetics data of Reference 3 have been convoluted into this newer data archive. Reminding the reader of the definition of dissociation energies, $D(X-Y) = \Delta H_f(X) + \Delta H_f(Y) - \Delta H_f(XY)$, it is clear that they can likewise also be derived numbers. The author believes that because the estimates in this chapter are comparatively inexact, the absence of direct literature citations in this chapter is legitimized.
8. Greenberg, A.; Winkler, R.; Smith, B.L.; Liebman, J.F. *J. Chem. Educ.* **1982**, *59*, 367.
9. (a) Beauchamp, J.L. *Annu. Rev. Phys. Chem.* **1971**, *22*, 527. (b) Lias, S.G.; Ausloos, P. "Ion-Molecule Reactions: Their Role in Radiation Chemistry". American Chemical Society: Washington, D.C., 1975, pp. 91–95.
10. This number, and a thorough discussion of the radical chemistry of boron-containing organic compounds, is found in Chapter IV of Ingold, K.U.; Roberts, B.P. "Free-Radical Substitution Reactions: Bimolecular Homolytic Substitutions ($S_H2$ Reactions at Saturated Multivalent Atoms)". Wiley-Interscience: New York, 1971.
11. Pauling, L. "The Nature of the Chemical Bond", 2nd ed. Cornell University Press: Ithaca, N.Y., 1940, pp. 138, 201–204.
12. Demuynck, J.; Fox, D.J.; Yamaguchi, Y.; Schaefer, H.F., III. *J. Am. Chem. Soc.* **1980**, *102*, 6204.
13. Bouma, W.J.; Nobes, R.H.; Radom, L. *Org. Mass Spectrom.* **1982**, *17*, 315.
14. Huber, K.P.; Herzberg, G. "Molecular Spectra and Molecular Structure Constants of Diatomic Molecules". Van Nostrand Reinhold: New York, 1979.
15. This number for HNCO is from Okabe, H. *J. Chem. Phys.* **1971**, *53*, 3507.
16. Poppinger, D.; Radom, L.; Pople, J.A. *J. Am. Chem. Soc.* **1977**, *97*, 7806.
17. Pauling, L. "The Nature of the Chemical Bond", 2nd ed. Cornell University Press: Ithaca, New York, 1944, pp. 138, 194–198.
18. Liebman, J.F.; Politzer, P.; Rosen, D.C. In "Applications of Electrostatic Potentials in Chemistry", Politzer, P.; and Truhlar, D.G., Eds.; Plenum Press: New York, 1981, pp. 295–308.
19. This conclusion is derived by using various qualitative models of carbenes (see Chapter 3 of Volume 1 by Liebman and Simons) to the electron deficient $HO_2^+$. In the particular, note that $HO_2^+$, HNO, and HCF are isoelectronic.
20. Quantum chemical calculations at the MP4–SDTQ/6–311+G(2d,p)//HF/6–31G(d) level suggest that the triplet state of $HO_2^+$ is more stable than the singlet by 8 kcal/mol. Using the isodesmic reaction $HOOH + {}^3HO^+ \rightarrow {}^3HOO^+ + H_2O$, quantum chemical calculations at this level and experimental thermochemical data for HOOH, $HO^+$, and $H_2O$ reproduce the experimental heat of formation of $HO_2^+$ (hence the proton affinity of $O_2$) to within 1 kcal/mol (J.E. Del Bene and J.F. Liebman, unpublished results).

**CHAPTER 3**

# Ionic Hydrogen Bonds

# Part I. Thermochemistry,
##    Structural Implications, and
##    Role in Ion Solvation

## Michael Meot-Ner (Mautner)

**National Bureau of Standards, Gaithersburg, Maryland**

## CONTENTS

© 1987 VCH Publishers, Inc.
  MOLECULAR STRUCTURE AND ENERGETICS, Vol. 4

# 1. INTRODUCTION

The interactions of ions with solvent molecules have profound effects on the energetics of ionic processes such as the protonation and deprotonation of aqueous bases and acids. Much quantitative knowledge has been gained since the early 1970s, when it became possible to study proton transfer equilibria in the gas phase, and thus to compare relative acidities and basicities in the absence and presence of ion solvation. The results showed that, for example, the differences between the relative gas phase heats of protonation of alkylamines, which range over 21 kcal/mol between ammonia and trimethylamine, are attenuated in water to a range of less than 2 kcal/mol due mostly to differential ion solvation. Given the extensive influence of solvation on ionic processes, a detailed understanding of ion–solvent interactions is needed.

The interaction between an onium ion and a hydrogen-bonding solvent may be broken down to several components, as shown in Figure 3-1. The most structure-specific component is the interaction between the protonated species $BH^+$ and the solvent molecules directly attached to it. Next, additional solvent molecules attach by hydrogen bonds to the inner solvent molecules. These interactions involve conventional ionic hydrogen bonds where a proton bridges between two $n$-donor centers, that is, bonds that may be said to be of the $nH^+ \cdot \cdot \cdot n$ type. As we shall see, these interactions are structure specific in that they are sensitive to molecular properties of the ion and the solvent.[1]

The ion may also interact with polar solvent molecules through hydrogen atoms on its alkyl substituents. These interactions involve unconventional

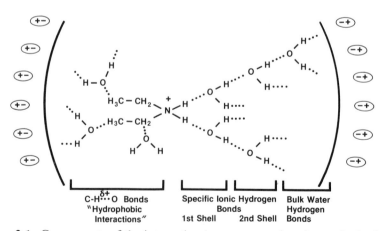

**Figure 3-1.** Components of the interaction between an onium ion and a hydrogen-bonding solvent.

hydrogen bonds, where at least one of the atoms attached to the proton is not a lone pair $n$ donor (eg, bonds of the $CH^{\delta+} \cdots O$ type).

Beyond the solvent molecules in its intimate vicinity, the ion interacts with the bulk solvent as a dielectric medium. The physical forces in these interactions, such as the energy required to form a cavity in the solvent, and the polarization of the bulk solvent by the ionic charge, depend only on the size of the ion and the dielectric properties of the solvent.

The specific, structure-dependent ion–solvent interactions can be studied in the gas phase[2], where the stepwise attachment of several solvent molecules is observed using clustering equilibrium studies:

$$AH^+ \cdot (n-1)H_2O + H_2O \rightleftarrows AH^+ \cdot nH_2O \qquad (3\text{-}1)$$

In these experiments the solvent molecules attach by strong conventional $nH^+ \cdots n$ bonds. The weaker, unconventional bonds are harder to study. Nevertheless, information can be obtained indirectly from the solvation of quaternary onium and carbonium ions, where only unconventional hydrogen bonds are possible.

The results of mass spectrometric ion–solvent clustering studies, and the insights obtained on the role of ionic hydrogen bonds in ion solvation, will be the subject of this chapter. The results of related ab initio studies will be reviewed in Chapter 4.

Beyond the interactions of simple, monofunctional onium ions and solvent molecules, ionic hydrogen bonds also occur in structurally complex polyfunctional species. The thermochemistry in such systems is indicative of the occurrence of intramolecular and of multiple ionic hydrogen bonds.[3] These systems will also be reviewed, since they serve as models for ionic interactions in biological systems such as enzymes and other peptides.

# 2. STRUCTURALLY SIMPLE SYSTEMS: MONOFUNCTIONAL IONS AND LIGANDS

## A. Conventional $nH^+ \cdots n$ Ionic Hydrogen Bonds: Correlations with Proton Affinity

The attachment of an $n$-donor ligand B to a protonated $n$-donor $AH^+$ results in partial proton transfer to B. The overall interaction results in a strong ionic hydrogen bond, whose dissociation energy ($\Delta H_D^\circ$) is typically 15–30 kcal/mol. These ionic hydrogen bonds are stronger by about an order of magnitude than hydrogen bonds between neutral molecules.

The efficiency of the partial proton transfer, or the sharing of the proton between the components of the dimer, is optimal when the proton affinities (PAs) of A and B are equal, and decreases in efficiency as the proton affinity difference $\Delta PA = PA(A) - PA(B)$ increases, since then the stronger base A

TABLE 3-1. Proton Affinities, Dissociation Energies, Atomic Charges, and Geometries Related to $R_3N-H^+ \cdots OH_2$ Complexes

| Complex | $\Delta PA$ (kcal/ mol)[a] | $\Delta H_D^\circ$ (kcal/ mol)[b] | $q(H^+)$[c] | $r(\text{Å})$[d] $(N-H^+)$ | $(H^+ \cdots OH_2)$ |
|---|---|---|---|---|---|
| $NH_4^+ \cdots OH_2$ | 37.5 | 19.9 | 0.470 | 1.081 | 1.512 |
| $CH_3NH_3^+ \cdots OH_2$ | 52.6 | 16.8 | 0.451 | 1.065 | 1.548 |
| $(CH_3)_2NH_2^+ \cdots OH_2$ | 58.3 | 15.0 | 0.440 | 1.053 | 1.582 |
| $(CH_3)_3NH^+ \cdots OH_2$ | 62.6 | 14.5 | 0.433 | 1.040 | 1.660 |

[a] $\Delta PA = PA(R_3N) - PA(H_2O)$. Proton affinities from Reference 13.
[b] Meot-Ner (Mautner), M. *J. Am. Chem. Soc.* **1984**, *106*, 1265.
[c] Atomic charge on the hydrogen atom before the formation of the complex; ab initio STO3–21G results, Reference 7.
[d] STO3–21G optimized result, Reference 7.

tends to retain the proton.[4] For example, in the addition of $H_2O$ to the oxonium ions $H_2OH^+$, $CH_3OH_2^+$, and $(CH_3)_2OH^+$, as $\Delta PA$ increases from 0 to 15.4 and 23.7, the dissociation energies of the complexes decrease from 31.6 to 24.1 and 22.6 kcal/mol, respectively.[5]

The inverse relationship between $\Delta PA$ and $\Delta H_D^\circ$ was first investigated experimentally by Yamdagni and Kebarle,[4] and was justified theoretically by Desmeules and Allen[6] for a series proton-bonded complexes of second- and third-row hydrides. These ab initio calculations showed that in the complexes $A-H^+ \cdots B$, the stretching of the $A-H^+$ bond increases and the length of the $B \cdots H^+$ distance decreases as the PA of B approaches that of A. Similarly, in a more regularly varying series, the attachment of $H_2O$ to protonated amines, Ikuta and Imamura[7] found a regular relation between $\Delta PA$, $\Delta H_D^\circ$, and the $N-H^+$ and $H^+ \cdots O$ distances (Table 3-1).

To explore more extensively the relationship between $\Delta PA$ and $\Delta H_D^\circ$, we recently investigated large series of complexes involving oxygen and nitrogen $n$-donors.[5] For each type of bond (Equations 3-2 to 3-4), a simple linear relationship is observed over a wide range of $\Delta PA$ (from 0 to 50 kcal/mol) and $\Delta H_D^\circ$ (from 30 to 15 kcal/mol); see, for example, Figure 3-2. The following equations are obtained for $\Delta H_D^\circ$ in kilocalories per mole:

$$OH^+ \cdots O \qquad \Delta H_D^\circ = 30.4 - 0.30\,\Delta PA \qquad (3\text{-}2)$$

$$NH^+ \cdots O \qquad \Delta H_D^\circ = 28.3 - 0.23\,\Delta PA \qquad (3\text{-}3)$$

$$NH^+ \cdots N \qquad \Delta H_D^\circ = 23.2 - 0.25\,\Delta PA \qquad (3\text{-}4)$$

where $OH^+$ is any protonated oxygen function and $NH^+$ is a protonated amine or pyridine. When the ligand is a nitrile, somewhat stronger bonding occurs, and the correlations are[8]:

$$OH^+ \cdots NCR \qquad \Delta H_D^\circ = 30.9 - 0.43\,\Delta PA \qquad (3\text{-}5)$$

$$NH^+ \cdots NCR \qquad \Delta H_D^\circ = 35.3 - 0.35\,\Delta PA \qquad (3\text{-}6)$$

$$RCNH^+ \cdots NCR' \qquad \Delta H_D^\circ = 28.1 - 0.37\,\Delta PA \qquad (3\text{-}7)$$

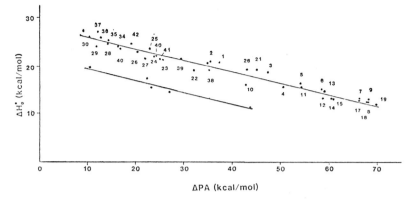

**Figure 3-2.** Correlation between $\Delta H_D^\circ$ and $\Delta PA$ for $NH^+ \cdots O$ bonds. (Reproduced by permission from Reference 5.) (The lower line is for $-NH^+ \cdots$ S-dimers). The complexes indicated by numbers are of the following types: A) $BH^+ \cdot H_2O$ where the number and the corresponding B are as follows: 1. $NH_3$, 2. $CF_3CH_2NH_2$, 3. $CH_3NH_2$, 4. $C_2H_5NH_2$, 5. $(CH_3)_2NH$, 6. $(CH_3)_3N$, 7. $(C_2H_5)_3N$, 8. $(n-C_3H_7)_3N$, 9. $(n-C_4H_9)_3N$, 10. 4-CNpy (py = pyridine), 11. py, 12. 2,6-$(CH_3)_2$ py, 13. 4-$CH_3$py, 14. 2-$i$-$C_3H_7$py, 15. 2-$t$-$C_4H_9$py, 16. 2,6-$(C_2H_5)_2$py, 17. 2,6-$(i$-$C_3H_7)_2$py, 18. 2,6-$(t$-$C_4H_9)_2$py, 19. 4-$N(CH_3)_2$py; B) Complexes $CH_3NH_3^+ \cdot A^+$ where A is as follows: 20. $H_2O$, 21. $CF_3CH_2OH$, 22. $CH_3OH$, 23. $C_2H_5OH$, 24. n-$C_3H_7OH$, 25. $n$-$C_4H_9OH$, 26. $t$-$C_4H_9OH$, 27. $(CH_3)_2O$, 28. $(C_2H_5)_2O$, 29. $(n$-$C_3H_7)_2O$, 30. $(n$-$C_4H_9)_2O$, 31. $(n$-$C_6H_{11})_2O$, 32. $CH_3OCH_2CH_2OCH_3$, 33. $CH_3OCH_2CH_2CH_2OCH_3$, 34. $CH_3COCH_3$, 35. $C_2H_5COCH_3$, 36. $C_2H_5COC_2H_5$, 37. $n$-$C_3H_7COC_2H_5$, 38. HCOOH, 39. $CF_3COOC_2H_5$, 40. $CH_3COOH$, 41. $HCOOCH_3$, 42. HCOO-$n$-$C_4H_9$.

In all cases, these correlations predict the dissociation energies of the ionic hydrogen bonds to within $\pm 2$ kcal/mol, which is as accurate as the experimental measurements.

In agreement with correlations 3-2 to 3-7, the strongest hydrogen bonds are observed in symmetric dimers where $\Delta PA = 0$. Thus, in all the oxonium dimer ions $(H_2O)_2H^+$, $(R_2O)_2H^+$, and $(R_2CO)_2H^+$, the dissociation energy H $= 30 \pm 2$ kcal/mol, regardless of the identity of R; but, for example, in $MeC(NMe_2)OH^+ \cdot H_2O$, where $\Delta PA = 50.3$ kcal/mol, $\Delta H_D^\circ$ has decreased to 18.3 kcal/mol. Similarly, in the symmetric dimers $(R_3N)_2H^+$, regardless of the identity of R, and also in $(Xpyridine)_2H^+$, $\Delta H_D^\circ = 24 \pm 2$ kcal/mol; but in $Et_3N^+$ where $\Delta PA = 28.9$ kcal/mol, $\Delta H_D^\circ$ has decreased to 16.3 kcal/mol.

Strong ionic hydrogen bonds are also observed in negative ion systems. In early studies, Yamdagni and Kebarle[9-11] measured the interaction energies betwen $OH^-$, $F^-$, $CCl_3^-$, $Cl^-$, and $Br^-$, and the ligands HOH, HF, $HCCl_3$, HCl, and HBr. The dissociation energies in these systems ranged between 13 and 22 kcal/mol, and it was shown that in series with a constant proton acceptor $B^-$ and a series of donors RH, a linear relationship of $\Delta H_D^\circ$ to the acidity of RH applies. Similarly, a linear relationship was observed when the ligand is kept constant and a series of ions with varying basicities is used. A more extensive set of alcohol–alkoxide complexes also lead to linear relationships of this kind[12]:

$$\Delta H_D^\circ(RO^- \cdots HOMe) = 146.2 - 0.44\, \Delta H_{acid}^\circ(ROH) \qquad (3\text{-}8)$$

and, when the ion is constant, but the proton donor changes,

$$\Delta H_D^\circ(ROH \cdots {}^-O\text{-}CH_2\text{-}t\text{-}Bu) = -160.2 + 0.37\, \Delta H_{acid}^\circ(ROH) \qquad (3\text{-}9)$$

In the case of such hydrogen-bonded complexes of negative ions, the best general relationships are obtained by the use of double-parameter equations that take into account separately the effects of the proton-acceptor strength of the ion and the proton-donor strengths of the neutrals. For example, for the alcohol–alkoxide complexes Caldwell and co-workers[12] found:

$$\Delta H_D^\circ(ROH \cdots {}^-OR')$$
$$= 15 - 0.40\, \Delta H_{acid}^\circ(ROH) + 0.31\, \Delta H_{acid}^\circ(R'OH) \qquad (3\text{-}10)$$

More recent measurements in our laboratory for OH ... $O^-$ type bonds involving $H_2O$, alcohols and carboxylic acids yield the correlation equation 3–11:

$$\Delta H_D^\circ(OH \cdots O^-) = 28.1 - 0.29\Delta\Delta H^\circ{}_{acid} \qquad (3\text{-}11)$$

Interestingly, for dimers with similar acidity or basicity differences, the strength of OH $\cdots$ $O^-$ bonds is remarkably similar (within 3 kcal/mol) to $OH^+$ $\cdots$ O bonds.

The empirical relations (3-2 to 3-11) are useful in several ways. First, they help to predict, together with further empirical relations concerning higher clustering, the thermochemistry of ion clustering—for example, in atmospheric chemistry and in analytical chemical ionization mass spectrometry. Together with further empirical relations, as we shall show presently, these relations can also estimate the bulk solvation energy of many onium ions and can indicate special effects for ions that deviate from these relations. Furthermore, Equations 3-2 to 3-11 can be used to estimate interactions between ionized and neutral groups in environments such as the interiors of proteins and enzymes, where direct measurements are not possible.

Although the ab initio studies of Desmeules and Allen[6] as well as of Ikuta and Imamura[7] show approximately linear $\Delta E_D$ versus $\Delta PA$ correlations, there is no obvious physical reason for the linear nature of these relations over the wide ranges observed. However, it is interesting to note that the interaction energies correlate with the calculated charge densities on the bonding proton, $q(H^{\delta+})$ at least in the $R_3NH^+$ $\cdots$ $OH_2$ series.[7] This trend can be related[8] to the $\Delta H_D^\circ/\Delta PA$ correlations (equating for now $\Delta H_D^\circ$ with $\Delta E_D$) by Equation 3-12.

$$\frac{dE}{dPA} = \frac{dE}{dq}\frac{dq}{dPA} \qquad (3\text{-}12)$$

The $dq/dPA$ term can be obtained from the calculated partial charges in the $R_3NH^+$ series and the experimental proton affinities of the amines. The relation is linear in this series. The electrostatic ion–dipole and ion-induced dipole components of the $dE/dq$ term are calculated from the electrostatic expressions for these terms, Equation 3-13.

$$-\frac{dE}{dq} = \frac{\mu_0}{r^2} + \frac{\alpha q}{r^4} \tag{3-13}$$

where $\mu_0$ is the dipole moment and $\alpha$ is the spherically averaged polarizability of the ligand, and $r$ is the ion–ligand separation. For polar ligands the $\mu$ term is dominant, and therefore the relation between $E$ and $q$ is approximately linear. Therefore, the two terms on the right-hand side of Equation 3-12 being constant, $dE/dq$ is also constant. These combined theoretical and electrostatic considerations justify an approximately linear $\Delta H_D^{\circ}/\Delta PA$ relation, at least in this series where the ion varies and the ligand is constant. The same arguments also predict a larger slope in the $\Delta H_D^{\circ}/\Delta PA$ correlation for the $R_3NH^+ \cdot \cdot \cdot HCN$ series, where $\mu$ is large, than for the same ions with the less polar ligands $H_2O$ and $NH_3$. This prediction is borne out experimentally, the slopes being 0.34, 0.26, and 0.25, respectively.

## B. Unconventional Ionic Hydrogen Bonds: $CH^{\delta+} \cdot \cdot \cdot O$

As we shall see below, the bulk solvation enthalpies of onium ions indicate that $CH^{\delta+} \cdot \cdot \cdot OH_2$ interactions make a significant contribution. Other indications of $CH^{\delta+} \cdot \cdot \cdot O$ interactions appear in ion clustering studies. For example, Grimsrud and Kebarle[14] observed the $(Me_2O)_3H^+$ cluster, and ab initio calculations showed that the dissociation energies of structures 1 and 2 are both about 8–10 kcal/mol.[15] $CH^{\delta+} \cdot \cdot \cdot O$ interactions are also indicated in complexes of $Me_3NH^+$ with polyethers, which are more stable by 4–10 kcal/mol than analogous complexes of monoethers, presumably due to multiple $CH^{\delta+} \cdot \cdot \cdot O$ interactions.

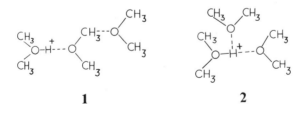

|   1   |   2   |

Further examples of analogous $CH^{\delta+} \cdot \cdot \cdot N$ interactions are indicated in the acetonitrile trimer $CH_3CN-H^+ \cdot \cdot \cdot NCCH_3 \cdot \cdot \cdot NCCH_3$.[16,17] $CH^{\delta+} \cdot \cdot \cdot X$ interactions are also indicated in crystals.[18,19]

In the examples above, the occurrence of $CH^{\delta+} \cdot \cdot \cdot O$ bonding is ambig-

TABLE 3-2. Thermochemistry of Unconventional Ionic Hydrogen Bonds[a]

| Complex | | $\Delta H_D^\circ$ (kcal/mol) | $\Delta S_D^\circ$ (cal/mol K) |
|---|---|---|---|
| $CH^{\delta+} \cdots O$ | | | |
| $Me_4N^+$ | $H_2O$ | 9.0 | 21.5 |
| | MeOH | 9.8 | 23.2 |
| | $Me_2CO$ | 14.6 | 24.7 |
| | $Me(OCH_2OCH_2)_3OMe$ | 24.2 | 33.8 |
| | $CH_3CONHCH_2COOCH_3$ | 20.1 | 29.4 |
| | $(CH_3CO-Gly-OCH_3)$ | | |
| | $MeNH_2$ | 8.7 | 17.4 |
| $NH^+ \cdots \pi$ | | | |
| $NH_4^+$ | $C_6H_6$ | 19.3 | 23.3 |
| | $C_6H_5F$ | 14.4 | 18.0 |
| | $1,3,5\text{-}(CH_3)_3C_6H_3$ | 21.8 | 21.1 |
| | $C_2H_4$ | $\approx 10$ | |
| $Me_3NH^+$ | $C_6H_6$ | 15.9 | 27.7 |
| $CH^{\delta+} \cdots \pi$ | | | |
| $Me_4N^+$ | $C_6H_6$ | 9.4 | 20 |

[a]Data from References 20 and 23.

uous because of the possibility of alternative structures such as **1** or **2**. To study such interactions unambiguously, we generated quaternary ammonium ions by gas phase reactions and studied the association of these ions with oxygen and nitrogen ligands, yielding clusters such as $Me_4N^+ \cdots OH_2$.[20] In these complexes, of course, the only possible interactions are of the $CH^{\delta+} \cdots X$ type. The results, as listed in Table 3-2, showed interaction energies of 9–15 kcal/mol with monodentate ligands. With amides and with polydentate ligands, stronger interactions (18–24 kcal/mol) are observed. The structures of some of the complexes and the nature of the interactions will be discussed below.

The $CH^{\delta+} \cdots O$ interactions we find in the gas phase are substantially stronger than the 2-kcal/mol interaction energies indicated in solution. However, in solution onium ions hydrogen-bond to water through the protonated functional group, and this may substantially delocalize the ionic charge. As a model, we are interested in the mutual effects of solvent molecules in $H_2O \cdots H_3COH_2^+ \cdots OH_2$, which is discussed in Chapter 4.

## C. $CH^{\delta+} \cdots O$ Complexes Versus Covalent Condensation: Interactions of Carbonium Ions with *n*-Donors

$CH^{\delta+} \cdots O$ hydrogen bonds should be important also in the solvation of carbonium ions. As will be shown later, theoretical calculations show that $CH_3^+$, $i\text{-}C_3H_7^+$, and $t\text{-}C_4H_9^+$ bond strongly to $H_2O$.

Apart from their role in solvation, the possibility of significant $CH^{\delta+}$

$\cdots$ O interactions raises an interesting question in relation to the gas phase association of carbonium ions with $n$-donor molecules. Such processes can lead to the formation of a covalent $C-O$ bond (reaction a) leading to condensation products.

$$R_3C^+ + OH_2 \rightarrow R_3COH_2^+ \tag{a}$$
$$R_3C^+ + OH_2 \rightarrow H_2O \cdot R_3C^+ \tag{b}$$

Such reactions may contribute to the prebiotic synthesis of organic molecules in ionized planetary atmospheres. Surprisingly, however, thermochemical considerations and ab initio calculations (Chapter 4) show that in some cases $CH^{\delta+} \cdots O$ bonded clusters may be comparable in stability to the condensation products.

To investigate this question, we examined recently the thermochemical information available on association reactions of carbonium ions with oxygen and nitrogen bases, and with ethylene and benzene, along with calculated or estimated values of $\Delta H°$ and $\Delta S°$ for reactions a and b.[21] The results showed that the association of carbonium ions with amines always results in condensation, leading to protonated ammonium ions, similar to reaction a. The reactions of $CH_3^+$ and $C_2H_5^+$ with $H_2O$ also lead to condensation, but the association of $t$-$C_4H_9^+$ and possibly of $i$-$C_3H_7^+$ with $H_2O$ leads to a cluster ion. These results were also confirmed by collisional dissociation studies on the association products. Similarly, the association of oxocarbonium ions with $H_2O$ and $CH_3OH$ leads to cluster ions, rather than protonated acetals and hemiacetals (reaction d).

$$\text{(c)}$$
$$\text{(d)}$$

Other cases of thermochemistry that is suggestive of clustering rather than condensation are the association of $t$-$C_4H_9^+$ with HCN, of $HCO^+$ with $H_2$, and of $CH_3NH_3^+$ and $C_2H_5NH_3^+$ with $CO_2$ (to produce clusters rather than protonated glycine and alanine). One factor in favor of clustering is that in these processes a loose bond is formed, leading to an entropy term that is more favorable by 10–15 cal/mol K than the entropy change for condensation reactions.

The energetics of competitive and successive condensation and clustering reactions are examined in Chapter 4 by ab initio calculations on the series $CH_3^+$, $H_2O \cdots CH_3^+$, $CH_3OH_2^+$, $CH_3OH_2^+ \cdots OH_2$, and $H_2O \cdots CH_3OH_2^+ \cdots OH_2$, as well as in the comparative clustering and condensation of $H_2O$ with $C_2H_5^+$, $i$-$C_3H_7^+$, and $t$-$C_4H_9^+$.

## D. Unconventional Ionic Hydrogen Bonds: $CH^{\delta+} \cdots X^-$ and $NH^+ \cdots \pi$

Ionic hydrogen bonds of another kind that also involve $CH^{\delta+}$ centers occur in the complexing of polar molecules such as $CH_3CN$ and $(CH_3)_2CO$ to negative ions.[22] Here the methyl hydrogens possess partial positive charge because of the electronegative nature of the substituent. In the complexes $Cl^-$ is therefore hydrogen-bonded to methyl hydrogens; that is, $CH^{\delta+} \cdots$ $Cl^-$ bonds are formed, with dissociation energies of 13–15 kcal/mol. $CH^{\delta+}$ $\cdots Cl^-$ hydrogen bonds also seem to be present in the complexes of $Cl^-$ with alkylbenzenes.

Another type of unconventional bond in which the proton does not bridge between two $n$-donor centers appears in complexes of onium ions with $\pi$ donors. As observed in Table 3-2, the interaction energy in a prototype of an $NH^+ \cdots \Pi$ complex, $NH_4^+ \cdots C_6H_6$, is fairly large, 19.3 kcal/mol. Interaction energies between several protonated amines and benzene derivatives are in the range of 13–22 kcal/mol, and even $C_2H_4$ bonds to $NH_4^+$ by 10 kcal/mol.

The complexes of ammonium ions with aromatic molecules constitute an interesting case of clustering that is more favorable than condensation (reaction e). Here the clustering products:

$$\Delta H^\circ = -19.3 \text{ kcal/mol} \qquad (e)$$
$$\Delta H^\circ = -16 \text{ kcal/mol} \qquad (f)$$

are more stable because the competing condensation reaction would destroy the aromatic nucleus.

Ab initio results on the structures of $NH^+ \cdots \pi$ complexes are discussed in Chapter 4.

# 3. EFFECTS OF STRUCTURAL COMPLEXITY

The preceding discussion dealt with clustering of structurally simple ions—that is, organic ions with one functional group, and without substantial steric hindrance. The introduction of structural complexity has marked thermochemical effects on gas phase ionic properties such as the enthalpies and entropies of protonation and clustering. These thermochemical effects can be assigned to phenomena such as intramolecular or multiple hydrogen bonding and steric effects. These effects were reviewed recently,[3] and are summarized here only briefly.

## A. Multiple Hydrogen Bonding

Polyfunctional $n$-donor ligands can form polydentate, multiply hydrogen-bonded complexes with onium ions. Correspondingly, the dissociation ener-

TABLE 3-3. Thermochemical Effects of Multiple Hydrogen Bonding

| Ion | Complex | $\Delta H_D^\circ$ (kcal/mol) | $\Delta S_D^\circ$ (cal/ mol K) | $\Delta H_{MB}^\circ$ (kcal/mol)[a] | $\Delta H_{MB}^\circ/n_{MB}$ (kcal/mol)[b] |
|---|---|---|---|---|---|
| | AmineH$^+$ · polyether[a] | | | | |
| | c-C$_6$H$_{11}$NH$_3^+$ · | 21.9 | 31.9 | 0 | |
| | | 29.4 | 35.5 | 8 | 8 |
| 3 | | 39.7 | 44.6 | 15 | 8 |
| 4 | 18-crown-6 | 46 | 38 | 21 | 10 |
| | AmineH$^+$ · ester[c] | | | | |
| | CH$_3$NH$_3^+$  CH$_3$COOCH$_3$ | 23.5 | 24.8 | 0 | |
| 5 | CH$_3$COO$i$-C$_3$H$_7$ | 30.0 | 35.2 | 6 | 6 |
| | AmineH$^+$ · amide[d] | | | | |
| 6 | CH$_3$NH$_3^+$ · HCONH$_2$ | 30.0 | 30.0 | 6 | 6 |
| 7 | CH$_3$CONH(CH$_3$)COOCH$_3$ (CH$_3$CO−Ala−OCH$_3$) | 40.1 | 35.1 | 14 | 7 |
| | Nucleic bases[e] | | | | |
| | PyridineH$^+$ · pyridine | 23.7 | 28 | 0 | |
| | AdenineH$^+$ · adenine | 30.3 | 39 | 7 | 7 |
| | CytosineH$^+$ · cytosine | 38.3 | 37 | 14 | 7 |

[a]Excess stabilization due to multiple bonding.
[b]$n_{MB}$ = number of extra hydrogen bonds.
[c]Reference 24.
[d]Reference 5.
[e]Reference 25.

gies and entropies of these complexes are significantly higher than in monodentate analogues (Table 3-3). Ions **3** and **4** show proposed polydentate structures with polyether ligands, including two or three hydrogen bonds. Comparison with model monodentate ligands allows an estimate of the extra stabilization due to multiple bonding ($\Delta H_{MB}^\circ$). We are also interested in the stabilization per extra hydrogen bond (beyond the first one); that is, $\Delta H_{MB}^\circ/(n_{HB} - 1)$. These values are shown in Table 3-1.

Similarly, complexes of CH$_3$NH$_3^+$ with large esters and with amides show enhanced stability, which may be assigned to multiple bonding, such as in ions **5** and **6**. We note that the stabilizing effects of the second and third hydrogen bonds are all in the range of 6–8 kcal/mol each.

Polydentate complexes are of biochemical interest in that the multiple polar ligand groups create an enzymelike environment (see Chapter 5). In this regard it is of particular interest that the interaction of the peptidelike alanine derivative CH$_3$CONHCH(CH$_3$)COOCH$_3$ with the ammonium ion CH$_3$NH$_3^+$ shows an extra stabilization of 14 kcal/mol. This effect indicates

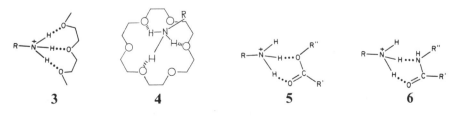

two extra hydrogen bonds with 7 kcal/mol each. Therefore, even a small peptidelike unit can bond to an onium ion by three hydrogen bonds, such as in ion **7**.

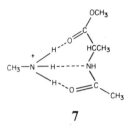

**7**

While we examined only polydentate complexes with ammonium ions, such polydentate structures are also observed between crown ethers and $H_3O^+$ or $CH_3OH_2^+$.[26]

In the cases above a single protonated group interacts with a polydentate ligand. A different type of multiple interaction seems to occur between protonated and neutral nucleic bases, where several pairs of functional groups can interact to form structures that may be ascribed to adenine$_2$H$^+$ and cytosine$_2$H$^+$ (ions **8, 9**).[27] In these dimers, of the total interaction energies of 30.3 and 38.3 kcal/mol, respectively, 7 and 15 kcal/mol can be ascribed to the second or second plus third hydrogen bonds, that is, again about 7 kcal/mol for each additional hydrogen bond. Interestingly, the configuration of hydrogen bonds that can be drawn in the dimer ions is analogous to hydrogen bonds in the nucleic base pairs in DNA—for example, cytosine$_2$H$^+$ (**9**) and cytosine · guanine (**10**).

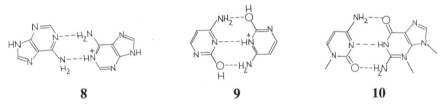

**8**                         **9**                         **10**

The additional hydrogen bonds formed in the complexes above are all of the conventional $nH^+ \cdots n$ type. However, multiple interactions involving the unconventional $CH^{\delta+} \cdots O$ type are also observed. For example, while the dissociation energy of the complex $Me_3NH^+ \cdot Et_2O$ is 19.5 kcal/mol, the dissociation energies of the complexes of $Me_3NH^+$ with $CH_3OCH_2CH_2OCH_3$ and $CH_3(OCH_2CH_2)OCH_3$ are 26.7 and 32.8 kcal/mol, indicating multiple interaction contributions of 7 and 10 kcal/mol in the polydentate complexes.[24] Since here only one proton is available for conventional hydrogen bonding, the added stabilization may be ascribed to $CH^{\delta+} \cdots O$ bonding. Even more clear-cut is the case in the complexing of $Me_4N^+$ with polyethers (Table 3-2), where the strong bonding energy clearly suggests multiple $CH^{\delta+} \cdots O$ interactions.

In summary, polydentate ligands complex strongly with onium ions. In general, each additional (second and third) hydrogen bond in the complexes adds about 7 kcal/mol to the dissociation energy of the complexes. Indeed, this rule may be used to estimate the number of hydrogen bonds in a complex.

## B. Intramolecular Ionic Hydrogen Bonds

The proton affinities of flexible, acyclic, polyfunctional compounds are generally substantially higher than those of monofunctional analogues, and large negative entropies of protonation of the polyfunctional compounds are also observed. The thermochemical effects, first observed by Yamdagni and Kebarle in diamines,[4] can be explained by the formation of intramolecular hydrogen bonds, which stabilize the ions and lead to an entropy loss upon cyclization. Similar effects have been observed in triamines,[28] amino alcohols,[28,29] polyethers and crown ethers,[30,31] diketones,[30] the amino acid derivative $CH_3CONHCH(CH_3)COOCH_3$,[25] and even in diphenylbutane[32] (Table 3-4). The significance of intramolecular bonding becomes evident when one considers the thermochemistry of the two proton transfer reactions g and h:

$$CH_3NH_3^+ + CH_3(CH_2)_3NH_2 \rightarrow CH_3(CH_2)_3NH_3^+ + CH_3NH_2 \qquad (g)$$
$$\Delta H° = -3.4 \text{ kcal/mol} \qquad \Delta S° \approx 0 \text{ cal/mol K}$$

$$CH_3NH_3^+ + NH_2(CH_2)_3NH_2 \rightarrow \dot{N}H_2(CH_2)_3\dot{N}H_3^+ + CH_3NH_2 \qquad (h)$$
$$\Delta H° = -19.2 \text{ kcal/mol} \qquad \Delta S° = -13.2 \text{ cal/mol K}$$

The observed strength of intramolecular bonds, as evaluated by comparison with monofunctional analogues, ranges from 2 to 30 kcal/mol in oxygen compounds and from 5 to 17 kcal/mol in diamines. In most cases, the intramolecular $OH^+ \cdots O$ and $NH^+ \cdots N$ bonds are significantly weaker than the corresponding intermolecular bonds in analogous protonated dimers. The weakening of the bond as expressed by the bond-weakening factor $\Delta H_{WF}°$ may be assigned to geometrical constraints imposed by the hydrocarbon chain, and possibly by intramolecular polarization of the neutral electron donor function by the protonated group. In general, difunctional ethanes give the least efficient intramolecular bonds. The bond strength increasingly approaches the optimal value as the chain length increases. For example, in the protonated polyethers $CH_3O(CH_2)_2OCH_3H^+$, $CH_3O(CH_2)_3OCH_3H^+$, and $CH_3(OCH_2CH_2)_2OCH_3H^+$ strength of the intramolecular bond increases from 7 to 16 to 22 kcal/mol, and the bond-weakening factors decrease correspondingly from 23 to 14 and 8 kcal/mol. In parallel, as the strength of the bond increases, so does the rigidity of the hydrogen-bonded ring. This is shown by the increase in $-\Delta S_{IHB}°/n$, where $n$ is the number of rotations about single bonds that are transformed into vibrations when the ring is formed (Table 3-4).

The presence of internal hydrogen bonds decreases the strength of interaction with external solvent molecules and vice versa. Thus, the attachment energy of $H_2O$ to $H_2N(CH_2)_3NH_3^+$ is only 11.4 kcal/mol, compared with 15.1 kcal/mol for $CH_3CH_2CH_2NH_3^+$. On the other hand, the strength of the inter-

TABLE 3-4. Thermochemical Parameters of Internal Hydrogen Bonds

| Compounds | $-\Delta H^\circ_{IHB}$ (kcal/mol)[a] | $-S^\circ_{IHB}$ (cal/mol K) | $-\Delta H^\circ_{BWF}$ (kcal/mol)[b] | $-\Delta S^\circ_{IHB}/n_6$ (cal/mol K)[c] |
|---|---|---|---|---|
| **Amines[d]** | | | | |
| $H_2NH^+\cdots NH_2$ (ring) | 7 | 8 | 17 | 2.6 |
| $H_2NH^+\cdot NH_2$ (ring) | 14 | 15 | 9 | 3.7 |
| $H_2N\cdots H\overset{+}{N}H \quad NH_2$ (ring) | 17 | 15 | 6 | 3.7 |
| **Ethers[e]** | | | | |
| $O\cdots H^+\cdots O$ (ring) | 7 | 4 | 23 | 1.4 |
| $O\cdots H^+\cdots O\cdots O$ (ring) | 30 | 19 | 0 | 3.2 |
| crown-type $O, O, H^+, O, O$ (ring) | 18 | 3 | 12 | |
| **Diketones[e]** | | | | |
| $OH^+\cdots O$ (1,3-diketone) | 2 | 4 | 28 | 2.0 |
| $OH^+\cdots\cdots O$ (diketone) | 6 | 7 | 24 | 2.3 |
| **Amino acid derivative[f]** | | | | |
| $CH_3C(OH^+\cdots O)C-CH_3,\ NH-CH(CH_3)$ | 7 | 14 | | 4.6 |
| **Intramolecular charge transfer complexes[g]** | | | | |
| diaryl $(CH_2)_4$ complex | 7 | 7 | 12 | 1.4 |
| $(CH_2)_4$ with $CH_2-CH_3$ groups | ≈3 | 7 | | 1.4 |

[a] $\Delta H^\circ_{IHB}$ (IHB = internal hydrogen bond) derived from comparison of protonation of monofunctional and polyfunctional compounds.
[b] $\Delta H^\circ_{BWF}$ (BWF = bond weakening factor) derived from comparison of intramolecular bonds to intermolecular bonds in dimers where geometry can optimize.
[c] Entropy of cyclization per single bond incorporated into the ring.
[d] Reference 28.
[e] Reference 30.
[f] Reference 25.
[g] Reference 32.

nal bond in the monohydrated $H_2N(CH_2)_3NH_3^+ \cdot H_2O$ ion is decreased by 3.7 kcal/mol, from 14.2 to 10.5 kcal/mol, compared with the nonhydrated ion; and in the fourfold hydrated ion the strength of the internal bond is further decreased to 8.4 kcal/mol. Indeed, sufficient further solvation, such as in bulk water, can completely displace the intramolecular bond.

## C. Steric Hindrance

Bulky groups introduced about the hydrogen-bonding center, such as in 2,6-dialkyl pyridines, can interfere with the hydrogen bond in two ways. Steric hindrance can keep the hydrogen bond extended, thus weakening the bond (an enthalpy effect); or the bond can achieve optimal distance, but at the cost of substantial constraints on the rotational freedom of the substituents (an entropy effect). An examination of the dimerization of 2,6-alkylpyridines and of trialkylamines showed that the association energies in the symmetric dimers did not change with increasing steric hindrance from pyridine$_2$H$^+$ to $(2,6-(i-Pr)_2$pyridine$)_2$H$^+$ and from $(Me_3N)_2$H$^+$ to $(n-Bu_3N)_2$H$^+$, but remained constant at $24 \pm 2$ kcal/mol. However, the entropies of dimer formation become increasingly negative, from 28 to 57 cal/mol K.

Similar trends are observed also in the gas phase hydration of the hindered alkylpyridinium ions by one $H_2O$ molecule. Here too, only a small variation in $\Delta H°$ is observed, consistent with the proton affinity correlations, but the steric hindrance leads to increasingly large entropy changes. An especially large entropy of association, $\Delta S° = -41$ cal/mol K, is observed in $2,6(t-Bu)_2$pyridineH$^+ \cdots OH_2$, where the rotations of both $t$-butyl groups, as well as of the $H_2O$ molecule become hindered, causing a steric entropy contribution of $-13$ cal/mol K. Indeed, comparison with solution thermochemistry shows that the steric effect of entropy on the bulk solvation of hindered pyridinium ions can be completely accounted for by the first $H_2O$ molecule.

The general conclusion about steric hindrance is that as long as the approach to optimal hydrogen bonding length is possible, the bond will not weaken, but the steric crowding may result in substantial negative entropy terms.

# 4. HIGHER ION–SOLVENT CLUSTERS

Up to this point, we have considered the attachment of a single ligand or solvent molecule to onium ions in the gas phase. However, for practical purposes our interest is in the condensed phases, where the ions interact with many neighboring solvent molecules. We shall turn therefore to the clustering of several solvent molecules onto onium ions, which is equivalent to the formation of the first solvent shells. From there, we shall proceed to a comparison of this partial solvation with the complete bulk hydration of monofunctional ions.

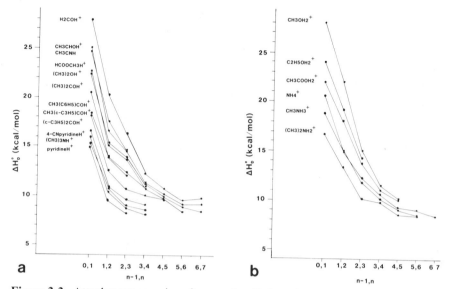

**Figure 3-3.** Attachment energies of successive $H_2O$ molecules in the stepwise hydration of (*a*) monoprotonic ions and (*b*) polyprotonic ions. (Reproduced by permission from Reference 34.)

Beyond the attachment of the first solvent molecule, pulsed high-pressure mass spectrometry can be used also to observe the attachment of several additional solvent molecules, usually up to a total of three to eight solvent molecules in a cluster. Generally, in each step the charge in the cluster becomes more diffuse, and the attachment energy of the next solvent molecule decreases (Figure 3-3). In the limit, the attachment of energies of the solvent molecules to the cluster approach the condensation energy of water (ie, about 10 kcal/mol). This limit is common to all clusters, and therefore the differential, ion-specific effects occur before this limit is reached.

Somewhat unexpectedly, in most clusters the observed limiting value, which is close to the condensation energy of water, is closely approached already at the fourth to sixth solvent molecule. This is surprising, since the condensation enthalpy of water reflects the formation of two hydrogen bonds as a water molecule is transferred from the gas phase into the bulk liquid. If the growing cluster behaves like the liquid in this respect, this would mean that: (a) the effects of the ionic charge on the attachment energy become negligible as the charge becomes diffuse, which is plausible, and (b) the attaching water molecule forms two hydrogen bonds to the cluster, which would suggest structures other than those conventionally assumed for $BH^+ \cdot 3H_2O$ or $BH^+ \cdot 4H_2O$, such as ions **11** or **12** below. Rather, structures involving cyclic hydrogen-bonded arrangements may be indicated.

Since the common lower limit of clustering energies is reached after the first four to six molecules, most significant specific interactions take place upon the attachment of the first four to six solvent molecules. Fortunately, the attachment energies in these clusters are within the range that is accessible to high-pressure mass spectrometric equilibrium measurements.

In fact, the fractional decrease of the dissociation energy with increasing cluster size shows an unexpectedly regular and uniform trend from ion to ion.[34] For our purposes, these trends can be expressed in terms of the relation of the integrated solvation energy by $n$ solvent molecules $\Delta H^{\circ}_{0,n}$ to the monomolecular solvation energy $\Delta H^{\circ}_{0,1}$. Thus, for example, we can examine the ratio $\Delta H^{\circ}_{4/1} = \Delta H^{\circ}_{0,4}/\Delta H^{\circ}_{0,1}$ in the clustering of 26 oxonium and carbonium ions with $H_2O$ molecules. The results show that this ratio is constant, between 2.7 and 3.0 for 24 of the 27 clusters for which data are available, even though the data include monoprotonic oxonium ions $R_2OH^+$ and $R_2COH^+$, polyprotonic oxonium ions $ROH_2{}^+$, monoprotonic and polyprotonic ammonium ions, and even the clustering of onium ions with nonaqueous solvents $CH_3OH$, $NH_3$, and $H_2S$.

The empirical observation that in the hydration of oxonium and monoprotonic ammonium ions the ratio $\Delta H^{\circ}_{0/4}$ is $2.8 \pm 0.1$, and in the hydration of polyprotonic ammonium ions $3.1 \pm 0.1$, can be combined with Equations 3-2 and 3-3, which predict the value of $\Delta H^{\circ}_{0,1}$ on the basis of the proton affinity difference $PA(B) - PA(H_2O) = PA(B) - 166.5$ kcal/mol. The resultant equations (3-14 to 3-16) predict the fourfold hydration energies of onium ions $BH^+$ on the basis of one variable, the proton affinity $PA(B)$.

*Oxonium Ions*

$-OH^+ \cdots 4H_2O$ clusters:

$$- \Delta H^{\circ}_{0,4} = 224 - 0.84 PA(B) \qquad (3\text{-}14)$$

*Protonated Primary and Secondary Amines*

$-NH^+ \cdots 4H_2O$ clusters:

$$- \Delta H^{\circ}_{0,4} = 206 - 0.71 PA(B) \qquad (3\text{-}15)$$

*Protonated Tertiary Amines and Pyridines*

$-NH^+ \cdots 4H_2O$ clusters:

$$- \Delta H^{\circ}_{0,4} = 186 - 0.64 PA(B) \qquad (3\text{-}16)$$

For the 20 clusters for which data are available, the average difference between the predicted and experimental values of $\Delta H^{\circ}_{0,4}$ is $\pm 2$ kcal/mol, which is comparable to the error in the experimental data.

In the addition of several solvent molecules to onium ions, an interesting structural question arises with respect to whether the formation of a complete inner solvent shell and the start of the next shell are energetically distinguishable. For example, in the hydration of a monoprotonic ion $R_2OH^+$ the first $H_2O$ molecule constitutes the first shell, the next two $H_2O$ molecules the second shell, and so on (ion **11**), while in the hydration of $H_3O^+$, the first three $H_2O$ molecules constitute the first shell, and the fourth molecule starts the second shell (ion **12**).

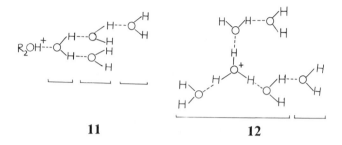

**11**                    **12**

If the energy distinction between shells is significant, a plot of $\Delta H^\circ_{n-1,n}$ versus $n$ should reveal a discontinuity between the two shells. Inspection of Figure 3-3 shows that such an effect is noticeable in a large drop from $\Delta H^\circ_{0,1}$ to $\Delta H^\circ_{1,2}$ and again from $\Delta H^\circ_{2,3}$ to $\Delta H^\circ_{3,4}$, corresponding to the transition from the first to the second and from the second to the third shell for the hydration of the monoprotonic ions $H_2COH^+$, $CH_3CHOH^+$, $(CH_3)_2COH^+$ and $CH_3CNH^+$. Similarly, Kebarle and co-workers[35,36] observed discontinuities in passing from the first to the second shell upon adding the fourth water molecule to $H_3O^+ \cdots nH_2O$. This was confirmed in our laboratory. More unexpectedly we also found that three $H_2O$ molecules constitute a shell for the solvation of $OH^-$, possibly by hydrogen bonds to the three oxygen lone pairs of $HO^-$. Altogether, out of 42 clustering systems, 10 show indications of distinct shell filling, including $NH_4^+ \cdot 4NH_3$ and $Na^+ \cdot 4NH_3$.

In contrast, $NH_4^+ \cdot nH_2O$ shows no distinct shell filling. In such cases, the distinction between consecutive solvent shells may be also diffuse if the thermochemistry for starting an outer shell is comparable to, or more favorable than, completing an inner shell. For example, the attachment energy of the fifth $H_2O$ molecule in the hydration of $NH_4^+$ is 10 kcal/mol, which is close to the common limiting value of 9 kcal/mol in the large hydrates. It is quite possible that in such large clusters the energies of a variety of structures such as **13–15** are very close.

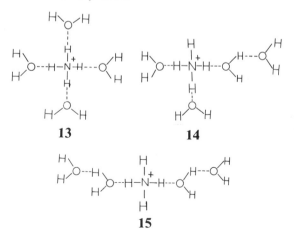

**13**                    **14**

**15**

In this case, the relative entropies may be dominant. The entropy term may favor the open-shell clusters **14** and **15,** since rotational freedom of the solvent molecules in the closed-shell structure **13** may be restricted. Furthermore, it is possible that several different structures are present in the equilibrium cluster population.

In that case, the observed van't Hoff plots are in fact composites, and the derived thermochemical values are intermediate between the true values corresponding to the various clusters.[37]

## 5. THE BULK SOLVATION OF ONIUM IONS

### A. The Relation Between Four-Molecular and Bulk Hydration

The partial solvation of ions by the first several solvent molecules serves as a model for the specific interactions of onium ions with the inner solvent spheres. It is of interest to compare trends in this specific, intimate solvation with trends in the bulk solvation of the same ions.

Fortunately, the differential enthalpies of solvation of ions can be obtained by combining the thermochemistries of protonation in the gas phase and in solution in a Born cycle:

$$NH_4^+)_g + (B)_g \longrightarrow (BH^+)_g + NH_3)_g$$
$$\downarrow \qquad \downarrow \qquad \downarrow \qquad \downarrow$$
$$NH_4^+)_{aq} + (B)_{aq} \longrightarrow BH^+)_{aq} + NH_3)_{aq}$$

Using this method, Taft and co-workers obtained the differential heats and entropies of solvation, that is, of transfer from gas phase to bulk water, for more than 60 onium ions.[1]

It may be expected that the stepwise solvation of ions will increasingly mimic the bulk solvation properties as the number of attached solvent molecules increases. Indeed, Kebarle,[38] Castleman and colleagues,[39,40] and Klots[41] have examined the relation between the $n$-fold clustering energies $\Delta H^\circ_{0,n}$ and the bulk hydration energies of alkali metal ions, halide ions, and also some onium ions. The general conclusion is that more than 80% of the differences between the bulk hydration enthalpies in each series of ions may be accounted for by the interactions with the first four to six water molecules. This is demonstrated for several ions in Figure 3-4. The results show that for most ions, the relative fourfold hydration enthalpies simulate the relative bulk hydration enthalpies surprisingly closely.

For a quantitative comparison of partial four-molecule hydration and complete bulk hydration, we examine the difference between the enthalpies of the two processes, that is, the differential term $\Delta\Delta H^\circ_{solv}$ (Equation 3-17a).

$$\Delta\Delta H^\circ_{solv} = \Delta H^\circ_{g \to aq}(BH^+) - \Delta H^\circ_{0,4} \qquad (3\text{-}17a)$$

Since the fourfold hydration enthalpies contain most of the specific ion–molecule hydrogen bonding interactions, the differential term $\Delta\Delta H^\circ_{solv}$

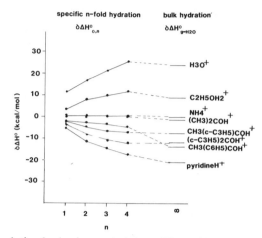

**Figure 3-4.** Cumulative hydration enthalpies $\Delta H^\circ_{0,n}$ and bulk hydration enthalpies of onium ions related to $NH_4^+$

should reflect the nonspecific electrostatic and intermolecular forces that occur beyond the fourfold solvation. These terms should be related to the ionic radius, and therefore we may expect a significant variation of $\Delta\Delta H^\circ_{solv}$ for the ions shown in Table 3-5. Surprisingly, however, for all the aliphatic oxonium ions, ranging in size from $H_3O^+$ to $Et_2OH^+$ and $Me(c\text{-}C_6H_{11})COH^+$, as well as for ammonium ions ranging from $NH_4^+$ to $Et_3NH^+$, the nonspecific solvation enthalpy is constant. That is, the empirical rule expressed by Equation 3-17b applies to all the 36 aliphatic ions in Table 3-5:

$$\Delta H^\circ_{g\rightarrow aq}(BH^+) - \Delta H^\circ_{0.4} = -26.8 \pm 2 \text{ kcal/mol} \qquad (3\text{-}17b)$$

Since the bulk solvation and intermolecular forces are expected to vary substantially with molecular size, the degree of substitution, and the number of protons available for hydrogen bonding, the constancy of $\Delta\Delta H^\circ_{solv}$ indicates compensating variation of the various physical factors.

As we noted, the overall ion–solvent interaction may be first decomposed to two factors: $XH^+ \cdots OH_2$ ionic hydrogen bonding, and other ion–bulk solvent physical forces. Since the specific hydrogen-bonding interactions are included in the addition of four water molecules to $BH^+$, the remaining ion–bulk solvent forces are represented by the solvation energy of the fourfold hydrated ion. This energy, that is, $\Delta H^\circ_{g\rightarrow aq}(BH^+ \cdot 4H_2O)$ can be calculated from a cycle that, starting with the bare ion $BH^+$, consists of the evaporation of four $H_2O$ molecules, the clustering of these $H_2O$ molecules onto $BH^+$, and the solvation of the cluster (Equation 3-18).

$$BH_g^+ + water_{aq} \rightarrow BH_g^+ + 4H_2O_g + (water - 4H_2O)_{aq}$$
$$\downarrow \qquad\qquad\qquad \downarrow$$
$$BH_{aq}^+ + water_{aq} \leftarrow BH^+ \cdot 4H_2O_g + (water - 4H_2O)_{aq}$$
$$\Delta H^\circ_{g\rightarrow aq}(BH^+ \cdot 4H_2O) = \Delta H^\circ_{g\rightarrow aq}(BH^+) - 4\Delta H^\circ_{vap,H_2O} - \Delta H^\circ_{0,4}(BH^+)$$

$$(3\text{-}18)$$

The solvation of the $BH^+ \cdot 4H_2O$ clusters is analyzed in the following sections. We present here the details of a model that uses the cycle 3–26 and physical models for the cavity and dielectric contributions to obtain the hydrophobic solvation of ions.

The solvation of the $BH^+ \cdot 4H_2O$ clusters, which is discussed below, is of interest mostly as a model for finding the hydrophobic energies. However, the model has been extended further to find the contribution of the various factors to the solvation of the $BH^+$ ions themselves, which is, of course, of more basic interest. The results give quantitatively the contribution of the cavity, dielectric, hydrophobic and hydrogen bonding factors to the solvation of protonated oxygen and nitrogen bases, ranging in size from $H_3O^+$ to $i\text{-}Pr_2COH^+$ and from $NH_4^+$ to $Et_3NH^+$. The model shows a regular and reasonable increase in the hydrophobic solvation energies with increasing alkyl substitution, and of the ionic hydrogen bonding contribution with increasing number of protic hydrogens. These results are summarized in the Appendix.

## B. The Contributions of Cavity, Dielectric Charging, and Intramolecular Interactions to the Hydration of $BH^+ \cdot 4H_2O$

The physical forces involved in the solvation of the cluster $BH^+ \cdot 4H_2O$ involve the following factors: the creation of a cavity in the solvent to accommodate the cluster, the dielectric charging of the solvent when the ionic cluster is introduced, and intermolecular interactions between the cluster and the rest of the solvent (Equation 3-19).

$$\Delta H^\circ_{g \rightarrow aq}(BH^+ \cdot 4H_2O) = \Delta H^\circ_{cav} + \Delta H^\circ_{dielectric} + \Delta H^\circ_{intermolecular} \qquad (3\text{-}19)$$

The cavity term is accounted for by the surface tension of the liquid at the cavity surface (Equation 3-20a). For the dielectric charging term, the Born equation is sometimes used, but improved accuracy is obtained by a double-shell model that refers to a cavity of radius $a$ within a concentric sphere of radius $b$. Outside the sphere the dielectric constant is that of the bulk medium ($\epsilon_0$), while within the layer of thickness $(b - a)$ it is assigned a lower value, $\epsilon_{loc}$, taken as 2 for water. The electrostatic energy in water is given by Equation 3-20b. Equations 3-20 yield free energies. From these we obtain enthalpies using Equation 3-21.

$$\Delta G^\circ_{cavity} = 4\pi r^2 \sigma_L \qquad (3\text{-}20a)$$

where $r$ = radius of the ion or cluster
$\sigma_L$ = surface tension of liquid water

$$\Delta G^\circ_{dielectric} = \frac{-q^2 e^2}{2}\left[\frac{\epsilon_a - 1}{-a\epsilon_a} + \frac{\epsilon_b - 1}{b\epsilon_b}\right] \qquad (3\text{-}20b)$$

where $\epsilon_a = \dfrac{2}{1 + (1 - \epsilon_{loc})(1 - \epsilon_b)a}{b\epsilon_b}$

TABLE 3-5. Thermochemistry of Hydration of Onium Ions[a] $BH^+$ and of Clusters $BH^+ \cdot 4H_2O$

Column groups: columns **Exp[c]** and **Calc[d]** fall under $-\Delta H^{\circ}_{g \to aq}(BH^+)$; columns $-\Delta H^{\circ}_{g \to aq}$[f] through $r(BH^+\cdot 4H_2O)$[l] fall under **$BH^+ \cdot 4H_2O$**; columns **Exp[m]** and **Calc[n]** fall under $-\Delta H^{\circ}_{g \to aq}(B)$.

| Ions | PA(B)[b] | Exp[c] | Calc[d] | $-\Delta H^{\circ}_{0,4}$[e] | $-\Delta H^{\circ}_{g \to aq}$[f] | $-\Delta H^{\circ}_{diel}$[g] | $\Delta H^{\circ}_{cav}$[h] | $\Delta H^{\circ\prime i}_{hydroph}$ | $\Delta H^{\circ ii}_{hydroph}$ | $\Delta H_{hydrophob}/mCH$ | $-T\Delta S_{g \to aq}$[k] | $r(BH^+\cdot 4H_2O)$[l] | Exp[m] | Calc[n] |
|---|---|---|---|---|---|---|---|---|---|---|---|---|---|---|
| **Water, alcohols, and ethers** | | | | | | | | | | | | | | |
| $H_3O^+$ | 166.5 | 110.3 | 111.1 | 84.1(83.7) | 68.2 | 49.4 | 16.1 | 0 | | 0 | 10 | 2.74 | 10.5 | |
| $MeOH_2^+$ | 181.9 | 97.2 | 98.2 | 71.0(72.8) | 68.2 | 47.1 | 18.0 | 9.1 | | 3.0 | 11 | 2.91 | 10.6 | 9.1 |
| $EtOH_2^+$ | 188.3 | 95.4 | 92.8 | 65.9 | 71.5 | 45.2 | 19.9 | 16.2 | | 3.2 | 5 | 3.05 | 12.6 | 9.1 |
| $Me_2OH^+$ | 192.1 | 88.7 | 89.6 | 62.6(62.1) | 68.1 | 45.0 | 20.0 | 18.1 | | 3.0 | 16 | 3.06 | 9.4 | 9.5 |
| $MeEtOH^+$ | 196.4 | 85.5 | 86.0 | 59.0 | 68.5 | 44.3 | 20.8 | 20.0 | | 2.5 | 16 | 3.12 | 10.2 | 10.1 |
| $Et_2OH^+$ | 200.2 | 82.7 | 82.8 | 55.8(56.8) | 68.9 | 41.9 | 23.6 | 25.6 | | 2.6 | 18 | 3.32 | 11.1 | 10.7 |
| **Aldehydes and ketones** | | | | | | | | | | | | | | |
| $C_6H_5CHOH^+$ | 200.2 | 79.8 | 77.8* | 55.8(55.8) | 66.0 | 40.8 | 25.2 | 25.5 | | 4.2 | 16 | 3.44 | 11.6 | 8.3* |
| $Me_2COH^+$ | 196.7 | 85.1 | 85.8 | 58.8(57.1) | 68.3 | 44.1 | 21.0 | 20.2 | | 3.4 | 14 | 3.14 | 10.1 | 10.2 |
| $MeEtCOH^+$ | 199.8 | 83.1 | 83.2 | 56.2 | 68.9 | 42.6 | 22.8 | 24.2 | | 3.0 | 16 | 3.27 | 11.2 | 11.1 |
| $Me(i\text{-}Pr)COH^+$ | 201.1 | 79.8 | 82.1 | 55.1 | 66.7 | 41.2 | 24.5 | 25.0 | | 2.5 | 15 | 3.38 | 11.6 | 13.1 |
| $Me(c\text{-}Pr)COH^+$ | 205.1 | 79.3 | 78.6 | 51.7(50.2) | 69.6 | 42.1 | 23.4 | 26.0 | | 3.2 | 16 | 3.31 | 12.0 | 10.8 |
| $(c\text{-}Pr)_2COH^+$ | 210.7 | 75.0 | 74.0 | 47.0(45.7) | 70.0 | 40.8 | 25.7 | 30.2 | | 3.0 | 17 | 3.47 | 14.7 | 12.6 |
| $Me(t\text{-}Bu)COH^+$ | 202.3 | 79.0 | 81.1 | 54.0 | 67.0 | 40.1 | 26.2 | 28.2 | | 2.3 | 16 | 3.50 | 12.4 | 13.7 |
| $Me(c\text{-}C_6H_{11})COH^+$ | 202.4 | 79.8 | 81.0 | 54.0 | 67.8 | 38.8 | 28.3 | 32.4 | | 2.3 | 16 | 3.64 | 12.9 | 13.4 |
| $Me(C_6H_5)COH^+$ | 205.4 | 73.2 | 73.5* | 51.4(53.4) | 63.8 | 39.7 | 26.8 | 25.9 | | 3.2 | 13 | 3.54 | 13.1 | 12.1* |
| $(i\text{-}Pr)_2COH^+$ | 204.9 | 78.1 | 78.8 | 51.8 | 68.3 | 39.1 | 27.8 | 31.4 | | 2.2 | 12 | 3.60 | 12.9 | 14.0 |
| **Esters** | | | | | | | | | | | | | | |
| $Me(OMe)COH^+$ | 197.8 | 83.5 | 84.8 | 57.8 | 67.7 | 43.4 | 21.9 | 21.2 | | 3.5 | 17 | 3.20 | 9.8 | 10.4 |
| $Et(OMe)COH^+$ | 200.2 | 82.9 | 82.8 | 55.8 | 69.1 | 41.9 | 23.6 | 25.8 | | 3.2 | 18 | 3.32 | 10.9 | 10.2 |
| **Sulfoxide** | | | | | | | | | | | | | | |
| $Me_2SOH^+$ | 211.3 | 73.7 | 73.5 | 46.4 | 69.3 | 42.3 | 23.2 | 25.2 | | 4.2 | 11 | 3.29 | 17.0 | 15.5 |
| **Amide** | | | | | | | | | | | | | | |
| $Me(NMe_2)COH^+$ | 216.8 | 67.5 | 68.8 | 41.8(47.3) | 67.7 | 41.6 | 24.0 | 25.2 | | 2.8 | | 3.35 | 16.1 | 16.4 |
| **Ammonia and primary amines** | | | | | | | | | | | | | | |
| $NH_4^+$ | 204.0 | 86.8 | 88.2 | 61.2(62.5) | 67.6 | 48.4 | 16.9 | 0 | | 0 | 7 | 2.81 | 8.5 | 10.1 |
| $MeNH_3^+$ | 214.1 | 80.5 | 81.0 | 53.9(54.0) | 68.6 | 46.2 | 18.8 | 9.7 | | 3.2 | 9 | 2.97 | 11.1 | 12.2 |
| $EtNH_3^+$ | 217.0 | 80.0 | 78.9 | 51.9(56.6) | 70.1 | 44.4 | 20.7 | 14.8 | | 3.0 | 11 | 3.12 | 13.0 | 12.6 |
| $CF_3CH_2NH_3^+$ | 202.5 | 86.5 | 89.2 | 62.2(63.8) | 66.3 | 43.8 | 21.4 | 12.2 | | | | 3.16 | 10.3 | 13.8 |
| $n\text{-}PrNH_3^+$ | 217.9 | 79.1 | 78.3 | 51.2 | 69.9 | 42.8 | 22.5 | 18.0 | | 2.6 | 11 | 3.24 | 13.3 | 12.7 |
| $i\text{-}PrNH_3^+$ | 218.6 | 77.9 | 77.8 | 50.8 | 69.1 | 42.8 | 22.5 | 17.2 | | 2.5 | 11 | 3.24 | 13.3 | 12.8 |
| $n\text{-}BuNH_3^+$ | 218.4 | 79.5 | 77.9 | 50.9 | 70.6 | 41.4 | 24.3 | 21.8 | | 2.4 | 13 | 3.37 | 14.1 | 13.0 |
| $t\text{-}BuNH_3^+$ | 220.8 | 76.7 | 76.2 | 49.2 | 69.5 | 41.4 | 24.2 | 20.8 | | 2.3 | 13 | 3.37 | 14.1 | 13.0 |

| Ions | $PA(B)$[b] | $-\Delta H^{\circ}_{g\to aq}(BH^+)$ Exp[c] | Calc[d] | $-\Delta H^{\circ}_{0,4}$[e] | $-\Delta H^{\circ}_{g\to aq}$[f] | $-\Delta H^{\circ}_{diel}$[g] | $\Delta H^{\circ}_{cav}$[h] | $\Delta H^{[vi]}_{hydroph}$ | $\Delta H^{\circ}_{hydroph}$ | $\Delta H^{\circ}_{hydrophob}/n_{CH}$ | $-T\Delta S_{g\to aq}$[k] | $r(BH^+\cdot 4H_2O)$[l] | $-\Delta H^{\circ}_{g\to aq}(B)$ Exp[m] | Calc[n] |
|---|---|---|---|---|---|---|---|---|---|---|---|---|---|---|
| c-C$_6$H$_{11}$NH$_3$‡ | 221.2 | 77.8 | 76.0 | 49.4 | 70.4 | 39.8 | 26.4 | 25.4 | 2.3 | | | 3.52 | 14.9 | 13.1 |
| C$_6$H$_5$NH$_3$‡ | 209.5 | 79.4 | 79.2* | 57.2 | 64.2 | 41.0 | 24.8 | 16.5 | 3.3 | | | 3.41 | 12.9 | 11.8* |
| **Secondary and tertiary amines** | | | | | | | | | | | | | | |
| Me$_2$NH$_2$‡ | 220.6 | 74.3 | 76.4 | 49.4 (49.4) | 66.9 | 44.4 | 20.7 | 16.2 | 2.7 | | 9 | 3.11 | 13.2 | 15.4 |
| (CH$_2$)$_4$NH$_2$‡ | 225.2 | 73.1 | 73.1 | 46.1 | 69.0 | 42.4 | 23.1 | 22.7 | 2.8 | | 10 | 3.28 | 15.2 | 15.7 |
| (CH$_2$)$_5$NH$_2$‡ | 226.2 | 72.3 | 72.4 | 45.2 | 69.1 | 41.8 | 24.8 | 25.8 | 2.6 | | 11 | 3.41 | 15.6 | 16.2 |
| Et$_2$NH$_2$‡ | 225.6 | 72.1 | 72.8 | 45.8 | 68.3 | 41.4 | 24.4 | 24.2 | 2.4 | | 12 | 3.38 | 15.3 | 16.0 |
| n-Pr$_2$NH$_2$‡ | 227.5 | 72.3 | 71.5 | 44.4 | 69.9 | 38.6 | 28.6 | 32.8 | 2.3 | | 13 | 3.65 | 17.2 | 16.2 |
| Me$_3$NH$^+$ | 225.1 | 66.6 | 68.9 | 41.9 (44.3) | 66.7 | 42.8 | 22.4 | 23.8 | 2.6 | | 11 | 3.24 | 13.2 | 13.6 |
| (CH$_2$)$_4$NMeH$^+$ | 228.7 | 65.1 | 66.6 | 39.6 | 67.5 | 41.1 | 24.8 | 28.7 | 2.6 | | 9 | 3.41 | 15.1 | 14.6 |
| (ring) NH$^+$ | 232.1 | 66.1 | 64.4 | 37.5 | 70.6 | 39.6 | 27.0 | 35.4 | 2.7 | | | 3.55 | 18.0 | 13.8 |
| Et$_3$NH$^+$ | 232.3 | 64.2 | 64.3 | 37.3 | 68.9 | 39.2 | 27.4 | 34.5 | 2.3 | | 13 | 3.58 | 16.7 | 14.7 |
| **Pyridines** | | | | | | | | | | | | | | |
| 4-CF$_3$pyrH$^+$ | 212.8 | 71.0 | 71.8* | 49.8 | 63.2 | 40.5 | 25.7 | 25.8 | 4.2 | | 9 | 3.46 | 11.6 | 10.8* |
| pyrH$^+$ | 220.8 | 65.6 | 65.7* | 44.7 (41.1) | 62.9 | 42.2 | 23.2 | 21.3 | 3.7 | | 10 | 3.29 | 11.9 | 11.1* |
| 4-MepyrH$^+$ | 225.2 | 64.8 | 63.8* | 41.8 | 64.4 | 41.0 | 24.9 | 25.8 | | | | 3.41 | 13.2 | 10.4* |
| **Proton and metal ions** | | | | | | | | | | | | | | |
| H$^+$ | 271.6 | | | 240 | 73 | | | | | | 9 | | | |
| Li$^+$ | 133.9 | | | 97 | 79 | | | | | | 10 | | | |
| Na$^+$ | 107.8 | | | 76 | 74 | | | | | | 8 | | | |
| K$^+$ | 87.6 | | | 60 | 70 | 51.0 | 15.1 | 15.4 | | | 5 | 2.65 | | |
| Rb$^+$ | 81.6 | | | 53 | 71 | 50.3 | 15.4 | | | | 4 | 2.69 | | |
| Cs$^+$ | 73.8 | | | 48 | 68 | 49.2 | 16.2 | | | | 4 | 2.76 | | |

[a] Values for $PA$ and $\Delta H^{\circ}$ in kilocalories per mole; $r$ in ångström units.

[b] Reference 13.

[c] Values relative to NH$_4^+$ from Reference 1. Absolute values are obtained using 86.8 kcal/mol for NH$_4$‡, based on Reference 41 but corrected using $PA(NH_3) = 204$ kcal/mol (Reference 13).

[d] Equations 3-25 to 3-27. Values for aromatic compounds (*) are adjusted by +5 kcal/mol (see text).

[e] Equations 3-14 to 3-16. Values in parentheses are experimental values (Reference 34).

[f] Equation 3-18, using $\Delta H^{\circ}_{0,4}$ values from Equations 3-14 to 3-16.

[g] Equation 3-21; see also Reference 42. Using $r(BH^+\cdot 4H_2O)$ from this table.

[h] Equation 3-20, using $r(BH^+\cdot 4H_2O)$ from this table.

[i] Data for CH$^{\circ+}$ . . . from Equation 3-22.

[j] $n_{CH}$ = number of hydrogen atoms on alkyl + aryl substituents.

[k] Using data from Reference 1.

[l] Value of $r$ calculated using the method of Reference 42 and supplementary material thereof.

[m] From values of Reference 1 relative to NH$_3$, and using $\Delta H^{\circ}_{g\to aq}(NH_3) = -8.5$ kcal/mol.

[n] Equations 3-28 to 3-30, using $\Delta H^{\circ}_{prot.aq}(NH_3) = -12.5$ kcal/mol. Values for aromatic compounds (*) are adjusted by +5 kcal/mol.

$$\epsilon_b = \frac{\epsilon_0}{\epsilon_{loc}}$$

$$\Delta H^\circ = \Delta G^\circ - T\left(\frac{\partial \Delta G^\circ}{\partial T}\right) \qquad (3\text{-}21)$$

The temperature coefficients are obtained numerically by calculating $\Delta G^\circ_{cavity}$ at 20 and 25°C using $\sigma_L$ at these temperatures, and by calculating $\Delta G^\circ_{dielectric}$ at 20 and 25°C using $\epsilon_0$ as the only temperature-dependent variable.

To apply Equations 3-20, the ions must be treated as spherical entities, and an ionic radius must be assigned. Ford and Scribner examined several possibilities and found the best predictive values when the radii used were those of a sphere having the same total volume as the ion in question.[42] We use the ionic volumes and radii used by these authors, for the solvation of the $BH^+$ ions in the Appendix. For the clusters, we use radii modified to reflect the volumes of the $BH^+ \cdot 4H_2O$ clusters. For this purpose, we add 16.1 $Å^3$ for each water molecule hydrogen-bonded to oxonium ions and 17.0 $Å^3$ for each water molecule hydrogen-bonded to ammonium ions. The results are summarized in Table 3-5.

Using the calculated cavity and dielectric charging terms, the contribution of the intermolecular forces can be obtained from Equation 3-19. To analyze the nature of these forces, we may consider the structures of the clustered ions such as $NH_4^+ \cdot 4H_2O$ and $Et_3NH^+ \cdot 4H_2O$.

## C. The Solvation of Alkyl Groups: A Compensation Variation of Solvation Terms

Intermolecular interactions with water of a clustered ion such as **16** or **17** will include hydrogen bonds from the inner four water molecules to the surrounding solvent, that is, OH · · · O bonds, and interactions of the alkyl groups with solvent molecules, which we term $CH^{\delta+}$ · · · O bonds, or

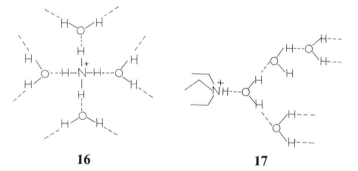

**16**                    **17**

hydrophobic solvation. The energies attributed to the total $CH^{\delta+}$ · · · O interactions may be due to electrostatic, dispersion, and van der Waals interactions of the alkyl groups with water, as well as structure-making effects of the alkyl groups on the solvent. We cannot evaluate these factors separately, but we can estimate the overall contribution of the $CH^{\delta+}$ · · · O interactions.

For this purpose, we find from Equation 3-19 $\Delta H^\circ_{\text{intermolecular}}$ and subtract the energy due to the residual hydrogen bonding (RHB) of the first four water molecules to the rest of the solvent. We must therefore first estimate the contribution of these OH $\cdots$ O hydrogen bonds. For this purpose we use the clusters $H_3O^+ \cdot 4H_2O$, where seven such "residual bonds" (eg, see ions **16** and **17**) yield a total intermolecular interaction of 35.0 kcal/mol, that is, 5.0 kcal/mol per bond, and $NH_4^+ \cdot 4H_2O$, where eight bonds yield a total of 36.1 kcal/mol, or 4.5 kcal/mol per bond. Therefore, we use Equation 3-22 to calculate the total contribution of $CH^{\delta+} \cdots$ O bonds (where $n$ is the number of hydrogen bonding sites on the ion, ie, $n = 1$ for $R_2OH^+$, 2 for $ROH_2^+$, etc, and, by simple arithmetic, $4 + n$ is the total number of OH $\cdots$ O bonds from the water molecules in $BH^+ \cdot 4H_2O$ to the rest of the aqueous solvent). Applying Equation 3-22 to all the ions assumes that the strength of these outer hydrogen bonds is not affected significantly by the identity of the central ion. This assumption is justified by the trends observed in clustering (Figure 3-3).

For oxonium ions:

$$-\Delta H^\circ_{\text{hydrophobic}} = -\Delta H^\circ_{\text{intermolecular}} - 5.0(4 + n) \text{ kcal/mol} \qquad (3\text{-}22a)$$

For ammonium ion:

$$-\Delta H^\circ_{\text{hydrophobic}} = -\Delta H^\circ_{\text{intermolecular}} - 4.5(4 + n) \text{ kcal/mol} \qquad (3\text{-}22b)$$

As noted above, Equation 3-17b implies a compensating variation of the bulk solvation factors (ie, charging plus cavity) vs the intermolecular terms with increasing alkyl substitution. To interpret this finding, we note again that the $\Delta\Delta H^\circ_{\text{solv}}$ term represents, after the addition of 4 $\Delta H^\circ_{\text{vap,H}_2\text{O}}$, the solvation enthalpy of the four-fold hydrated cluster (Equation 3-18). Therefore, for all alkylated onium ions $\Delta H^\circ_{\text{g}\to\text{aq}}(BH^+ \cdot 4H_2O)$ turns out be constant, and Equation 3-17b can be rewritten as follows:

$$\Delta H^\circ_{\text{g}\to\text{aq}}(BH^+ \cdot 4H_2O) = -68.8 \pm 2 \text{ kcal/mol} \qquad (3\text{-}23)$$

Again, considering that the various terms that enter into the solvation of the clustered ions varies significantly, the constancy of this term throughout the diverse series in Table 3-5 is remarkable. For example, it is easy to see that the ions **16** and **17** are different in all important aspects. Thus, in the solvation of **16** versus **17**, the charging plus cavity terms contribute $-31.5$ versus $-11.8$ kcal/mol and the OH $\cdots$ O bonds $-36.0$ versus $-22.5$ kcal/mol. The total contribution of these terms stabilizes $NH_4^+ \cdot 4H_2O$ more than $Et_3NH^+ \cdot 4H_2O$ by 34 kcal/mol. This is about exactly compensated for by the $CH^{\delta+} \cdots$ O interactions between the ethyl groups and water in the solvation of $Et_3NH^+ \cdot 4H_2O$. Similar fortuitous cancellations occur throughout the ions in Table 3-5.

Table 3-5 shows that the contribution of each hydrophobic $CH^{\delta+} \cdots$ O interaction in the hydration of alkylammonium ions is 2.4–3.2 kcal/mol. As expected, there is a tendency toward the lower values in the larger ions where the charge is more diffuse. Similar trends are observed in oxonium ions.

It is of interest to compare the solvation of alkyl groups of ions with the corresponding neutral solutes. For this purpose it is advantageous to examine solutes where dipole–dipole and hydrogen bonding forces are minimal with respect to the total $CH^{\delta+} \cdots O$ interaction. Therefore, we examine $Me_3N$ and $Et_3N$. For a treatment consistent with that of the ions, we consider the terms that contribute to $\Delta H^{\circ}_{g \to aq}(B)$ as follows:

$$\Delta H^{\circ}_{g \to aq}(B) = \Delta H^{\circ}_{cav} + \Delta H^{\circ}_{HB} + \Delta H^{\circ}_{hydrophobic} \qquad (3\text{-}24)$$

$$\text{For } Me_3N: \quad -13.2 \ = \ 14 \ - \ 3 \ + \Delta H^{\circ}(CH \cdots O)$$

$$\text{For } Et_3N: \quad -16.7 \ = \ 20 \ - \ 3 \ + \Delta H^{\circ}(CH \cdots O)$$

For neutral solutes the charging term is absent; the cavity-forming energy is calculated as above, using the equivalent radius of B. We assign about 3 kcal/mol to the hydrogen bond $R_3N \cdot H_2O$. This treatment yields $-24$ kcal/mol for the total hydrophobic interaction for $Me_3N$ and $-34$ kcal/mol for $Et_3N$, which is almost identical with $-24$ and $-35$ kcal/mol for the respective ions. In other words, for both the neutrals and the ions we obtain $-2.7$ and $-2.3$ kcal/mol per hydrophobic interaction in the hydration of the two compounds. Evidently, the presence of the ionic charge does not make the hydration of the alkyl groups significantly larger. This lack of ionic effect may be due to the large degree of charge dispersion in these ions both among the alkyl groups and to the solvent through the $NH^+ \cdots OH_2$ hydrogen bond. The similarity of the hydrophobic solvation of B and $BH^+$ is noted throughout Table A3-1 in the Appendix.

Similar to the enthalpies, the entropies of solvation, $\Delta S^{\circ}_{g \to aq}(BH^+)$ also become, in general, more negative with increasing alkyl substitution. This trend is consistent with increasing structure-making effect due to the increasing number of $CH^{\delta+} \cdots OH_2$ interactions. Using the data of Taft,[1] it is seen that the incremental entropies with added alkyl substitution are again comparable in the ions and the corresponding neutrals. This supports the conclusion above that alkyl group–water interactions of ions and corresponding neutrals are similar.

As a further comment on the enthalpies of ion hydration, we note that the alkali metal ions $K^+$, $Rb^+$, and $Cs^+$ also yield the same value of $-69 \pm 2$ kcal/mol for $\Delta H^{\circ}_{g \to aq}(BH^+ \cdots 4H_2O)$ as the aliphatic onium ions.

## D. Deficient Solvation of Aromatic Ions

In contrast to the aliphatic ions, the aromatic onium ions in Table 3-5 (ie, ions containing benzene or pyridine moieties) show values of $\Delta H^{\circ}_{g \to aq}(BH^+ \cdot 4H_2O)$ that are smaller, by an average of 5 kcal/mol, than that expected from Equation 3-23. This could be due to decreased specific solvation of the aromatic ions because, for example, of charge delocalization in the aromatic nucleus. This is, indeed, the implicit interpretation given by

Ford and Scribner,[42] who developed separate estimation parameters for "localized" and "delocalized" (carbonyl and pyridinium) ions. In this case, Equations 3-14 to 3-16 would not be used to estimate $\Delta H_{0,4}^{\circ}$ in Equation 3-18. However, data on the clustering of pyridinium ions shows that they fit the $\Delta H_{0,1}^{\circ}$ versus $\Delta PA$ correlations; that is, Equation 3-16 is applicable. Therefore the gas phase hydrogen-bonding data do not suggest that the clustering of charge-delocalized ions would be weaker than for aliphatic ions of similar acidities. Consequently, the small values of $\Delta H_{g\rightarrow aq}^{\circ}(BH^{+}\cdot 4H_{2}O)$ for ions with aromatic substituents should be attributed to deficient bulk solvation rather than to deficient specific solvation.

The decreased bulk solvation of aromatic ions is not due to the cyclic nature of the substituents, since the aliphatic cyclic ions are solvated normally (ie, in accordance with Equation 3-23). In line with the preceding considerations, it may be assumed that the deficient solvation of aromatic ions results from the hydrogen deficiency of the substituents, which decreases the number of $CH^{\delta+}\cdots OH_{2}$ interactions. In fact, comparing the total $CH^{\delta+}\cdots O$ interactions of anilineH$^{+}$ versus c-$C_{6}H_{11}NH_{3}^{+}$, pyridineH$^{+}$ versus piperidineH$^{+}$, and Me($C_{6}H_{5}$)COH$^{+}$ versus Me(c-$C_{6}H_{11}$)COH$^{+}$ shows that total $CH^{\delta+}\cdots O$ interactions contribute more by 9.9, 4.5, and 6.5 kcal/mol to the solvation enthalpies of the saturated versus the aromatic analogues, respectively.

While the total $CH^{\delta+}\cdots O$ interactions of the aromatic ions are deficient, the individual $CH^{\delta+}\cdots O$ interaction per phenyl or pyridine hydrogen atom is stronger than in the aliphatic ions. This can result from efficient charge delocalization from the protonated group to the aromatic ring. The deficient solvation of aromatic ions is caused therefore not by a weakening of the $CH^{\delta+}\cdots O$ bonds in these ions, but by the smaller number of hydrogens available for such interactions.

## E. A Simple Empirical Rule for the Hydration Enthalpies of Onium Ions

We have observed that the enthalpies of solvation of onium ions by one and by four $H_{2}O$ molecules can be predicted on the basis of $PA(B)$ (Equations 3-14 to 3-16). We can combine these equations and predict the enthalpies of solvation in bulk water on the basis of $PA(B)$, by means of Equations 3-25 to 3-27.

*Oxygen Compounds*

$$-\Delta H_{g\rightarrow aq}^{\circ}(BH^{+}) = 251 - 0.84PA(B) \qquad (3\text{-}25)$$

*Primary and Secondary Amines*

$$-\Delta H_{g\rightarrow aq}^{\circ}(BH^{+}) = 233 - 0.71PA(B) \qquad (3\text{-}26)$$

*Tertiary Amines*

$$-\Delta H_{g\rightarrow aq}^{\circ}(BH^{+}) = 213 - 0.64PA(B) \qquad (3\text{-}27)$$

In all cases, these equations predict the solvation enthalpies with a standard deviation better than $\pm 1.5$ kcal/mol (Table 3-5). This is only slightly less accurate than the experimental values, which are derived from a Born–Haber cycle and therefore contain the combined errors in $PA(B)$, $\Delta H^\circ_{g\rightarrow aq}(B)$, and $\Delta H_{protonation,aq}(B)$.

For onium ions containing a benzene or pyridine moiety, $\Delta H^\circ_{g\rightarrow aq}(BH^+)$ is obtained by adding $+5$ kcal/mol to the result calculated from Equations 3-25 to 3.27.

Table 3-5 contains the data for 42 onium and ammonium ions. These are all the ions for which data are given by Taft,[1] except for four esters, where the data seem suspect.[25] The agreement between the experimental and predicted values can be seen in Table 3-5.

Furthermore, using the predicted values of $\Delta H^\circ_{g\rightarrow aq}(BH^+ \cdot 4H_2O)$, Equations 3-25 to 3-27 can be used to predict other entities in the Born–Haber cycle.

$$
\begin{array}{ccc}
& -PA(B) & \\
B_g + H^+_g & \longrightarrow & BH^+_g \\
\Big\downarrow \Delta H^\circ_{g\rightarrow aq}(B) & \Big\downarrow \Delta H^\circ_{g\rightarrow aq}(H^+) & \Big\downarrow \Delta H^\circ_{g\rightarrow aq}(BH^+) \\
& \Delta H^\circ_{prot,aq}(B) & \\
B_{aq} + H^+_{aq} & \longrightarrow & BH^+_{aq}
\end{array}
$$

For example, given only $PA(B)$ and the aqueous enthalpy of protonation of a compound, the heat of solvation (from gas phase to aqueous solution) of the compound can be predicted using Equations 3-28 to 3-30.

*Oxygen Compounds*

$$\Delta H^\circ_{g\rightarrow aq}(B) = 20.6 - 0.16 PA(B) - \Delta H^\circ_{prot,aq}(B) \qquad (3\text{-}28)$$

*Primary and Secondary Amines*

$$\Delta H^\circ_{g\rightarrow aq}(B) = 36.6 - 0.29 PA(B) - H^\circ_{prot,aq}(B) \qquad (3\text{-}29)$$

*Tertiary Amines*

$$\Delta H^\circ_{g\rightarrow aq}(B) = 58.6 - 0.36 PA(B) - \Delta H^\circ_{prot,aq}(B) \qquad (3\text{-}30)$$

Equations 3-28 to 3-30 are obtained from the Born–Haber cycle and using Equations 3-25 to 3-27 for $\Delta H^\circ_{g\rightarrow aq}(BH^+)$. Again, these equations predict the solvation enthalpies with standard deviations better than $\pm 1.5$ kcal/mol.

This error is substantially larger than in experimental heats of solvation, which are usually measured to better than $\pm 0.2$ kcal/mol.

For compounds containing a benzene or pyridine moiety, $\Delta H^\circ_{g \to aq}(B)$ is obtained by adding $+5$ kcal/mol to the values calculated from Equations 3-28 to 3-30.

The agreement between the experimental and predicted values of $\Delta H^\circ_{g \to aq}(B)$ can be seen in Table 3-5.

# 6. IMPLICATIONS CONCERNING CONDENSED-PHASE ION-CHEMISTRY

## A. The Interaction of Water with Protic and Alkyl Groups

In summary, we shall review the trends observed in the hydration of onium ions and consider some implications concerning the structure of water about the ions. Finally, we shall examine some implications of the gas phase data concerning the condensed-phase biochemistry of structurally complex ions.

The enthalpy of hydration of $BH^+$ by one water molecule is inversely related to the proton affinity of B. This trend is explained qualitatively by the variation of charge density on the protic hydrogens of $(BH^+)$ which ranges from $-14$ to $-32$ kcal/mol.

The enthalpy of hydration by four water molecules, $\Delta H^\circ_{0,4}$ turns out to be a constant multiple of the one-molecular hydration enthalpy $\Delta H^\circ_{0,1}$. (This ratio may be denoted in short as $\Delta H^\circ_{4/1}$.) The four-molecular hydration enthalpies range from $-37$ to $-84$ kcal/mol. After the fourth water molecule, $\Delta H^\circ_{n-1,n}$ levels off to about $9 \pm 1$ kcal/mol for all onium ions. Therefore, the specific ionic hydrogen-bonding interactions are accounted for by the first four water molecules.

Furthermore, the four-molecular solvation enthalpies reproduce closely the relative bulk hydration enthalpies. The same effect is expressed alternatively by the statement that the bulk solvation enthalpies of the fourfold hydrated ions are constant, $-68.8 \pm 2$ kcal/mol for all aliphatic onium ions.

As the combined result of the trends above, the bulk solvation enthalpies of all aliphatic onium ions can be predicted on the basis of $PA(B)$ alone.

All the foregoing correlations are verified statistically for a fairly comprehensive set of oxonium and ammonium ions. Nevertheless, the correlations are not only unexpected, but they run contrary to expectation. This is true for the linear $\Delta H^\circ_{0,1}$ versus $\Delta PA(B)$ relation, where one may have expected rather an asymptotic decrease of $\Delta H^\circ_{0,1}$ with increasing $\Delta PA$. Also, for $\Delta H^\circ_{4/1}$ one may have expected, rather than the actual constant value, a strong dependence on the number of protic hydrogens and a larger proportional decrease of $\Delta H^\circ_{n,n-1}$ with $n$ when the first attachment energy is large. Furthermore, especially unexpected is the constant value of

$\triangle H^{\circ}_{g \to aq}(BH^{+} \cdot 4H_2O)$, where one may have expected larger bulk solvation energies for small polyprotonic ions versus large monoprotonic ions.

Evidently, all the observed trends need theoretical interpretation. Nevertheless, relations between the clustering and bulk solvation energies can yield some physical insight.

In relation to the specific hydrogen bonding, we observed that $\triangle H^{\circ}_{4/1}$ is approximately constant for all the onium ions. These include ions with from one to four protic hydrogens. The constant ratio includes all, which suggests that the number of protic sites is not directly effective in determining the overall specific ionic hydrogen bonding in the fourfold solvated clusters. In other words, in these $BH^{+} \cdot 4H_2O$ clusters ion–water and water–water hydrogen bonding is involved to varying degrees, depending on $n$ (see, eg, ions **16** with no OH $\cdots$ O bonds and **17** with three OH $\cdots$ O bonds). The common value of $\triangle H^{\circ}_{4/1}$ shows that the effect of the ion on the total hydrogen bonding in $BH^{+} \cdots 4H_2O$ is expressed equally well whether the water molecules are directly hydrogen bonded to it or the ion's effect is transmitted through OH $\cdots$ O hydrogen bonds.

Overall hydrogen bonding enhances more the hydration of small polyprotonic ions than of large ions with small $n$. The cavity and dielectric charging terms also favor the solvation of the smaller ions. The constant value of $\triangle H^{\circ}_{g \to aq}(BH^{+} \cdot 4H_2O)$ results from a compensation of these terms by the structure-making hydrocarbon–water interactions, which contribute up to $-35$ kcal/mol to the solvation of the ions we examined.

It appears that the solvation energies of the hydrocarbon groups of onium ions, although substantial, are similar to those in the corresponding neutrals. The lack of ionic effects may be attributed to the delocalization of the protonic charge onto the solvent attached to the protonated functional group. In this respect, it should be considered that total proton affinity of four water molecules clustered together in $(H_2O)_4H^{+}$ is 235 kcal/mol, which is higher even than the proton affinity of $Et_3N$. Therefore, the proton may be pulled away substantially from the protonated bases in aqueous solution, leaving the charge densities on the alkyl groups similar to that in neutrals. This would also explain why each $CH^{\delta+} \cdots$ O interaction contributes only about 2 kcal/mol in bulk water, compared with 7–10 kcal/mol in the gas phase.

The effect of alkyl group hydration also accounts for the observed inefficient solvation of hydrogen-deficient, benzene- or pyridine-substituted ions. Since the clustering properties of these ions obey the appropriate correlations, their inefficient solvation must be a bulk solvation effect.

## B. Ion–Molecule Complexes in Biochemistry

Many biological processes involve protonated forms of structurally complex molecules and macromolecules. Some of the systems observed in the gas phase may serve as models for the interactions of the ionic intermediates with polar $n$-donors in their environment. The gas phase model is especially pertinent in environments such as the interiors of enzymes, where the bulk solvent is excluded.

In enzymes, polar groups provide an environment resembling crown ethers, which stabilizes ionic reaction intermediates.[46] Ionic states of substrates possess groups such as $-NH_3^+$ and $COO^-$. To compete with aqueous environments, such intermediates must be stabilized by the enzyme by more than 70 kcal/mol. Interactions of the positive moiety of such intermediates with polar neighbors is modeled by the $RNH_3^+ \cdots$ polyether and $RNH_3^+ \cdots CH_3CO\text{-}Ala-OCH_3$ complexes discussed above (Table 3-3 and ions **3, 4,** and **7**). The experimental complexing energies in these model systems range up to 45 kcal/mol. This demonstrates that the ion–dipole interactions can provide a substantial part of the needed stabilization. However, the model also shows that other stabilizing factors such as the interactions of ions with ionic groups, the interactions of the negative ionic residue of the intermediate with the enzyme, and possibly entropy factors must also play a substantial role in the enzymic stabilization of ionic intermediates.

The complexing of $RNH_3^+$ by polyethers also provides a model for the interaction of neurotoxins, containing a triad of oxygen atoms, with a protonated lysine residue in nerve membranes. According to a recent model,[47] this interaction prevents the lysine residue from closing a channel for cation transport through the nerve membrane. The observed complexing energies of $RNH_3^+$ with crown ethers suggest that the multiply hydrogen-bonded structures such as **3** and **4** can indeed stabilize strongly such ion–toxin complexes.

Also of interest to neurochemistry is the complexing of the neurotransmitter acetylcholine with polar binding groups in neuroreceptors. Acetylcholine is a quaternary ammonium compound, and its complexing to a protein can be modeled by the complexing of $Me_4N^+$ to the peptidelike derivative $CH_3CO\text{-}Ala-OCH_3$. We observe that these binding energies can exceed 20 kcal/mol even with this small peptidelike derivative, and stronger bonding probably can be obtained by further multiple bonding to a large peptide.

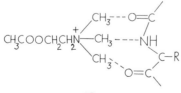

**18**

# 7. CONCLUSIONS

A comprehensive investigation of hydrogen bonding of protonated organic species to neutral ligands has been accomplished during the past two decades. From these investigations we can now derive some predictive generalizations concerning the ionic hydrogen bond.

1. Ionic hydrogen bonds between $n$-donors are very strong, with $\triangle H_D^\circ$ up to 32 kcal/mol.

2. There exist linear correlations between the hydrogen bond dissociation energies and the proton affinity difference of the components of the dimers, over wide ranges of $\triangle H_D^\circ$ and $\triangle PA$.

3. Unconventional $CH^{\delta+} \cdots X$ and $BH^+ \cdots \Pi$ type hydrogen bonds can be significant, with $\triangle H_D^\circ$ of 8–20 kcal/mol.

4. In polyfunctional species, intramolecular and multiple hydrogen bonding occurs. Usually, a second or third hydrogen bond in a complex contributes each $7 \pm 1$ kcal/mol to the stability of the complex.

5. Steric hindrance introduces substantial negative entropies of clustering.

6. The ratio between four-molecular solvation and monomolecular solvation energies, $\triangle H_{0,4}^\circ / \triangle H_{0,1}^\circ$, is constant, being 2.8–3.1 in the solvation of most onium ions.

7. The relative four-molecular solvation energies, $\triangle H_{0,4}^\circ$, reproduce closely the relative bulk solvation energies of alkylated onium ions; in other words, $\triangle H_{g \rightarrow aq}^\circ (BH^+ \cdot 4H_2O)$ is constant for most onium ions. This relation holds for a large variety of ions with varying sizes and protic hydrogens. The trend reflects a fortuitous compensation of electrostatic, hydrogen-bonding, and hydrophobic interactions.

8. The hydrophobic interactions of alkyl groups of ions are comparable to the interactions of the respective neutrals.

9. Ionic hydrogen bonding contributions to ion hydration vary regularly, increasing by 10 kcal/mol for each protic hydrogen.

In summary, for structurally simple ions, the strengths of the hydrogen bonds correlate simply with proton affinity differences. For polyfunctional ions, intramolecular and multiple bonding and steric hindrance cause significant effects. Finally, the properties of higher order clustering allow the prediction and quantitative analysis of the hydration energies of organic ions.

# REFERENCES

1. Taft, R.W. *Prog. Phys. Org. Chem.* **1983**, *14*, 247. Review article with numerous useful references.
2. Kebarle, P. *Annu. Rev. Phys. Chem.* **1977**, *28*, 445.
3. Meot-Ner (Mautner), M. *Acc. Chem. Res.* **1984**, *17*, 186.
4. Yamdagni, R.; Kebarle, P. *J. Am. Chem. Soc.* **1973**, *95*, 3504.
5. Meot-Ner (Mautner), M. *J. Am. Chem. Soc.* **1984**, *106*, 1257.
6. Desmeules, P.J.; Allen, L.C. *J. Chem. Phys.* **1980**, *72*, 4731.
7. Ikuta, S.; Imamura, M. *Chem. Phys.* **1984**, *90*, 37.
8. Speller, C.V.; Meot-Ner (Mautner), M. *J. Phys. Chem.* Submitted.
9. Yamdagni, R.; Kebarle, P. *J. Am. Chem. Soc.* **1977**, *93*, 7139.
10. Yamdagni, R.; Payzant, J.D.; Kebarle, P. *Can. J. Chem.* **1973**, *51*, 2507.

11. French, M.A.; Ikuta, S.; Kebarle, P. *Can. J. Chem.* **1981**, *60*, 1907.
12. Caldwell, G.; Rozeboom, M.D.; Kiplinger, J.P.; Bartmess, J.E. *J. Am. Chem. Soc.* **1984**, *106*, 4660.
13. Lias, S.G.; Liebman, J.F.; Levin, R.D. *J. Phys. Chem. Ref. Data* **1984**, *13*, 695.
14. Grimsrud, E.P.; Kebarle, P. *J. Am. Chem. Soc.* **1978**, *100*, 4694.
15. Hirao, K.; Jano, J.; Yamabe, S. *Chem. Phys. Lett.* **1982**, *87*, 181.
16. Meot-Ner (Mautner), M. *J. Am. Chem. Soc.* **1978**, *100*, 4694.
17. Hirao, K.; Yamabe, S.; Jano, J. *J. Phys. Chem.* **1982**, *86*, 2626.
18. DeBoer, J.A.A.; Reinhardt, D.N.; Harkima, S.; van Hummel, G.J.; de Jong, F. *J. Am. Chem. Soc.* **1982**, *104*, 4073.
19. Taylor, R.; Kennard, O. *J. Am. Chem. Soc.* **1982**, *104*, 5063.
20. Meot-Ner (Mautner), M.; Deakyne, C.A. *J. Am. Chem. Soc.* **1985**, *107*, 469.
21. Meot-Ner (Mautner), M.; Ross, M.M.; Campagna, J.E.; *J. Am Chem. Soc.* **1985**, *107*, 4839.
22. French, M.A.; Ikuta, S.; Kebarle, P. *Can. J. Chem.* **1982**, *60*, 1907.
23. Deakyne, C.A.; Meot-Ner (Mautner), M. *J. Am. Chem. Soc.* **1985**, *107*, 474.
24. Meot-Ner (Mautner), M. *J. Am. Chem. Soc.* **1983**, *105*, 4912.
25. Meot-Ner (Mautner), M. *J. Am. Chem. Soc.* **1984**, *106*, 278.
26. Sharma, R.B.; Kebarle, P. *J. Am. Chem. Soc.* **1984**, *106*, 3913.
27. Meot-Ner (Mautner), M. *J. Am. Chem. Soc.* **1979**, *101*, 2396.
28. Meot-Ner (Mautner), M.; Hamlet, P.; Hunter, E.P.; Field, F.H. *J. Am. Chem. Soc.* **1980**, *102*, 6393.
29. Houriet, R.; Ruefenacht, H.; Carrupt, P.A.; Vogel, P.; Tichy, M.J. *J. Am. Chem. Soc.* **1983**, *105*, 3417.
30. Meot-Ner (Mautner), M. *J. Am. Chem. Soc.* **1983**, *105*, 4906.
31. Sharma, R.B.; Blades, A.T.; Kebarle, P. *J. Am. Chem. Soc.* **1984**, *106*, 510.
32. Meot-Ner (Mautner), M.; Sieck, L.W. *J. Am. Chem. Soc.* **1981**, *103*, 5342.
33. Meot-Ner (Mautner), M.; Sieck, L.W. *J. Am. Chem. Soc.* **1983**, *105*, 2956.
34. Meot-Ner (Mautner), M. *J. Am. Chem. Soc.* **1984**, *106*, 1265.
35. Payzant, J.D.; Cunningham, A.J.; Kebarle, P. *Can. J. Chem.* **1973**, *51*, 3242.
36. Lau, Y.K.; Ikuta, S.; Kebarle, P. *J. Am. Chem. Soc.* **1982**, *104*, 1462.
37. Meot-Ner (Mautner), M.; Speller, C.V.; *J. Chem. Phys.* Submitted.
38. Kebarle, P. In "Modern Aspects of Electrochemistry", Vol. 9, Conway, B.E.; and Bockris, J. O'M., Eds.; Plenum Press: New York, 1974.
39. Keesee, R.G.; Lee, N.: Castleman, A.W. *J. Chem. Phys.* **1980**, *73*, 2195.
40. Lee, N.; Keesee, R.G.; Castleman, A.W. *J. Colloid Interface Sci.* **1980**, *75*, 555.
41. Klots, C.E. *J. Phys. Chem.* **1981**, *85*, 3585.
42. Ford, G.P.; Scribner, J.D. *J. Org. Chem.* **1983**, *48*, 2226.
43. Abraham, M.H.; Liszi, J. *J. Chem. Soc.*, Faraday Trans, 1, **1978**, *74*, 1604.
44. Beveridge, D.L.; Schnuelle, G.W. *J. Phys. Chem.* **1975**, *79*, 2562.
45. The values of $\triangle\triangle H_{solv}$, as well as $\triangle S^{\circ}_{g \rightarrow aq}$ calculated from $\delta\triangle G^{\circ}_{g \rightarrow aq}$ and $\delta\triangle H^{\circ}_{g \rightarrow aq}$ of Reference 1, deviate from the expected trend by 2–13 kcal/mol and 10–40 cal/mol K, respectively, suggesting that the $\triangle H^{\circ}_{g \rightarrow aq}$ values as given are too small (not sufficiently negative) by about: $i$-Pr(OMe)COH$^+$, 4; c-Pr(OMe)COH$^+$, 2; $t$-Bu(OMe)COH$^+$, 7, and C$_6$H$_t$(OCH$_5$)COH$^+$, 7 kcal/mol. The $\triangle H^{\circ}$ values are suspect rather than the $\triangle G^{\circ}$ values, since the latter fit the estimation schemes of Reference 22.
46. Warshel, A. *Acc. Chem. Res.* **1981**, *14*, 284.
47. Kosower, E.M. *FEBS Lett.* **1983**, *163*, 161.

# CHAPTER 4

# Ionic Hydrogen Bonds

# Part II. Theoretical Calculations

## Carol A. Deakyne

Air Force Geophysics Laboratory, Hanscom Air Force Base, Massachusetts.

## CONTENTS

## 1. INTRODUCTION

The hydrogen bond is an intermolecular interaction between an electron-deficient hydrogen and an unshared pair of electrons on an electronegative atom or the electrons in a double or triple bond. Ab initio studies of hydrogen-bonded complexes ($-A-H \cdots B-$) have made important contributions to the theory of these intermolecular interactions. (For recent reviews, see References 1–5 and references cited therein.) The systems investigated

MOLECULAR STRUCTURE AND ENERGETICS, Vol. 4

comprise neutral and ionic dimers, neutral and ionic clusters, infinite chains, and complexes involving large organic molecules and ions.[1−5] Most of the research has involved dimers, especially neutral dimers, but in recent years more work has focused on the other systems. Equilibrium geometries, total interaction energies and the components of the total interaction energies,[3,6] charge rearrangements, and spectroscopic properties of hydrogen-bonded complexes have been probed in the studies.[1−5] The A · · · B distances, A−H distances, hydrogen bond directionality, and the linearity of hydrogen bonds are among the structural details that have been examined. Spectroscopic properties investigated include vibrational constants, electronic transitions, and NMR chemical shifts.[1−5]

The following general characteristics of hydrogen bonds have been derived from theoretical calculations[1−5] and from experimental studies.[1,7,8]

1. Neutral hydrogen bonds generally range in energy from 3 to 10 kcal/mol, whereas ionic hydrogen bonds are generally in the 10–35 kcal/mol range (see Table 4-1, cf Chapter 3).

2. The strength of a hydrogen bond is related to the acidity of the proton donor and the basicity of the electron donor.

3. The A−H bond lengthens upon hydrogen bond formation, leading to a decrease in the A−H stretching frequency $\nu_{AH}$ and the A−H force constant $k_{AH}$.

TABLE 4-1. Comparison of the Calculated ($\Delta E$) and Experimental ($\Delta H°$) Hydrogen Bond Energies of Some (B)AH, (B)AH$^+$, and (AH)B$^-$ Complexes

| Complex | $\Delta E^a$ (kcal/mol) | Ref. | $\Delta H°$ (kcal/mol) | Ref. |
|---|---|---|---|---|
| (HF)HF | 8.00 | 9 | 4.3 | 5 |
| (HF)H$_2$F$^+$ | 40.70 | 10 | | |
| (HF)F$^-$ | 40.24 | 11$^b$ | 37 | 14 |
| (H$_2$O)H$_2$O | 8.20 | 9 | 3.6 | 5 |
| (H$_2$O)H$_3$O$^+$ | 44.70 | 10 | 32 | 15 |
| (H$_2$O)OH$^-$ | 36 | 12a | | |
| (NH$_3$)NH$_3$ | 4.13 | 9 | 4.5 | 16 |
| (NH$_3$)NH$_4^+$ | 32.00 | 10 | 25 | 17 |
| (H$_2$O)HF | 13.40 | 9 | 6.2 | 5 |
| (HF)H$_2$O | 4.41 | 9 | | |
| (HF)H$_3$O$^+$ | 24.12 | 10 | | |
| (H$_2$O)F$^-$ | 39.4 | 13 | 23.3 | 18 |
| (NH$_3$)H$_2$O | 8.93 | 9 | | |
| (H$_2$O)NH$_3$ | 4.07 | 9 | | |
| (H$_2$O)NH$_4^+$ | 27.72 | 10 | 17 | 17 |
| (NH$_3$)HF | 16.30 | 9 | | |
| (HF)NH$_3$ | 3.64 | 9 | | |
| (HF)NH$_4^+$ | 17.78 | 10 | | |

$^a$Calculated using the 4–31G basis set.
$^b$Calculated using an extended basis set.

4. As the A—H bond lengthens, the A · · · B bond shortens. This is also reflected in the pertinent frequencies and force constants.

5. The H · · · B distance lies between the H—B distance in the protonated base and the sum of the H and B van der Waals radii.

6. The A · · · B and A—H bond lengths are highly dependent on the nature of the proton donor and acceptor molecules, on steric effects, and on the number of hydrogen bonds to which a given proton donor or electron donor contributes.

7. There is an energetic preference for linear or nearly linear hydrogen bonds. Although A—H · · · B angles of 180° are rare in crystals, when the observed distribution of A—H · · · B angles is corrected for the number of possible spatial configurations with A—H · · · B = $\phi + \delta\phi$, the "corrected" distribution is consistent with linearity.

8. Gas phase hydrogen-bonded complexes tend to form hydrogen bonds along the directions of lone pairs. In the crystalline state, there is a significantly smaller preference for hydrogen bonding in the directions of $sp^3$-hybridized lone pairs than $sp^2$-hybridized lone pairs. For example, N—H · · · O=C hydrogen bonds form along the lone pair directions, while N—H · · · O bonds do not.

9. The polarity of the A—H bond increases when a complex is formed. The magnitude of the dipole increment $\Delta\mu$ brought about by complexation is usually larger than expected from vectorial addition of the components. To a first approximation, $\Delta\mu$ has the direction of the hydrogen bond.

10. When hydrogen bonds are formed, the hydrogen-bonded proton loses electrons and A and B gain electrons. The remaining atoms in the proton donor molecule show a net gain of electrons, while the remaining atoms in the proton acceptor molecule show a net loss of electrons. Overall, electron density is transferred from the proton acceptor to the proton donor.

11. There is a nonadditivity effect, that is, a mutual enhancement or diminution of hydrogen bond strengths and A · · · B distances, in larger clusters and condensed phases.

## 2. COMPUTATIONAL METHODS

The classes of basis sets primarily used in calculations of hydrogen-bonded systems are minimal basis sets, split valence or double-zeta (DZ) basis sets, polarized basis sets (DZ + P), and large basis sets, for example, triple-zeta or doubly polarized bases. To date, the majority of calculations, especially for more extensive systems, have been carried out utilizing minimal and double-zeta bases.[1-5] There are several sources of errors in the geometries, interaction energies, energy components, and molecular properties obtained with these small basis sets.[1,5,19,20] They are the poorly reproduced multipole moments (ie, monopole, dipole, quadrupole, etc) of the monomers, the incorrect representation of the long-range tail of the wave function, and the basis set superposition error (BSSE),[21-23] that is, the nonphysical charge delocalization and stabilization of the supersystem caused by the different sizes of the monomer and supersystem basis sets.

Part of the basis set dependence of the properties above is removed by correcting for BSSE.[3,5,20,23,24] This has been done[25] via the counterpoise method,[26] whereby the total energies of the subunits are obtained in the supermolecule basis. Sokalski and co-workers[19,24] find that the rest of the basis set dependence is primarily due to the electrostatic portion of the self-consistent field (SCF) energy (see Section 3, "Energy Components"). To eliminate the latter dependence, Sokalski and co-workers[19] recommended the use of "uniform quality" basis sets[27] to improve charge distributions calculated with small bases.[28] Others[29] have suggested scaling the electrostatic term.

Amelioration of the improper description of the long-range tail of the wave function can be achieved[19] by the use of the "overlap matched atomic orbitals,"[30] which yield intermolecular overlap integrals closer to the exact ones.[19,31]

Sokalski and associates[19] believe that a calculation combining all these factors will lead to interaction energies for small bases comparable to those for more extended bases. It has been shown that binding energies and equilibrium geometries computed with polarized[1-5,24,32,36] and larger[3,5,37-40] basis sets are reasonable, even though BSSE can be significant[24] and monomer multiple moments can be incorrect for DZ + P basis sets.[1,3]

The size of the basis set, hence the amount of CPU time, may be reduced further by employing minimal all-valence basis sets with effective core-model potentials[19,23,24] or core-deficient[20,41] basis sets. The all-valence $3^s3^p$ MODPOT[42] and 3-21G[43] bases have been suggested as possibilities.[19,20,24]

If the hydrogen bond energy is computed using single-determinant theory, two other sources of error limit the quantitative accuracy of the results, namely, the inability of single-determinant wave functions to account for intramolecular correlation,[1,5] and dispersion effects.[44] An exact computation of these effects requires extended basis sets[45] and extensive and time-consuming configuration interaction (CI) calculations.[5] (The amount of computer time required for an SCF computation is roughly proportional to $n^4$, where $n$ is the number of basis functions. For a CI calculation, it is roughly proportional to $n^5$.) However, the dispersion energy can be reliably estimated empirically.[46] When the estimated dispersion energy is added to the near Hartree–Fock energy, quantitative accuracy is approached.[5]

## 3. ENERGY COMPONENTS

The most widely used decomposition scheme for analysis of the origin of hydrogen bonding is the scheme developed by Morokuma and Kitaura[6,47] based on SCF calculations. This approach partitions the SCF interaction energy as follows[6,47]:

$$\Delta E = ES + PL + EX + CT + MIX = E_{AB} - E_A - E_B \qquad (4\text{-}1)$$

where $E_{AB}$ is the total energy of the complex; $E_A$ and $E_B$ are the total energies of the monomers A and B, respectively; $ES$ is the electrostatic energy, the interaction energy between the undistorted charge distributions of the isolated monomers; $PL$ is the polarization energy, the attractive interaction energy between the permanent charges and higher multipoles of one molecule and the induced multipoles of the other; $EX$ is the exchange repulsion, the short-range repulsion resulting from the overlap of the charge densities of molecules A and B; $CT$ is the charge transfer of electrons between monomers; and $MIX$ is a coupling term that takes into account the higher order interactions between the other components. The perturbation theory approach developed by Kolos and co-workers[3] partitions the interaction energy into components identical to or closely related to these.

Using this scheme, Umeyama and Morokuma[29] and Kollman[48] analyzed the interaction energy of a number of intermolecular complexes. They drew the following general conclusions:

1. The import of the various energy components is dependent on the type of hydrogen bond (ie, weak or strong) and on the A · · · B distance.

2. At the equilibrium geometry, all the components can be important.

3. For weak neutral complexes, although the largest attractive component is $ES$, $CT$ makes a significant contribution to the total interaction energy.

4. Hydrogen bonds in strong neutral complexes are essentially electrostatic.

5. The hydrogen bonds in strong and weak neutral complexes differ because the magnitude of $CT$ does not vary much from complex to complex, whereas the magnitude of $ES$ does.

6. Ionic complexes have large $ES$ contributions and can have sizable $CT$ contributions to the total complexation energy.

7. Although $PL$ is not essential for binding, changes in $PL$ do account for some substituent effects on the hydrogen bond energy.

8. The electrostatic energy is often the key factor in determining the directionality and relative energies of hydrogen-bonded complexes.

9. The exchange repulsion controls A · · · B bond lengths.

10. Both $ES$ and $EX$ are responsible for the preference for linear or nearly linear hydrogen bonds.

Most of the foregoing conclusions[29,48] are based on calculations carried out using the 4–31G[49] basis set. The magnitudes of the individual energy components are highly basis set dependent.[19,20,24,29,33] However, the primary effect of adding polarization functions is to decrease the values of $ES$ and the computed total interaction energy (which are generally overestimated with the 4–31G basis[29,48] for the reasons cited below), such that the observed trends are essentially unchanged.

Part of the basis set dependence is removed by correcting for BSSE.[19,20,24,33] The expression for $\Delta E$ becomes[24]:

$$\Delta E = ES + PL + EX + CT + MIX = E_{AB} - E_{A(B)} - E_{B(A)} \qquad (4\text{-}2)$$

where $E_{AB}$ is the total energy of the complex and $E_{A(B)}$ and $E_{B(A)}$ are the energies of monomers A and B in the supermolecule basis set; $ES$ and $PL$ are the same as in Equation 4-1; and $EX$ and $CT$ are now free of BSSE. Overall, correcting for BSSE generally increases the exchange energy, decreases the charge transfer energy, and decreases the total energy.[24] Nevertheless, for a series of complexes with the same proton donor (except for $CH_4$, since the uncorrected 4–31G relative dimerization energies of the (B)$CH_4$ complexes do not correlate with the basicity of B) or the same electron donor, the qualitative trends are unaffected.[20,24,33] When complexes with different proton donors and different electron donors are compared, there are sometimes changes in relative energies.[20,24] In the complexes studied by Sokalski and co-workers,[24] however, the relative energies are altered more by accounting for dispersion effects than by correcting for BSSE.

The basis set dependence of the corrected energies has been found mainly to result from the atomic monopole–monopole term of the electrostatic energy.[19,23,24,50] It is, therefore, related to the quality of the charge distribution.[19,23,24]

An expression similar to the energy decomposition equation can be written for the charge redistribution generated by complex formation.[6,47] From their analysis of a variety of hydrogen-bonded complexes, Umeyama and Morokuma[29] attribute (a) the loss of electron density on and adjacent to the hydrogen-bonded proton to $EX$ and $PL$, (b) the gain of electron density between the hydrogen-bonded proton and the proton acceptor to $CT$ and $PL$, and (c) the alternate loss and gain of electron density on the atoms of the proton donor and proton acceptor not directly bonded to the hydrogen-bonded proton to $PL$. Clearly, even though $PL$ generally makes the smallest contribution to $\Delta E$, it makes the largest contribution to the charge redistribution.[29]

The redistribution of electron density that occurs when a hydrogen bond is formed depends on the basis set employed.[23] The primary effect of removing BSSE is that the $CT$ term becomes smaller and more consistent.[23] Thus, the trends reported by Umeyama and Morokuma[29] are not changed.

# 4. MODELS OF HYDROGEN BONDING

L. C. Allen[51] has proposed a model of hydrogen bonding based on ab initio calculations of second- and third-row hydrides. The model focuses on the properties of the electron and proton donor that determine hydrogen bond strengths. The monomer properties that were selected are the A—H bond dipole and $\Delta I$, the difference in the first ionization potential of the electron donor and that of the noble gas atom in its row. The model is reasonably successful in rationalizing the dimerization energy, internuclear separation, directionality, dimer–dipole moment, charge transfer, infrared intensity enhancement, and stretching force constants.

Since the electrostatic energy component gives a good description of many energetic, structural, and spectral features of hydrogen-bonded complexes,[29,48] electrostatic models of the hydrogen bond have been propounded, also. Buckingham and Fowler's model[52] assigns point charges to each monomer atom and embeds them in hard spheres to take the short-range repulsions into account. The multipoles employed are the point charge, point dipole, and point quadrupole. They are obtained by distribution multipole analysis[53] of the SCF charge densities.[53,54] The predicted angular features of the complexes are in qualitative agreement with experiment.[52] Kollman[48] has used the observation[13] that the interaction energy is related to the molecular electrostatic potential ($V$) of the proton donor and acceptor to construct an expression for $\Delta E$ based on the product $V_A V_B$. The directionality of the complex is determined by finding the angular variable that yields the most negative electrostatic potential at the given reference point. The model is able to predict qualitatively the energies and shapes of hydrogen-bonded complexes.[48]

The molecular electrostatic potential has proven to be a useful tool for probing molecular interactions.[48,55–59] However, its calculation from an ab initio wave function is not feasible for large molecules. Consequently, in recent years several research groups have derived analytical expressions for $V$.[55–58,60–64] Their strategy is to employ many-center multipole expansions.[55–57,60–62] Expansions over atomic centers[55,56] and expansions centered on the charge centers[57,60,61] and on the centroids[62] of localized molecular orbitals have been developed. Most of these expansions are truncated at the quadrupole term.

In other studies,[58,63,64] the many-centered multipole expansion is truncated at the monopole term. One approach is to describe the molecular charge distribution in terms of directly transferable and orthogonal fragments.[58,61,63] Using atomic charges, obtained by Mulliken population analysis[65] or by fitting accurate electrostatic potentials,[64] is another approach.

To obtain a more complete analytical representation of $\Delta E$, an empirical expression for the electrostatic energy has been combined with empirical expressions for other energy components.[56,58,62,66–68] For example, the Kollman group[64,66] has concentrated on the repulsive and dispersion contributions, whereas Pullman's group[56,62,67,68] has included equations for all the constituents.

## 5. IONIC HYDROGEN BONDS

The general properties of hydrogen-bonded complexes reviewed above are based on calculations on conventional neutral and ionic systems, where both A and B are $n$-donors. Recently, a number of theoretical investigations have been carried out on unconventional ionic hydrogen bonds. Here the term "hydrogen bonding" indicates only that the interaction involves partially charged hydrogen atoms and polar or polarizable ligands. It does not imply that there is a significant amount of $n$-donation into the bond. It is of interest to compare the general properties of the unconventional ionic supersystems to those of the conventional supersystems. Since most of the reviews cited previously[1-5,29,48,61] concentrate on neutral complexes, a summary of some representative studies of both conventional and unconventional ionic complexes is presented below.

For the most part, minimal or split valence basis sets were utilized in the studies. The 4–31G basis[49] was the one most widely employed, usually without correcting for BSSE. The sources of error in the absolute magnitudes of the results obtained at this level of calculation were mentioned in Sections 2 and 3, "Computational Methods" and "Energy Components." However, although the absolute magnitudes are often incorrect, relative values along a series of related molecules are reproduced quite well.[10,12,29,48,69-72]

In addition, for most of the calculations on the supermolecules, the monomer geometries were fixed at their optimum or experimental values and only a few bond lengths and bond angles between them were optimized. This procedure is valid provided the hydrogen bond is not symmetric.[10,73]

## A. Conventional Ionic Hydrogen Bonds

### a. Dimers

Desmeules and Allen[10] have examined 21 (B)AH$^+$ complexes, where A, B = $NH_3$, $OH_2$, FH, $PH_3$, $SH_2$, and ClH at the 4–31G basis set level.[49] (See Table 4-2 for the hydrogen bond energies and Scheme I for the $(NH_3)NH_4^+$

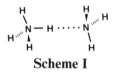

**Scheme I**

system.) The A $\cdots$ B distance, the A—H distance, and the angle between the A $\cdots$ B bond and the symmetry axis of B were optimized for the asymmetric systems. For the symmetric systems, a complete optimization was performed. The charge density difference plots obtained for the complexes demonstrate that conventional ionic and neutral hydrogen bonds exhibit the same pattern of charge redistributions.

Desmeules and Allen[10] drew the following conclusions from this work. First, for a given electron donor, the binding energy is inversely related to the H $\cdots$ B separation. This is also true for conventional neutral hydrogen bonds.[1-5] Second, for a given proton donor, the dimerization energy is directly related to the proton affinity of the electron donor and to the change in charge on the hydrogen-bonded proton (cf Chapter 3, this volume, and

**TABLE 4-2.** Calculated Hydrogen Bond Energies ($\Delta E$) and Experimental Hydrogen Bond Enthalpies ($\Delta H°$) of (B)AH$^+$ Dimers: A, B = NH$_3$, H$_2$O, HF, PH$_3$, H$_2$S, HCl

| Complex | $\Delta E$ (kcal/mol)[b] | $\Delta H°$ (kcal/mol) | Ref. |
|---------|--------------------------|------------------------|------|
| (NH$_3$)NH$_4^+$ | 32.00 | 25 | 17 |
| (H$_2$O)NH$_4^+$ | 27.72 | 17 | 17 |
| (HF)NH$_4^+$ | 17.78 | | |
| (PH$_3$)NH$_4^+$ | 16.47 | | |
| (H$_2$S)NH$_4^+$ | 13.55 | | |
| (HCl)NH$_4^+$ | 8.79 | | |
| (H$_2$O)H$_3$O$^{+a}$ | 44.70 | 32 | 15 |
| (HF)H$_3$O$^+$ | 24.12 | | |
| (PH$_3$)H$_3$O$^+$ | 25.75 | | |
| (H$_2$S)H$_3$O$^+$ | 21.18 | 25 | 74 |
| (HCl)H$_3$O$^+$ | 13.05 | | |
| (HF)H$_2$F$^{+a}$ | 40.70 | | |
| (HCl)H$_2$F$^+$ | 25.51 | | |
| (H$_2$O)PH$_4^+$ | 20.56 | | |
| (HF)PH$_4^+$ | 12.36 | | |
| (PH$_3$)PH$_4^+$ | 11.97 | 12 | 75 |
| (H$_2$S)PH$_4^+$ | 9.53 | | |
| (HCl)PH$_4^+$ | 6.08 | | |
| (HF)H$_3$S$^+$ | 17.14 | | |
| (H$_2$S)H$_3$S$^+$ | 16.47 | 13 | 76 |
| | | 15 | 74 |
| (HCl)H$_3$S$^+$ | 9.51 | | |
| (HCl)H$_2$Cl$^{+a}$ | 16.67 | | |

[a]Symmetric hydrogen bond.
[b]Reference 10.

Chapter 7, Volume 2). The latter quantity was estimated from the difference maps. Third, if the H · · · B distance increases, the A · · · B distance increases, also. The correlation between the two is linear for a specific row of proton donors. Divergence from the linear relationships indicates that the hydrogen bond is bent (at large A · · · B separation) or that the hydrogen bond is predominantly covalent (at small A · · · B separation).

Ikuta and Imamura[77] used the 3–21G[43] and, for the amines only, the 6–31G*[78] basis sets to probe the connection between the interaction energy ($\Delta E_{0,1}$) and the proton affinity (PA) of the base and between $\Delta E_{0,1}$ and the enthalpy of solution ($\Delta H_{sol}$). They studied B, BH$^+$, and (H$_2$O)BH$^+$ for the following bases: NH$_3$, CH$_3$NH$_2$, (CH$_3$)$_2$NH, (CH$_3$)$_3$N, C$_5$H$_5$N (py), 4−CN−py, and 4−CH$_3$−py [see Table 4-3 and Scheme II for (H$_2$O)pyH$^+$]. Either optimized geometries (ammonia systems) or standard geometries (pyridine systems, except for the N−H bond length) were utilized for the subunits. Partial optimizations of the complexes were carried out. There is excellent agreement between the 6–31G* $\Delta E_{0,1}$ values and the $\Delta H_{0,1}$ values. The 3–

TABLE 4-3. Calculated Hydrogen Bond Energies ($\Delta E$), Experimental Hydrogen Bond Enthalpies ($\Delta H°$), Experimental Enthalpies of Solution ($\Delta H_{sol}$), and Proton Affinities ($PA$) of $(H_2O)R_3NH^+$ and $(H_2O)4 - X - PyH^+$

| Complex | $\Delta E$ (kcal/mol)[a] | | $\Delta H°$ (kcal/mol) | | $\Delta H_{sol}$ (kcal/mol) | | $PA$ (kcal/mol) | | | |
| | | | | | | | Calculated[a] | | Experimental | |
| | 3-21G | 6-31G* | Ref. 79 | Ref. 80 | Ref. 81 | Ref. 82 | 3-21G | 6-31G* | Ref. 83 | Ref. 84 |
|---|---|---|---|---|---|---|---|---|---|---|
| $(H_2O)NH_4^+$ | 33.9 | 19.5 | 17.3 | 19.9 | −83.8 | −77.5 | 226.9[c] | 218.4[c] | 202.3 | 202.3 |
| $(H_2O)CH_3NH_3^+$ | 30.1 | 16.8 | 15.0 | 16.8 | −78.0 | −71.3 | 237.0[c] | 228.5[c] | 211.4 | 211.3 |
| $(H_2O)(CH_3)_2NH_2^+$ | 27.7 | 15.2 | | 15.0 | −72.9 | −63.6 | 243.5[c] | 235.2[c] | 217.8 | 217.9 |
| $(H_2O)(CH_3)_3NH^+$ | 25.9 | 14.6 | 14.8 | 14.5 | −65.6 | | 248.5 | | 221.6 | 222.1 |
| $(H_2O)4-CN-PyH^+$ | 25.1 | | 15.9 | | | −66.7[b] | 229.5 | | 207.9 | 207.6 |
| $(H_2O)PyH^+$ | 23.6 | | 15.0 | | | −62.7[b] | 243.0 | | 217.7 | 218.1 |
| $(H_2O)4-CH_3-PyH^+$ | 22.7 | | 14.3 | | | −61.2[b] | 247.9 | | 220.9 | |

[a]Values calculated with the two basis sets shown; Reference 77.
[b]The magnitude of $\Delta H_{sol}$ (PyH$^+$) is from reference 82. $\Delta H_{sol}$ (substituted PyH$^+$) = $\Delta H_{sol}(PyH^+) + \delta\Delta H_{sol}$. The $\delta\Delta H_{sol}$ values are from reference 81.
[c]Reference 85.

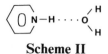

**Scheme II**

21G $\Delta E_{0,1}$ values are 1.5–2 times larger than the $\Delta H_{0,1}$ values,[77] primarily due to BSSE.[20]

For a given set of ions, the change in charge on the hydrogen-bonded proton ($+\delta(H)$) and $-\Delta H_{sol}$ are directly related to $\Delta E_{0,1}$.[77] The plots of $\Delta E_{0,1}$ versus $+\delta(H)$ are linear for both systems. The plot of $\Delta E_{0,1}$ versus $-\Delta H_{sol}$ is linear for the pyridinium ions[77,81] and curved for the ammonium ions.[77] The point for $(CH_3)_3NH^+$ deviates the most from the other data, which is explained by the lower value of $\Delta E_{0,1}$ for $(CH_3)_3NH^+$ compared to the other ammonium ions and by the limited number of hydrogen bonds it can form.

As the H $\cdots$ O bond length of the $(H_2O)BH^+$ cations decreases, $\Delta E_{0,1}$ increases in a linear manner. There is also a negative linear correlation between $\Delta E_{0,1}$ and the proton affinity of the corresponding base for the pyridines; the graph for the amines may be gently curved. However, the relationship between $\Delta E_{0,1}$ and $\Delta PA$ is linear for both systems (cf Part I of this discussion, in Chapter 3). The graphs of $\Delta E_{0,1}$ versus $PA$ and $-\Delta H_{sol}$ versus $PA$ have different slopes, indicating that successive hydrogen bond energies also become larger as the proton affinity of the base becomes smaller. From these data, Ikuta and Imamura[77] find that $\Delta E_{0,1}$ is not a uniform fraction of $-\Delta H_{sol}$ for the ammonium ions and the differences in the $\Delta E_{0,1}$ values ($\delta\Delta E_{0,1}$) are not a uniform fraction of the differences in the $-\Delta H_{sol}$ values ($-\delta\Delta H_{sol}$) for both sets of ions.

The interaction between the formate anion and water has been investigated by Berthod and Pullman.[33] They employed two basis sets, Dunning's[86] (9s, 5p/4s) set of Gaussian functions contracted into a [5s, 3p/3s] basis and a (7s, 3p/3s) basis set contracted to [2s, 1p/1s].[87] A set of d functions on C and O and a set of p functions on H were added to Dunning's basis. The geometrical parameters for $H_2O$ and $HCOO^-$ were taken from average values observed in crystals. They considered three structures of the dimer. The most stable has the $H_2O$ and $HCOO^-$ in the same plane with the water bridged to the two anionic oxygens (see Scheme III). The computed total energies were corrected for BSSE and added to an estimated value of the

**Scheme III**

dispersion energy to obtain more quantitative results for the hydrogen bond energy. The components of the binding energy were evaluated also.

The data show[33] that the minimal basis set reproduces the structures and relative energies of the more extensive basis set quite well. In fact, when the interaction energies are corrected for BSSE, even the magnitudes of $\Delta E$ are close for the two basis sets: $\Delta E$ for the bridged structure calculated with the extended basis set is 19.2 kcal/mol (without BSSE) and 17.0 kcal/mol (with BSSE). For the minimal basis set, the values are 22.1 and 16.5 kcal/mol, respectively. However, the trends in the dimerization energies are not changed by accounting for either BSSE or the dispersion effects. The electrostatic energy is the dominant term in the binding energy, and the relative values of $\Delta E_{EL}$ parallel those of $\Delta E$.

A number of calculations have been carried out to resolve whether the geometry of $(H_2O)OH^-$ is symmetric or asymmetric.[12a,73,88-91] The most complete study has been undertaken by McMichael Rohlfing and co-workers[73] who performed a global geometry optimization using a correlated wave function. The earlier investigations were carried out with limited basis sets,[12a,88,91] with only partial geometry optimizations,[12a,88-91] and/or without including correlation effects.[12a,88,90] All the work shows that the asymmetric structure is preferred at the SCF level.[12a,73,88-91] However, a symmetric hydrogen bond is more stable when correlation is included[73,89,91] (Table 4-4).

TABLE 4-4  Total Energies (Hartrees) for $(H_2O)OH^-$ at the Basis Set Levels Indicated

| Basis | Method[a] | Energy | Ref. |
|---|---|---|---|
| 4−31G (sym) | HF-SCF | −151.20325 | 73 |
| 4−31G (asym) | HF-SCF | −151.20358 | 73 |
| 6−31G* (sym) | HF-SCF | −151.39176 | 73 |
| 6−31G* (asym) | HF-SCF | −151.39340 | 73 |
| 6−31G** (sym) | HF-SCF | −151.41105 | 73 |
| 6−31G** (asym) | HF-SCF | −151.41228 | 73 |
| 6−31G** (sym) | MP2 | −151.82012 | 73 |
| O(11,7,1/5,4,1) (sym) H(6,1/3,1) | HF-SCF | −151.49523 | 90 |
| O(11,7,1/5,4,1) (asym) H(6,1/3,1) | HF-SCF | −151.49774 | 89 |
| O(11,7,1/5,4,1) (sym) H(6,1/3,1) | CISD | −151.91454 | 89 |
| O(11,7,1/5,4,1) (asym) H(6,1/3,1) | CISD | −151.91518[b] | 89 |
| O(9,5/4,2) (sym) H(4,1/2,1) | HF-SCF | −151.45295 | 91 |
| O(9,5/4,2) (asym) H(4,1/2,1) | HF-SCF | −151.45341 | 91 |
| O(9,5/4,2) (sym) H(4,1/2,1) | CISD | −151.72689 | 91 |
| O(9,5/4,2) (asym) H(4,1/2,1) | CISD | −151.72642 | 91 |

[a]HF = Hartree–Fock, SCF = self-consistent field, MP = Møller–Plesset, CISD = configuration interaction, single and double substitutions.
[b]The authors state that they expect a complete geometry optimization to remove the remaining asymmetry in their structure.

McMichael Rohlfing and colleagues[73] also examined whether the symmetric and asymmetric ions contain linear or bent hydrogen bonds and whether they are planar or nonplanar. They found that a nonplanar symmetric form with a linear hydrogen bond is most stable (SCF and post-SCF levels), whereas a nonplanar asymmetric form with a nonlinear hydrogen bond is most stable (SCF level). (Incorporation of correlation decreases the angles and lengthens the bonds.) Nevertheless, the potential energy surface is so flat with respect to both proton motion and torsional changes in the dihedral angle that zero-point energy and environmental effects will determine the conformation of the anion.[73]

Ikuta[39] has looked at the basis set dependence of the geometries and hydrogen bond energies of $(H_2O)OH^-$ and $(H_2O)OCH_3^-$. Four basis sets were employed to compute the interaction energies: 6–31G*,[78] 6–311G*,[92] 6–311+G**,[92,93] and [5s 4p1d/3s1p].[89,94] Configuration interaction with double substitutions (CID)[95] and Møller–Plesset theory to third order (MP3)[96] were applied to calculate the electron correlation contribution to the interaction energy of $(H_2O)OH^-$. The asymmetric structure of monohydrated $OH^-$ and the structure of monohydrated $CH_3O^-$ were completely optimized, assuming linear hydrogen bonds, with the 6–31G*[39,88] and 6–31+G*[39,93] bases.

Inclusion of the diffuse p function in the basis set lengthens the H · · · O bond, shortens the O−H bond, and increases the OH · · · OH$^-$ bond angle in $(H_2O)OH^-$.[39,88] Thus, adding the diffuse function decreases the covalent character of the OH · · · O bond.[10,73] All the other bond lengths and bond angles in both ions are very similar for both bases.

It is necessary to include the diffuse p function in the basis set or to use an extensive basis set such as [5s4p1d/3s1p] to obtain an accurate value for the $(H_2O)OH^-$ dimerization energy (Table 4-5). The electron correlation contribution raises this value by 2.5 (CID) and 3.4 (MP3) kcal/mol. A similar electron correlation effect has been reported elsewhere.[89,98,99]

In contrast, the 6–31G* dimerization energy for $(H_2O)OCH_3^-$ is quite close (within 3 kcal/mol) to the results calculated with the larger basis sets (Table 4-5). This suggests that extensive basis sets may be less essential for bigger anions.

In an earlier article, Ikuta[100] found that $\delta\Delta E_{0,1}$ is 42% of $\delta\Delta H_{sol}$ for $NH_4^+$ versus $CH_3NH_3^+$ and that 90% of $\delta\Delta H_{sol}$ was accounted for after solvation by three water molecules. Similar results were obtained for $NH_4^+$ and $H_3O^+$.[101] A comparison[39] of the $\Delta E_{0,1}$ values for $OH^-$ and $CH_3O^-$ shows that $\delta\Delta E_{0,1}$ is only 25% of the $\delta\Delta H_{sol}$, implying that a larger number of solvent molecules may be required to account for $\delta\Delta H_{sol}$ for the anions.

Caldwell and co-workers[72] have computed the interaction energies and structures of $(H_2O)OH^-$, $(H_2O)C_2H^-$, $(CH_3OH)OCH_3^-$, and $(C_2H_2)C_2H^-$ with the 4–31+G basis set[102] and of $(C_2H_5OH)OCH_3^-$ and $(C_2H_5OH)OC_2H_5^-$ with the 4–31G basis set.[49] The 4–31+G $\Delta E$ values are within about 3 kcal/mol of the experimental values they reported for each of the anions above. Even the 4–31+G difference in the bond strengths of

TABLE 4-5. Calculated Hydrogen Bond Energies ($\Delta E$) and Experimental Hydrogen Bond Enthalpies ($\Delta H°$) of (AH)B$^-$ Ions

| Complex | Basis set | Method | $\Delta E$ (kcal/mol) | Ref. | $\Delta H°$ (kcal/mol) | Ref. |
|---|---|---|---|---|---|---|
| (H$_2$O)OH$^-$ | 4−31G | HF-SCF | 40.8 | 88 | 25.0 | 97 |
| | 4−31+G | HF-SCF | 28.3 | 72 | | |
| | 6−31G*//4−31G$^a$ | HF-SCF | 34.5 | 88 | | |
| | 6−31G* | HF-SCF | 34.6 | 39 | | |
| | 6−311G** | HF-SCF | 33.6 | 39 | | |
| | 6−311+G** | HF-SCF | 22.8 | 39 | | |
| | [5s4p1d/3s1p] | HF-SCF | 24.6 | 39 | | |
| | [5s4p1d/3s1p] | MP3 | 28.0 | 39 | | |
| | [5s4p1d/3s1p] | CID | 27.1 | 39 | | |
| (H$_2$O)OCH$_3^-$ | 4−31G | HF-SCF | 30.5 | 88 | 19.9 | 72 |
| | 6−31G*//4−31G$^a$ | HF-SCF | 24.5 | 88 | | |
| | 6−31G* | HF-SCF | 25.2 | 39 | | |
| | 6−311G** | HF-SCF | 23.7 | 39 | | |
| | 6−311+G** | HF-SCF | 21.7 | 39 | | |
| | [5s4p1d/3s1p] | HF-SCF | 22.3 | 39 | | |
| (H$_2$O)C$_2$H$^-$ | 4−31+G | HF-SCF | 17.9 | 72 | $\simeq$14.9 | 72 |
| (CH$_3$OH)OH$^-$ | 4−31G | HF-SCF | 42.4 | 88 | | |
| | 6−31G*//4−31G$^a$ | HF-SCF | 36.7 | 88 | | |
| (CH$_3$OH)OCH$_3^-$ | 4−31G | HF-SCF | 31.2 | 88 | 21.8 | 72 |
| | 4−31+G | HF-SCF | 25.9 | 72 | | |
| | 6−31G*//4−31G$^a$ | HF-SCF | 25.7 | 88 | | |
| (CH$_3$OH)OC$_2$H$_5^-$ | 4−31G | HF-SCF | 29.9 | 72 | 20.2 | 72 |
| (C$_2$H$_5$OH)OC$_2$H$_5^-$ | 4−31G | HF-SCF | 30.2 | 72 | 20.6 | 72 |
| (C$_2$H$_2$)C$_2$H$^{-b}$ | 4−31+G | HF-SCF | 10.6 | 72 | | |

$^a$Total energies obtained at the 6−31G* level by using 4−31G optimized geometries.
$^b$Not observed experimentally; Reference 72.

(H$_2$O)OH$^-$ and (CH$_3$OH)OCH$_3^-$ is the correct size. This difference and the difference in the bond strengths of (H$_2$O)OH$^-$ and (H$_2$O)OCH$_3^-$ are substantially overestimated at the 4−31G and 6−31G* levels.[72,88]

Caldwell and co-workers[72] have compared the 4−31G binding energies of (CH$_3$OH)OCH$_3^-$, (C$_2$H$_5$OH)OCH$_3^-$ and (C$_2$H$_5$OH)OC$_2$H$_5^-$ with the experimental dissociation enthalpies, also. Although the magnitudes of the $\Delta E$'s are approximately 50% too high, the order of the $\Delta E$'s is correct.

These results agree with those of Ikuta[39] by demonstrating that small basis sets yield reliable interaction energies for anionic complexes. The combined data imply that polarization functions are not required, provided diffuse functions are added to the basis.[39,72]

A systematic and consistent examination of the barrier to proton transfer in (NH$_3$)$_2$H$^+$, (H$_2$O)$_2$H$^+$, and (H$_2$O)(NH$_3$)H$^+$ has been performed by Scheiner and co-workers.[99,103−106] These investigators have probed the effect of polarization functions,[99] electron correlation,[99,104] bond stretching,[99,103,105,106] and angle deformations[103,105,106] on the barrier height. They have also analyzed the charge redistributions that occur as the proton moves along the hydrogen bond axis via Mulliken[65] population analysis and electron density difference maps.[105,106] The results obtained for each cation have been compared and contrasted.

The rigid molecule approximation, where all the nuclei except the central proton are held fixed throughout the transfer, was employed in all the calculations. The transfer barrier was computed at several different heavy-atom (N,O) separations.

Some of the general conclusions drawn from this work are as follows:

1. Inclusion of polarization functions raises the proton transfer barrier, while taking electron correlation effects into account significantly lowers it (Table 4-6). The overall outcome is that 4–31G[49] barriers are in excellent agreement with those from much more sophisticated treatments.
2. Intrapair correlation effects increase the height of the barrier and interpair correlation effects decrease it.
3. At small values of the heavy-atom separation $R$, single-well proton transfer potentials are observed; at large separations double-well potentials are observed.
4. The barrier to proton transfer is increased when $R$ is increased and when the symmetry axis of the proton donor and/or the proton acceptor is twisted away from the intermolecular axis. The barrier is lowered by allowing $R$ to relax at each distance of the central proton from the proton-donating molecule.
5. As the heavy-atom separation increases, the equilibrium position of the hydrogen-bonded proton becomes closer to the proton donor. For any given $R$, inclusion of electron correlation moves the equilibrium position of the central proton further from the proton donor. The reverse holds when polarization functions are included in the basis set. However, both lengthen the equilibrium intermolecular distance.
6. The barrier height is correlated with the total charge on the electron-donating group, the orbital energies of all the lone pairs on the N or O elec-

TABLE 4-6. Calculated Barriers to Proton Transfer[a]

| $(\alpha_1,\alpha_2)$[b] | Method | Ref. | NH $\cdots$ O | OH $\cdots$ O | NH $\cdots$ N | OH $\cdots$ N |
|---|---|---|---|---|---|---|
| (0,0) | HF/4–31G | 106 | 29.1 | 16.9 | 11.4 | 2.7 |
| (0,0) | HF/DZP | 99 | | | 14.6 | |
| (0,0) | POL–CI | 99 | | | 7.3 | |
| (0,20) | HF/4–31G | 106 | 30.9 | 18.6 | 12.9 | 3.3 |
| (20,20) | HF/4–31G | 106 | 32.4 | 19.9 | 14.0 | 4.3 |
| (20,−20) | HF/4–31G | 106 | 33.5 | 20.5 | 15.1 | 4.6 |

[a] $R$ = 2.95 Å; energies in kilocalories/mol.
[b] $\alpha_1$ and $\alpha_2$ are the angles between the A $\cdots$ B axis and the rotation axis of the proton-donating molecule and the proton-accepting molecule, respectively.

tron-donating atoms, and the $X-H$ and $H \cdots X$ ($X = N,O$) overlap populations. All these properties are calculated before proton transfer.

7. The difference maps confirm the results obtained from the Mulliken population analysis. Both methods demonstrate that as the hydrogen-bonded proton moves along the hydrogen bond axis, there is a charge transfer from the proton acceptor to the proton donor accompanied by a polarization of charge (ie, electron) density perpendicular to the hydrogen bond axis. In net, all the atoms in the proton-donating group gain electron density, whereas all the atoms in the proton-accepting group lose it.

8. The relative barrier heights and variations in charge redistributions for the different dimers can be explained on the basis of electronegativities.

Several of the foregoing observations have also been noted by other researchers: namely, the effect of including electron correlation on the proton transfer barrier[107] and on the equilibrium distance between the central proton and the proton donor[89,107] and the effect of increasing $R$ on the proton transfer barrier[107,108] and on the shape of the potentials.[108]

### b. Clusters

Numerous investigations of the properties of cluster ions have been undertaken in recent years. Among them are the studies of $(NH_3)_nH^+$, $n = 1-6$[36]; $(HCN)_nH^+$, $n = 1-5$[69]; $(CH_3OH)_nH^+$, $n = 1-4$[70]; $(H_2O)_nH^+$, $n = 1-6$[12]; $(H_2O)_nOH^-$, $n = 1-4$[12a,34]; $(H_2O)_nNO_2^-$, $n = 1-3$[109]; $(H_2O)_nNO_3^-$, $n = 1-3$.[109] The basis sets used in the various calculations are 3–21G,[36,43] 3–21G*,[36,43] 4–31G,[12,49,69,70] 6–31G,[34,109,110] 6–31G*,[34,78] and 6–31G**.[36,78] The groups of Hirao[36,69] and Sapse[34] have obtained global optimizations for $(NH_3)_nH^+$, $n = 1-5$ and $(HCN)_nH^+$, $n = 1-3$, and for $(H_2O)_nOH^-$, $n = 1-4$, respectively. The remaining complexes were partially optimized.

For the most part, these researchers have concentrated on determining the structures of the ions, the binding energies as successive solvent molecules are added (Table 4-7), the dominant contributions to the binding energy, and the charge redistributions that occur upon hydrogen bond formation. However, Newton and Ehrenson[12a] have computed $O-H$ stretching force constants and frequencies and have examined models for proton transfer, also.

The most stable geometry of the $(HCN)_nH^+$ systems is a linear chainlike structure.[69] The $(CH_3OH)_nH^+$ cations have a zigzag framework.[70] Chain structures, preferably with branching, rather than cyclic ones are favored by the $(H_2O)_nH^+$ and $(H_2O)_nOH^-$ ions.[12] However, cyclic geometries become more competitive as $n$ gets larger. Only branched structures were considered for the $(NH_3)_nH^+$ complexes.[36] [See Scheme IV for $(H_2O)_4H^+$.]

The $(H_2O)_nNO_3^-$ anions are most stable when both water hydrogens interact with nitrate oxygens (ie, with a bridged structure).[109] Nevertheless, the supermolecules with only one of the hydrogens bonded to a nitrate oxygen (ie, with a linear structure) are only about 1 kcal/mol less stable. $(H_2O)NO_2^-$ has a bridged hydrogen bond, $(H_2O)_2NO_2^-$ has two linear hydro-

TABLE 4-7. Calculated Hydrogen Bond Energies ($\Delta E$) and Experimental Hydrogen Bond Enthalpies ($\Delta H°$) of Some Conventionally Hydrogen-Bonded Cluster Ions

| Complex | Method | $\Delta E$ (kcal/mol) | Ref. | $\Delta H°$ (kcal/mol) | Ref. |
|---|---|---|---|---|---|
| $(NH_3)_2H^+$ | HF/3–21G | 37.12 | 36 | 25.4 | 111 |
| | | | | 24.8 | 17 |
| | HF/3–21G* | 32.20 | 36 | | |
| | HF/6–31G** | 26.21 | 36 | | |
| $(NH_3)_3H^+$ | HF/3–21G | 26.59 | 36 | 17.3 | 111 |
| | | | | 17.5 | 17 |
| | HF/3–21G* | 24.63 | 36 | | |
| | HF/6–31G** | 20.62 | 36 | | |
| $(NH_3)_4H^+$ | HF/3–21G | 21.56 | 36 | 14.2 | 111 |
| | | | | 13.8 | 17 |
| | HF/3–21G* | 20.37 | 36 | | |
| $(NH_3)_5H^+$ | HF/3–21G | 17.20 | 36 | 11.8 | 111 |
| | | | | 12.5 | 17 |
| $(NH_3)_6H^+$ | HF/3–21G | 11.50 | 36 | | |
| $(HCN)_2H^+$ | | | | | |
| Asymmetrical | HF/4–31G | 28.8 | 69 | 30.0 | 112 |
| Asymmetrical | CI/4–31G | 30.2 | 69 | | |
| Symmetrical | HF/4–31G | 28.3 | 69 | | |
| Symmetrical | CI/4–31G | 30.3 | 69 | | |
| $(HCN)_3H^+$ | | | | | |
| Linear | HF/4–31G | 18.8 | 69 | 13.8 | 112 |
| T-shaped | HF/4–31G | 8.7 | 69 | | |
| $(HCN)_4H^+$ | HF/4–31G | 14.5 | 69 | 11.8 | 112 |
| $(HCN)_5H^+$ | HF/4–31G | 11.4 | 69 | 9.2 | 112 |
| $(CH_3OH)_2H^+$ | HF/4–31G | 36.4 | 70 | 33.1 | 15 |
| $(CH_3OH)_3H^+$ | | | | | |
| Linear | HF/4–31G | 27.0 | 70 | 21.3 | 15 |
| T-shaped | HF/4–31G | 12.9 | 70 | | |
| $(CH_3OH)_4H^+$ | HF/4–31G | 16.6 | 70 | 16.1 | 15 |
| $(H_2O)_2H^+$ | | | | | |
| Straight chain | HF/4–31G | 37 | $12^a$ | 31.6 | 15 |
| Ring | HF/4–31G | 21 | $12^a$ | | |
| $(H_2O)_3H^+$ | | | | | |
| Straight chain | HF/4–31G | 26 | $12^a$ | 19.5 | 15 |
| Ring | HF/4–31G | 10 | $12^a$ | | |
| $(H_2O)_4H^+$ | | | | | |
| Branched chain | HF/4–31G | 22 | $12^a$ | 17.5 | 15 |
| Straight chain | HF/4–31G | 17 | $12^a$ | | |
| Ring | HF/4–31G | 11 | $12^a$ | | |
| $(H_2O)_5H^+$ | | | | | |
| Branched chain | HF/4–31G | 16 | $12^a$ | 15.3 | 15 |
| Straight chain | HF/4–31G | 13 | $12^a$ | | |
| Ring | HF/4–31G | 12 | $12^a$ | | |
| $(H_2O)_6H^+$ | | | | | |
| Branched chain | HF/4–31G | 15 | $12^a$ | 13.3 | 15 |
| $(H_2O)OH^-$ | | | | | |
| Straight chain | HF/4–31G | 36 | $12^a$ | 34.5 | 113 |
| Straight chain | HF/6–31G | 38.9 | 34 | | |
| Ring | HF/4–31G | 25 | $12^a$ | | |

CAROL A. DEAKYNE

TABLE 4-7. (continued)

| Complex | Method | $\Delta E$ (kcal/mol) | Ref. | $\Delta H°$ (kcal/mol) | Ref. |
|---|---|---|---|---|---|
| $(H_2O)_2OH^-$ | | | | | |
| Straight chain | HF/4–31G | 26 | 12[a] | 23.1 | 113 |
| Straight chain | HF/6–31G | 29.4 | 34 | | |
| Ring | HF/4–31G | 15 | 12[a] | | |
| $(H_2O)_3OH^-$ | | | | | |
| Branched chain | HF/4–31G | 19 | 12[a] | 18.4 | 113 |
| Branched chain | HF/6–31G | 24.3 | 34 | | |
| Straight chain | HF/4–31G | 16 | 12[a] | | |
| Ring | HF/4–31G | 17 | 12[a] | | |
| $(H_2O)_4OH^-$ | | | | | |
| Branched chain | HF/4–31G | 13 | 12[a] | | |
| Branched chain | HF/6–31G | 1.9 | 34 | | |
| Ring | HF/4–31G | 18 | | | |
| $(H_2O)NO_2^-$ | | | | | |
| Bridged | HF/6–31G | 23.4 | 109 | 14.3 | 114 |
| Linear | HF/6–31G | 18.5 | 109 | | |
| $(H_2O)_2NO_2^-$ | HF/6–31G | 15.9 | 109 | 12.9 | 114 |
| $(H_2O)_3NO_2^-$ | HF/6–31G | 17.7 | 109 | 10.4 | 114 |
| $(H_2O)NO_3^-$ | | | | | |
| Bridged | HF/6–31G | 18.5 | 100 | 12.3 | 114 |
| Linear | HF/6–31G | 17.2 | 109 | | |
| $(H_2O)_2NO_3^-$ | | | | | |
| Bridged | HF/6–31G | 16.3 | 109 | | |
| Linear | HF/6–31G | 15.3 | 109 | | |
| $(H_2O)_3NO_3^-$ | | | | | |
| Bridged | HF/6–31G | 14.5 | 109 | | |
| Linear | HF/6–31G | 13.7 | 109 | | |

[a]Corrected for zero-point energies.

gen bonds, and $(H_2O)_3NO_2^-$ has one bridged and two linear hydrogen bonds.[109]

Other more general results and conclusions from these studies are as follows:

1. The decreases in successive solvation enthalpies are reproduced satisfactorily by small basis sets.[12,34,36,69,70,109]
2. The X—H and H · · · X (X = N,O) bond distances calculated with and without polarization functions in the basis are quite similar regardless of the size of the complex.[34,36]
3. As solvent molecules are added to the inner shell, the X—H bonds shorten and the H · · · X bonds lengthen, thereby weakening the hydrogen bonds.[12,34,36,69,70,109] This is primarily due to electrostatic[12,34,69,70,109] and steric effects.[8] As $n$ increases, however, the X—H overlap populations increase,[12]

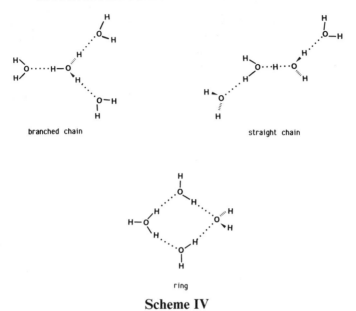

branched chain                    straight chain

ring

**Scheme IV**

the H · · · X overlap populations decrease,[12] and the charge transferred from each added solvent molecule decreases.[12,36,69,70] Thus, delocalization effects are also important.[12,36,69,70]

4. Hydrogen bonds between inner and outer shell molecules are longer and weaker than those between inner shell molecules and the central ion for the same reasons cited above.[12,34,36,69,70]

5. When an inner shell solvent molecule acts as both a proton donor and a proton acceptor, the hydrogen bond between the central ion and the molecule becomes stronger, at the expense of the other hydrogen bonds involving the central ion.[12,69,70] This is reflected in the changes in the X—H and H · · · X bond distances.[12,69,70]

6. As the size of the complex gets larger, the charge on the central ion decreases.[12,34,36,69,70,109]

7. Branched-chain structures are preferred over straight-chain structures mainly for electrostatic reasons.[12]

8. There is a monotonic relationship between calculated X—H bond lengths and stretching force constants.[12]

9. Symmetric intermediates facilitate aqueous (ie, liquid $H_2O$) proton transfers.[12]

## B. Unconventional Ionic Hydrogen Bonds

### a. $CH^{\delta+}$ · · · X Complexes

Two types of structure for complexes containing $CH^{\delta+}$ · · · X hydrogen bonds have been considered. The first, termed a linear conformation, is analogous to the structure of a complex containing a conventional ionic hydrogen bond, with X the proton-accepting atom and C the proton-donating atom. The second type is termed a cavity conformation. In the 1-Me-cavity conformation, the X attaches to the cavity generated by three hydrogens of a single methyl group. In the 3-Me-cavity conformation, the X is also attached to a cavity generated by three hydrogens, but each hydrogen is from a different methyl group.

Hirao and associates have investigated the geometries and complexation energies of $(CH_3CN)_nH^+$, where $n = 1-3$,[69] $((CH_3)_2O)_nH^+$, where $n = 1-3$,[70] and $(CH_3CN)_nX^-$, where X = F, Cl and $n = 1-4$.[71] In each case, the 4–31G basis set[49] was used (plus a diffuse function on the anions)[115] and a partial geometry optimization of the clusters was carried out.

For $((CH_3)_2O_nH^+$ and $(CH_3CN)_nH^+$, when the third solvent molecule is added, there are no longer any acidic terminal hydrogens with which it can form a conventional ionic hydrogen bond. Yet, $\Delta E_{2,3}$ has been measured experimentally and is nonzero for both ions.[15,112] One possibility for the structure of the $n = 3$ ion is that it is T-shaped; that is, the third solvent molecule interacts with the central proton. Two other possibilities are the linear and the 1-Me-cavity conformations. (See Scheme V for $(CH_3CN)_3H^+$.)

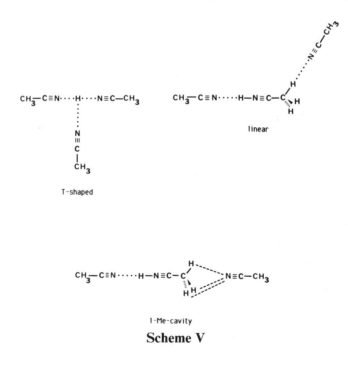

**Scheme V**

The methyl hydrogens are more positive in $((CH_3)_2O)_2H^+$ and $(CH_3CN)_2H^+$ than they are in $(CH_3)_2O$ and $CH_3CN$, respectively, as a result of the charge redistributions induced by the formation of the hydrogen bonds.

TABLE 4-8. Calculated Hydrogen Bond Energies ($\Delta E$) and Experimental Hydrogen Bond Enthalpies ($\Delta H°$) of the $C-H^{\delta+} \cdots X$ Bonds in $((CH_3)_2O)_3H^+$, $(CH_3CN)_3H^+$, $(CH_3CN)_nF^-$, and $(CH_3CN)_nCl^-$

| Complex | $\Delta E$ (kcal/mol) | Ref. | $\Delta H°$ (kcal/mol) | Ref. |
|---|---|---|---|---|
| $((CH_3)_2O)_3H^+$ | | | | |
| T-shaped | 8.1 | 70 | 10.1 | 15 |
| Linear | 7.7 | 70 | | |
| l-Me-cavity | 7.1 | 70 | | |
| $(CH_3CN)_3H^+$ | | | | |
| T-shaped | 8.3 | 69 | 9.3 | 112 |
| Linear | 11.9 | 69 | | |
| l-Me-cavity | 11.0 | 69 | | |
| $(CH_3CN)F^-$ | | | | |
| Linear | 18.2 | 71[a] | 16.0 | 116 |
| | 18.0 | 71[b] | | |
| l-Me-cavity | 16.1 | 71[a] | | |
| | 15.6 | 71[b] | | |
| $(CH_3CN)_2F^-$ (linear) | 15.5 | 71[a] | 12.9 | 116 |
| $(CH_3CN)_3F^-$ (linear) | 13.6 | 71[a] | 11.7 | 116 |
| $(CH_3CN)_4F^-$ (linear) | 9.3 | 71[a] | 10.4 | 116 |
| $(CH_3CN)Cl^-$ | | | | |
| Linear | 10.3 | 71[a] | 13.4 | 116 |
| | 10.0 | 71[b] | | |
| l-Me-cavity | 11.5 | 71[a] | | |
| | 10.8 | .71[b] | | |
| $(CH_3CN)_2Cl^-$ (l-Me-cavity) | 10.3 | 71[a] | 12.2 | 116 |
| $(CH_3CN)_3Cl^-$ (l-Me-cavity) | 8.9 | 71[a] | 10.6 | 116 |
| $(CH_3CN)_4Cl^-$ (1-Me-cavity) | 7.2 | 71[a] | 6.2 | 116 |

[a]SCF calculation.
[b]SCF + CI calculation.

The data show (Table 4-8) that for $((CH_3)_2O)_3H^+$ the T-shaped structure is most stable, but for $(CH_3CN)_3H^+$ the linear structure is. However, for both ions the hydrogen bond energies of all three conformations are quite close in magnitude and are comparable to the hydrogen bond enthalpies. Hirao and co-workers[69] rationalize the destabilization of the $(CH_3CN)_3H^+$ T-shaped complex in terms of the large exchange repulsion between the $\pi$-electron density on the nitrogens of the $(CH_3CN)_2H^+$ moiety and the $\sigma$-electron density on the nitrogen of the approaching $CH_3CN$. Thus, even though the central proton has the largest positive charge, the T-shaped conformation is the least stable.

$(CH_3CN)_nF^-$ and $(CH_3CN)_nCl^-$, $n = 1-4$, anions have been observed experimentally also.[116] For these ions, all the hydrogen bonds are of the unconventional type. Yamabe and Hirao[71] examined linear and 1-Me-cavity conformations for both $(CH_3CN)F^-$ and $(CH_3CN)Cl^-$. In addition, a model with the anion interacting with the cyano carbon was studied for

(CH₃CN)F⁻ only ($R_{CF} = 3.16$ Å, see Scheme VI). The latter complex was found to be appreciably less stable than the former two, which was again ascribed to a large exchange repulsion. The linear conformation is slightly favored for (CH₃CN)F⁻, but the 1-Me-cavity conformation is slightly

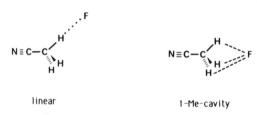

linear                    1-Me-cavity

**Scheme VI**

favored for (CH₃CN)Cl⁻ at both SCF and SCF + CI levels of calculation (Table 4-8).[71] The computed binding energies are in reasonable agreement with the experimental enthalpies. Based on these results, for $n > 1$, only linear conformations were considered for (CH₃CN)$_n$F⁻ and only 1-Me-cavity conformations were considered for (CH₃CN)$_n$Cl⁻.

Yamabe and Hirao[71] attribute the difference in the preferred geometries of the singly solvated ions to the smaller size and greater basicity of F⁻. One consequence of this difference in geometries is that although the interactions in both complexes are dominated by the electrostatic contribution, the covalent contribution is also important for the F⁻ complexes.[71] This observation on the nature of the linear conformation is supported by the bent (following Walsh's rule on AH₂–type species) rather than 180° H · · · F⁻ · · · H angle[71] in (CH₃CN)₂F⁻ and by the shorter H · · · X⁻ distance and larger charge transfer in the linear conformation compared to the 1-Me-cavity conformation of a given ion.[69–71]

Meot-Ner and Deakyne[117] have studied CH^{δ+} · · · B interactions in clusters of quaternary ammonium ions with various solvent molecules experimentally and theoretically. Calculations were carried out on five of the supermolecules employing the 3–21G basis: (H₂O)Me₄N⁺, (H₂O)₂Me₄N⁺, (CH₃OH)Me₄N⁺, (CH₃NH₂)Me₄N⁺, and (CH₃Cl)Me₄N⁺. For all the dimers, 1-Me-cavity conformations and 3-Me-cavity conformations were investigated (see Scheme VII). The linear conformation was examined also for (H₂O)Me₄N⁺. Since upon geometry optimization it rotated to the 3-Me-cavity conformation, it was not considered for the other ions. The NC · · · X bond (X = N, O, Cl) was assumed to be linear in the 1-Me-cavity complexes, and only the C · · · X bond distance and the orientation of the lone

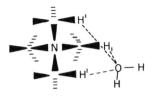

3-Me-cavity

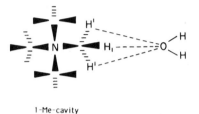

1-Me-cavity

**Scheme VII**

pairs were optimized. This approach is similar to the one adopted by Hirao and co-workers.[69-71] The N · · · X distance, the H' · · · X distance, and the orientation of the lone pairs were optimized in the 3-Me-cavity complexes.

The most stable placement of each solvent moelcule in the 3-Me-cavity dimers is to the right of and below the $H_1$ and H' hydrogens.[117] However, the interaction of the water and methanol molecules with $Me_4N^+$ differs from the interaction of the methyl amine and methyl chloride molecules. The lone pairs on the oxygen in $H_2O$ and $CH_3OH$ are oriented such that these molecules overlap more strongly with the H' hydrogens than with the $H_1$ hydrogen. In contrast, the positioning of the lone pair on the $CH_3NH_2$ nitrogen and the lone pairs on the $CH_3Cl$ chlorine leads to essentially equally strong interactions between these molecules and the three hydrogens. In the 1-Me-cavity dimers, although the electron-donating atoms are equidistant from each hydrogen, the orientations of the lone pairs generate unequal X · · · H overlaps for X = N and O. These atoms interact more effectively with $H_1$ than with the H' atoms.

As the results in Table 4-9 demonstrate, the dimer 3-Me-cavity conformations are universally more stable than the 1-Me-cavity conformations. Both types of complex are stabilized primarily by electrostatic effects, but charge transfer effects and polarization effects are also influential. In fact, the relative interaction energies along the series of 3-Me- and of 1-Me-cavity complexes is directly related to the N · · · X, X · · · H', and X · · · $H_1$ separations, the magnitudes of the X · · · H' and X · · · $H_1$ overlap populations, and the charge transferred to the $Me_4N^+$ ion. The lower stability of the 1-Me-cavity conformation of a given ion compared to its 3-Me-cavity

TABLE 4-9. Calculated Solvation Energies ($\Delta E$) and Experimental Solvation
Enthalpies ($\Delta H°$) of the $Me_4N^+$ Complexes[a]

| Complex | | $-\Delta E$ (kcal/mol) | $-\Delta H°$ (kcal/mol) |
|---|---|---|---|
| $(H_2O)Me_4N^+$ | (3-Me) | 17.9 | 9.0 |
| $(H_2O)Me_4N^+$ | (l-Me) | 11.9 | |
| $(H_2O)_2Me_4N^+$ | (water–water) | 17.0 | 9.4 |
| $(H_2O)_2Me_4N^+$ | (two-cavity) | 15.8 | |
| $(CH_3OH)Me_4N^+$ | (3-Me) | 17.5 | 9.8 |
| $(CH_3OH)Me_4N^+$ | (l-Me) | 11.5 | |
| $(CH_3NH_2)Me_4N^+$ | (3-Me) | 13.8 | 8.7 |
| $(CH_3NH_2)Me_4N^+$ | (l-Me) | 9.9 | |
| $(CH_3Cl)Me_4N^+$ | (3-Me) | 9.0 | 6.5 |
| $(CH_3Cl)Me_4N^+$ | (l-Me) | 6.3 | |

[a]Reference 117.

conformation correlates with the same properties and, in addition, with the
amount of polarization of the $Me_4N^+$ group.

Two models were tested for $(H_2O)_2Me_4N^+$.[117] The first model has the two
water molecules hydrogen-bonded to each other (water–water model). The
$O \cdots O$ distance was assumed to be the same as it is in $(H_2O)_2$. The second
model has the added water molecule in another 3-Me-cavity site placed at
the optimal solvent position in $(H_2O)Me_4N^+$ (two-cavity model). No geometry reoptimization was done in either case.

The data show (Table 4-9) that the water–water model is the more stable
of the two and that $\Delta E_{1,2}$ is essentially equivalent to $\Delta E_{0,1}$. The latter result
agrees with the experimental observation.[117] This trend in binding energies
is unusual for hydrogen-bonded complexes. However, any solvent that can
form only the two-cavity complex, such as acetone, would be expected to
show the usual drop in complexation energy for the second step. This is also
in accordance with the experimental data.

The stabilization energy associated with the attachment of the second
water originates from three main sources[117]: the formation of the water–
water conventional hydrogen bond, the further polarization of $Me_4N^+$ and
the first $H_2O$ (making the H′ atoms more positive and the oxygen more negative), and the extra charge transferred to $Me_4N^+$.

**b. $CH^{\delta+} \cdots O$ Complexes Versus Condensation: Interactions of Carbonium Ions
with n-Donors**

When n-donor molecules such as $H_2O$ react with carbonium ions in the gas
phase, two types of reaction can occur:

$$R_3C^+ + OH_2 \rightarrow R_3COH_2^+ \qquad (4\text{-}3)$$
$$R_3C^+ + OH_2 \rightarrow (H_2O)CR_3^+ \qquad (4\text{-}4)$$

In reaction 4-3 a protonated alcohol is formed; in reaction 4-4 a hydrogen-bonded cluster ion is formed. Meot-Ner and Deakyne[118] and Hiraoka and Kebarle[119] have looked at reactions between carbonium ions and n-donors utilizing experimental[118,119] and theoretical[118] techniques. The ions investigated via ab initio molecular orbital theory[118] are $CH_3OH_2^+$, $C_2H_5OH_2^+$, $i$-$C_3H_7OH_2^+$, $t$-$C_4H_9OH_2^+$, $(H_2O)CH_3OH_2^+$, $(H_2O)_2CH_3OH_2^+$, $(H_2O)CH_3^+$, $(H_2O)C_2H_5^+$, $(H_2O)i$-$C_3H_7^+$, and $(H_2O)t$-$C_4H_9^+$. The structures of the protonated alcohols were completely optimized,[118,120] and the structures of the complexes were partially optimized[118] by use of the 3–21G basis set.[43] Energies were then computed at the 4–31G level[49] utilizing the 3–21G structures (ie, 4–31G//3–21G calculations were performed). Electron correlation, zero-point energy, polarization, and BSSE effects were taken into account for several of the systems (Table 4-10).

Five conformations of the complexes were considered: the linear, 1-Me-cavity, 1'-Me-cavity, 2-Me-cavity, and 2'-Me-cavity conformations. The first two were examined for each carbonium ion; the last was studied only for $(H_2O)t$-$C_4H_9^+$. The 1'-Me-cavity conformation was investigated for $(H_2O)CH_3^+$ and $(H_2O)C_2H_5^+$, and the 2-Me-cavity conformation was investigated for $(H_2O)i$-$C_3H_7^+$ and $(H_2O)t$-$C_4H_9^+$. The 2-Me-cavity dimers have the water interacting with two of the cation hydrogens, each from a different methyl group. The two hydrogens are on the charge-bearing carbon in the 1'-Me-cavity supermolecules. The water attaches to the cavity created by three hydrogens, two from the same methyl group and one from a different methyl group, in the 2'-Me-cavity cation (see Scheme VIII).

Two hydration sites were examined for monohydrated, protonated methanol, $(H_2O)CH_3OH_2^+$, namely, the $H_2O$ group and the $CH_3$ group. Only linear conformations were considered. Conformations 1a and 1b have been studied for $(H_2O)_2CH_3OH_2^+$; 1c and 1d are currently being investigated (see Scheme IX).

The parameters varied in the linear conformers are the C−H (or O−H) and H · · · O bond lengths, the CH · · · O (or OH · · · O) bond angle, and the angle between the H−O−H bisector and the OH · · · O bond. The C · · · O separation, the angle between the H−O−H bisector and the H−C−H bisector (1'-Me-cavity complexes), the angle between the H−O−H bisector and the C · · · O bond (1-Me-cavity and 2'-Me-cavity complexes), and the CH · · · O and H · · · OH angles (2'-Me-cavity complexes) were optimized in the other four conformations.

The binding energies obtained for the various protonated alcohols and cluster ions are presented in Table 4-10. If no binding energy is included in the table for a given conformation of a specific complex, it is not a minimum on the potential energy surface of that complex. Jorgensen and co-workers found similar relative stabilization energies for the protonated alkyl fluorides and alkyl chlorides that they studied.[121–123]

Clearly, even when the cluster ions are much less stable than the conden-

TABLE 4-10. Calculated ($\Delta E$) and Experimental ($\Delta H°$) Hydrogen Bond Energies of $R_3COH_2^+$ and $(H_2O)CR_3^+$

| Complex | Method | $\Delta E^a$ (kcal/mol) | $ZPE^{a,b}$ (kcal/mol) | $\Delta E + \Delta ZPE^{a,e}$ (kcal/mol) | $\Delta H°$ (kcal/mol) |
|---|---|---|---|---|---|
| $CH_3OH_2^+$ | 3–21G//3–21G | 81.47(70.17)[c] | 41.90 | 62.74 | 66[d] |
| | 4–31G//3–21G | 66.93(63.05)[c] | | 55.62 | |
| | 6–31G*//3–21G | 60.03 | | 52.60 | |
| | MP2/4–31G//HF/3–21G | 77.81(73.93)[c] | | 66.50 | |
| | MP2/6–31G*//HF/3–21G | 74.94 | | 67.51 | |
| $(H_2O)CH_3^+$ (linear) | 3–21G//3–21G | 28.73(24.50)[c] | 37.00 | 21.97 | |
| | 4–31G//3–21G | 23.31(22.36)[c] | | 19.83 | |
| | 6–31G*//3–21G | 17.08 | | 14.55 | |
| | MP2/4–31G//HF/3–21G | 24.74(23.79)[c] | | 21.26 | |
| | MP2/6–31G*//HF/3–21G | 20.07 | | 17.54 | |
| $(H_2O)CH_3^+$ (1'-Me-cavity) | 3–21G//3–21G | 22.24(18.06)[c] | 36.61 | 15.92 | |
| | 4–31G//3–21G | 19.60(18.56)[c] | | 16.42 | |
| | 6–31G*//3–21G | 16.19 | | 14.05 | |
| | MP2/4–31G//HF/3–21G | 20.23(19.19)[c] | | 17.05 | |
| | MP2/6–31G*//HF/3–21G | 17.69 | | 15.55 | |
| $(H_2O)CH_3OH_2^+$ (linear to O—H) | 3–21G//3–21G | 42.26 | | | |
| | 4–31G//3–21G | 34.33 | | | |
| $(H_2O)CH_3OH_2^+$ (linear to C—H) | 3–21G//3–21G | 17.58 | | | |
| | 4–31G//3–21G | 14.13 | | | |
| $(H_2O)_2CH_3OH_2^+$ (1b) | 3–21G//3–21G | 32.29 | | | |
| | 4–31G//3–21G | 27.46 | | | |
| $C_2H_5OH_2^+$ | 3–21G//3–21G | 61.32 | | | |
| | 4–31G//3–21G | 46.64 | | | |
| | MP2/4–31G//HF/3–21G | 54.30 | | | |
| $(H_2O)C_2H_5^+$ (linear) | 3–21G//3–21G | 24.45(20.37)[c] | | | 37[d] |
| | 4–31G//3–21G | 19.97 | | | |
| | MP2/4–31G//HF/3–21G | 20.59 | | | |

TABLE 4-10. (continued)

| Complex | Method | $\Delta E^a$ (kcal/mol) | $ZPE^{a,b}$ (kcal/mol) | $\Delta E + \Delta ZPE^{a,e}$ (kcal/mol) | $\cdot \Delta H^\circ$ (kcal/mol) |
|---|---|---|---|---|---|
| $(H_2O)C_2H_5^+$ (l'-Me-cavity) | 3–21G//3–21G | 20.28 | | | |
| | 4–31G//3–21G | 17.26 | | | |
| | MP2/4–31G//HF/3–21G | 17.59 | | | |
| $(H_2O)C_2H_5^+$ (l-Me-cavity) | 3–21G//3–21G | 13.86 | | | |
| | 4–31G//3–21G | 11.56 | | | |
| | MP2/4–31G//HF/3–21G | 13.06 | | | |
| $i$-$C_3H_7OH_2^+$ | 3–21G//3–21G | 47.50 | | | $22.8^d$ |
| | 4–31G//3–21G | 32.27 | | | |
| | MP2/4–31G//HF/3–21G | 39.31 | | | |
| $(H_2O)i$-$C_3H_7^+$ (linear) | 3–21G//3–21G | 21.70 | | | |
| | 4–31G//3–21G | 17.45 | | | |
| | MP2/4–31G//HF/3–21G | 18.29 | | | |
| $(H_2O)i$-$C_3H_7^+$ (2-Me-cavity) | 3–21G//3–21G | 19.48 | | | |
| | 4–31G//3–21G | 14.07 | | | |
| $(H_2O)i$-$C_3H_7^+$ (l-Me-cavity) | 3–21G//3–21G | 12.78 | | | |
| | 4–31G//3–21G | 10.39 | | | |
| $t$-$C_4H_9OH_2^+$ | 3–21G//3–21G | 38.24 | | | $11.2^{a,d}$ |
| | 4–31G//3–21G | 22.41 | | | |
| $(H_2O)t$-$C_4H_9^+$ (2-Me-cavity) | 3–21G//3–21G | 18.56 | | | |
| | 4–31G//3–21G | 13.09 | | | |
| $(H_2O)t$-$C_4H_9^+$ (l-Me-cavity) | 3–21G//3–21G | 11.94 | | | |
| | 4–31G//3–21G | 9.56 | | | |

[a] Reference 118.
[b] Zero-point energy.
[c] $\Delta E$ corrected for BSSE. The ZPE of $CH_3^+$ is 20.80 kcal/mol and of $H_2O$ is 13.67 kcal/mol.
[d] Reference 119.

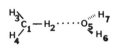

linear

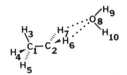

1'-Me-cavity

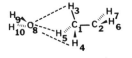

1-Me-cavity

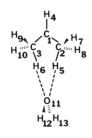

2-Me-cavity

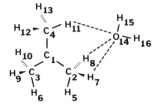

2'-Me-cavity

**Scheme VIII**

sation product, there is a substantial $CH^{\delta+} \cdots O$ interaction. The $CH^{\delta+} \cdots F$ and $CH^{\delta+} \cdots Cl$ energies are quite large, also.[121,122] Examining the uncorrected 4–31G//3–21G $\Delta E$ values in Table 4-10 more closely leads to the conclusion that the reactions between $CH_3^+$ and $H_2O$, $C_2H_5^+$ and $H_2O$, and $i$-$C_3H_7^+$ and $H_2O$ yield the protonated alcohols, in agreement with Hiraoka and Kebarle's[119] predictions based on their experimental data. In contrast, the disparity in the dissociation energies of protonated $t$-butanol and $(H_2O)t$-$C_4H_9^+$ of about 9 kcal/mol suggests that the formation of the cluster ion may be competitive with the formation of the condensation product in this case. Comparing the uncorrected 4–31G//3–21G $\Delta E$ values with the $\Delta H°$ values shows that the $\Delta E$'s of the protonated alcohols are generally overestimated by about 9–10 kcal/mol, due in part to zero-point

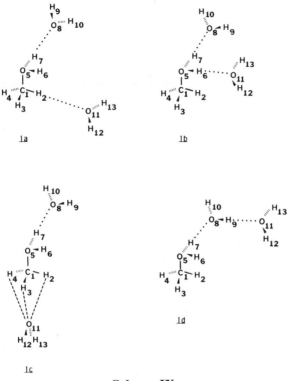

**Scheme IX**

energy effects. On the other hand, at this level of calculation $CH^{\delta+}\cdots O$ calculated interaction energies generally range from being 2–3 kcal/mol too low to being 2–3 kcal/mol too high.[69–71] This implies that the difference in the stabilities of the two products $\delta\Delta E$ should be decreased by at least 5–6 kcal/mol. If the proposed decrease is correct, the difference in the $\Delta S$ values for the two reactions ($\sim$10–15 cal/mol K) would make their $\Delta G$ values comparable. However, accounting for electron correlation effects, polarization effects, zero-point energy effects, and BSSE leads to an overall increase in the $\delta\Delta E$ for $CH_3OH_2^+$ and both $(H_2O)CH_3^+$ complexes. The source of this increase is the inclusion of electron correlation effects in the calculations. Correcting for BSSE, polarization effects, and zero-point energy effects decreases $\delta\Delta E$. Since the MP2/4–31G//HF/3–21G results parallel the MP2/6–31G*//HF/3–21G results, although the increase in $\delta\Delta E$ is larger for the latter, only MP2/4–31G//HF/3–21G calculations were carried out for the other ions. Assuming that the pattern above is repeated throughout the series of ions, and the preliminary results for the $C_2H_5^+$ and $i$-$C_3H_7^+$ species indicate that it will be, the theoretical data suggest that the reaction between $H_2O$ and $t$-$C_4H_9^+$ yields the protonated alcohol, also. Additional work is in progress to augment the preliminary results.

Other observations to note from Table 4-10 include the following:

1. The MP2/4–31G//HF/3–21G and MP2/6–31G*//HF/3–21G $\Delta E$ + $\Delta$ZPE values are in excellent agreement with the experimental $\Delta H°$ value for $CH_3OH_2^+$.
2. The BSSE for the 4–31G//3–21G stabilization energies of the hydrogen-bonded complexes is similar for all the clusters and is 1.0 kcal/mol or less.
3. Enlarging the basis set lowers the interaction energy of the supermolecules and including correlation effects raises it. This pattern has been reported also by Scheiner and co-workers.[4,99]

For both $(H_2O)CH_3OH_2^+$ and $(H_2O)_2CH_3OH_2^+$, as expected, the conformer with the added water molecule hydrogen-bonded to the $OH_2$ moiety of protonated methanol is the most stable.[118] In fact, **1a** is not a minimum on the potential energy surface of $(H_2O)_2CH_3OH_2^+$. The charge distributions obtained from Mulliken population analysis[65] show that with the addition of just two water molecules to $CH_3OH_2^+$, the positive charge on the methyl hydrogens drops from an average value of 0.319 to an average value of 0.271. The total positive charge on the methyl group decreases from 0.632 to 0.514. Both quantities are rather rapidly approaching their magnitudes in the uncomplexed methanol molecule (ie, 0.191 and 0.304, respectively[120]). These data support the explanation presented by Meot-Ner (in Chapter 3) to account for (a) the observed weak $CH^{\delta+} \cdots O$ interactions between alkyl groups of onium ions and the bulk solvent and (b) the observed similarity of the size of these interactions for onium ions and the corresponding neutrals.

The trend in the stabilization energies of the protonated alcohols correlates with the $C-O$ bond length, with the $C-O$ bond overlap population, and with the amount of charge transferred from the cation to the $H_2O$ group, in particular to the water hydrogens. The last point is consistent with Hiraoka and Kebarle's[119] and Jorgensen's[123] rationalization of the decrease in $\Delta H°$ as the size of the alkyl group increases. They proposed that the attachment of a water or HCl molecule to the larger alkyl cations is not as instrumental in delocalizing the positive charge as it is in the smaller ones. Jorgensen has related the trend in interaction energies to the energy of the lowest unoccupied molecular orbital of the carbonium ion, also.[123]

The relative binding energies along the series of linear unconventional hydrogen-bonded supermolecules are directly related to the charge on the interacting proton before formation of the complex $q_H$, the loss of electron density on the water hydrogens $\Delta q_{H,H_2O}$, the gain in electron density on the electron-donating and proton-donating atoms, the charge transferred to the $H_2O$, the $H \cdots O$ bond length, and the $H \cdots O$ overlap population.[118] They are indirectly related to the $C-H$ overlap population and the $H \cdots O$ bond length.[118] In comparison, the differences in the complexation energies of the 1-Me-cavity and the 1'-Me-cavity dimers correlate with only $q_H$, the charge gained on the electron-donating atom, and the $C-H$ and $H$

$\cdots$ O overlap populations (1'-Me-cavity dimers); $\Delta q_{H,H_2O}$, the change in charge density on the proton-donating atom, and the charge transferred to $H_2O$ have essentially equivalent values in all the 1-Me-cavity and 1'-Me-cavity complexes. For the 2-Me-cavity conformers of $(H_2O)i$-$C_3H_7^+$ and $(H_2O)t$-$C_4H_9^+$, $\delta\triangle E$ depends only on $q_H$. For these cations, all the other parameters have virtually equal magnitudes.[118]

The most stable conformation of a specific cluster ion is the one with the largest $\Delta q_{H,H_2O}$, the largest charge transfer, the largest H $\cdots$ O overlap population, and the smallest H $\cdots$ O separation.[118] These findings are in agreement with those mentioned above.[12,69-71]

Clearly, the magnitudes of the interaction energies of $CH^{\delta+} \cdots$ X ionic hydrogen bonds can be appreciable.[15,69-71,112,116,117] However, not unexpectedly, binding energies are smaller and H $\cdots$ X bonds are longer for supermolecules containing unconventional ionic hydrogen bonds than for their conventional counterparts with the same value of $n$.[10,12,69,70] Still, the properties exhibited by linear conformations of the former systems, though less pronounced, are very similar to those exhibited by the latter systems. Examples of analogous properties are the charge redistributions brought about by hydrogen bond formation, the dominance of the electrostatic contribution to $\triangle E$, and the structural changes that occur as solvent molecules are added.[10,12,69,70,118]

Two major differences were found in the properties of the cavity complexes and the conventional complexes; otherwise, they are quite similar also.[10,12,69,70,117,118] First, the polarization of the subunits can be more extensive in the unconventional systems (2-Me-cavity and 3-Me-cavity conformations).[117,118] Second, in the 1-Me-cavity and 1'-Me-cavity supermolecules, the proton-donating atom loses electron density rather thans gains it[69,70,117,118] and the interacting protons sometimes gain electron density rather than lose it.[117,118] In the 3-Me-cavity dimers, the proton-donating atom of $Me_4N^+$ gains more electron density in the weaker complexes than in the stronger ones.[117]

### c. NH$^+$ $\cdots$ $\pi$ Complexes

Hydrogen bonds with a $\pi$-electron-donating group have been studied theoretically and experimentally by Deakyne and Meot-Ner.[124] The three systems for which calculations were carried out are $(C_2H_4)NH_4^+$, $(C_6H_6)NH_4^+$, and $(C_6H_5F)NH_4^+$. Partially optimized structures of the complexes were obtained with the STO–3G basis.[125] Then 3–21G//STO–3G energies were computed for each ion. 3–21G//3–21G structures were determined also for $(C_2H_4)NH_4^+$. The 3–21G//3–21G and 3–21G//STO–3G geometries and interaction energies are in excellent agreement; therefore, only the 3–21G//STO–3G energies are reported, since they were calculated for all the systems investigated.

The two conformations of $(C_2H_4)NH_4^+$ shown in Scheme X were examined,[124] one with the hydrogen of $NH_4^+$ pointing toward the ethylene double bond (T-shaped) and one with two of the $NH_4^+$ hydrogens straddling the

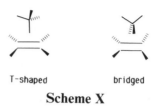

T-shaped        bridged

**Scheme X**

double bond (bridged). (The conformation with two of the $NH_4^+$ hydrogens pointing toward the ethylene carbons is essentially equal in energy to the bridged conformation.) Only the distance between the nitrogen and the center of the C=C bond was varied.

The T-shaped structure is more stable (Table 4-11), but it is about 21 kcal/mol higher in energy than $C_2H_5NH_3^+$.[124] The latter ion is the condensation product produced in the reaction between $C_2H_4$ and $NH_4^+$. This result corresponds reasonably well with the experimental value of about 16 kcal/mol.[124]

The T-shaped conformer is preferred over the bridged conformer for several reasons. The most favorable approach of the hydrogen-bonded proton electrostatically is along the center of the C=C bond.[13] More charge is transferred from $C_2H_4$ to $NH_4^+$ and $NH_4^+$ polarizes the charge density of $C_2H_4$ more extensively in the T-shaped dimer. The H · · · C distance is shorter and the overlap between the ethylene carbons and the hydrogen-bonded proton(s) is greater in the T-shaped model. The average absolute values of the molecular orbital shifts are larger for the T-shaped structure.

When the complexes are formed,[124] the $\pi$-electron density on the ethylene carbons decreases. However, the net negative charge on the carbons increases due to the shift in charge density from the ethylene hydrogens to

TABLE 4-11. Calculated Hydrogen Bond Energies ($\Delta E$) and Experimental Hydrogen Bond Enthalpies ($\Delta H°$) of $(C_2H_4)NH_4^+$, $(C_6H_6)NH_4^+$, and $(C_6H_5F)NH_4^{+a}$

| Complex | $\Delta E$ (kcal/mol) | $\Delta H°$ (kcal/mol) |
|---|---|---|
| $(C_2H_4)NH_4^+$ (T-shaped) | 8.9 | 10 |
| $(C_2H_4)NH_4^+$ (bridged) | 6.5 | |
| $(C_6H_6)NH_4^+$ (**2a**) | 15.2 | 19.3 |
| $(C_6H_6)NH_4^+$ (**2b**) | 16.3 | |
| $(C_6H_6)NH_4^+$ (**2c**) | 14.7 | |
| $(C_6H_6)NH_4^+$ (**2d**) | 14.2 | |
| $(C_6H_5F)NH_4^+$ (**2a′**) | 11.0 | 14.4 |
| $(C_6H_5F)NH_4^+$ (**2b′**) | 11.4 | |
| $(C_6H_5F)NH_4^+$ (**2c′**) | 10.4 | |
| $(C_6H_5F)NH_4^+$ (**2d′**) | 10.3 | |
| $(C_6H_5F)NH_4^+$ (**3**) | 16.8 | |

[a]Reference 118.

the carbons. Every atom in $NH_4^+$ (except for the nitrogen in the bridged structure) including the hydrogen-bonded proton(s) gains electron density. The charge change on the interacting proton(s) distinguishes these complexes from conventional hydrogen-bonded complexes. This is also true for some 1-Me-cavity and 1'-Me-cavity supermolecules,[117,118] as noted previously, and for some $(C_6H_6)NH_4^+$ and $(C_6H_5F)NH_4^+$ $\pi$ systems.[124]

Overall, the charge shifts between the subunits and the charge redistributions within the subunits are fairly small. This suggests once again that the dimers are stabilized primarily by electrostatic effects.

Structures and energies were determined for four conformations of $(C_6H_6)NH_4^+$.[124] Three conformations have the nitrogen over the center of the ring. Conformers 2a, 2b, and 2c have one, two, and three hydrogens, respectively, pointing toward the ring. For illustrative purposes, only conformer 2b is shown in Scheme XI. (The total energies of 2b and 2c are independent of the positioning of the $NH_4^+$ hydrogens, ie, whether the hydrogens are directed toward the carbons or between them.) Conformer 2d has the nitro-

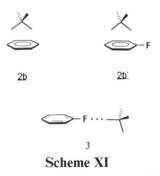

2b                2b'

3

**Scheme XI**

gen over the center of a C—C bond with one hydrogen directed toward the bond. Analogous conformations, designated 2a', 2b', 2c', and 2d' were investigated for $(C_6H_5F)NH_4^+$. Again, only conformer 2b' is depicted in Scheme XI. Three different 2d' dimers are possible. Since they all have energies within 1 kcal/mol of each other, only the results for the most stable one ($NH_4^+$ over the $C_{2,5}-C_{3,6}$ bond) are tabulated (Table 4-11). The $NH^+ \cdots F$ complex 3 (Scheme XI) was examined also for $(C_6H_5F)NH_4^+$.

The distance between the center of the ring or the center of the C—C bond was varied in 2a–2d and 2a'–2d'. For 3 the hydrogen bond was assumed to be linear and only the H $\cdots$ F separation was optimized.

The dissociation energies of the $(C_6H_6)NH_4^+$ dimers are uniformly too low (Table 4-11).[124] They are also very close in magnitude and could be reordered at a higher level of calculation. This is true for the $\pi$ conformers of $(C_6H_5F)NH_4^+$ as well. The similarity in the $\Delta E$ values indicates that the interaction energy is relatively independent of the orientation of $NH_4^+$. This supports the suggestion, based on the experimental results,[124] that the forces binding $NH_4^+$ and $K^{+126}$ to benzene are comparable; that is, electrostatic effects are predominant in the $\pi$ complexes. In fact, Castora and associates[127] have advanced the general rule that $K^+$ and $NH_4^+$ bind in essentially the same way to any species.

Structures **2b** and **2b'** are the most stable $\pi$ conformers of $(C_6H_6)NH_4^+$ and $(C_6H_5F)NH_4^+$, respectively, for essentially the same reasons cited above for the T-shaped conformer of $(C_2H_4)NH_4^+$.[124] The only dissimilarities are that the maximum polarization of the charge density of $C_6H_6(C_6H_5F)$ by $NH_4^+$ does not occur in **2b** (**2b'**) and the most favorable line of approach of $NH_4^+$ is along the $C_6$ axis.

The difference in the experimental dissociation energies of $(C_6H_6)NH_4^+$ and $(C_6H_5F)NH_4^+$ is approximately 5 kcal/mol.[124] If $(C_6H_5F)NH_4^+$ is a $\pi$ complex, then this difference, which is reproduced well by the calculations, can be explained as follows. The highest occupied molecular orbitals (HOMOs) of benzene are a degenerate pair of $\pi$ orbitals. The highest occupied $\pi$-molecular orbital of $C_6H_5F$ is nearly equivalent in energy to the $C_6H_6$ HOMOs. This is consistent with the general observation that any number of hydrogens can be replaced by fluorines without affecting the energy of the $\pi$ HOMO significantly.[128] In contrast, the second highest occupied $\pi$-molecular orbital of $C_6H_5F$ is considerably lower in energy than the $C_6H_6$ HOMOs. This makes the $\pi$ electrons of $C_6H_5F$ more tightly bound than those of $C_6H_6$, which causes a weaker interaction between $C_6H_5F$ and $NH_4^+$ than that between $C_6H_6$ and $NH_4^+$.[124] However, if **3** is the most stable structure of $(C_6H_5F)NH_4^+$, as is found in this work,[124] more accurate calculations are required to explain the drop in $\triangle H°$ for the $C_6H_5F$ dimer.

In summary, $CH^{\delta+} \cdots X$ and $NH^+ \cdots X$ interaction energies can be quite large, and conventionally and unconventionally hydrogen-bonded complexes exhibit many, but not all, of the same properties.

## ACKNOWLEDGMENTS

The support of the Holy Cross Data Processing Center and of the U.S. Air Force Geophysics Laboratory Information Resource Center is gratefully acknowledged.

## REFERENCES

1. Kollman, P. In Modern Theoretical Chemistry, Vol. 4, "Applications of Electronic Structure Theory", Schaefer, H.F., III, Ed.; Plenum Press: New York, 1977, Chapter 3, pp. 109–152.
2. Schuster, P. *Angew. Chem. Int. Ed. Engl.* **1981**, *20*, 546.
3. Kolos, W. In "New Horizons of Quantum Chemistry", Vol. 4, Symposium IV, Lowdin, P.-O.; and Pullman, B., Eds.; Reidel: Boston, 1983, pp. 243–278.
4. Scheiner, S. *Stud. Phys. Theor. Chem.* **1983**, *26*, 462.
5. Beyer, A.; Karpfen, A.; Schuster, P. *Top. Curr. Chem.* **1984**, *120*, 1.
6. Kitaura, K.; Morokuma, K. *Int. J. Quantum Chem.* **1976**, *10*, 325.

7. Zeegers-Huyskens, T.; Huyskens, P. *Mol. Interactions* **1981**, *2*, 1.
8. Taylor, R.; Kennard, O. *Acc. Chem. Res.* **1984**, *17*, 320.
9. Topp, W.C.; Allen, L.C. *J. Am. Chem. Soc.* **1974**, *96*, 5291.
10. Desmeules, P.J.; Allen, L.C. *J. Chem. Phys.* **1980**, *72*, 4731.
11. Noble, P.N.; Kortzeborn, R.N. *J. Chem. Phys.* **1970**, *52*, 5375.
12. (a) Newton, M.D.; Ehrenson, S. *J. Am. Chem. Soc.* **1971**, *93*, 4971. (b) Newton, M.D. *J. Chem. Phys.* **1977**, *67*, 5535.
13. Kollman, P.; McKelvey, J.; Johansson, A.; Rothenberg, S. *J. Am. Chem. Soc.* **1975**, *97*, 955.
14. Harrell, S.A.; McDaniel, D.H. *J. Am. Chem. Soc.* **1964**, *86*, 4497.
15. Grimsrud, E.P.; Kebarle, P. *J. Am. Chem. Soc.* **1973**, *95*, 7939.
16. Lowder, J.E. *J. Quant. Spectrosc. Radiat. Transfer* **1970**, *10*, 1085.
17. Payzant, J.D.; Cunningham, A.J. *Can. J. Chem.* **1973**, *51*, 3242.
18. Arshadi, M.; Yamdagni, R.; Kebarle, P. *J. Phys. Chem.* **1970**, *74*, 1475.
19. Sokalski, W.A.; Hariharan, P.C.; Kaufman, J.J. *J. Comput. Chem.* **1983**, *4*, 506.
20. Hobza, P.; Zahradnik, R. *Chem. Phys. Lett.* **1981**, *82*, 473.
21. Kestner, N.R. *J. Chem. Phys.* **1968**, *48*, 252.
22. Kolos, W. *Theor. Chim. Acta* **1979**, *51*, 219.
23. Sokalski, W.A. *J. Chem. Phys.* **1982**, *77*, 4529.
24. Sokalski, W.A.; Hariharan, P.C.; Kaufman, J.J. *J. Phys. Chem.* **1983**, *87*, 2803.
25. Sokalski, W.A.; Roszak, S.; Hariharan, P.C.; Kaufman, J.J. *Int. J. Quantum Chem.* **1983**, *23*, 847.
26. Boys, S.F.; Bernardi, F. *Mol. Phys.* **1970**, *19*, 558.
27. Mezey, P.G.; Csizmadia, I.G. *Can. J. Chem.* **1977**, *55*, 1181.
28. Mezey, P.G.; Haas, E.C. *J. Chem. Phys.* **1982**, *77*, 870.
29. Umeyama, H.; Morokuma, K. *J. Am. Chem. Soc.* **1977**, *99*, 1316.
30. Cusachs, L.C.; Trus, B.L.; Caroll, D.G.; McGlynn, S.P. *Int. J. Quantum Chem. Symp.* **1967**, *1*, 423.
31. Murrell, J.N.; Shaw, G. *Mol. Phys.* **1968**, *15*, 325.
32. Dill, J.D.; Allen, L.C.; Topp, W.C.; Pople, J.A. *J. Am. Chem. Soc.* **1975**, *97*, 7220.
33. Berthod, H.; Pullman, A. *J. Comput. Chem.* **1981**, *2*, 87.
34. Sapse, A.M.; Osorio, L.; Snyder, G. *Int. J. Quantum Chem.* **1984**, *26*, 223, 231.
35. Wright, L.R.; Borkman, R.F. *J. Chem. Phys.* **1982**, *77*, 1938.
36. Hirao, K.; Fujikawa, T.; Konishi, H.; Yamabe, S. *Chem. Phys. Lett.* **1984**, *104*, 184.
37. Clementi, E.; Kistenmacher, H.; Kolos, W.; Romano, S. *Theor. Chim. Acta* **1980**, *55*, 257.
38. Clementi, E.; Kolos, W.; Lie, G.C.; Ranghino, G. *Int. J. Quant. Chem.* **1980**, *17*, 377.
39. Ikuta, S. *J. Comput. Chem.* **1984**, *5*, 374.
40. Lister, D.G.; Palmieri, P. *J. Mol. Struct.* **1977**, *39*, 295.
41. Mezey, P.G.; Ramachandra Rao, C.V.S. *J. Comput. Chem.* **1980**, *1*, 134.
42. Popkie, H.E.; Kaufman, J.J. *Int. J. Quantum Chem.* **1976**, *10*, 47.
43. (a) Binkley, J.S.; Pople, J.A.; Hehre, W.J. *J. Am. Chem. Soc.* **1980**, *102*, 939. (b) Gordon, M.S.; Binkley, J.S.; Pople, J.A.; Pietro, W.J.; Hehre, W.J. ibid, **1982**, *104*, 2797.
44. Margenau, H.; Kestner, N.R. "The Theory of Intermolecular Forces". Pergamon Press: Oxford, 1969.
45. Sokalski, W.A.; Chojnacki, H. *Int. J. Quantum Chem.* **1978**, *13*, 679.
46. Ahlrichs, R.; Penco, P.; Scoles, G. *Chem. Phys.* **1977**, *19*, 119.
47. Morokuma, K. *J. Chem. Phys.* **1971**, *55*, 1236.
48. Kollman, P. *J. Am. Chem. Soc.* **1977**, *99*, 4875.
49. Ditchfield, R.; Hehre, W.J.; Pople, J.A. *J. Chem. Phys.* **1971**, *54*, 724.
50. Karlstrom, G.; Sadlej, A.J. *Theor. Chim. Acta* **1982**, *61*, 1.
51. Allen, L.C. *J. Am. Chem. Soc.* **1975**, *97*, 6921.
52. Buckingham, A.D.; Fowler, P.W. *J. Chem. Phys* **1983**, *79*, 6426.
53. Stone, A.J. *Chem. Phys. Lett.* **1981**, *83*, 233.
54. Fowler, P.W. *Mol. Phys.* **1982**, *47*, 355.
55. (a) Rein, R. *Adv. Quantum Chem.* **1973**, *7*, 335. (b) Rein, R. In "Intermolecular Interactions: From Diatomics to Biopolymers", Pullman, B., Ed.; Wiley: New York, 1978, pp. 307–362.
56. Claverie, P. In "Intermolecular Interactions: From Diatomics to Biopolymers", Pullman, B., Ed.; Wiley: New York, 1978, pp. 69–305.

57. Scrocco, E.; Tomasi, J. *Top. Curr. Chem.* **1973**, *42*, 95.
58. Bonaccorsi, R.; Ghio, C.; Tomasi, J. *Int. J. Quantum Chem.* **1984**, *26*, 637.
59. Pullman, A.; Pullman, B. In "Chemical Applications of Atomic and Molecular Electro-static Potentials", Politzer, P.; and Truhlar, D.G., Eds.; Plenum Press: New York, 1981, pp. 381–405.
60. Bonaccorsi, R.: Cimiraglia, R.; Scrocco, E.; Tomasi, J. *Theor. Chim. Acta* **1974**, *33*, 97.
61. Tomasi, J. *Mol. Interactions* **1982**, *3*, 119.
62. Lavery, R.; Etchebest, C.; Pullman, A. *Chem. Phys. Lett.* **1982**, *85*, 266.
63. Bonaccorsi, R.; Scrocco, E.; Tomasi, J. *J. Am. Chem. Soc.* **1977**, *99*, 4546.
64. Singh, U.C.; Kollman, P.A. *J. Comput. Chem.* **1984**, *5*, 129.
65. Mulliken, R.S. *J. Chem. Phys.* **1955**, *23*, 1833.
66. Douglas, J.E.; Kollman, P.A. *J. Am. Chem. Soc.* **1980**, *102*, 4295.
67. Gresh, N.; Claverie, P.; Pullman, A. *Int. J. Quantum Chem. Symp.* **1979**, *13*, 243.
68. Gresh, N.; Claverie, P.; Pullman, A. *Int. J. Quantum Chem.* **1982**, *22*, 199.
69. Hirao, K.; Yamabe, S.; Sano M. *J. Phys. Chem.* **1982**, *86*, 2626.
70. Hirao, K.; Sano, M.; Yamabe, S. *Chem. Phys. Lett.* **1982**, *87*, 181.
71. Yamabe, S.; Hirao, K. *Chem. Phys. Lett.* **1981**, *84*, 598.
72. Caldwell, G.; Rozeboom, M.D.; Kiplinger, J.P.; Bartmess, J.E. *J. Am. Chem. Soc.* **1984**, *106*, 4660.
73. McMichael Rohlfing, C.; Allen, L.C.; Cook, C.M.; Schlegel, H.B. *J. Chem. Phys.* **1983**, *78*, 2498.
74. Hiraoka, K.; Kebarle, P. *Can. J. Chem.* **1977**, *55*, 24.
75. Long, J.W.; Franklin, J.L. *J. Am. Chem. Soc.* **1974**, *96*, 2320.
76. Meot-Ner (Mautner), M; Field, F.H. *J. Am. Chem. Soc.* **1977**, *99*, 998.
77. Ikuta, S.; Imamura, M. *Chem. Phys.* **1984**, *90*, 37.
78. Hariharan, P.C.; Pople, J.A. *Theor. Chim. Acta* **1973**, *28*, 203.
79. Davidson, W.R.; Sunner, J.; Kebarle, P. *J. Am. Chem. Soc.* **1979**, *101*, 1675.
80. Meot-Ner (Mautner), M. *J. Am. Chem. Soc.* **1984**, *106*, 1257, 1265.
81. Arnett, E.M.; Chawla, B.; Bell, L.; Taagepera, M.; Hehre, W.J.; Taft, R.W. *J. Am. Chem. Soc.* **1977**, *99*, 5729.
82. Taft, R.W. *Prog. Phys. Org. Chem.* **1983**, *14*, 247.
83. Aue, D.H.; Bowers, M.T. In "Gas Phase Ion Chemistry", Vol. 2, Bowers, M.T., Ed.; Academic Press: New York, 1979, pp. 1–51.
84. Kebarle, P. *Annu. Rev. Phys. Chem.* **1977**, *28*, 445.
85. Smith, S.F.; Chandrasekhar, J.; Jorgensen, W.L. *J. Phys. Chem.* **1982**, *86*, 3308.
86. Dunning, T.H., Jr. *J. Chem. Phys.* **1970**, *53*, 2823.
87. Pullman, A.; Berthod, H.; Gresh, N. *Int. J. Quantum Chem. Symp.* **1976**, *10*, 59.
88. Jorgensen, W.L.; Ibrahim, M. *J. Comput. Chem.* **1981**, *2*, 7.
89. Roos, B.O.; Kraemer, W.P.; Diercksen, G.H.F. *Theor. Chim. Acta* **1976**, *42*, 77.
90. Kraemer, W.P.; Diercksen, G.H.F. *Theor. Chim. Acta* **1972**, *23*, 398.
91. Stogard, A.; Strich, A.; Almlof, J.; Roos, B. *Chem. Phys.* **1975**, *8*, 405.
92. Dunning, T.H., Jr.; Hay, P.J. In Modern Theoretical Chemistry, Vol. 3, "Methods of Electron Structure Theory", Schaefer, H.F., III, Ed.; Plenum Press: New York, 1977, Chapter 1, pp. 1–28.
93. Krishnan, R.; Binkley, J.S.; Seeger, R.; Pople, J.A. *J. Chem. Phys.* **1980**, *72*, 650.
94. Salez, C.; Veillard, A. *Theor. Chim. Acta* **1968**, *11*, 441.
95. Roos, B.O.; Siegbahn, P.E.M. In Modern Theoretical Chemistry, Vol. 3, "Methods of Electronic Structure Theory", Schaefer, H.F., III, Ed.; Plenum Press: New York, 1977, Chapter 7, pp. 277–318.
96. (a) Møller, C.; Plesset, M.S. *Phys. Rev.* **1934**, *46*, 618. (b) Pople, J.A.; Binkley, J.S.; Seeger, R. *Int. J. Quantum Chem. Symp.* **1976**, *10*, 1.
97. Mackay, G.I.; Bohme, D.K. *J. Am. Chem. Soc.* **1978**, *100*, 327.
98. Diercksen, G.H.F.; Kraemer, W.P.; Roos, B.O. *Theor. Chim. Acta* **1975**, *36*, 249.
99. Scheiner, S.; Harding, L.B. *J. Am. Chem. Soc.* **1981**, *103*, 2169.
100. Ikuta, S. *Chem. Phys. Lett.* **1983**, *95*, 604.
101. Ikuta, S. *Mass Spectrosc.* **1982**, *30*, 297.
102. Chandrasekhar, J.; Andrade, J.G.; Schleyer, P.R. *J. Am. Chem. Soc.* **1981**, *103*, 5609.
103. Scheiner, S. *J. Am. Chem. Soc.* **1981**, *103*, 315.
104. Scheiner, S.; Harding, L.B. *Chem. Phys. Lett.* **1981**, *79*, 39.

105. Scheiner, S. *J. Phys. Chem.* **1982,** *86,* 376.
106. Scheiner, S. *J. Chem. Phys.* **1982,** *77,* 4039.
107. Meyer, W.; Jakubetz, W.; Schuster, P. *Chem. Phys. Lett.* **1973,** *21,* 97.
108. Delpuech, J.-J.; Serratrice, G.; Strich, A.; Veillard, A. *Mol. Phys.* **1971,** *29,* 849.
109. Howell, J.M.; Sapse, A.M.; Singman, E.; Snyder, G. *J. Phys. Chem.* **1982,** *86,* 2345.
110. Hehre, W.J.; Ditchfield, R.; Pople, J.A. *J. Chem. Phys.* **1972,** *56,* 2257.
111. Tang, I.N.; Castleman, A.W., Jr. *J. Chem. Phys.* **1975,** *62,* 4576.
112. Meot-Ner (Mautner), M. *J. Am. Chem. Soc.* **1978,** *100,* 4694.
113. De Paz, M.; Guidoni, A.G.; Friedman, L. *J. Chem. Phys.* **1969,** *51,* 3748.
114. Payzant, J.D.; Yamdagni, R.; Kebarle, P. *Can. J. Chem.* **1971,** *49,* 3308.
115. Keil, F.; Ahlrichs, R. *J. Am. Chem. Soc.* **1976,** *98,* 4787.
116. Yamdagni, R.; Kebarle, P. *J. Am. Chem. Soc.* **1972,** *94,* 2940.
117. Meot-Ner (Mautner), M.; Deakyne, C.A. *J. Am. Chem. Soc.* **1985,** *107,* 469.
118. Meot-Ner (Mautner), M.; Deakyne, C.A. In preparation.
119. Hiraoka, K.; Kebarle, P. *J. Am. Chem. Soc.* **1977,** *99,* 360.
120. Whiteside, R.A.; Frisch, M.J.; Binkley, J.S.; DeFrees, D.J.; Schlegel, H.B.; Raghavachari, K.; Pople, J.A. Carnegie–Mellon Quantum Chemistry Archive, Department of Chemistry, Carnegie–Mellon University, Pittsburgh, PA 15123.
121. Jorgensen, W.L. *J. Am. Chem. Soc.* **1978,** *100,* 1057.
122. Cournoyer, M.E.; Jorgensen, W.L. *J. Am. Chem. Soc.* **1984,** *106,* 5104.
123. Jorgensen, W.L. *J. Am. Chem. Soc.* **1978,** *100,* 1049.
124. Deakyne, C.A.; Meot-Ner (Mautner), M. *J. Am. Chem. Soc.* **1985,** *107,* 474.
125. Hehre, W.J.; Stewart, R.F.; Pople, J.A. *J. Chem. Phys.* **1969,** *51,* 2657.
126. Sunner, J.; Nishizawa, K.; Kebarle, P. *J. Phys. Chem.* **1981,** *85,* 1814.
127. Castora, F.J.; Liebman, J.; Meot-Ner (Mautner), M. Unpublished work.
128. Liebman, J.; Politzer, P.; Rosen, D.C. in "Chemical Applications of Atomic and Molecular Electrostatic Potentials", Politzer, P.; and Truhlar, D.G., Eds.; Plenum Press: New York, 1981, pp. 295–308.

# CHAPTER 5

# The Design and Action of a Totally Synthetic Serine Protease Partial Model

Howard Edan Katz

AT&T Bell Laboratories, Murray Hill, New Jersey

## CONTENTS

## 1. INTRODUCTION

The design and synthesis of enzyme models is one of the prime challenges of organic chemistry. Such models make possible the observation in enzymatic reactions of mechanistic details that are often obscured in the enzymes themselves. Furthermore, the increased understanding of enzyme action that is gained by the study of enzyme models might lead to the tailoring of artificial enzymes to carry out reactions other than those associated with biochemical processes.

Enzymes generally catalyze reactions by virtue of their ability to recognize

© 1987 VCH Publishers, Inc.
MOLECULAR STRUCTURE AND ENERGETICS, Vol. 4

reagents from solutions and hold them in reactive orientations.[1] To this end, enzymes must contain a binding site for substrates, and also one or more functional groups suitably located to react with the bound substrates. In addition, many enzymes require low molecular weight cocatalysts to function, and these must also be readily available to the reacting complex. Similarly, a successful model of an enzyme also would need to incorporate a binding site, reactive functionality, and cofactor availability (where appropriate) into a single chemical system.

Extensive work on enzyme modeling has focused on organic compounds that could serve as binding sites for smaller chemical entities. For example, crown ethers[2] and cryptands[3] are compounds that rely on the interaction of convergent polar groups with a guest of opposite polarity to achieve favorable binding energies in organic media. In contrast to crowns and crypts, the naturally occurring cyclodextrins[4] introduce a lipophilic cavity into an otherwise hydrophilic solvent system and are therefore able to form complexes with nonpolar molecules. The behavior of cyclodextrins has inspired the synthesis of man-made hydrophobic macrocycles such as cyclophanes,[5] calixarenes,[6] and cavitands.[7]

Only a few researchers have successfully combined binding sites with reactive functional groups to produce molecules with enzymelike kinetic behavior. Many of these molecules have been cyclodextrins that are modified by the addition of auxiliary reactive species. The reactive species have included imidazole[8] and pyridoxamine,[9] thereby modeling phosphodiesterases and transaminases, respectively. Crown ethers have also been employed as binding units, in conjunction with thiols[10] (cysteine protease models), dihydropyridines[11] (hydrogenase models), and flavins.[12] Recently, a cyclophane was shown to act as a catalytic binding site for an appended pyridoxamine.[13]

The serine proteases are a class of enzymes that catalyze the hydrolysis of acyl linkages such as peptides and esters. The key features[14] of these enzymes (Figure 5-1) are a substrate binding site and a nucleophilic hydroxymethyl group (on a serine residue) that attacks the carbonyl of a tightly bound substrate, giving an acylated enzyme. This acyl enzyme is subsequently hydrolyzed, releasing the acyl unit as a carboxylic acid and regenerating the free enzyme. While the gross mechanism of the reaction is well known, many details are not yet established,[15] particularly the roles of intramolecular general base catalysts. Initial attempts to model the activity of the serine proteases have been based on semisynthetic cyclodextrins,[16] whose hydroxymethyl groups are acylated by substrates at remarkably accelerated rates relative to simple acyl hydrolysis.

This chapter is concerned with the development of an organic compound that models some of the attributes of the serine proteases. Two aspects of the chemistry of this compound are explored in some detail: the derivation and properties of its binding site, and the kinetics of the reaction of a hydroxymethyl group attached to the binding site, in the presence of a complementary substrate–buffer system. Although this account is somewhat

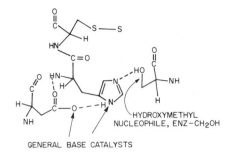

(a)  ENZ–CH$_2$OH  +  RCOX  $\rightleftharpoons$  ENZ–CH$_2$OH·RCOX

(b)  ENZ–CH$_2$OH·RCOX $\rightleftharpoons$ ENZ–CH$_2$OCOR + HX

(c)  ENZ–CH$_2$OCOR + H$_2$O  $\longrightarrow$  ENZ–CH$_2$OH + RCOOH

**Figure 5-1.** Active site structure and gross mechanism of action of a serine protease (chymotrypsin). (See Reference 14.)

simplified, enough information is presented so that the activity of the model compound may be understood as a consequence of its particular molecular structure.

While much of this chapter is narrowly focused on the properties of a few organic substances, the presentation also outlines a rational approach to the design and evaluation of synthetic enzyme models in general. A prototypical binding compound is chosen, and its structure is modified to optimize its affinity for a proposed substrate. Next, a reactive functionality is appended to the binding site so that the functionality is well positioned to react with the bound substrate. Finally, a kinetic analysis is performed on the reaction between the substrate and the functionalized binder. The kinetics are used to judge the effectiveness of the enzyme model based on two criteria: an accelerated rate of reaction compared to a nonbinding model, and the progression of the reaction through a model–substrate complex that is at a free energy minimum relative to the starting materials.

When either the model or the substrate is present in excess, the passage of the reaction through a stable complex results in saturation kinetics,[17] whereby the rate of the reaction is unaffected by an increase in the concentration of the excess reagent. Saturation kinetics are an important feature of enzyme-catalyzed reactions. Potential enzyme models that are poor substrate binders do not exhibit saturation phenomena at reasonable concentrations, and thus may be distinguished kinetically from strong substrate binders, even when their rates of reaction are comparable.

## 2. DESIGN OF THE BINDING SITE

For the serine protease model to operate in organic media, a binding site that would hold substrates through polar interactions was selected. Because the substrates were to contain ammonium groups (as in protonated amino acid esters), the incorporation of an ammonium group receptor into the model was desirable. Previous work on cysteine protease analogues[10] took advantage of crown ethers to bind these ammonium substituents via pole–dipole attractions and hydrogen bonds. (Chapters 3 and 4 in this volume treat strong hydrogen bonding.)

However, a functioning serine protease model, containing a hydroxymethyl nucleophile, would need to meet two requirements not applicable to cysteine protease models, which employed thiol nucleophiles. First, the binding of the ammonium ion would have to be strong enough to enable complexation to be favored when the pH is greater than the $pK_a$ of the ammonium group. (Crown ether–alkylammonium complexes are generally dissociated by added bases.) Only at such a pH would it be possible to partially deprotonate, and thereby unmask, the hydroxymethyl nucleophile before the reaction. Thiols, on the other hand, are fractionally deprotonated[18] at a pH well below the $pK_a$[19] of amino esters. It has been shown[20] that this deprotonation is necessary for the model thiolysis reactions to proceed efficiently, and by analogy, one would expect that oxynucleophiles would need to be deprotonated if they were to function in a similar way.

Second, the oxygen nucleophile would have to be carefully oriented with respect to the bound substrate carbonyl. This is less important in the case of thiols due to the greater atomic radius of sulfur as compared to oxygen and the greater polarizability of sulfur compounds[21] as compared to analogous oxygen compounds. Simple crown ethers would not possess the ammonium binding power or the rigidity necessary to meet these requirements; therefore, an improved binding site was sought.

The formal insertion of aromatic rings into crown frameworks is one method of generating synthetic cation binders of various shapes. Table 5-1 lists a series of macrocyclic polyethers containing six oxygen atoms, varying in the number of $CH_2OCH_2$, $ArOCH_3$, and $ArOCH_2$ units incorporated in the macrorings. Some of these have been discussed at great length in previous publications,[2] mostly regarding their abilities to bind alkali metal cations. Less attention has been paid to their affinities for alkylammonium compounds. The free energies of association $(-\Delta G_A)$ given in Table 5-1 are a measure of the relative affinities of the hosts for unhindered alkylammonium guests.

The first two entries are derivatives of 18-crown-6. These prototypical hosts are fair alkylammonium binders, mainly because they offer six Lewis basic oxygen atoms for bonding to the ammonium protons of the guest. However, the organizing of the conformationally flexible crown ether for binding (thereby freezing out many degrees of freedom) must occur at the time the guest interacts with the host, which is expected to result in an extremely unfavorable entropy of complexation.

A small degree of conformational rigidity is imparted upon the crown ether when a biaryloxy unit replaces part of the aliphatic structure of the macrocycle, as in **3** and **4**. Unfortunately, the potential improvement in the

TABLE 5-1. Free Energies of Association $(-\Delta G_A)$ of Methylammonium with Macrocyclic Polyethers[a]

| Compound | Structure | $-\Delta G_A$ (kcal/mol) | Ref. |
|---|---|---|---|
| 1 | | 7.4 | 22 |
| 2 | | 8.1 | 22 |
| 3 | | 7.1 | 23 |
| 4 | | 5.6 | 23 |
| 5 | | 8.2 | 23 |
| 6 | | 6.2 | 24 |
| 7 | | 5.4 | 24 |
| 8 | | 9.0 | 25 |
| 9 | | <1 | 2 |

TABLE 5-1. (continued)

| Compound | Structure | $-\Delta G_A$ (kcal/mol) | Ref. |
|----------|-----------|--------------------------|------|
| 10 | 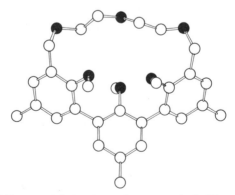 | 5.1 | 27 |

[a]Values of $-\Delta G_A$ are for methylammonium picrate in chloroform-d at 25°C, determined according to Reference 22.

binding free energy is not realized because of the lowered basicity of the aryl oxygens[28] relative to the aliphatic ether oxygens they replace. Thus, the polar interactions between the host and alkylammonium ions would be less favorable enthalpically, and the overall free energy of association is actually lower than in **1** and **2**.

The situation is dramatically different in the case of compound **5**. Because of the presence of the triaryl unit, the uncomplexed molecule exists in a semirigid conformation not very different from the one that is necessary for alkylammonium complexation (see Figure 5-2). Five of the six oxygen atoms in the free host converge about the cavity, spaced so that all three protons of a potential $RNH_3^+$ guest may be hydrogen bonded. The only major reorientation required to achieve the binding conformation is the redirection of the central $CH_2OCH_2$ oxygen toward the cavity. This single conformational change, along with some adjustments in bond angles, represents a much less extensive reorganization than that which is required for compounds **1–4** to bind alkylammonium salts. As a result, **5** is a superior binder of alkylammonium ions compared to **1–4**, even though three of the oxygens in **5** are of lower basicity. The improvement in binding entropy,

**Figure 5-2.** X-Ray crystal structure of macrocycle **5**. (From Reference 2.)

reduction of binding strain, or possible trifurcation in the $Ar-OCH_3$ hydrogen bonds must override the unfavorable enthalpic effect caused by the less basic oxygen atoms.

Macrocycles **6** and **7**, which are still more rigid than **5**, are nevertheless poorer binders of alkylammonium ions. This is probably because the favored conformation of **6** and **7** does not allow for the lone pairs of three oxygen atoms to engage simultaneously in hydrogen bonding, according to CPK model studies. Altering this favored conformation introduces strain, which detracts from the otherwise thermodynamically favorable interaction between the cycles and the guest ion. Molecular model studies and NMR spectroscopy both indicate that rotation about the aryl–aryl bonds is extremely hindered in **6** and **7**, with the methoxy groups locked in an alternating up-down-up-down arrangement.

Pentaaryl compound **8** exhibits the strongest alkylammonium ion affinity of all the compounds in this survey. Most of its framework consists of rigid subunits, and CPK models indicate that it possesses a cavity shaped like an oval saucer. One unstrained conformation allows the $CH_2OCH_2$ oxygen and two aryl oxygens to participate in hydrogen bonding to a guest alkylammonium, although uncomplexed **8** may also exist with its $CH_2OCH_2$ oxygen directed away from the cavity, analogous to uncomplexed **5**. As in **5, 6,** and **7,** the aryl–oxygen bonds of **8** point in alternating directions with respect to the best plane of the molecule. The organic substituent on a complexed alkylammonium cation is in a hydrophobic environment determined by the hydrocarbon framework of the macrocycle, where it may enjoy a degree of special stabilization.

The last two derivatives, **9** and **10,** are the most rigid in the series. However, the lone pairs on these oxygens converge about a very small region of space in which there is no room to insert an alkylammonium group. Therefore, these three compounds would be of little use in designing a strong binder for alkylammonium ions.

An analysis of Table 5-1 suggests that the shape of compound **8** appears optimal for alkylammonium binding relative to the other entries in the table. The importance of this "shape" is brought out when one considers how much less basic are the oxygen atoms in **8** than in **1** or **3**. This "shape" effect seems to outweigh a presumably great difference in Lewis-type electron-donating ability between **1** and **8**.

Although compounds **1–10** have been useful in studying binding ability as a function of molecular structure, rather than functionality,[27] there was no intrinsic reason to "settle" for hydrogen-bonding groups as weakly basic as anisyl oxygens[28] in designing an ideal alkylammonium binder. It was desirable to find a structural unit that could replace some of the anisyl groups in compounds shaped like **8** but would not alter the gross structure of the molecule and would carry a more strongly hydrogen-bonding functionality than the anisole oxygen. Preferably, the new group would not act as a strong base or nucleophile.

The tetrahydro-2(1*H*)-pyrimidinone (hereafter referred to as "cyclic urea") group fulfilled this need. The carbonyl oxygen of this heterocyclic unit is much more basic[29] than an anisole oxygen would be, since the carbonyl oxygen lies at the negative end of a dipole while the anisole oxygen has reduced negative character due to $\pi$ donation to the benzene ring. However, like the anisyl unit, the cyclic urea is a six-membered ring, with $sp^2$-hybridization at the atoms that would be closest to the cavity in an actual binder.

Both cation binding studies[30,31] and X-ray crystallography[30] have provided experimental evidence that cyclic urea groups are nearly isostructural with anisoles when incorporated into macrocyclic ligands. The greater electron density of the urea oxygens sometimes contributes to generally greater cation binding ability in oligourea hosts than in all-anisyl analogues.[32] Occasionally, the greater conformational flexibility in an aryl–urea linkage relative to an anisyl–anisyl linkage is detrimental to binding in that more organization of the host must occur during the binding step, probably leading to a more negative binding entropy.

One consistently favorable effect of the substitution of anisole units by cyclic ureas in polydentate ligands is the strengthening of alkylammonium binding. This is illustrated by the data in Table 5-2. For each of three pairs of semirigid methylammonium binders, the replacement of anisole by urea increases the binding free energy. The very best alkylammonium binders synthesized, such as 14, have been those in which three anisyl components are replaced by three ureas, poised for the formation of hydrogen bonds to each of the three acidic protons on the guest. It must be pointed out, however, that the remaining anisole units in such binders do contribute electron density and/or conformational definition to the macrocycles. This is demonstrated by compounds 15 and 16, which display reduced affinities for methylammonium relative to 14.

The X-ray crystal structure of the t-butylammonium complex of 14 (Figure 5-3) illustrates the convergence of the three urea oxygens toward the guest nitrogen atom, so that all three hydrogens bonded to the nitrogen may be accommodated by unstrained hydrogen bonds. The lone pair electrons of the $ArOCH_3$ groups are also directed toward the cationic nitrogen atom in the complex.

The extraordinary ability of compound 14 to complex alkylammonium guests arises from two optimal attributes. Host 14 is shaped like the superior polyether binder 8 and is therefore likely to be preorganized to a similar degree. This host also incorporates three hydrogen-bonding cyclic urea units in orientations well suited for cooperative binding to the three acidic protons of alkylammonium compounds. Although temperature-dependence NMR experiments have shown that uncomplexed 14 exists in several conformations (due to changes in the relative positions of the methoxy and urea groups) that interconvert with significant activation energies, CPK models reveal that the binding conformer, in which all three ureas are syn to the best plane of the macrocycle, is relatively unstrained.

Having approached in 14 an ideal shape and functionality for an alkylam-

**TABLE 5-2.** Comparison of $-\Delta G_A$ of Methylammonium with Macrocycles Containing Cyclic Urea and Anisole Units

| Compound | Structure | $-\Delta G_A$ (kcal/mol) | Ref. |
|---|---|---|---|
| **5** | | 8.4 | 23 |
| **11** | | 9.1 | 30 |
| **8** | | 9.0 | 25 |
| **12** | | 10.6 | 25 |
| **13** | | 12.0 | 30 |
| **14** | | 14.4 | 30 |
| **15** | | 9.0 | 33 |
| **16** | | 7.4 | 33 |

[a]Values of $-\Delta G_A$ are for methylammonium picrate in chloroform-d at 25°C, determined according to Reference 22.

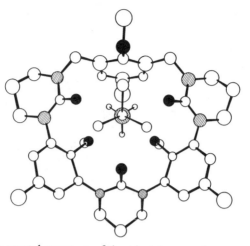

**Figure 5-3.** X-Ray crystal structure of the *t*-butylammonium complex of **14**. (From Reference 30.)

monium binding site, the next step was to employ this binding site in a molecule that would both complex and transesterify an amino acyl derivative in a manner analogous to serine proteases. Because the position of the amino nitrogen and its attached organic residue in such an amino acyl complex was predictable, with the substrate triply hydrogen bonded to the cyclic ureas, it was possible to use CPK models to assess possible locations of the hydroxymethyl nucleophile relative to the few conceivable orientations of the substrate carbonyl. It was correctly anticipated that the synthetically accessible[34] compound **17** (Figure 5-4) would provide a hydroxymethyl group properly positioned to attack the carbonyl of a bound amino acyl derivative. The final section of this chapter will present a kinetic analysis of

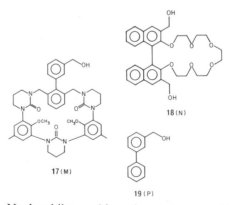

**Figure 5-4.** Nucleophiles used in serine protease model reactions.

the reaction between **17** and the substrate *l*-alanine *p*-nitrophenyl ester. Nucleophiles **18**[35] and **19**[34] will be examined as well, for comparison.

# 3. KINETICS OF A SERINE PROTEASE MIMIC

The reaction[34] used to model the acyl enzyme formation step of the serine protease catalytic cycle is illustrated in Figure 5-5. A large excess of nucleophile **17** ("M"), **18** ("N"), or **19** ("P"), was treated with the substrate *l*-alanine *p*-nitrophenyl ester ("SL", 0.1 mM in CDCl₃), buffered with diisopropyl ethyl amine ("B") and its perchlorate salt ("HB⁺"), to give the acylated host ("ROS") and *p*-nitrophenol ("LH"). The appearance of LH over time was monitored by UV spectroscopy, giving rise to data that were used to calculate rate constants ($k_{obs}$) for the psuedo-first-order formation of LH. The rates obtained are listed in Table 5-3.

A cursory analysis of the data leads to the generalization that the complexing nucleophiles M and N are efficiently esterified under the reaction conditions employed, while the noncomplexing P did not react at a measurable rate. It might then be naively concluded that M and N act almost equally well as serine protease models. If speed of reaction were the only criterion for evaluating potential enzyme models, we might ask what advantage there was in designing the idealized binding site of M, as opposed to making do with a traditional crown ether.

However, more careful examination of the rates as a function of nucleophile and buffer concentration reveals that M and N are acylated via different mechanisms. In particular, it will be shown that only one of the mechanisms leads to saturation kinetics, which are commonly associated with many enzymatic processes. The "saturation" mechanism will be the distinguishing feature of the superior model for the enzyme.

The elementary steps that need to be considered are presented in Table 5-4, where ROH is any hydroxymethyl nucleophile. It is assumed that com-

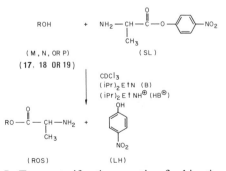

**Figure 5-5.** Transesterification reaction for kinetic analysis.

TABLE 5-3. Rates of Acylation of Hydroxymethyl Compounds by *l*-Alanine *p*-Nitrophenyl Ester (0.1 mM in CDCl$_3$) at 25°C[a]

| Run | Nucleophile | (ROH) | Concentration (mM) | [B] (mM) | [HB$^+$] (mM) | $10^3 k_{obs}$ (min$^{-1}$) |
|-----|-------------|-------|--------------------|----------|---------------|------------------------------|
| 1 | M | (17) | 1.4 | 1 | 1 | 2.8 |
| 2 | M | (17) | 1.4 | 3 | 3 | 2.2 |
| 3 | M | (17) | 1.4 | 3 | 1 | 9.1 |
| 4 | M | (17) | 1.4 | 6 | 1 | 17 |
| 5 | M | (17) | 0.7 | 6 | 1 | 19 |
| 6 | M | (17) | 1.4 | 10 | 1 | 24 |
| 7 | N | (18) | 1.6 | 3 | 1 | 13[b] |
| 8 | N | (18) | 1.6 | 6 | 1 | 9[b] |
| 9 | N | (18) | 1.6 | 10 | 1 | 10[b] |
| 10 | N | (18) | 3.2 | 6 | 1 | 18[b] |
| 11 | P | (19) | 16 | 6 | 1 | <0.046 (0.001) |
| 12 | P | (19) | 32 | 6 | 1 | <0.043 (0.001) |

[a]Reference 34.
[b]Not corrected for the statistical factor of two CH$_2$OH groups per molecule.

pletely undissociated alcohol ROH is too weak a nucleophile to be acylated by a *p*-nitrophenyl ester under these reaction conditions, but instead must first be converted to the alkoxide RO$^-$ to react (see Section 2); this assumption is also compatible with the data in Table 5-3. It is also assumed that unprotonated substrate does not form kinetically significant complexes with the macrocyclic ligands. The values of the rate constants depend on the nucleophile ROH, except in reactions III and IV.

Three special cases of nucleophiles must be considered here: noncomplexing, weakly complexing, and strongly complexing. The rate of product

TABLE 5-4. Elementary Steps in the Serine Protease Model Reaction Mechanism

$$\text{I. ROH + B} \quad \underset{k_{-1}}{\overset{k_1}{\rightleftharpoons}} \quad \text{RO}^- + \text{HB}^+ \quad \left[ K_1 = \frac{k_1}{k_{-1}} \right]$$

$$\text{II. RO}^- + \text{SL} \quad \overset{k_2}{\longrightarrow} \quad \text{ROS} + \text{L}^-$$

$$\text{III. SL + HB}^+ \quad \underset{k_{-3}}{\overset{k_3}{\rightleftharpoons}} \quad \text{HSL}^+ + \text{B} \quad \left[ K_3 = \frac{k_3}{k_{-3}} \right]$$

$$\text{IV. HSL}^+ + \text{ROH} \quad \underset{k_{-4}}{\overset{k_4}{\rightleftharpoons}} \quad \text{ROH·HSL}^+ \quad \left[ K_4 = \frac{k_4}{k_{-4}} \right]$$

$$\text{V. ROH·HSL}^+ + \text{B} \quad \underset{k_{-5}}{\overset{k_5}{\rightleftharpoons}} \quad \text{RO}^-\text{·HSL}^+ + \text{HB}^+ \quad \left[ K_5 = \frac{k_5}{k_{-5}} \right]$$

$$\text{VI. RO}^-\text{·HSL}^+ \quad \overset{k_6}{\longrightarrow} \quad \text{ROS} + \text{LH}$$

$$\text{VII. L}^- + \text{HB}^+ \quad \overset{k_7}{\longrightarrow} \quad \text{LH} + \text{B}$$

formation by a noncomplexing alcohol is given by Equation 5-1, assuming that $k_1 \gg k_2$ (justified by UV spectroscopy of the reaction mixture).

$$\frac{d[LH]}{dt} = k_2[RO^-][SL] \tag{5-1}$$

Also assuming that the equilibrium of reaction I is established instantaneously, or that $RO^-$ exists at an infinitesimal steady state concentration (Equation 5-2), Equation 5-1 may be rewritten as Equation 5-3.

$$[RO^-] = \frac{k_1[B][ROH]}{k_{-1}[HB^+] + (k_2[SL])} \qquad k_{-1} \gg k_2 \tag{5-2}$$

$$\frac{d[LH]}{dt} = \frac{k_2 K_1[B][ROH][SL]}{[HB^+]} \tag{5-3}$$

Since the reaction medium was buffered, $[B]/[HB^+]$ is constant, and the overall rate is bimolecular.

For an alcohol that forms a complex with the substrate, Table 5-3 indicates that reaction VI occurs much faster than reaction II. (Cycles including **17** and **18** whose cavities were occupied by $Na^+$ also underwent reaction II very slowly, since the faster reaction VI was inhibited.[34]) Therefore, the rate of product formation by a complexing nucleophile is given by Equation 5-4.

$$\frac{d[LH]}{dt} = k_6[RO^- \cdot HSL^+] \tag{5-4}$$

Treating $RO^- \cdot HSL^+$ as a species present at an infinitesimal steady state concentration (Equation 5-5), Equation 5-4 is transformed into Equation 5-6.

$$[RO^- \cdot HSL^+] = \frac{k_5[ROH \cdot HSL^+][B]}{k_{-5}[HB^+] + (k_6)} \qquad k_{-5} \gg k_6 \tag{5-5}$$

$$\frac{d[LH]}{dt} = \frac{k_6 K_5[ROH \cdot HSL^+][B]}{[HB^+]} \tag{5-6}$$

Because reactions III and IV are rapid equilibria,[36] Equation 5-7 holds as well.

$$\frac{[ROH \cdot HSL^+]}{[SL]} = \frac{K_4 K_3[ROH][HB^+]}{[B]} \tag{5-7}$$

Under the experimental conditions, all the terms on the right-hand side of Equation 5-7 are constants, with the following approximate values: $K_3 = 10^{-3}$ (from Reference 37), $[ROH] = 10^{-3}$, and $[HB^+]/[B] = 10^{-1}$. Thus, Equation 5-7 may be approximated by Equation 5-8.

$$\frac{[ROH \cdot HSL^+]}{[SL]} \approx 10^{-7} K_4 \tag{5-8}$$

The relatively weak alkylammonium binder $N$ $(-\Delta G_A$ for $CH_3NH_3^+$ in $CDCl_3$ of 6.9 kcal/mol) has $K_4 \approx 10^5$.[38] Thus, $ROH \cdot HSL^+$ is present in small concentrations relative to [SL] and may be treated as a steady state species (Equation 5-9). Substituting Equation 5-9 into Equation 5-6 gives the final rate expression as Equation 5-10.

$$[ROH \cdot HSL^+] = K_4[HSL^+][ROH] = \frac{K_4 K_3[SL][ROH][HB^+]}{[B]} \tag{5-9}$$

$$\frac{d[LH]}{dt} = k_6 K_5 K_4 K_3[SL][ROH] \tag{5-10}$$

On the other hand, the stronger alkylammonium binder M $(-\Delta G_A$ for $CH_3NH_3^+$ in $CDCl_3$ of 12.7 kcal/mol) has $K_4 \approx 10^9$.[34] For this host, reactions III and IV are not kinetically significant because virtually all the SL is instantly[36] converted to $ROH \cdot HSL^+$. *Therefore, $ROH \cdot HSL^+$ is not amenable to a steady state treatment in this case.* The appropriate kinetic expression for M is simply Equation 5-6.

According to this kinetic analysis, the transacylation reaction of strong binder M is expected to show a first-order dependence on specific base ([B]/[HB$^+$]) and zero-order dependence on [ROH]. In contrast, weak binder N should react with first-order dependence on [ROH] and zero-order dependence on specific base. The data in Table 5-3 are consistent with these predictions.

In discussing models of enzymatic reactions, it is appealing to speak of "rate accelerations," that is, the rate of the model reaction compared to an "analogous" uncatalyzed reaction. Unfortunately, one cannot obtain dimensionless ratios of rate constants when the reactions being compared are of differing molecularity.[39] In the present analysis, rate equations 5-3 and 5-10 are bimolecular, whereas rate equation 5-6 is unimolecular.

One solution to this dilemma is to treat the bimolecular reactions as pseudo-first order, assuming that rate equation 5-11 holds for all three nucleophiles.

$$\frac{d[LH]}{dt} = k_{obs}([SL]_o - [LH]) \tag{5-11}$$

The pseudo-first-order rate constant $k_{obs}$ is equal to the measured rates listed in Table 5-3. Using Equations 5-6, 5-10, and 5-3, $k_{obs}$ can be defined for M, N, and P in Equations 5-12, 5-13, and 5-14, respectively. (Note again that $K_4$, $K_5$, and $k_6$ are different for M and N.)

$$k_{obs} = \frac{k_6 K_5[B]}{[HB^+]} \text{ for M} \qquad k_M' = \frac{k_6 K_5[B]}{[HB^+][M]} \tag{5-12}$$

$$k_{obs} = k_6 K_5 K_4 K_3[N] \text{ for N} \qquad k_N' = k_6 K_5 K_4 K_3 \tag{5-13}$$

$$k_{obs} = \frac{k_2 K_1[B][P]}{[HB^+]} \text{ for P} \qquad k_P' = \frac{k_2 K_1[B]}{[HB^+]} \tag{5-14}$$

Dividing $k_{obs}$ by [ROH] gives a second-order rate constant $k'$ for the reaction shown in Figure 5-5, which is a measure of the relative efficiency of each nucleophile at transesterifying $l$-alanine $p$-nitrophenyl ester.

Runs 11 and 12 in Table 5-3 set a conservative upper limit of $10^{-4}$ M$^{-1}$ min$^{-1}$ for $k_P'$. Using runs 5 and 8, the rate accelerations for nucleophiles M and N are calculated in Equations 5-15 and 5-16.

$$\frac{k_M'}{k_P'} = \frac{k_6 K_5}{k_2 K_1 [M]} \approx 3 \times 10^5 \tag{5-15}$$

$$\frac{k_N'}{k_P'} = \frac{k_6 K_5 K_4 K_3 [HB^+]}{k_2 K_1 [B]} \approx 6 \times 10^4 \tag{5-16}$$

Note that under this treatment the apparent rate acceleration for M would increase if a smaller excess of M were employed. Similarly, the calculated acceleration for N would increase under conditions more acidic than those used in the present experiment. (This increase would be limited by the experimental difficulties in observing the reaction at all when [RO$^-$] is vanishingly small.)

A different method for calculating the rate accelerations is to treat all the reactions as though they were in fact bimolecular, and did not proceed through complexes. The applicable rate law would then be Equation 5-17.

$$\frac{d(LH)}{dt} = k''[ROH][SL] \tag{5-17}$$

For this treatment to apply to M, the substitution in Equation 5-18 must be carried out on equation 5-6.

$$[ROH \cdot HSL^+] = \frac{K_4 K_3 [ROH][HB^+][SL]}{[B]} \tag{5-18}$$

The result is coincidentally the same as Equation 5-10. Using Equation 5-10 for M and N, and Equation 5-3 for P, the respective second-order rate constants $k''$ may be calculated. (For N and P, $k'' = k'$.)

$$k_M'' = k_6 K_5 K_4 K_3 \tag{5-19}$$

$$k_N'' = k_6 K_5 K_4 K_3 \tag{5-20}$$

$$k_P'' = \frac{k_2 K_1 [B]}{[HB^+]} \tag{5-21}$$

Equations 5-12 to 5-14 can be used to introduce $k_{obs}$ into the expressions for $k''$.

$$k_M'' = \frac{k_{obs}[HB^+]K_4 K_3}{[B]} \approx 3 \times 10^3 \text{ M}^{-1} \text{ min}^{-1} \tag{5-22}$$

$$k_N'' = \frac{k_{obs}}{[N]} \qquad \approx 6 \text{ M}^{-1} \text{ min}^{-1} \tag{5-23}$$

$$k_P'' = \frac{k_{obs}}{[P]} \qquad \approx 10^{-4}\,M^{-1}\,min^{-1} \qquad (5\text{-}24)$$

Taking $K_4$ for M as $10^9$ and $K_3 \approx 10^{-3}$, the data in runs 5, 8, and 12 allow the estimation of numerical values for $k''$. The "second-order" rate accelerations for M and N are roughly $3 \times 10^7$ and $6 \times 10^4$, respectively, compared to P. (A previously reported[34] value of $k_M''/k_P'' \approx 10^{11}$ did not include the factor $K_3$.)

The value for the rate acceleration by M is drastically different in the two treatments. The reason for this is illustrated in the energy level diagrams in Figure 5-6, where the separated reagents are considered to be in standard states.[40] The rate constant $k_M'$ reflects the first-order reaction that proceeds experimentally from the complex $M\cdot HSL^+$ and is therefore a measure of $\Delta G_1^\ddagger$. On the other hand, $k_M''$ is obtained assuming that the reacting species are the separate entities M and SL, and thus $k_M''$ is a measure of $\Delta G_2^\ddagger$.

The calculated rate accelerations express the degree to which the activation energy for a complexing nucleophile is less than $\Delta G_4^\ddagger$ ($\Delta G_4^\ddagger$ is the same using either treatment). The rate acceleration for M appears greater when the second-order treatment is used, because $\Delta G_2^\ddagger$ is less than $\Delta G_1^\ddagger$. The rate

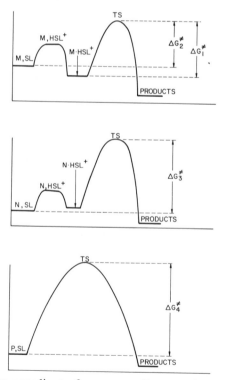

**Figure 5-6.** Reaction coordinate–free energy diagrams for serine protease model reactions.

ratio $k''_M/k''_P$ reflects the lowering of the free energy of the transition state TS(M) relative to TS(P), without taking into account the lowering of the reactant energy in the M system upon complexation.[40] The experimentally observed additional activation energy $\Delta G^{\ddagger}_1 - \Delta G^{\ddagger}_3$ is accounted for in the rate constant $k'_M$, so that the apparent rate acceleration is less using the first-order treatment. Since $\Delta G^{\ddagger}_3$ is the same, whether the first-order or second-order treatment is used, the derived rate acceleration for N is of the same order of magnitude either way.

The experimental rates and the mathematical analysis both reveal that M exhibits saturation kinetics[17] in its reaction with SL. As long as one of the two reagents is present in sufficient excess, the rate is independent of the concentration of excess reagent. Saturation kinetics are an important attribute of enzymatic reactions, and it is especially noteworthy that only the excellent ammonium binder M reacted in this enzymelike manner; N did not. Based on this kinetic test, as well as on the observed rate accelerations, M may be regarded as a partial serine protease mimic. (It is conceivable that N would display saturation kinetics if [N] or [HB$^+$]/[B] were impractically high.)

However, at least two additional qualities must be conferred on an M-based system before it will act as a full serine protease model. First, the "extra activation energy," $\Delta G^{\ddagger}_1 - \Delta G^{\ddagger}_2$, needs to be reduced, so that the overall transacylation can proceed more rapidly. Second, and most important, the transesterified substrate must be made more easily hydrolyzable than the uncomplexed substrate would be, so that true catalytic turnover will occur, as it does in actual enzymes. It is hoped that the incorporation of general bases and other hydrophilic groups onto the macrocyclic framework will provide the necessary kinetic improvements.[41]

## 4. SUMMARY

An optimal binding site for substrates containing an ammonium substituent was designed. The union of this binding site with a hydroxymethyl-containing residue led to a compound that acted as a serine protease mimic in that it was acylated by amino esters via a saturation mechanism and at accelerated rates. This compound is thus a promising basis for the synthesis of a true model for the serine proteases. In the future, one might expect that by maximizing the binding strength of other host–guest couples and devising kinetic tests similar to those presented here, new enzymelike reactions of totally synthetic reagents will be uncovered.

## ACKNOWLEDGMENTS

We are most grateful to Prof. D. J. Cram and Dr. R. C. Helgeson for helpful discussions. We also thank Prof. Cram and Mrs. J. Hendrix for supplying needed reprints and X-ray crystal structural diagrams.

# REFERENCES

1. Dickerson, R.E.; Geis, I. "The Structure and Action of Proteins". Benjamin: Menlo Park, Calif., 1969.
2. Cram, D.J.; Trueblood, K.N. *Top. Curr. Chem.* **1981**, *98*, 43–106.
3. Lehn, J.-M. *Acc. Chem. Res.* **1978**, *11*, 49–57.
4. (a) Tabushi, I. *Acc. Chem. Res.* **1982**, *15*, 66–72. (b) Breslow, R. *Science* **1982**, *218*, 532–537. (c) Bender, M.L.; Komiyama, M. "Cyclodextrin Chemistry". Springer-Verlag: New York, 1978.
5. Tabushi, I.; Yamamura, K.; Nonoguchi, H.; Hirotsu, K.; Higuchi, T. *J. Am. Chem. Soc.* **1984**, *106*, 2621–2625.
6. Gutsche, C.D. *Acc. Chem. Res.* **1983**, *16*, 161–170.
7. Cram, D.J. *Science* **1983**, *219*, 1117–1183.
8. Breslow, R.; Bovy, P.; Hersh, C.L. *J. Am. Chem. Soc.* **1980**, *102*, 2115–2117.
9. Breslow, R.; Hammond, M.; Lauer, M. *J. Am. Chem. Soc.* **1980**, *102*, 421–422.
10. (a) Chao, Y.; Cram, D.J. *J. Am. Chem. Soc.* **1976**, *98*, 1015–1017. (b) Matsui, T.; Koga, K. *Tetrahedron Lett.* **1978**, 1115–1118. (c) Lehn, J.-M.; Sirlin, C. *J. Chem. Soc. Chem. Commun.* **1978**, 949–951.
11. (a) Kellogg, R.M. *Top. Curr. Chem.* **1982**, *101*, 111–145. (b) Behr, J.-P.; Lehn, J.-M. *J. Chem. Soc. Chem. Commun.* **1978**, 143–146.
12. Shinkai, S.; Ishikawa, Y.; Shinkai, H.; Tsuno, T.; Makishima, H.; Ueda, K.; Manabe, O. *J. Am. Chem. Soc.* **1984**, *106*, 1801–1808.
13. Winkler, J.; Coutouli-Argyropoulou, E.; Leppkes, R.; Breslow, R. *J. Am. Chem. Soc.* **1983**, *105*, 7198–7199.
14. Blow, D.M.; Birktoft, J.J.; Hartley, B.S. *Nature* **1969**, *221*, 337–340.
15. (a) Hunkapiller, M.W.; Smallcombe, S.H.; Whitaker, D.R.; Richards, J.H. *Biochemistry* **1973**, *12*, 4732–4743. (b) Bachovchin, W.W.; Roberts, J.D. *J. Am. Chem. Soc.* **1978**, *100*, 8041–8047. (c) Banacky, P.; Linder, B. *Biophys. Chem.* **1981**, *13*, 223–231. (d) Kollman, P.A.; Hayes, D.M. *J. Am. Chem. Soc.* **1981**, *103*, 2935–2941. (e) Hamilton, S.E.; Zerner, B. ibid, **1981**, *103*, 1827–1831. (f) Kossiakoff, A.A.; Spencer, S.A. *Biochemistry* **1981**, *20*, 6462–6474.
16. Trainor, G.L.; Breslow, R. *J. Am. Chem. Soc.* **1981**, *103*, 154–158.
17. Fersht, A. "Enzyme Structure and Mechanism". W.H. Freeman: San Francisco, 1977, p. 85.
18. Bordwell, F.G.; Hughes, D.L. *J. Org. Chem.* **1982**, *47*, 3224–3232.
19. Hay, R.W.; Morris, P.J. *J. Chem. Soc. B* **1970**, 1577–1582.
20. Chao, Y.; Weisman, G.R.; Sogah, G.D.Y.; Cram, D.J. *J. Am. Chem. Soc.* **1979**, *101*, 4948–4958.
21. Pearson, R.J. *Surv. Prog. Chem.* **1969**, *5*, 1.
22. Helgeson, R.C.; Weisman, G.R.; Toner, J.L.; Tarnowski, T.L.; Chao, Y.; Mayer, J.M.; Cram, D.J. *J. Am. Chem. Soc.* **1979**, *101*, 4928–4941.
23. Koenig, K.E.; Lein, G.M.; Stuckler, P.; Kaneda, T.; Cram, D.J. *J. Am. Chem. Soc.* **1979**, *101*, 3553–3566.
24. Artz, S.P.; Cram, D.J. *J. Am. Chem. Soc.* **1984**, *106*, 2160–2171.
25. Katz, H.E.; Cram, D.J. *J. Am. Chem. Soc.* **1984**, *106*, 4977–4987.
26. Cram, D.J.; deGrandpre, M.; Knobler, C.B.; Trueblood, K.N. *J. Am. Chem. Soc.* **1984**, *106*, 3286–3292.
27. Cram, D.J.; Lein, G.M.; Kaneda, T.; Helgeson, R.C.; Knobler, C.B.; Maverick, E.; Trueblood, K.N. *J. Am. Chem. Soc.* **1981**, *103*, 6228–6232 (for $Li^+$ and $Na^+$).
28. Levitt, L.S.; Levitt, B.W. *Z. Naturforsch. B* **1979**, *34*, 614–620.
29. (a) Arnett, E.M.; Mitchell, E.J.; Murty, T.S.S.R. *J. Am. Chem. Soc.* **1974**, *96*, 3875–3889. (b) Mitsky, J.; Jaris, L.; Taft, R.W. ibid, **1972**, *94*, 3442–3445.
30. (a) Cram, D.J.; Dicker, I.B.; Lein, G.M.; Knobler, C.B.; Trueblood, K.N. *J. Am. Chem. Soc.* **1982**, *104*, 6827–6828. (b) Cram, D.J.; Dicker, I.B.; Knobler, C.B.; Trueblood, K.N. ibid, **1982**, *104*, 6828–6830.
31. Cram, D.J.; Dicker, I.D. *J. Chem. Soc. Chem. Commun.* **1982**, 1219–1221.
32. Cram, D.J.; Lein, G.M.; Kaneda, T.; Helgeson, R.C.; Knobler, C.B.; Maverick, E.; Trueblood, K.N. *J. Am. Chem. Soc.* **1981**, *103*, 6228–6232.
33. Nolte, R.J.M.; Cram, D.J. *J. Am. Chem. Soc.* **1984**, *106*, 1416–1420.

34. (a) Cram, D.J.; Katz, H.E. *J. Am. Chem. Soc.* **1983,** *105,* 135–137. (b) Cram, D.J.; Katz, H.E. ibid, **1984,** *106,* 4987–5000.
35. Cram, D.J.; Helgeson, R.C.; Koga, K.; Kyba, E.P.; Madan, K.; Sousa, L.R.; Siegel, M.G.; Moreau, P.; Gokel, G.W.; Timko, J.M.; Sogah, G.D.Y. *J. Org. Chem.* **1978,** *43,* 2758–2772.
36. Anthonsen, T.; Cram, D.J. *J. Chem. Soc. Chem. Commun.* **1983,** 1414–1416.
37. (a) Werber, M.W.; Shalitin, Y. *Bioorganic Chem.* **1973,** *2,* 202–220. (b) Williams, A.W.; Young, G.T. *J. Chem. Soc. Perkin Trans. 1* **1972,** 1194–1200; also Reference 19. The absolute p$K_a$ values would be different in $CDCl_3$, but the *ratios* of the values should be the same.
38. Lingenfelter, D.S.; Helgeson, R.C.; Cram, D.J. *J. Org. Chem.* **1981,** *46,* 393–406.
39. Schowen, R.L. *Chem. Eng. News* **1983,** *39,* 47.
40. Schowen, R.L. In "Transition States of Biochemical Processes", Gandour, R.D.; and Schowen, R.L., Eds.; Plenum Press: New York, 1978, pp. 77–114.
41. The synthesis of a compound similar to **17** with an appended imidazole has been completed, and its acyl transfer kinetics have been probed: Cram, D.J.; Lam, P.Y.-S., *J. Am. Chem. Soc.* **1986,** *108,* 839–841.

CHAPTER 6

# Empirical Resonance Energies of Acyl and Carbonyl Derivatives

## Philip George

University of Pennsylvania, Philadelphia, Pennsylvania

## Charles W. Bock and Mendel Trachtman

Philadelphia College of Textiles and Science, Philadelphia, Pennsylvania

## CONTENTS

© 1987 VCH Publishers, Inc.
  MOLECULAR STRUCTURE AND ENERGETICS, Vol. 4

# 1. INTRODUCTION: HISTORICAL PERSPECTIVE

It is one of the tenets of organic chemistry that certain molecules show a chemical reactivity significantly different from what would be expected on the basis of their classical valence bond structures. For more than a century it has been known that benzene lacks the facile additivity expected of 1,3,5-cyclohexatriene, and that in acyl derivatives RC(X)=O and carbonyl derivatives XYC=O, the bonding of X, Y, and O to the same carbon atom modifies, if not changes entirely, the reactivity otherwise expected of the C−X, C−Y, and C=O functional groups. In these cases the "whole" is by no means the "sum of the parts" (in terms of the formal bonding) but something rather different.

For more than half a century it has been recognized that the key molecular property responsible for these contrasts in reactivity is an alteration in the electronic structure that results in benzene, the acyl and the carbonyl derivatives having an enhanced stability relative to reference molecules with the same formal bonding. Against a background of quantum mechanical valence bond theory, Pauling and Sherman[1] in 1933 evaluated this enhanced stability, designated as an empirical resonance energy, in the following way.

The enthalpy of atomization $\Delta H_a^\circ$ for the molecule in question is first evaluated from its enthalpy of formation from the elements, $\Delta H_f^\circ$. For example, in the case of benzene, we have:

$$\Delta H_a^\circ(\text{benzene}) = 6\,\Delta H_f^\circ(\text{H}) + 6\,\Delta H_f^\circ(\text{C}) - \Delta H_f^\circ(\text{benzene}) \qquad (6\text{-}1)$$

where $\Delta H_f^\circ(\text{H})$ and $\Delta H_f^\circ(\text{C})$ are the enthalpies of formation of atomic hydrogen and atomic carbon, respectively. The empirical resonance energy $RE$ is then obtained from the expression:

$$RE(\text{benzene}) = \Delta H_a^\circ(\text{benzene}) - [6E(\text{C}-\text{H})$$
$$+ 3E(\text{C}-\text{C}) + 3E(\text{C}=\text{C})] \qquad (6\text{-}2)$$

where $E(\text{C}-\text{H})$, $E(\text{C}-\text{C})$, and $E(\text{C}=\text{C})$ are the thermochemical bond energies for C−H, C−C, and C=C derived from $\Delta H_f^\circ$ data for paraffins and olefins. The $\Delta H_f^\circ$ values are for the gas phase species.

Following the same procedure in the case of acyl derivatives—for example, acetic acid—yields:

$$\Delta H_a^\circ(\text{acetic acid}) = 4\,\Delta H_f^\circ(\text{H}) + 2\,\Delta H_f^\circ(\text{C})$$
$$+ 2\,\Delta H_f^\circ(\text{O}) - \Delta H_f^\circ(\text{acetic acid}) \qquad (6\text{-}3)$$

and

$$RE(\text{acetic acid}) = \Delta H_a^\circ(\text{acetic acid}) - [3E(\text{C}-\text{H})$$
$$+ E(\text{C}-\text{C}) + E(\text{C}-\text{O}) + E(\text{C}=\text{O}) + E(\text{O}-\text{H})] \qquad (6\text{-}4)$$

where $E(\text{O}-\text{H})$, $E(\text{C}-\text{O})$, and $E(\text{C}=\text{O})$ are additional thermochemical bond energies derived from $\Delta H_f^\circ$ data for $H_2O$, ethers, and ketones.

Thus the following values were obtained: benzene 37.3 kcal/mol; nine monocarboxylic acids, mean value 28 kcal/mol; eight dicarboxylic acids, mean value per COOH group 27 kcal/mol; methyl and esters of monocarboxylic acids, the same value as the parent acids to within 5 kcal/mol; formamide and acetamide, 21 kcal/mol; oxamide 24 kcal/mol per $CONH_2$ group; urea 37 kcal/mol; and the dimethyl and diethyl esters of carbonic acid, 41 and 42 kcal/mol, respectively.[1,2] The values for the acyl derivatives are of the same order of magnitude as that for benzene. Yet over the years the stabilization in benzene has attracted far more interest, undoubtedly because it is the parent molecule typifying "aromatic" structures in general. However, in biological chemistry, the reactions of acyl derivatives especially play a role in many diverse catabolic and anabolic processes, whereas those in which the benzene ring is involved are restricted to no less important but more specialized metabolic pathways.

To redress the balance to some extent, the empirical resonance energy of acyl and carbonyl derivatives will now be subjected to the same scrutiny that has been brought to bear on the empirical resonance energy of benzene and benzenoid hydrocarbons in recent years.[3-8]

There are two interrelated themes: first, the method for evaluating $RE$ using bond energies is shown to be equivalent to calculating $\Delta H°$ for a particular reaction involving the molecule in question and the self-same set of reference molecules from which the bond energies are evaluated; and second, matching structural elements in the reference molecules as far as possible with those in the molecule in question gives a value for $RE$ that can be regarded as the most appropriate measure of its inherent stabilization.

However, before proceeding, it may be noted that there is a fundamental distinction between "empirical resonance energies" and "resonance energies," which involve a specific conceptual approach to chemical bonding. The former are essentially phenomenological, derived from experimental $\Delta H_f°$ data, hence can be evaluated from semiempirical or ab initio calculations of $\Delta H_f°$ or total molecular energies. Some examples will be given later. The latter are not strictly phenomenological and have been defined in various ways—utilizing properties of individual bonds inferred from calculations using, for example, the $\pi$-electron approximation, model structures with idealized bonding, or hypothetical reference states such as an infinitely large cyclic polyene.[9-17] In the case of benzene, interest has been focused almost exclusively on this sort of resonance energy, and there has been no comparable treatment of acyl and carbonyl derivatives. It may also be noted that insofar as an empirical resonance energy serves to evaluate the stabilization of a certain type of molecule in relation to other reference molecules, the more general term "stabilization energy" could just as well be employed. The stabilization energy, for example, of an unbranched alkane relative to a branched alkane or to a cycloalkane is simply the isomerization energy. But, in view of the long-standing usage for benzene, acyl, and carbonyl derivatives, the term "empirical resonance energy" is used in this chapter.

## 2. EMPIRICAL RESONANCE ENERGIES AS REACTION ENTHALPIES: BENZENE

The correlation between $RE$ for benzene calculated via bond energies and $\Delta H°$ for an underlying reaction has been demonstrated in 10 particular cases in a recent article,[8] and, not unexpectedly, the $\Delta H°$ $(RE)$ values were found to vary widely: from 4.4 to 64.2 kcal/mol. Four of these cases have been chosen to illustrate the differing extents to which the structural elements are matched in reactants and products, and thus to provide a basis for judging the rather more complicated matching that is at issue when heteroatoms are involved in the case of acyl and carbonyl derivatives.

1. With $E(C-H)$, $E(C-C)$, and $E(C=C)$ calculated from $\Delta H_f°$ for $CH_4$, $CH_3-CH_3$, and $CH_2=CH_2$,[18] the underlying reaction is:

$$\text{benzene} + 6CH_4 \rightarrow 3CH_3-CH_3 + 3CH_2=CH_2 \qquad (6\text{-}5)$$

and $RE = 64.2$ kcal/mol.

2. With $E(C-H)$ and $[E(C-C) + E(C=C)]$ calculated from $\Delta H_f°$ for $CH_4$ and $CH_3-CH=CH_2$,[18] the underlying reaction is:

$$\text{benzene} + 3CH_4 \rightarrow 3CH_3-CH=CH_2 \qquad (6\text{-}6)$$

and $RE = 48.1$ kcal/mol.

3. With $E(C-H)$, $E(C-C)$, and $E(C=C)$ calculated from $\Delta H_f°$ for $CH_4$, cyclohexane, and cyclohexene,[18] the underlying reaction is:

$$\text{benzene} + 2 \text{ cyclohexane} \rightarrow 3 \text{ cyclohexene} \qquad (6\text{-}7)$$

and $RE = 35.8$ kcal/mol.

4. With $E(C-H)$, $E(C-C)$, and $E(C=C)$ calculated from $\Delta H_f°$ for $CH_4$, $CH_2=CH-CH=CH_2$, and $CH_2=CH_2$,[18] the underlying reaction is:

$$\text{benzene} + 3CH_2=CH_2 \rightarrow 3CH_2=CH-CH=CH_2 \qquad (6\text{-}8)$$

and $RE = 21.6$ kcal/mol.

In cases 3 and 4, $CH_4$ in italics has been included to solve the simultaneous equations for the bond energies. However the terms in the enthalpy of atomization, $\Delta H_a°(CH_4)$, cancel out in the final expression, which is equivalent to $\Delta H°$ for the underlying reaction, as they should do, since $CH_4$ does not appear as a reactant or a product.

## 3. MATCHING THE BONDING: BENZENE

The aforementioned reactions are all *isodesmic;* that is, there is retention of the number of bonds of a given formal type, but with a change in their relation to one another.[19] In other words, the number of C−H bonds, C−C bonds, and C=C bonds is the same in reactants and products. Reaction 6-5, in which the bonds between the heavy atoms are separated from one another, giving two heavy-atom product molecules containing the same bonds, has been termed a bond separation reaction.[19]

But within the broad category of isodesmic reactions there is a progressive improvement in specific matching of the bonding as one goes from reaction 6-5 to 6-8. Using the connectivity notation,[20] C3 and C4, to describe carbon atoms bonded to three and four nearest neighbor atoms, respectively, there are three kinds of carbon–carbon single bond C4−C4, C4−C3, and C3−C3, together with the carbon–carbon double bond C3=C3. In addition, there are differences in the carbon–hydrogen bonding according to the number of hydrogen atoms bonded to each kind of carbon. In reactions 6-5 and 6-6 neither the carbon–carbon nor the carbon–hydrogen bonding is matched, although in reaction 6-6 the mismatch is less in the latter. In reaction 6-7 the carbon–hydrogen bonding is matched, but there remains a mismatch in the carbon–carbon bonding; whereas in reaction 6-8 both carbon–hydrogen and carbon–carbon bonding are now matched.

This particular subclass of isodesmic reactions in which the bonding is matched, not only by the number of formal bonds of each type but also according to more specific structural features, has been termed *homodesmotic* to emphasize the sameness of the bonding.[4,5] In such reactions extraneous energy contributions that otherwise arise from changes in the bonding state of the carbon and/or the carbon–hydrogen bonding are clearly going to be reduced—and for this reason their $\Delta H°$ values can be considered to be a more appropriate measure of the stabilization energy.

In reaction 6-8 the stabilization in benzene is assessed, inter alia, relative to three times whatever the stabilization is in *trans*-1,3-butadiene, the reference molecule containing the necessary C3−C3 structural element. The use of the rotamer in which one $H_2C=CH$ plane is twisted 90° relative to the other $CH=CH_2$ plane, such that $\pi$ delocalization is absent,[21] has been proposed as an alternative to make allowance, as best one can in principle, for the stabilization in the planar trans structure.[6] (Cyclooctatetraene has been used as a model for attached unconjugated double bonds by Liebman and Van Vechten and by Stevenson; see Chapter 8 in Volume 2 and Chapter 2 in Volume 3 in this series.)

Quite apart from this sort of consideration, it should be noted that there is no unique homodesmotic reaction. Reaction 6-8 is just the most simple reaction of this type that can be set up for benzene, and, analogous to isodesmic bond separation,[19] it may be termed a homodesmotic group separation.[22] Other examples utilizing larger reference molecules, which in general may be termed homodesmotic fission reactions, are discussed later in connection with issues that arise when molecules containing heteroatoms are involved.

## 4. EMPIRICAL RESONANCE ENERGIES AS REACTION ENTHALPIES: ACYL AND CARBONYL DERIVATIVES

The reactions that underlie the evaluation of empirical resonance energies for acyl and carbonyl derivatives using bond energies can be identified by employing a procedure similar to that developed for benzene.[7,8] Adopting Pauling and Sherman's choice of reference molecules[1,2] (ie, methane, ethane, dimethylether, acetone, and water), the bond energies are given by the following expressions:

$$E(C-H) = \tfrac{1}{4}\Delta H_a^\circ(CH_4) \tag{6-9}$$

$$E(C-C) = \Delta H_a^\circ(CH_3CH_3) - \tfrac{3}{2}\Delta H_a^\circ(CH_4) \tag{6-10}$$

$$E(C-O) = \tfrac{1}{2}\Delta H_a^\circ(CH_3OCH_3) - \tfrac{3}{4}\Delta H_a^\circ(CH_4) \tag{6-11}$$

$$E(C=O) = \Delta H_a^\circ(CH_3COCH_3) - 2\Delta H_a^\circ(CH_3CH_3) + \tfrac{3}{2}\Delta H_a^\circ(CH_4) \tag{6-12}$$

$$E(O-H) = \tfrac{1}{2}\Delta H_a^\circ(H_2O) \tag{6-13}$$

Using current $\Delta H_f^\circ$ values,[18] $E(C-H) = 99.28$, $E(C-C) = 78.82$, $E(C-O) = 81.14$, $E(C=C) = 183.48$, and $E(O-H) = 110.78$ kcal/mol. Hence, taking acetic acid as the example, with $\Delta H_f^\circ(CH_3COOH) = -103.27$ kcal/mol, $\Delta H_a^\circ(CH_3COOH) = 772.59$ kcal/mol, and so, from Equation 6-4, we have:

$$RE(CH_3COOH) = 772.59 - [297.84 + 78.82$$
$$+ 81.14 + 183.48 + 110.78] = 20.5 \text{ kcal/mol}$$

Alternatively, substituting the general expressions for the bond energies in Equation 6-4 and collecting like terms, it follows that:

$$RE(CH_3COOH) = \Delta H_a^\circ(CH_3COOH) + \Delta H_a^\circ(CH_3CH_3)$$
$$- \Delta H_a^\circ(CH_3COCH_3) - \tfrac{1}{2}\Delta H_a^\circ(CH_3OCH_3) - \tfrac{1}{2}\Delta H_a^\circ(H_2O)$$

Consideration of the makeup of the right-hand side of this equation leads to the conclusion that it is the enthalpy change, $\Delta H^\circ$, for the reaction:

$$CH_3COOH + CH_3CH_3 \rightarrow CH_3COCH_3 + \tfrac{1}{2}CH_3OCH_3 + \tfrac{1}{2}H_2O \tag{6-14}$$

Calculated from the $\Delta H_f^\circ$ data, $\Delta H^\circ$ is found to be 20.5 kcal/mol, identical with the $RE$ value above. The determination of the empirical resonance energy indirectly from bond energy terms is thus equivalent to the direct calculation of a reaction enthalpy—which has to be so because the same thermochemical data are employed in both cases.

If acetaldehyde is used instead of acetone to calculate $E(C=O)$, the underlying reaction for acetic acid is:

$$CH_3COOH + CH_4 \rightarrow CH_3CHO + \tfrac{1}{2}CH_3OCH_3 + \tfrac{1}{2}H_2O \tag{6-15}$$

with $\Delta H^\circ$, $RE = 30.6$ kcal/mol.

Likewise if methane, ethane, methylamine, acetone, and ammonia are used as reference molecules in the determination of the empirical resonance energy of urea, the underlying reaction is:

$$(NH_2)_2CO + 2CH_3CH_3 \rightarrow CH_3COCH_3 + 2CH_3NH_2 \qquad (6\text{-}16)$$

with $\Delta H°$, $RE = +36.0$ kcal/mol.

The procedure adopted by Liebman and Greenberg[23] for evaluating a resonance energy—namely, as the difference between the experimental $\Delta H_f°$ value and a value derived from data for model (ie, reference) compounds—is also equivalent to the determination of a reaction enthalpy. For example, in the case of $N,N$-dimethylacetamide, if this value is based on data for acetone, trimethylamine and ethane, that is:

$$\Delta H_f°(CH_3CON(CH_3)_2)$$
$$= \Delta H_f°(CH_3COCH_3) + \Delta H_f°((CH_3)_3N) - \Delta H_f°(CH_3CH_3) \qquad (6\text{-}17)$$

the underlying reaction is:

$$CH_3CON(CH_3)_2 + CH_3CH_3 \rightarrow (CH_3)_2CO + (CH_3)_3N \qquad (6\text{-}18)$$


## 5. MATCHING THE BONDING: ACYL AND CARBONYL DERIVATIVES

For molecules containing heteroatoms, N, O, halogens, S, and so on, the criteria for an isodesmic reaction need no elaboration beyond that necessary for hydrocarbons to require matching of the formal single, double, and triple bonding according to the kind of hetero-atom bonds in reactant and product. But in the case of homodesmotic reactions, more narrowly defined criteria are called for. For hydrocarbon reactions, the criteria alluded to above require the following in reactant and product molecules:

i. Equal numbers of each type of carbon atom (C4, C3, and C2).
ii. Equal numbers of each type of carbon–carbon bond (C4−C4, C4−C3, C3−C3, C3=C3, etc).
iii. Equal numbers of each type of carbon atom with 0, 1, 2, 3 (and 4) hydrogen atoms attached.

With heteroatoms present, the corresponding matching of the additional structural elements consists of:

iv. Equal numbers of each type of heteroatom (N3, N2, N1, O2, O1, etc).
v. Equal numbers of each type of carbon–heteroatom bond (eg, C4−O2, C3−O2, C3=O1, C4−N3, C3−N3).
vi. Equal numbers of each type of heteroatom with 0, 1, 2 (and 3) hydrogen atoms attached.

Two points are to be noted in passing. First, according to these extended criteria it is nevertheless possible for a reaction to be homodesmotic with respect to carbon yet isodesmic with respect to the heteroatom, for example:

$$2CH_3OH \rightarrow CH_3OCH_3 + H_2O \qquad (6\text{-}19)$$

Second, these criteria do not encompass reactions in which there is direct bonding between heteroatoms: in such cases additional criteria are required.

Hence, while the reactions above for acetic acid and urea (6-14, 6-15, and 6-16) are isodesmic, they are not homodesmotic. In reaction 6-14 there is a mismatch in both the carbon–carbon and oxygen–hydrogen bonding, in reaction 6-15 in both the carbon–hydrogen and oxygen–hydrogen bonding, and in reaction 6-16 in both the carbon–carbon and carbon–nitrogen bonding. Reaction 6-18, which underlies Liebman and Greenberg's procedure[23] for evaluating the resonance energy of $N,N$-dimethylacetamide, is also isodesmic but not homodesmotic, because again there is a mismatch in both the carbon–carbon and carbon–nitrogen bonding.

In view of the profound influence of the adoption of a homodesmotic reaction in place of an isodesmic reaction on the magnitude of $\Delta H°$ ($RE$) in the case of benzene—leading to a much lower value—it thus becomes important to establish whether the values for acyl and carbonyl derivatives would likewise be affected.

Homodesmotic reactions for formyl, acetyl, and carbonyl derivatives are set out below utilizing formaldehyde to provide the C3=O1 group, vinyl derivatives $CH_2=CHX$ and $CH_2=CHY$ to provide the C3−X and C3−Y groups, and ethylene, propene, and isobutene, as required, to complete the matching of the structural elements. These reference molecules are the smallest that can be used for this purpose, and the reactions thus come into the homodesmotic group separation category (see above).

*Formyl*

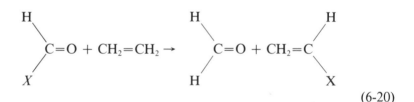

(6-20)

*Acetyl*

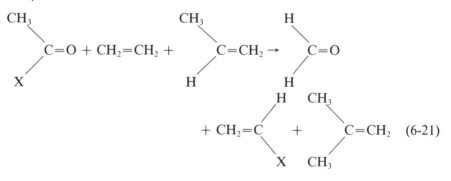

(6-21)

*Carbonyl*

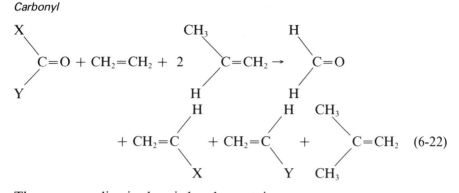

$$(6\text{-}22)$$

The corresponding isodesmic bond separations are:

*Formyl*

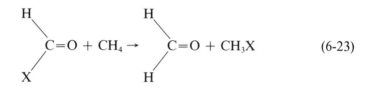

$$(6\text{-}23)$$

*Acetyl*

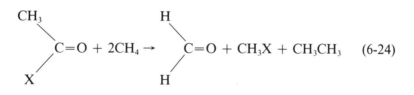

$$(6\text{-}24)$$

*Carbonyl*

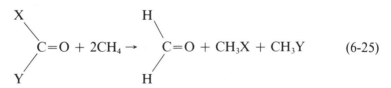

$$(6\text{-}25)$$

Both types of reaction are a little more complicated in the case of anhydrides because there are two carbonyl groups present. With formic anhydride, the isodesmic bond separation reaction is:

$$\begin{matrix} H-C \overset{O}{\underset{O}{\diagdown}} \\ H-C \overset{}{\underset{O}{\diagdown}} \end{matrix} O + 2CH_4 + H_2O \rightarrow 2\left( \overset{H}{\underset{H}{\diagup}} C=O \right) + 2CH_3OH \qquad (6\text{-}26)$$

while the corresponding reaction for acetic anhydride involves, in addition, fission of the C−C bonds:

$$CH_3-C(=O)O,\ CH_3-C(=O)O + 4CH_4 + 2H_2O \rightarrow 2\left(H,H\ C=O\right) + 2CH_3OH + 2CH_3CH_3 \quad (6\text{-}27)$$

The homodesmotic group separation reactions are:

$$H-C(=O)O,\ H-C(=O)O + 2CH_2{=}CH_2 \rightarrow 2\left(H,H\ C=O\right) + H-C(=CH_2)O,\ H-C(CH_2) \quad (6\text{-}28)$$

and

$$CH_3-C(=O)O,\ CH_3-C(=O)O + 2\ CH_3,H\ C{=}CH_2 + 2CH_2{=}CH_2 \rightarrow 2\ H,H\ C=O$$

$$+\ 2\ CH_3,CH_3\ C{=}CH_2 +\ H-C(=CH_2)O,\ H-C(CH_2) \quad (6\text{-}29)$$

Although appreciably larger than most of the molecules needed as reference, divinylether is nevertheless the smallest molecule containing O2 bonded to two C3's. With vinyl alcohol, instead the reactions are only isodesmic, for example:

$$H-C(=O)O,\ H-C(=O)O + 2CH_2{=}CH_2 + H_2O \rightarrow 2\left(H,H\ C=O\right) + 2CH_2{=}C(H)OH \quad (6\text{-}30)$$

since there is a mismatch with regard to the number of hydrogens bonded to O2. The $\Delta H°$ values for these reactions, calculated from the $\Delta H_f°$ data listed in Tables 6-1 and 6-2 are given in Table 6-3. The procedures used to estimate some of the $\Delta H_f°$ values are described in the Appendix. It can be

TABLE 6-1. $\Delta H_f^\circ$ at 298 K (kJ/mol) for the Reference Molecules

| Molecule | $\Delta H_f^\circ$ | Molecule | $\Delta H_f^\circ$ |
|---|---|---|---|
| $H_2O$ | −241.8 | $CH_3NH_2$ | −23.0 |
| | | $CH_2=CHNH_2$ | +57.1[a] |
| $CH_4$ | −74.5 | | |
| $CH_2=CH_2$ | +52.2 | $CH_3F$ | −233.9[b] |
| $CH_3-CH_3$ | −84.0 | $CH_2=CHF$ | −138.8 |
| $CH_3CH=CH_2$ | +20.2 | | |
| $(CH_3)_2C=CH_2$ | −16.9 | $CH_3Cl$ | −82.0 |
| | | $CH_2=CHCl$ | +37.3 |
| $H_2C=O$ | −108.7 | | |
| $CH_3OH$ | −201.6 | $CH_3Br$ | −37.2 |
| $CH_3CHO$ | −165.8 | $CH_2=CHBr$ | +79.3 |
| $CH_2=CHOH$ | −127.4[a] | | |
| $CH_2=CHOCH_3$ | −99.0[a] | $CH_3I$ | +15.4 |
| $CH_2=CHOC_2H_5$ | −140.9 | $CH_2=CHI$ | +132.6[a] |
| $(CH_2=CH)_2O$ | −14.0 | | |
| $(CH_3)_2C=O$ | −217.2 | $CH_3SH$ | −22.9 |
| | | $CH_2=CHSH$ | +82.5[a] |

[a]Estimated using Equation 6-48. See Appendix.
[b]Reference 24.

seen in Table 6-3 that as in the case of benzene, the values are consistently lower by a substantial amount for the homodesmotic reactions. For example, with F, OH, and $NH_2$ as the substituents, the average increments are 10.4 ± 2.6 kcal/mol for the formyl derivatives, 17.0 ± 2.6 kcal/mol for the acetyl derivatives, and 22.0 ± 5.2 for the carbonyl derivatives, the increments being smaller throughout with F as the substituent. The important

TABLE 6-2. $\Delta H_f^\circ$ at 298 K (kJ/mol), for Formyl, Acetyl, and Carbonyl Derivatives

| Molecule | $\Delta H_f^\circ$ | Molecule | $\Delta H_f^\circ$ | Molecule | $\Delta H_f^\circ$ |
|---|---|---|---|---|---|
| HCOF | −393.3[a] | $CH_3COF$ | −468.7 | $COF_2$ | −639.8 |
| HCOCl | −189.8[b] | $CH_3COCl$ | −242.8 | $COCl_2$ | −218.4 |
| HCOBr | −137.1[b] | $CH_3COBr$ | −190.4 | $COBr_2$ | −144.6 |
| HCOI | −72.5[b] | $CH_3COI$ | −126.1 | COFCl | −445.6 |
| HCOOH | −378.8 | $CH_3COOH$ | −432.1 | $CO(OH)_2$ | −615.1[c] |
| $HCOOCH_3$ | −355.5 | $CH_3COOCH_3$ | −410.0 | $CO(OC_2H_5)_2$ | −639.1 |
| $(HCO)_2O$ | −461.8[d] | $(CH_3CO)_2O$ | −567.3 | — | — |
| $HCONH_2$ | −189.1[e] | $CH_3COHN_2$ | −238.3 | $CO(NH_2)_2$ | −245.9 |
| HCOSH | −121.7[b] | $CH_3COSH$ | −175.0 | — | — |

[a]Reference 25.
[b]Estimated using Equation 6-49. See Appendix.
[c]Estimated using $\Delta H_f^\circ$ for the diethyl ester and the increment [$\Delta H_f^\circ$(ethyl acetate) − $\Delta H_f^\circ$(acetic acid)]; see Reference 28.
[d]Estimated using $\Delta H_f^\circ$ for acetic anhydride and the average increment [$\Delta H_f^\circ$(formyl−X) − $\Delta H_f^\circ$(acetyl−X)].
[e]References 26 and 27.

TABLE 6-3 $\Delta H°$ Values (kcal/mol)[a] for Isodesmic Bond Separation Reactions and for Homodesmotic Reactions

Isodesmic bond separation reactions are listed for formyl, acetyl, and carbonyl derivatives (ie, reactions 6-23, 6-24, and 6-25). In the homodesmotic reactions, the following carbonyl reference molecules are employed:

(i). Formaldehyde (ie, reactions 6-20, 6-21, and 6-22).
(ii). Acetaldehyde (ie, reactions 6-36, 6-38, and 6-40).
(iii). Acetone (ie, reactions 6-37, 6-39, and 6-41).

$\mathrm{H}\!\!-\!\!\overset{\displaystyle}{\underset{X}{C}}\!=\!O$

| X | iso | homo i | homo ii | homo iii |
|---|---|---|---|---|
| F | 29.9 | 22.4 | 16.4 | 13.0 |
| OH[b] | 34.2 | 21.6 | 15.6 | 12.2 |
| OMe[c] | 38.2 | 22.9 | 16.8 | 13.4 |
| Anhyd[c] | 55.5 | 30.1 | 18.1 | — |
| NH$_2$ | 31.5 | 20.4 | 14.4 | 11.0 |
| Cl | 17.6 | 15.8 | 9.8 | 6.4 |
| SH | 15.4 | 10.4 | 4.3 | 0.9 |
| Br | 15.7 | 13.3 | 7.3 | 3.9 |
| I | 12.8 | 10.6 | 4.6 | 1.2 |

$\mathrm{CH_3}\!\!-\!\!\overset{\displaystyle}{\underset{X}{C}}\!=\!O$

| X | iso | homo i | homo ii | homo iii |
|---|---|---|---|---|
| F | 45.7 | 31.5 | 25.5 | 22.1 |
| OH[b] | 44.7 | 25.5 | 19.5 | 16.1 |
| OMe[b] | 49.0 | 27.0 | 21.0 | 17.6 |
| Anhyd[c] | 76.1 | 37.6 | 25.6 | 18.8 |
| NH$_2$ | 41.0 | 23.3 | 17.3 | 13.9 |
| Cl | 28.0 | 19.6 | 13.6 | 10.2 |
| SH | 25.9 | 14.2 | 8.2 | 4.8 |
| Br | 26.2 | 17.1 | 11.1 | 7.7 |
| I | 23.4 | 14.5 | 8.5 | 5.1 |

$$\begin{matrix} X \\ & \!\!\!\!C{=}O \\ Y \end{matrix}$$

| X,Y | iso | homo i | homo ii | homo iii |
|---|---|---|---|---|
| F, F | 50.7 | 34.4 | 28.4 | 25.0 |
| OH, OH | 60.3 | 34.0 | 28.0 | 24.5 |
| OEt, OEt[d] | 80.7 | 33.2 | 27.2 | 23.8 |
| NH$_2$, NH$_2$ | 57.4 | 33.9 | 27.9 | 24.5 |
| Cl, Cl | 22.6 | 17.9 | 11.9 | 8.5 |
| F, Cl | 40.6 | 30.1 | 24.1 | 20.7 |
| Br, Br | 26.4 | 20.3 | 14.3 | 10.9 |

[a]Calculated from the $\Delta H_f^\circ$ data in Tables 6-1 and 6-2.
[b]Includes $\Delta H^\circ$ for the ester hydrolysis, +4.0 kcal/mol for methyl formate, and +4.3 kcal/mol for methyl acetate.
[c]Includes $\Delta H^\circ$ for the anhydride hydrolysis, −12.9 kcal/mol for formic anhydride and −13.2 kcal/mol for acetic anhydride.
[d]Includes $\Delta H^\circ$ for the hydrolysis of the diethyl carbonate and the isodesmic bond separation of the ethanol, +20.4 kcal/mol, that is:

$$(C_2H_5O)_2C{=}O + 2CH_4 + 2H_2O \rightarrow (HO)_2C{=}O + 2CH_3OH + 2CH_3CH_3$$

point is that there is a substantial difference, not that the values for the homodesmotic reactions are necessarily lower. A mismatch in bonding could, in principle, make either a positive or negative contribution to $\Delta H°$. It happens that the contribution serves to enhance the positive values of $\Delta H°$ in these particular cases.

Within each class of derivative the $\Delta H°$ values for both isodesmic and homodesmotic reactions are significantly larger when the atom bonded to C3=O1 is a first-row heteroatom (ie, F, O, or N), and there is a marked tendency for the values to decrease progressively in going from second-, to third-, to fourth-row heteroatoms (eg, Cl to Br to I). A similar conclusion is reached if empirical resonance energies are evaluated on the basis of bond energy assignments, although a different choice of reference molecules from which the bond energies have been derived does, of course, result in minor numerical differences. For example, using the bond energy assignments of Cox and Pilcher,[29] the $RE$ values for acetyl derivatives, given by the expression:

$$\Delta H_a° - [3E(C-H)_p^{co} + E(C_{co}-C) + E(C=O) + E(C-X)]$$

where $E(C=O)$ and $E(C-X)$ relate to aldehyde–ketone, and C4−X bonding with seven substituents, are as follows: F, 26.4; OH, 26.2; NH$_2$, 30.2; Cl, 13.0; SH, 13.6; Br, 11.4; and I, 10.0 kcal/mol, respectively.

For each substituent, especially those in which the atom directly bonded to C3=O1 is a first-row heteroatom, the $\Delta H°$ values for the homodesmotic reactions are significantly larger for the carbonyl as compared to the formyl derivatives (eg, 34.4 vs 22.4 kcal/mol for F, 34.0 vs 21.6 kcal/mol for OH, and 33.9 vs 20.4 kcal/mol for NH$_2$). Further substitution thus results in increased stabilization, in accord with the original empirical resonance energy calculations of Pauling and Sherman.[1]

The $\Delta H°$ values for the carboxylic acids, thioacids, and carboxylic acid anhydrides call for special comment in view of the important role played by thioacid derivatives and by anhydrides in metabolic reactions. In this context thioesters are often regarded as containing "activated" acyl groups, and anhydrides are regarded as having "high-energy" bonding of some kind. The $\Delta H°$ values, on the other hand (10.4 vs 21.6 and 14.2 vs 25.5 kcal/mol), indicate less stabilization in the thioacid compared to the carboxylic acid to the extent of about 11 kcal/mol, which suggests that relative to appropriate reference structures, the bonding is more ordinary. In like manner, the $\Delta H°$ values for the anhydrides are significantly less than twice the value for the corresponding acids: 30.1 compared to 2 × 21.6 and 37.6 compared to 2 × 25.5 kcal/mol. Thus the fission of the anhydride group to give two carboxylic acid groups is favored by an increase in stabilization to the extent of about 13 kcal/mol. This is a major factor responsible for the more favorable $\Delta H°$ of hydrolysis of anhydrides compared to esters. The values for formic anhydride and methyl formate calculated from the $\Delta H_f°$ data in Tables 6-1 and 6-2 are −12.9 compared to +4.0 kcal/mol, and for acetic anhydride and methyl acetate −13.2 compared to +4.3 kcal/mol.

It is very probable that the stabilization in mixed carboxylic–phosphoric acid anhydrides, and in phosphoric acid anhydrides, would likewise be significantly less than that for the component acids.

# 6. MORE EXTENSIVE MATCHING OF STRUCTURAL ELEMENTS

In the homodesmotic reactions above, formaldehyde has been used as the reference molecule to provide the C3=O1 group. Alternatives are, of course, acetaldehyde or acetone, and in fact bond energy calculations using these molecules to evaluate $E(C=O)$ are of long standing.[29,30] It is important therefore to establish whether these other reference molecules would be *inherently* a more suitable choice. Is there a matching of structural elements more extensive than that already covered by the homodesmotic reaction criteria? This is the issue, and such criteria, as given above, are in effect a matching of structural elements in reactants and products according to the connectivity of the heavy atoms, the bonding between them, and the number of hydrogen atoms bonded to each type. This kind of matching, however, like the less specific matching of formal bonding in isodesmic reactions, is piecemeal. The next step is to match the immediate structural environment about each type of heavy atom in its entirety (ie, the nearest neighbors, and after that the next nearest neighbors, etc). A similar approach has recently been used in defining a transferable valence force field for a series of molecules containing the carbonyl group.[31]

The matching of neighboring atoms around each type of heavy atom in reactants and products by both number and type of heavy atom will be designated "homoplesiotic," and, when the matching is by number alone, by "isoplesiotic." These terms are derived from the Greek adjective *plesios,* meaning near, close, neighboring. "Homoplesiotic" thus suggests sameness among neighbors (ie, matching by both number and kind), while "isoplesiotic" implies equality among neighbors (ie, in magnitude, hence by number alone).

A consequence of the homoplesiotic criteria is that matching at the nearest-neighbor level predetermines the number of H atoms bonded to each type of central atom, and so no additional specification is called for. However, although the homoplesiotic criteria may be met at the next-nearest-neighbor level, this does not necessarily entail an equal number of H atoms around each type of central atom at this level. Differences in chain branching in reactant and product molecules result in different numbers of H atoms. But for the present purpose at least, there is nothing further to be gained by imposing this additional constraint.

To demonstrate the extent of the matching that can be achieved, and thus put the matching for the reactions of acyl and carbonyl derivatives into better perspective, some further reactions of benzene will be presented first, together with the reaction energies ($\Delta E_T$), evaluated as the difference between the total molecular energies of the product and reactant species calculated ab initio.

Using all-trans, long-chain linear polyenes, a series of reactions can be set up analogous to the group separation reaction:

$$\text{benzene} + 3 \text{ ethylene} \rightarrow 3 \text{ butadiene} \quad \Delta E_T = 27.7 \qquad (6\text{-}8)$$

in which one polyene serves as the product while another, shorter by $-CH=CH-$, serves as the reactant; that is:

$$\text{benzene} + 3 \text{ butadiene} \rightarrow 3 \text{ hexatriene} \quad \Delta E_T = 26.2 \qquad (6\text{-}31)$$
$$\text{benzene} + 3 \text{ hexatriene} \rightarrow 3 \text{ octatetraene} \quad \Delta E_T = 25.7 \qquad (6\text{-}32)$$
$$\text{benzene} + 3 \text{ octatetraene} \rightarrow 3 \text{ decapentaene} \quad \Delta E_T = 25.4 \qquad (6\text{-}33)$$

Alternatively the CH groups from the benzene can be restructured as the chain CH groups in a single polyene molecule; for example:

$$\text{benzene} + \text{ethylene} \rightarrow \text{octatetraene} \quad \Delta E_T = 26.5 \qquad (6\text{-}34)$$
$$\text{benzene} + \text{butadiene} \rightarrow \text{decapentaene} \quad \Delta E_T = 25.8 \qquad (6\text{-}35)$$

The nearest-neighbor (NN) and next-nearest-neighbor (NNN) atoms around the various carbon atoms in ethylene, the linear polyenes and benzene are given in Table 6-4. Checking off the assignments for each carbon atom in the reactants and products, it appears that at the NN level of matching all the reactions 6-8 and 6-31 through 6-35 are homoplesiotic, but at the NNN level only reactions 6-31, 6-32, 6-33, and 6-35 are homoplesiotic. In reaction 6-8 the NNN atoms are not matched for the carbon atoms of either the CH or the $CH_2$ groups; in reaction 6-34 the NNN atoms are matched for four of the carbon atoms of the CH groups, but not for two, nor are they matched for the carbon atoms of the $CH_2$ groups.

Experimental $\Delta H_f^\circ$ data are not yet available beyond hexatriene to calculate $\Delta H^\circ$ values, but for purposes of comparison $\Delta E_T$ values can be employed, calculated from the total molecular energies obtained using the 6–31G basis set with full geometry optimization.[32] These values, in kilocalories per mole, are listed alongside the various reactions (eg, 6-35). It is gratifying to find that the more extensive matching of structural elements is

TABLE 6-4. Checklist of Nearest-Neighbor and Next-Nearest-Neighbor Atoms in Ethylene, Linear Polyenes $C_4$ to $C_{10}$, and Benzene

| Molecule | Reference carbon | Nearest-neighbors | Next-nearest-neighbors |
|---|---|---|---|
| Ethylene | C | C3,2H | 2H |
| Polyenes | $C_1$ | C3,2H | C3,H |
| | $C_2$ | 2C3,H | C3,3H |
| | $C_3$ | 2C3,H | 2C3,2H |
| | $C_4$ | 2C3,H | 2C3,2H |
| | $C_5$ | 2C3,H | 2C3,2H |
| Benzene | C | 2C3,H | 2C3,2H |

paralleled by trends in the $\Delta E_T$ values toward a lower limit for reactions that are homoplesiotic at both levels. But the increments are by no means negligible. The onset of matching at the NNN level in reaction 6-31 leads to a $\Delta E_T$ value 1.5 kcal/mol less than that for reaction 6-8, in which the matching is restricted to the NN level; and even with reactions 6-31 through 6-33, which are all matched at the NNN level, there is still a further progressive decrease of 0.8 kcal/mol.

Returning to the consideration of acyl and carbonyl derivatives, the homodesmotic reactions in which acetaldehyde and acetone furnish the $C3=O1$ structural elements are as follows:

*Formyl*

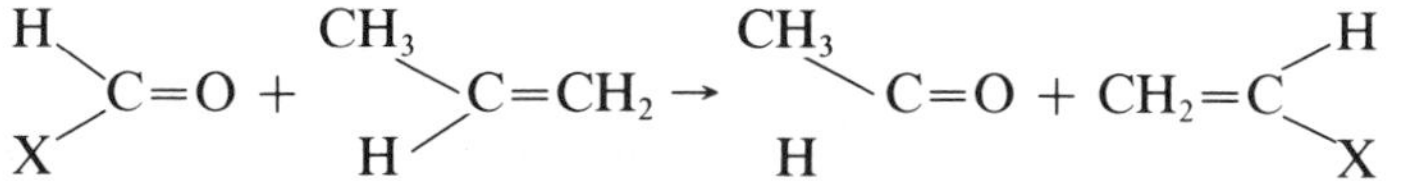

$$\underset{X}{\overset{H}{>}}C=O + \underset{H}{\overset{CH_3}{>}}C=CH_2 \rightarrow \underset{H}{\overset{CH_3}{>}}C=O + CH_2=C\underset{X}{\overset{H}{<}} \tag{6-36}$$

$$\underset{X}{\overset{H}{>}}C=O + \underset{CH_3}{\overset{CH_3}{>}}C=CH_2 \rightarrow \underset{CH_3}{\overset{CH_3}{>}}C=O + CH_2=C\underset{X}{\overset{H}{<}} \tag{6-37}$$

*Acetyl*

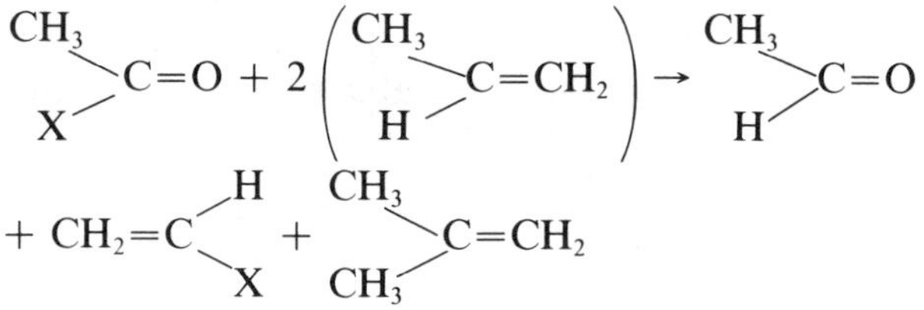

$$\underset{X}{\overset{CH_3}{>}}C=O + 2\left(\underset{H}{\overset{CH_3}{>}}C=CH_2\right) \rightarrow \underset{H}{\overset{CH_3}{>}}C=O \tag{6-38}$$

$$+ CH_2=C\underset{X}{\overset{H}{<}}\overset{CH_3}{} + \underset{CH_3}{\overset{CH_3}{>}}C=CH_2$$

$$\underset{X}{\overset{CH_3}{>}}C=O + \underset{H}{\overset{CH_3}{>}}C=CH_2 \rightarrow \underset{CH_3}{\overset{CH_3}{>}}C=O + CH_2=C\underset{X}{\overset{H}{<}} \tag{6-39}$$

*Carbonyl*

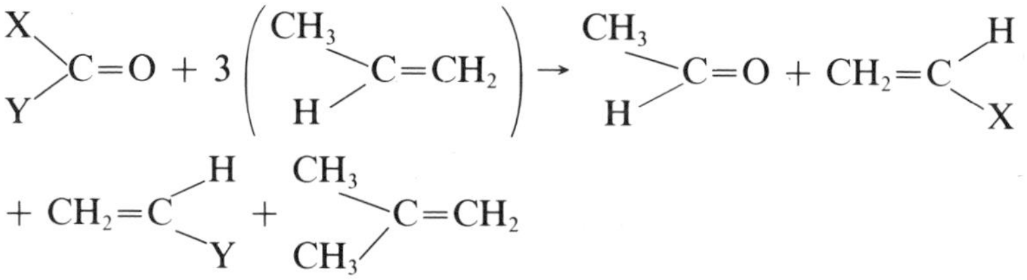

$$\underset{Y}{\overset{X}{>}}C=O + 3\left(\underset{H}{\overset{CH_3}{>}}C=CH_2\right) \rightarrow \underset{H}{\overset{CH_3}{>}}C=O + CH_2=C\underset{X}{\overset{H}{<}} \tag{6-40}$$

$$+ CH_2=C\underset{Y}{\overset{H}{<}}\overset{CH_3}{} + \underset{CH_3}{\overset{CH_3}{>}}C=CH_2$$

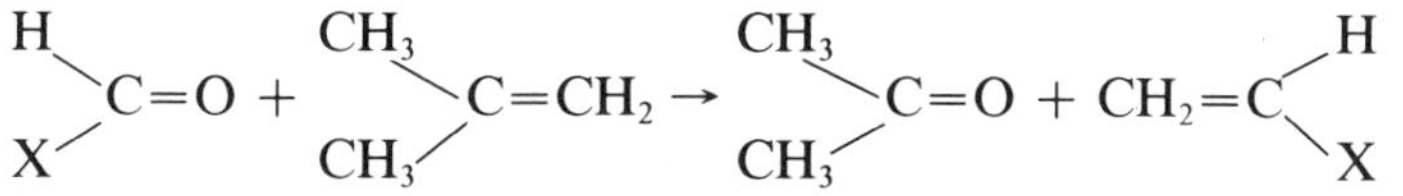

$$\underset{Y}{\overset{X}{>}}C=O + 2\left(\underset{H}{\overset{CH_3}{>}}C=CH_2\right) \rightarrow \underset{CH_3}{\overset{CH_3}{>}}C=O \tag{6-41}$$

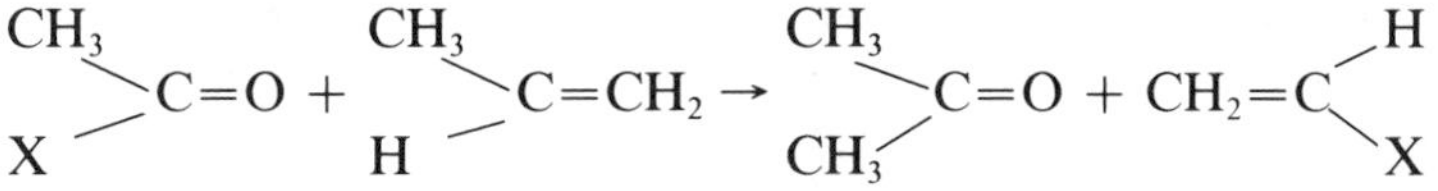

$$+ CH_2=C\underset{X}{\overset{H}{<}} + CH_2=C\underset{Y}{\overset{H}{<}}$$

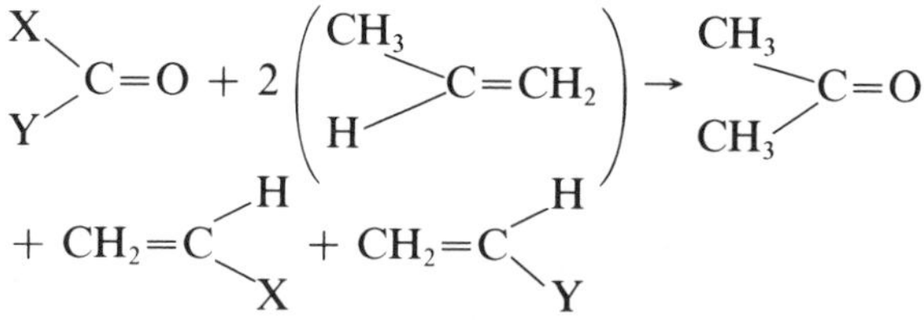

The $\Delta H°$ values for these reactions, calculated from the $\Delta H_f°$ data listed in Tables 6-1 and 6-2, are given in Table 6-3 along with those with formaldehyde as the reference molecule. The values using formaldehyde are consistently more positive than those using acetaldehyde by 6.0 kcal/mol, and those using acetone are consistently more positive by 9.4 kcal/mol (the increments for the anhydrides are twice as much because two C=O groups are involved). These increments are simply $\Delta H°$ for exchange reactions between the reference molecules, namely:

$$\underset{H}{\overset{CH_3}{>}}C=O + CH_2=CH_2 \rightarrow \underset{H}{\overset{H}{>}}C=O + CH_2=C\underset{CH_3}{\overset{H}{<}} \qquad (6\text{-}42)$$

and

$$\underset{CH_3}{\overset{CH_3}{>}}C=O + CH_2=CH_2 \rightarrow \underset{H}{\overset{H}{>}}C=O + CH_2=C\underset{CH_3}{\overset{CH_3}{<}} \qquad (6\text{-}43)$$

and are thus independent of the particular formyl, acetyl, or carbonyl derivative. Alternatively, these $\Delta H°$ values can be regarded as the empirical resonance energies of acetaldehyde and acetone with respect to formaldehyde as the reference C3=O1 structure.

To illustrate the sort of comparison that can be made between $\Delta H°$ values based on experimental $\Delta H_f°$ data and $\Delta E_T$ values obtained from total molecular energies calculated ab initio, the $E_T$ values reported by Radom and co-workers[33,34] have been used (Table 6-5). For a rigorous comparison, corrections would have to be made for heat capacity increments, 0 K $\rightarrow$ 298 K, and for zero-point energy contributions to the reaction energy, but these experimental quantities are unknown for several of the reactant and product species. These corrections would account in part for the relatively small differences between the two sets of values.

TABLE 6-5. Comparison of $\Delta H°$ Values (kcal/mol) for the Isodesmic Bond Separation and Homodesmotic Fission Reactions

The table lists reactions of formyl, acetyl, and carbonyl derivatives calculated from experimental $\Delta H_f°$ data, with $\Delta E_T$ values calculated from total molecular energies[a] obtained using the 4–31G basis set[33,34] and standard geometries.[20] Homodesmotic reactions i and iii are as described in Table 6-3.

| Substituent | $\underset{X}{\overset{H}{>}}C=O$ | | $\underset{X}{\overset{CH_3}{>}}C=O$ | | Substituents | $\underset{Y}{\overset{X}{>}}C=O$ | |
| --- | --- | --- | --- | --- | --- | --- | --- |
| | iso | homo i | iso | homo ii | | iso | homo iii |
| OH: $\Delta H°$ | 34.2 | 21.6 | 44.7 | 16.1 | OH, OH: $\Delta H°$ | 60.3 | 24.5 |
| $\Delta E_T$ | 30.1 | 21.9 | 42.1 | 16.6 | $\Delta E_T$ | 52.2 | 22.5 |
| NH$_2$: $\Delta H°$ | 31.5 | 20.4 | 41.0 | 13.9 | NH$_2$, NH$_2$: $\Delta H°$ | 57.4 | 24.5 |
| $\Delta E_T$ | 35.9 | 22.6 | 45.5 | 15.0 | $\Delta E_T$ | 64.6 | 24.7 |

[a]Conversion factor: 1 atomic unit (au) $\equiv$ 627.5095 kcal/mol.

TABLE 6-6. Checklist of Nearest-Neighbor and Next-Nearest Neighbor Atoms in Nine Molecules

Formaldehyde, formyl fluoride, propene, acetaldehyde, acetyl fluoride, isobutene, acetone, carbonyl fluoride, and vinyl fluoride are listed; for key to subscripts i, ii, and iii, see headnote to Table 6-3.

| Molecule | Reference atom | Nearest-neighbors | Next-nearest-neighbors |
|---|---|---|---|
| $H_2C{=}O$ | C3 | O1,2H | None |
| | O1 | C3 | 2H |
| $HC(F){=}O$ | C3 | O1,F,H | None |
| | O1 | C3 | F,H |
| | F | C3 | $O_1$,H |
| $CH_3C_{ii}H{=}C_iH_2$ | $C_i3$ | $C_{ii}3$,2H | C4,H |
| | $C_{ii}3$ | $C_i3$,C4,H | 5H |
| | C4 | $C_{ii}3$,3H | $C_i3$,H |
| $CH_3CHO$ | C3 | O1,C4,H | 3H |
| | O1 | C3 | C4,H |
| | C4 | C3,3H | O1,H |
| $CH_3C(F){=}O$ | C3 | O1,C4,F | 3H |
| | O1 | C3 | C4,F |
| | C4 | C3,3H | O1,F |
| | F | C3 | O1,C4 |
| $(CH_3)_2C_{ii}{=}C_iH_2$ | $C_i3$ | $C_{ii}3$,2H | 2C4 |
| | $C_{ii}3$ | $C_i3$,2C4 | 8H |
| | C4 | $C_{ii}3$,3H | $C_i3$,C4 |
| $(CH_3)_2C{=}O$ | C3 | O1,2C4 | 6H |
| | O1 | C3 | 2C4 |
| | C4 | C3,3H | O1,C4 |
| $F_2C{=}O$ | C3 | O1,2F | None |
| | O1 | C3 | 2F |
| | F | C3 | O1,F |
| $FC_{ii}H{=}C_iH_2$ | $C_i3$ | $C_{ii}3$,2H | F,H |
| | $C_{ii}3$ | $C_i3$,F,H | 2H |
| | F | $C_{ii}3$ | $C_i3$,H |

To check the extent to which matching occurs between NN and NNN atoms in the reactant and product species in the homodesmotic reactions above, formyl fuloride, acetyl fluoride, and carbonyl fluoride have been chosen as examples. It is understood that with a polyatomic group in place of the fluorine, additional NN and NNN atoms must be taken into account. The NN and NNN assignments for these molecules are given in Table 6-6, together with those for formaldehyde, propene, acetaldehyde, isobutene, acetone, and vinyl fluoride. In the formyl fluoride reaction with formaldehyde as the reference carbonyl structure (reaction 6-20), there is matching at the NN level for O1, F, and one C3, but not for the two other C3's: at the NNN level there is matching for two C3's, but not for O1, F, and the other C3. In the acetyl fluoride reaction with acetaldehyde as the reference carbonyl structure (reaction 6-38), there is matching at the NN level for O1, F, C4, and for two of the five C3's; at the NNN level there is matching for only one of the five C3's. In the carbonyl fluoride reaction with acetone as the reference carbonyl structure (reaction 5-41), there is also matching at the NN level for O1, F, and C4 and for two of the five C3's: but at the NNN level there is no matching at all. Hence none of these reactions is homoplesiotic at the NN (or the NNN) level, and the same would be true with polyatomic substituent groups.

In terms of the less exacting matching criteria, all three reactions are isoplesiotic at the NN level, but not at the NNN level.

These findings thus show that with respect to the matching of structural elements beyond that required for homodesmotic reactions, neither acetaldehyde nor acetone is inherently a better choice than formaldehyde as the reference carbonyl structure. Recognizing that this sort of justification is lacking, a compromise that could nevertheless be adopted entails taking formaldehyde as the reference molecule for the formyl and carbonyl derivatives, and acetaldehyde for acyl derivatives in general, since the derivatives are related to the reference molecules by H atom substitution as a common feature. This procedure would have a parallel in calculations using bond energies, where, for example, individual assignments are made for $E(C_{co}-H)_2$, $E(C_{co}-H)_1$ and $E(C_{co}-C)$.[29]

The presence of heteroatoms and the branched-chain structures of many of the reactant and product molecules are obviously the origin of the very limited matching in these acyl and carbonyl reactions compared to that in the benzene reactions. Even so, the presence of just one heteroatom can have a profound influence on the extent to which matching can be achieved. This can be illustrated by reactions analogous to that utilized by Liebman and Greenberg[23] to evaluate a resonance energy for $N,N$-dimethylacetamide (ie, reaction 6-18), in which the heteroatom structural elements are replaced by their carbon counterparts (ie, reactions 6-44, 6-45, and 6-46) and then comparing the classification of these reactions with that for the corresponding paraffin (ie, reaction 6-47[35]); see Table 6-7.

*N,N-Dimethylacetamide*

$$CH_3-C \overset{O}{\underset{N(CH_3)_2}{\lessgtr}} + CH_3CH_3 \rightarrow \overset{CH_3}{\underset{CH_3}{\diagdown}}C=O + (CH_3)_3N \qquad (6\text{-}18)$$

*1,1-Dimethylacetone (3-methyl-2-butanone)*

$$CH_3-C \overset{O}{\underset{CH(CH_3)_2}{\lessgtr}} + CH_3CH_3 \rightarrow \overset{CH_3}{\underset{CH_3}{\diagdown}}C=O + (CH_3)_3CH \qquad (6\text{-}44)$$

*N,N-Dimethyl-2-aminopropene*

$$CH_3-C \overset{CH_2}{\underset{N(CH_3)_2}{\lessgtr}} + CH_3CH_3 \rightarrow \overset{CH_3}{\underset{CH_3}{\diagdown}}C=CH_2 + (CH_3)_3N \qquad (6\text{-}45)$$

**TABLE 6-7.** Reaction Classifications

| | Reaction | Isodesmic | Homodesmotic | NN level | | NNN level | |
|---|---|---|---|---|---|---|---|
| | | | | Isoplesiotic | Homoplesiotic | Isoplesiotic | Homoplesiotic |
| 6-18 | *N,N*-Dimethylacetamide | Yes | No | Yes | No | No | No |
| 6-44 | 1,1-Dimethylacetone | Yes | Yes | Yes | No | No | No |
| 6-45 | *N,N*-Dimethyl-2-aminopropene | Yes | No | Yes | No | No | No |
| 6-46 | 2,3-Dimethyl-1-butene | Yes | Yes | Yes | No | No | No |
| 6-47 | 2,3-Dimethylbutane | Yes | Yes | Yes | Yes | Yes | Yes |

Matching criteria

*2,3-Dimethyl-1-butene*

$$CH_3-C\underset{CH(CH_3)_2}{\overset{CH_2}{\diagup}} + CH_3CH_3 \rightarrow \underset{CH_3}{\overset{CH_3}{\diagup}}C=CH_2 + (CH_3)_3CH \qquad (6\text{-}46)$$

*2,3-Dimethylbutane*

$$(CH_3)_2CH-CH(CH_3)_2 + CH_3CH_3 \rightarrow 2(CH_3)_3CH \qquad (6\text{-}47)$$

First, it may be noted in passing that the presence of the N atom in the main skeletal chain has a very marked effect: reactions 6-18 and 6-45 are only isodesmic, whereas reactions 6-44, 6-46 and 6-47 are homodesmotic. Second, the less restrictive nature of the isoplesiotic criteria is exemplified by the finding that all five reactions come into this category (like the acyl and carbonyl reactions discussed above). Third, the far more restrictive nature of the homoplesiotic criteria is exemplified by the finding that only the last reaction comes into this category. Moreover, matching at the NNN level also occurs only in this reaction. The onset of this extensive matching is attributable to all the heavy atoms being identical (ie, C4), just as in the benzene reactions, where the matching is comparable, all the heavy atoms are C3. But even though the heavy atoms are matched at the NNN level, the H atoms are not matched in the 2,3-dimethylbutane reaction (6-47), nor are they matched in the benzene reactions 6-8 and 6-34—only in reactions 6-31, 6-32, 6-33, and 6-35. Chain branching is the reason for the mismatch in the paraffin reaction, and in the benzene reactions matching occurs only when the polyene chains in the other reactant and product molecules reach a certain length.

# 7. SUMMARY

Empirical resonance energies calculated using bond energies based on thermochemical data are equivalent to evaluating $\Delta H°$ for a particular reaction involving the molecule in question and the self-same set of reference molecules from which the bond energies have been obtained. The reactant and product molecules in the reactions that underlie the original calculations of the resonance energies for benzene and for typical acyl and carbonyl derivatives, although matched in terms of formal single and double bonds, have been shown to be mismatched when the more specific bonding criteria for homodesmotic reactions are taken into account.

An appropriate choice of reference molecules, which results in the reactions meeting these more demanding criteria, does not, however, alter the conclusion that the resonance energies for typical acyl and carbonyl derivatives are of the same order of magnitude as that for benzene.

In setting up the homodesmotic reactions for the acyl and carbonyl derivatives, formaldehyde, acetaldehyde or acetone can be used to provide the

C3=O1 group, so the question arises as to whether one of these reference molecules is inherently a better choice. Again, the issue is whether there is a matching of structural elements more extensive than that already covered by the homodesmotic reaction criteria.

This question has been answered by considering the matching, not piecemeal in terms of individual bonds, but in terms of nearest-neighbor heavy atoms and their next-nearest-neighbor heavy atoms around each of the heavy atoms in the reactant and product molecules. Matching by number alone has been designated "isoplesiotic"; matching by both number and type, "homoplesiotic."

On this basis it appears that no distinction can be made between these carbonyl reference molecules—the matching is rudimentary, only isoplesiotic at the nearest-neighbor level in all three cases. If a choice needs to be made, it could be formaldehyde on the grounds that this is the simplest structure containing the necessary C3=O1 structural element. On the other hand, it could be formaldehyde for the formly and carbonyl derivatives, and acetaldehyde for acyl derivatives in general, since the derivatives are related to the reference molecules by H atom substitution as a common feature.

A comparison of the matching in benzene reactions with that in a fission reaction of $N,N$-dimethylacetamide, in the analogous reactions in which the heteroatom structural elements are replaced by their carbon counterparts, and in the corresponding reaction for the paraffin 2,3-dimethylbutane, leads to the conclusion that extensive matching of heavy atoms at the nearest-neighbor level, and particularly the next-nearest-neighbor level, is quite exceptional.

## APPENDIX

In calculating $\Delta H°$ for the various reactions experimental values of $\Delta H_f°$ for the formyl, acetyl, and carbonyl derivatives, for benzene, and for the reference molecules have been taken from a computer-analyzed tabulation of thermochemical data[18]; see Tables 6-1 and 6-2. References for a few additional experimental values are given in footnotes. The values are quoted in kilojoules per mole (kJ/mol), since this is the unit adopted in the tabulation. When needed, the conversion factor 4.184 kJ $\equiv$ 1 kcal has been employed.

However, to treat as many diverse structures as possible, some $\Delta H_f°$ values have had to be estimated.

In the case of the vinyl reference molecules, $CH_2=CHX$, where X is OH, $OCH_3$, $NH_2$, SH, and I, values have been calculated from the empirical equation:

$$\Delta H_f°(\text{vinyl}-X) = -30.55 + 1.0060 \, \Delta H_f°(\text{phenyl}-X) \quad (6\text{-}48)$$

based on experimental data for 12 pairs of derivatives with X = H, $CH_3$, $C_2H_5$, F, Cl, Br, CN, HC≡C, CHO, $CH_2OH$, $OC_2H_5$, and $OCOCH_3$, $\Delta H_f°$ for acrylaldehyde being obtained from that for crotonaldehyde with the increment for methyl group substitution derived from data for *trans*-1,3-pentadiene and *trans*-1,3-butadiene.[18] Equation 6-48 has a correlation coefficient of 0.9994. The mean difference $[\Delta H_f^\circ(\text{vinyl}-X) - \Delta H_f^\circ(\text{phenyl}-X)]$ is $-30.4 \pm 5.8$ kJ/mol, so the uncertainty in the values estimated for the vinyl derivatives above is probably about 6 kJ/mol.

In the case of the formyl derivatives, $HC(X)=O$, where X is Cl, Br, I, and SH, values have been calculated from the empirical equation:

$$\Delta H_f^\circ(\text{formyl}-X) = 54.16 + 1.0046\ \Delta H_f^\circ(\text{acetyl}-X) \qquad (6\text{-}49)$$

based on experimental data for seven pairs of derivatives with X = H, $CH_3$, $C_2H_5$, F, OH, $OCH_3$, and $NH_2$. This equation (6-49) has a correlation coefficient of 0.9998. The mean difference $[\Delta H_f^\circ(\text{formyl}-X) - \Delta H_f^\circ(\text{acetyl}-X]$ is $52.7 \pm 2.7$ kJ/mol, so the uncertainty in the values estimated for the formyl derivatives above is probably about 3 kJ/mol.

The procedures used to estimate $\Delta H_f^\circ$ for formic anhydride and for carbonic acid are outlined in footnotes to Table 6-2.

# REFERENCES

1. Pauling, L.; Sherman, J. *J. Chem. Phys.* **1933**, *1*, 606.
2. The values listed by Linus Pauling in "The Nature of the Chemical Bond" (Cornell University Press: Ithaca, N.Y., 1st ed., 1939; 2nd ed., 1940; 3rd ed., 1960) are the same for the acyl and carbonyl derivatives; that for benzene is given as 39 kcal/mol in the first two editions.
3. George, P. *Chem Rev.* **1975**, *75*, 85.
4. George, P.; Trachtman, M.; Bock, C.W.; Brett, A.M. *Theor. Chim. Acta* **1975**, *38*, 121.
5. George, P.; Trachtman, M.; Bock, C.W.; Brett, A.M. *J. Chem. Soc. Perkin Trans. 2* **1976**, 1222.
6. George, P.; Trachtman, M.; Bock, C.W.; Brett, A.M. *Tetrahedron* **1976**, *32*, 1357.
7. George, P.; Bock, C.W.; Trachtman, M. "The Evaluation of Stabilization Energies (Empirical Resonance Energies) for Benzene, Porphine and [18]Annulene from Thermochemical Data and from Ab Initio Calculations". In "The Biological Chemistry of Iron", Dunford, B.; Dolphin, D.; Raymond, K.; and Sieker, L., Eds.; Reidel: Boston, 1982, p. 273.
8. George, P.; Bock, C.W.; Trachtman, M. *J. Chem. Educ.* **1984**, *61*, 225.
9. Lo, D.H.; Whitehead, M.A. *Can. J. Chem.* **1968**, *46*, 2027, 2041.
10. Dewar, M.J.S.; de Llano, C. *J. Am. Chem. Soc.* **1969**, *91*, 789.
11. Hess, B.A., Jr.; Schaad, L.J. *J. Am. Chem. Soc.* **1971**, *93*, 2413.
12. Aihara, J.-I. *J. Am. Chem. Soc.* **1976**, *98*, 2750.
13. Gutman, I.; Milun, M.; Trinajstic, N. *J. Am. Chem. Soc.* **1977**, *99*, 1692.
14. Bauld, N.L.; Welsher, T.L.; Cessac, J.; Holloway, R.L. *J. Am. Chem. Soc.* **1978**, *100*, 6920.
15. Haddon, R.C. *J. Am. Chem. Soc.* **1979**, *101*, 1722.
16. Kollmar, H. *J. Am. Chem. Soc.* **1979**, *101*, 4832.
17. Herndon, W.C. *Isr. J. Chem.* **1980**, *20*, 270.
18. Pedley, J.B.; Rylance, J. "Sussex-N.P.L. Computer-Analysed Thermochemical Data: Organic and Organometallic Compounds". University of Sussex, Brighton, U.K., 1977.
19. Hehre, W.J.; Ditchfield, R.; Radom, L.; Pople, J.A. *J. Am. Chem. Soc.* **1970**, *92*, 4796.
20. Pople, J.A.; Gordon, M. *J. Am. Chem. Soc.* **1967**, *89*, 4253.
21. Mulliken, R.S. *Tetrahedron* **1959**, *6*, 68. In the absence of $\pi_x$-electron delocalization in the 90° conformer, $\pi_x$-electron nonbonded repulsion across the C—C bond and $\pi_y$-electron repulsion must be taken into account along with hyperconjugation effects. There is a further complication, as yet unresolved, in that the vinyl groups may no longer be precisely planar.

22. George, P.; Trachtman, M.; Bock, C.W. *J. Chem. Soc. Perkin Trans. 2* **1977,** 1036.
23. Liebman, J.F.; Greenberg, A. *Biophys. Chem.* **1974,** *1,* 222.
24. Stull, D.R.; Westruum, E.F., Jr.; Sinke, G.C. "The Chemical Thermodynamics of Organic Compounds". Wiley: New York, 1969.
25. Stull, D.R., Ed. "JANAF Thermochemical Tables", 2nd ed., NSRDS-NBS 37. Government Printing Office, Washington, D.C., 1971.
26. Wagman, D.D.; Evans, W.H.; Parker, V.B.; Halow, I.; Bailey, W.M.; Schum, R.H. "Selected Values of Chemical Thermodynamic Properties", National Bureau of Standards Technical Note No. 270-3. Government Printing Office, Washington, D.C., 1968.
27. Bauder, A.; Günthard, H.H. *Helv. Chim. Acta* **1958,** *41,* 670.
28. George, P.; Bock, C.W.; Trachtman, M. *J. Comput. Chem.* **1982,** *3,* 283.
29. Cox, J.D.; Pilcher, G. "Thermochemistry of Organic and Organometallic Compounds". Academic Press: New York, 1970.
30. Cottrell, T.L. "The Strengths of Chemical Bonds", 2nd ed.; Butterworths: London, 1958.
31. Hollenstein, H.; Günthard, H.H. *J. Mol. Spectrosc.* **1980,** *84,* 457.
32. Bock, C.W.; George, P.; Trachtman, M. *Theochem.* **1984,** *18,* 1.
33. Radom, L.; Hehre, W.J.; Pople, J.A. *J. Am. Chem. Soc.* **1971,** *93,* 289.
34. Radom, L.; Lathan, W.A.; Hehre, W.J.; Pople, J.A.P. *Aust. J. Chem.* **1972,** *25,* 1601.
35. This is the type of homodesmotic reaction that can be employed to assess the destabilization in branched-chain paraffins due to partial eclipsing of methyl (or larger) groups on adjacent C atoms. In this particular instance $\Delta H° = -1.3$ kcal/mol.[18] With more heavily substituted paraffins (eg, 2,2,3,3-tetramethylbutane), the corresponding reaction is:

$$(CH_3)_3C-C(CH_3)_3 + CH_3CH_3 \rightarrow 2(CH_3)_4C$$

and a larger negative value is obtained for $\Delta H°$: $-6.0$ kcal/mol, indicating considerably greater destabilization.

# CHAPTER 7

# Some Conformational Aspects of Linear Peptides Containing Proline

## Marc J. O. Anteunis and Jef J. M. Sleeckx

State University of Ghent, Ghent, Belgium

## CONTENTS

> *Statements in angular brackets such as this, « », are intended to supplement the text by offering details of arguments, additional examples, and extensions of the ideas presented. They may be omitted in an initial reading without loss of continuity.*

Proline (Pro) and its congeners are, among the 20 naturally occurring amino acids, of unique interest in that they have some of the most profound influences on the conformations of peptide segments. It is our aim to briefly review insights that have been obtained from both the theoretical and experimental fields, related to the peptide structuralization capabilities that Pro possesses. This problem is closely allied with the conformational feature of the five-membered pyrrolidine ring itself, but emphasis also must be placed on the spatial aspects of the peptide backbone. The reader who is not well versed in protein conformational terminology and coding should review Section 4 before beginning this chapter.

# 1. "PRO-SINGLE"

The isolated proline system ("Pro-single") is characterized by a certain flexibility due to the pseudorotational freedom of the pyrrolidine ring; nevertheless, the allowed $\phi$, $\psi$ values remain confined within certain limits. A preference for $\gamma^-/\gamma^+$ of the pyrrolidine ring conformations is widely accepted,[1-12,27] being characterized by $\phi$ values of $-60°/-75°$ for an L residue. However, in these conformations a relatively important ring angle variation along $\chi^5$ may be introduced without an appreciable increase in energy,[13,14] an aspect that usually has not been considered in theoretical calculation approaches. Therefore a much wider $\phi$ range should be allowed for (eg, $-40° < \phi < -90°$).

A cooperative effect correlating the pyrrolidine ring conformation with concomitant $\phi$ values (and therefore also with the cis–trans isomeric state)

≪Concomitantly with a displacement of the pseudorotational phase, the relative populations of respectively limiting $\gamma^-$ (or $N$-conformations) and $\gamma^+$ (or $S$-conformations) are typically affected in solution (see References 27–29, 97, and Section 3 for the solid state).≫

has been proposed,[3,8] although the conformational "transmission" effect is very small because of the ease of deformations along $\chi^5$ and/or deviations from planarity of the peptide bond (or $N$-hybridization state). Nevertheless, for cis-peptide bonds and out-of-register $\phi$ values (eg, $< -80°$), a change of the ring conformation from $\gamma^-/\gamma^+$ toward higher puckering in the $N-C^\alpha-C^\beta$ region

≪For a somewhat different behavior of Ac-Hyp-OH, however, see Reference 30.≫

seems to be typical.[3,8,26] Additional comments about solid state behavior are provided in Section 3.

A feature common to all imino acid residues is the possibility for cis-amide structures to occur, giving rise to the existence of cis- ($\omega = 0° \pm 15$) and trans- ($\omega = 180° \pm 15$) amide isomers of comparable energy content; the cis contents for Ac-Pro-NHMe have been calculated as 3.8 and 30% (References 16 and 3, respectively) and as 13.8% for the hydrated model.[16] In water, the amount of cis content has been found to be strongly dependent on the concentration[17] (eg, from $\leq$ 7% for 40 mM to 30% for 3 M solutions), presumably because the trans isomer, in contrast to the cis isomer,[18] occurs in a $\gamma$-turn conformation (Figure 7.1) that can't aggregate.[18] Finally the cis

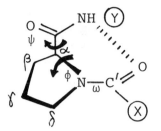

**Figure 7-1.** Conformational parameters of the proline residue in peptides.

$\rightleftharpoons$ trans isomerization is a slow process on the NMR time scale, $\Delta H^{\ddagger}_{cis-trans}$ amounting to 17 kcal/mol.

It is important to realize that deviation from planarity of the amide bond is relatively easy,[15,19–23] resulting in deviations of $\omega$ as large as 15°

≪In strained cyclic Pro-peptides and in the solid state, deviations of $\omega$ as high as 20° have been observed (see, eg, Reference 24).≫

at the cost of only 1 kcal/mol and in deviations of $\theta_N$ (for $\Delta\omega = 5$–15°)

≪$\theta_N$ is the torsion angle between $C'NC^{\alpha}$ and $C'NC^{\delta}$ and is a measure for the hybridization state at the amide nitrogen; see also page 222.≫

as large as 20–30° at the cost of the same increase in energy. This aspect is amply discussed in Section 3 in relation to data obtained from solid state parameters.

For the common amino acid residues, the parameter $\psi$, which describes the carboxamide rotational state, usually is not heavily constrained to certain values, covering a region from $-100°$ to $+180°$.[15] However, certain values are prohibited (eg, $\psi = 0°$ when in combination with $\phi < -140°$ and $\phi \simeq 0°$ and the span from $-100°$ to $-180°$ especially for concomitant $\phi$ values of $> -60°$; see also Figures 7.3 and 7.4 in Section 2). From theoretical calculations, $\psi$ for Pro-single seems to be confined to the following (see Reference 16) in the "hydrated" model [in apolar solvents, the trans′ isomer is always relatively more stable than the cis′ isomer and covers moreover a wider span in $\psi$ (eg, from $+70°$ to $+170°$)]:

$\psi = -50°$ (cis′) and $+170°$ (trans′), when preceded by a *cis*-peptide bond (cis isomer)

$\psi = -50°$ (cis′) and $+110°/+170°$ (trans′), in the trans isomer

or alternatively in the "naked" model (see Reference 3):

$\psi = -50°$ (cis′) and $+160°$ (trans′), for the cis isomer

$\psi = -40°$ (cis′) and $+60°$ and $+160°$ (trans′), for the trans isomer

From calculations, the order of stability would be: trans–trans' > cis–trans' ~ trans–cis' > cis–cis' (according to Reference 16 for the "hydrated" model) or cis–trans' ⌣ trans–cis' ≳ cis-cis'> trans-trans' (according to Reference 3). Hydration of the cis–trans', through better solvent access of the carboxamide function, would be a rather biasing aspect.[16]

Characteristically, the barrier for cis' ⇌ trans' isomerization would be high, according to Empirical Conformational Energy Program for Peptides (ECEPP) calculations for the hydrated model[16] ($\simeq$12 kcal/mol) but moderate (cis–cis' ⇌ cis–trans'; 7 kcal/mol) to low (trans–cis' ⇌ trans–trans'; 2.5 kcal/mol) for the vacuum model[3] (cf Reference 114).

The X-ray data[25a] for Ac-Pro-NH$_2$ are: $\phi = -80°$, $\psi = -14°$, and $\omega = -176°$; the pyrrolidine ring is close to a $\gamma^+$–$\beta^-$ ($^\gamma_\beta$T)-conformation, and $\theta_N$ is 6°. The analogous parameters of Ac-Pro-NHMe are almost identical.[25b] We discuss solid state data of Pro in Section 3.

# 2. "PRO-MIDDLE" IN (LINEAR) PEPTIDES

## A. General Considerations

In relatively good agreement with perturbative configuration using localized orbitals (PCILO) calculations,[15] frequently encountered conformations of Pro residues in globular proteins have approximate values of $-80° \leq \phi \leq -35°$, $-80° < \psi < 0°$ (cis') and $110° < \psi < 185°$ (trans'). Values of $\phi$(Pro) = 0° and $-110°$ and $\psi$(Pro) < $-120°$ have been observed only exceptionally. Exaggerated negative $\phi$ values result from a change of "regular" pyrrolidine $\gamma^-$/$\gamma^+$ ring conformations, to the $\beta^-$/$\alpha^+$ region, which may be detected by NMR spectroscopic means.[26] As will be discussed later (cf Table 7-2 in Section 2), such forms may be expected for L-Pro involved in certain typical peptide segments, restricted to bend positions of type II*[($i + 2$)], type V*[($i + 2$)], type IV[($i + 1$) and, exceptionally, ($i + 2$)], type VI[($i + 2$) (involved in cis-peptide bonding with the preceding residue)], or in a $\gamma$ corner.

≪A $\beta$ turn is frequently observed in cyclic peptides because it initiates pleated sheets that are especially favorable for ($4n + 2$) cyclic peptides[110,111] with Pro therefore often in the ($i + 2$)-position of a II* turn.[92] Also a $\gamma$ corner with Pro in the ($i + 1$)-position is a feasible structure for cyclopentapeptides.[112]≫

It is clear that such deformed rings must be compensated by stabilizing contributions other than intraresidual in origin. In this context it is interesting to speculate on the effect in substituting higher homologues such as pipecolic acid and congeners, which possess intrinsically highly negative $\phi$ values, for Pro.

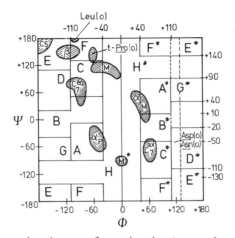

**Figure 7-2.** The Ramachandran conformational $\phi,\psi$ map showing the centrosymmetric distribution of regions A–H and antimeric regions A*–H* with the locations of low-energy minima for typical secondary structures. The locations of global (bold) and local energy minima for the 20 amino acid residues of proteins are as follows: $C_5$: **Phe**(0), **Lys**(0), **Arg**(0), **Cys**(0), **Val**(0), **Leu**(0), **Tyr**(0), **Gln**(0), Ala(1), Trp(1), Ser(1), Asp(1), Gly(2), His(2), Thr(2); $\beta$: Cys(0), Lys(0), Arg(0),Val(1), Ala(2); $C_7^{eq}$: **Gly**(0), **Ala**(0), **Val**(0), **Leu**(0), **Ile**(0), **Phe**(0), **His**(0), **Ser**(0), **Thr**(0), **Gln**(0), **Glu**(0), **Asn**(0), **Arg**(0), **Lys**(0), **Cys**(0), **Met**(0), Tyr(0), Trp(0); M: **Gln**(0), Leu(0), Ile(0), Phe(0), Trp(0), Tyr(0), Thr(0), Met(0), Val(1), Arg(1), Gly(2), Ala(2), His(2); $\alpha_p$: **t-Pro**(0), Phe(0), Tyr(0), His(0), Trp(0), Cys(0), Gln(1), Ser(2); M: Gly(2), Ala(2); $\alpha_M$: **Gln**(0), **Asn**(0), Asp(1), Ile(1), Phe(2), Glu(2), Ser(2); $C_7^{ax}$: **Gly**(0), **Tyr**(0), **Trp**(0), Ala(0), Phe(0), Ile(1), Lys(1), Arg(1), Val(2), Leu(2), His(2), Cys(2). Values in parentheses report the energy (kcal/mol) of the local minimum compared with the global minimum (PCILO calculations; according to Reference 15).

A myriad of a number of 10 to 100 low-energy minima (stabilized backbone conformations) is usually found for Ac-X-Pro-NHMe and Ac-Pro-X-NHMe. Dipeptides represent therefore a blend of low-energy structures in the $\phi$, $\psi$, $\omega$ conformational space and the normalized statistical weight of a particular architecture will seldom exceed 0.3.[1] Using the stereoalphabet[31] (conformational letter code) for the designation of specific conformations in the $\phi$, $\psi$ dimension (Figure 7-2), one can define the a priori allowed regions for Pro, being characterized by its constrained $\phi$, $\psi$ values. It then turns out that only regions A, C, and F can be occupied by L-Pro residues. For *cis*-peptide and out-of-register Pro residues, zones D and G may sometimes be considered.

It has become accepted practice to classify secondary structures of proteins into four classes: $\alpha$ helices, $\beta$-pleated sheets (both parallel and antiparallel), chain reversals (hairpins, U turns, $\beta$ turns, bends, coils, kinks), and random fragments. The first three structures mentioned bear specific features,[33] which for present purposes are the $(\phi,\psi)$ values which are characteristic for these classes. Table 7-1 lists some $\phi,\psi,\omega$ sets for representative structural units, while Table 7-2 gives the classification of the standard bends

**TABLE 7-1.** Approximate Torsion Angles $\phi$, $\psi$, and $\omega$ for Some Regular Structures[a]

| Structure | Torsion angles (°) | | | |
| | $\phi$ | $\psi$ | $\omega$ | Stereoalphabet[b] |
|---|---|---|---|---|
| 3.6$\alpha_P$-Helix ($\alpha$-poly-L-Ala) | −57 | −47 | +180 | · · ·−A−A−· · · |
| 3$_{10}\alpha$-helix | −50 | −26 | +180 | · · ·−$\overline{\text{A}}$−$\overline{\text{A}}$−· · · |
| Poly-*trans*-proline (PP II) | −77 | +146 | +180 | · · ·−F−F−· · · |
| Poly-*cis*-proline (PP I)[c] | −83 | +158 | 0 | · · ·−F−F·· · · |
| Polyglycine II | −80 | +150 | +180 | · · ·−F−F−· · · |
| Triple helix (collagen) | −51 | +153 | +180 | F |
| | −76 | +127 | | F |
| | −45 | +148 | | $\overline{\text{F}}_f$ |
| Parallel pleated sheet | −119 | +113 | +180 | · · ·−E$_e$−E$_e$−· · · |
| Antiparallel pleated sheet | −139 | +135 | −178 | · · ·−E−E−· · · |
| Fully extended | −180 | +180 | +180 | · · ·−$^e$E−$^e$E−· · · |

[a]According to Reference 52.
[b]We denote the occupied region in the $\phi,\psi$ conformational space by a capital letter code as introduced by Reference 31 (cf Figure 7-2). In addition, a lowercase superscript is used to indicate subregions at, respectively, the top right and top left (eg, A$^a$ and $^a$A) and a subscript to indicate subregions at, respectively, the bottom right and bottom left (eg, A$_a$ and $_a$A) of these regions. To indicate the center of the region, no scripts are used. An upper-middle and a bottom-middle region are represented, respectively, by a bar above and below the letter (eg, $\overline{\text{A}}$ and A).
[c]Calculated from Reference 53.

(References 1, 33; cf 52, 53) together with the torsion angles as estimated by Venkatalacham.[34]

It should be noticed that $\beta$ or $\alpha$ segments are defined by a sequence of three to four residues having *identical* ($\phi,\psi$) pairs.[35,36] An $\alpha$ segment is further characterized by a distance between the C$^\alpha$ atoms of the first and fourth residue ($R_{i,i+3}$) < 6 Å (eg, poly-Ala, 5.4 Å) and a $\beta$ fragment has a minimum of *two* successive $R_{i,i+3}$ of more than 7 Å (eg, fully extended 11 Å). A turn[41,108,109] is described in terms of four residues $i, \ldots, i + 3$, whereby *only* the *two* central residues $i + 1$ and $i + 2$ determine the bend conformation (Table 7-2) resulting in $R_{i,i+3}$ < 7 Å. It is furthermore a segment isolated by surrounding $\alpha$ or $\beta$ structural fragments. Thus bend type III has ($\phi,\psi$)$_{2,3}$ sets equivalent to those of the residues in a 3$_{10}$ helix, whereas ($\phi,\psi$)$_1$ and ($\phi,\psi$)$_4$ differ from these. Each bend type can therefore be characterized by a combination of two conformational coding letters. One exception is the $\gamma$ turn (an "isolated" $\gamma$ helix)[37,38] for which only one residue with its $\phi,\psi$ set must be considered (see also footnotes $g$ and $h$ of Table 7-2). This type of kink is especially important in cyclic oligopeptides.[39] For an extensive review of the $\gamma$ turn, see Reference 9.

An important point for further discussion relates to the propensity a certain residue may have to fit in one of these secondary structures. That is, allowance is made for a deviation as large as 50° in one (but only one) of the typical ($\phi,\psi$) value sets that define the bends in Table 7-2. To pinpoint the resulting "imperfection," one speaks then about "distorted" secondary

TABLE 7-2. Standard Bend Types[a] and Calculated[b] Torsion Angles

| Bend type[c] | Torsion angles (°) | | | | | | | Stereoalphabet[d] |
|---|---|---|---|---|---|---|---|---|
| | $\omega_{i+1}$ | $\phi_{i+1}$ | $\psi_{i+1}$ | $\omega_{i+2}$ | $\phi_{i+2}$ | $\psi_{i+2}$ | $\omega_{i+3}$ | |
| I | 180 | −60 | −30 | +180 | −90 | 0 | 180 | A B |
| I* | 180 | +60 | +30 | +180 | +90 | 0 | 180 | A*B* |
| II | 180 | −60 | +120 | +180 | +80 | 0 | 180 | C̄ B* |
| II* | 180 | +60 | −120 | +180 | −80 | 0 | 180 | C̄*B |
| III | 180 | −60 | −30 | +180 | −60 | −30 | 180 | A A |
| III* | 180 | +60 | +30 | +180 | +60 | +30 | 180 | A*A* |
| IV | A bend of type I–III* with two or more torsion angles differing by at least 40°, but with $R_{i,i+3} < 7$Å ("dump"). | | | | | | | $[D^{d}E_{e}]^{e}$ |
| V | 180 | −80 | +80 | +180 | +80 | −80 | 180 | C C* |
| V*[f] | 180 | +80 | −80 | +180 | −80 | +80 | 180 | C*C |
| VI[e] | 180 | — | — | 0 | — | — | 180 | |
| VII | A chain reversal produced by either $|\psi_{i+1}| \geq 140°$ and $|\psi_{i+2}| < 60°$, or $|\psi_{i+1}| < 60°$ and $|\phi_{i+2}| \geq 140°$. | | | | | | | — — |
| γ Turn[g] | 180 | +80 | −80 | 180 | | | | C* |
| γ* Turn[h] | 180 | −80 | +80 | 180 | | | | C |

[a]Reference 33.
[b]Reference 41.
[c]The types coded with an asterisk (I*–V*) are the backbone enantiomers of the corresponding types (roman numerals). Therefore they should especially be considered when dealing with D-residues and/or Gly.
[d]Enantiomeric structures and regions are indicated with asterisks; see footnote b of Table 7-1 for additional description of code.
[e]No $\phi,\psi$ values have been considered originally for this type.[1,33] A hairpin that matches this definition has been recognized as a "reverse open chain"[43,54] with local conformations of (−120,+90), 0, (−120,+120).
[f]Not yet experimentally observed.
[g]References 37 and 38. Also called a γ coil.[51] This is equivalent to a $C_{7}^{x}$ conformation involving an L-residue and is considered to be generally less likely,[33,44] unless for Gly, L-Tyr, L-Trp, and D-residues (see also Reference 15c). Experimental values would be $(\phi,\psi) = (70°/85°, -60°/-70°)$.
[h]This is equivalent to the reversed γ turn and to $C_{7}^{eq}$ involving an L-residue; see note g.

structures. Also the typical values reported in, for example, Table 7-2, are theoretical predictions,[34] and there are indications[40] that these and other proposals[41] may deviate somewhat from the observed mean value.

It is generally accepted, both from theoretical models[33,42] and experimental evidence[35,43,46–51] that each amino acid possesses a predisposition to occur in each of the established types of secondary structure. These spatial preferences are highly dependent on the individual amino acid residue, that is, on short-range, intraresidue interactions. Medium-range and long-range interresidue contributions must nevertheless be considered, and they are major contributors to the proclivity to specific tertiary structuralization.[15a] Medium-range interresidue interactions have major consequences in Pro-(di)peptide sequences,[1] as discussed in the next paragraphs.

Tertiary structuralization that does not depend on intraresidue effects results, for example, from the simple fact that typical hydrophobic regions tend to become buried in the inner cavity of a globular protein,[32,45,55,56] and conversely, polar amino acids occur at the surface. Thus also (Ala$_j$–Ala$_{j+3}$) or (Ala$_j$–Ala$_{j+1}$) favorable lipophilic interactions[57] and clustering become important (1.4 kcal/mol) in water.[51,58−61] Among the earliest recognized interresidue interactions are those involving H-bond bridging mechanisms, such as exist between side chains of, for example, Ser, Thr, Asp, and Asn with the backbone peptide at the $NH_{j−3}$ position.[62,63] These result in helix-disrupting, but bend-promoting effects.[64] Thus in Thr-Pro-Arg-Lys an out-of-register chain reversal was calculated to be 1.1 kcal/mol lower in energy[65] as a result of a H-bond bridging between Arg(NH) to Thr(C=O). Frequently the calculated energy minima in tri- and higher peptides are not simply com-binations of the dipeptide backbone energy minima, and new minima are formed.[1,15,45,61,65−67] Attractive and repulsive side chain–side chain interac-tions between residues that can approach each other in space to give con-tacts within 5–8 Å have been revealed by statistical treatments of X-rayed proteins.[68] Specific atomic (functional) interactions and charge–charge inter-actions are revealed,[69,71,75] and one interesting stabilization that has been proposed rests on a sulfur-$\pi$ attraction.[71−73] Finally, the vectorial summation of individual peptidic dipole moments may play an especially important role in the interplay of the peptide with its surroundings[16] (cf Reference 74). It is likely that the mechanism of conservation [isofunctional and isosteric replacements during the evolutionary selection of amino acid residues belonging to important proteins such as cytochromes and globulins (see Ref-erence 75)], also has medium- and long-range cooperative interactions at its origin. In other words, as a kind of feedback process in the relationship between sequence and tertiary structure, it can be understood that certain amino acids not only prefer specific emplacement in bends and $\alpha$ and $\beta$ regions, but also show some preference for occurring at adjacent regions (eg, especially from the $i − 4$ to $i + 8$ regions).[50,102]

## B. Statistical Emplacements of Proline in Linear Peptides

In the context of interresidue interactions with Pro as a participant, the fol-lowing statements may be of interest:

1. Proline is an ambivalent residue, being rather hydrophilic[68] but uncharged. It is therefore not surprising to find Pro often at the boundary of a native protein[32,42] characteristically involved in chain reversals. Also Pro is frequently surrounded by polar residues (especially Gln, Ser, Cys) or by itself ("autoassociation"[68]).
2. Proline is an excellent bend promoter,[42,76] and the probability that it will be located at an $(i + 1)$-position is especially high when the preceding pep-tide bond is trans.[50] A cis-Pro residue is not frequently found in linear pro-teins, but if it does occur, it is at the $(i + 2)$-position of bend type VI[50] or in bend type IV (a type that is in a way a bend-dump).
3. Typically, Pro prefers to occur at an N-terminal helical boundary[76] (but not at an N-terminal sheet boundary), and it has the highest probability among all natural amino acid residues of occurring there. Concerning tetra-

peptidic bend boundaries, again Pro is a remarkable participant,[50] with $\beta$-turn potentials that exceed 1.0 in the following order: $i + 6$ (18.8%), $i + 5$ (14.8%), $i + 7$ (11.4%), $i + 8$ (9.1%), $i - 4$ (8.5%), $i - 1$ (8.0%), $i - 2$ (6.8%), and $i - 3$ (5.7%). This can be viewed together with the autoassociation propensity noted in item 1.

4. The statistical distribution listed in item 3, may result because Pro is a typical $\alpha$-helix breaker[40,77] and a $\beta$-disruptor.[49] Depending on its sequential position, given the orchestrated context of a succession of similar residues,[70] Pro engages a peptidic chain into a conformation reversal.

## C. Occupation of Proline (Inner Residue) in the $\phi$, $\psi$-Conformational Space; Its Propensity for Bend Induction

Because of lack of strong long-range interresidue interactions and cooperative effects, the conformational space available to noncyclic small peptides is large. This results in a coexistence of blended conformations mostly characterized by the absence of (restricted) regions of minimum conformational energy that are separated by barriers low enough to prevent them from being interchanged slowly enough to permit observation by NMR spectroscopy. On the contrary, cis–trans isomerization is a slow process ($\Delta H^{\ddagger} \sim 20$ kcal/mol) (see References 6, 82, 83),

≪As an example, rates for polyproline (PP) conversion due to cis–trans isomerization as in PP I → PP II and the related zipper mechanism (in aqueous solutions) have been revealed in an elegant study by Torchia and Bovey.[78] For related observations in PP, see Reference 79 and for poly-D,L-Pro, Reference 80. Theoretical considerations are found in References 5 and 81.≫

also observed during prolidase hydrolysis[84] and unfolding processes.[85] As such, Pro could trigger the dynamic aspects of globular protein folding.[118] The expected[16] additional possibility that a slow isomerization of the type cis′ → trans′ ($\Delta H^{\ddagger}$ 8–14 kcal/mol) can take place, and especialy for Pro-Pro sequences,[3,7] may result in observable consequences.

≪Due to spectral complexity, these cis′–trans′ isomers have not yet been separately observed with certainty (cf Reference 80) in the NMR spectrum, although assumed by Reference 82; see subsection c, "Pro-Pro Sequences."≫

To a first approximation only, the computed minima for tetrapeptide models are simply combinations of the single-residue backbone energy minima, but especially for imino-containing oligopeptides there are typical and important deviations. It is nevertheless instructive to consider a priori the

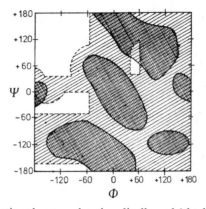

**Figure 7-3.** Conformational space, showing disallowed (shaded) regions for Ala residues; dashed lines: limits obtained from "hard sphere" approximation; solid lines: limits from PCILO quantum mechanics with 6 kcal/mol above the global minimum. (Adapted from Reference 15.)

location of global and local energy minima for each single residue as depicted earlier in Figure 7-2, and the allowed regions for typical residues such as Ala (Figure 7-3) and Gly and Pro (Figure 7-4) as the representative cases.

We will now briefly discuss the conformational aspects of Pro-containing linear peptides for some typical sequences obtained from model calculation and experimental evidence (Reference 1 and references cited therein). We

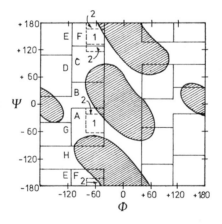

**Figure 7-4.** Conformational space, limiting disallowed (shaded) regions for Gly residues (solid lines: flexible model; PCILO calculations with 6 kcal/mol above the global minimum) and for Pro: zone 1, allowed for *cis*-Pro; zones 1 + 2, allowed for *trans*-Pro (adapted from Reference 15; the horizontal region has been widened to allow librations of $\phi = -60 \pm 20°$; see Section 1, "Pro-Single"). Stereoalphabet letters define specific regions A–H (cf Figure 7-2): A, $\alpha$ helix; B, bridge; C, $C_7^{eq}$; D, F, and G, neighbor; E, extended (Reference 31).

will place special stress on bend propensities, the most important feature for Pro-containing peptidic fragments.

In all cases considered, new minima at $\psi$ (Pro) = $-23 \pm 5°$ are found (region A, Figure 7-4), not present[1] in Pro-single (Figure 7-2). However in the hydrated model the very favorable $N_{i+3}H \rightarrow O_i$ interaction, which is at the basis of this behavior is disrupted, and $\psi$ invariably moves to $-48°$.[16]

### a. (Aaa)¹-Xxx-Pro-(Aaa)⁴

The presence of Pro severely limits the conformation of the preceding residue, and this is even more true[86–90] for the less flexible residues (see further discussions including Pro-Pro sequences). Due to steric congestion between $H^\delta$ (Pro) and NH (X), $|\psi|$ tends to become large in the preceding residue ($|\psi (X)| > 70°$), limiting the occupation for X to regions D, E, and F but occasionally A for Gly. As expected for the presence of an $\alpha$ breaker in the sequence, the helical region A generally becomes less important for X than in the single residue, and this tendency would be even more pronounced in the hydrated state[16] (intra-H-bond-disrupting solvent). With these restrictions in mind, together with the prerequisite of limited $\phi,\psi$ sets for Pro (Section 1), no high probability for typical bends (Table 7-2) may be expected with L-Pro$_{i+2}$. In principle, we may envision only types IV and VI and a $\gamma^*$ turn, and for Gly-Pro or D-Xxx-L-Pro in addition, types II*, V*, and a $\gamma$ turn, as well. The ECEPP calculation[1] discloses that a Gly-Pro sequence may form a II* bend,

≪A II* bend has been observed in exceptional cases in solution in a cyclopentapeptide[91] and is more frequently observed in cyclohexapeptides.[92] In solution of c/Gly¹-Pro²-Ser³-D-Ala⁴-Pro⁵/ there may be a II* turn present along 4–5, and a $\gamma$ turn at 2,[93] while in the solid the II* turn is located at 1–2.[24]≫

a IV bend, and a somewhat more unstable VI bend. None of these latter kinks however has high probability for preceding residues other than Gly.

The possibility for a type VI bend characterized by a *cis*-peptide bond deserves attention. *cis*-Peptide bonds occur relatively seldom, not only because of their instability when amino acid residues are involved, but because the sequence trans–cis–trans is highly improbable in a section of three peptide units. When X-Pro sequences are involved, however, up to 30% cis may be expected.[51] In this case a reverse open chain can create a local kink (Table 7-2, footnote *e*) together with an exceptionally high negative $\phi$ (Pro) value of about $-90°$.

Experimental data show indeed that cis ⇌ trans equilibria are in the range of prediction, and that in X-Pro sequences the amount of cis increases in L,L (but decreases in D,L) with bulkiness of X (eg, in the order Gly < Ala < Leu < Phe[94,95]). The cis content also seems to increase with the bulkiness of the residue following Pro[96] (see, however, Reference 97). The population of cis also increases substantially with the polarity of the medium. This is

TABLE 7-3. cis-Prolines Observed in Native Globular Proteins at Position $i + 2$ of Bends[a]

| Entry | Tetrapeptide section | $R_{i,i+3}$ (Å) | $(\phi,\psi)_{i+1}$ | $(\phi,\psi)_{i+2}$ | Region | SR[b] | Allowed ECEPP[c] (kcal/mol) |
|---|---|---|---|---|---|---|---|
| 1 | Glu$^{72}$-Leu-Pro-Asn$^{75}$ | 6.1 | $-49,+136$ | $-96,-11$ | F$\bar{\text{A}}$ | + | (2.48) |
| 2 | Lys$^{91}$-Tyr-Pro-Asn$^{94}$ | 5.1 | $-39,+132$ | $-89,+11$ | F$\underline{\text{B}}$ | + | (2.48) |
| 3 | Gly$^{166}$-Tyr-Pro-Gly$^{169}$ | 5.6 | $-96,+145$ | $-86,+13$ | F$\underline{\text{B}}$ | − | (2.48) |
| 4 | Tyr$^{115}$-Lys-Pro-Asn$^{118}$ | 5.6 | $-53,+133$ | $-83,+14$ | F$\underline{\text{B}}$ | + | (2.48) |
| 5 | Ser$^{93}$-Leu-Pro-Tyr$^{96}$ | — | $-82,+150$ | $-73,+156$ | FF | + | — |
| 6 | Thr$^{49}$-Leu-Pro-Gly$^{52}$ | 6.1 | $-113,+156$ | $-68,-29$ | E/FA | + | (0.66) |
| 7 | Gly$^{112}$-Asn-Pro-Tyr$^{115}$ | 4.2 | $-150,+105$ | $-64,+162$ | DF | − | (0.66) |
| 8 | Gln$^{6}$-Ser-Pro-Ser$^{9}$ | — | $-145,+139$ | $-62,174$ | E$\bar{\text{F}}$ | − | (2.55) |

[a]Adapted from Reference 3.
[b]Allowed for an occupation of the $(i + 1)$ residue as a single residue (cf Figure 7-2).
[c]According to ECEPP calculations[1]; values given are the relative minima for X = Ala and prefixed $\phi$ (Pro) $= -75°$.

expected if, for example, hydration is allowed for.[16] A rationale invoking intramolecular electrostatic interactions has been offered,[94] more specifically a stabilizing attraction between $(^{\delta+}C=O)_{j+1}$ and the preceding $(C=O^{\delta-})_j$ carboxamide function,[98] which is possible only in the trans $j$ bond and with the highest contribution in nonpolar surroundings.

Only 10 bends with a cis-Pro residue in the $(i + 2)$-position have been recognized among 30 native proteins, and for eight of them peptide torsions were available.[3] Table 7-3 discloses that they occupy a region in the conformational space that fits for an $(i + 1)$ single residue (Figure 7-2) and/or that they are predicted from ECEPP calculations to be reasonably stable. It is noteworthy, however, that next to $\phi,\psi$ sets typical for Pro as in poly-Pro (Table 7-1 and entries 5, 7, and 8 in Table 7-3), the $\phi$ values are often exaggeratedly low [eg, those encountered in the $(i + 2)$-position of bends I, II*, IV, V*, or VII (entries 1–4)]. In considering the two $(\phi,\psi)_i$ sets, entries 1–4 might best be classified as bends VII, were it not for the cis-peptide nature. It is clear that here long-range attractive contributions must compensate for a relatively strained state of Pro, occurring in a pyrrolidine ring conformation with high pucker in the $C^\alpha - C^\beta$ region. These are situations that might be observed

≪An $\alpha^+$ form displays in the $^1$H NMR spectrum an apparent triplet for H$^\alpha$ (Pro) with $\Sigma J^3 \backsim 8$ Hz.[3,26] Also cis-Pro can be identified through $^{13}$C NMR spectroscopy.[99] Finally, unusually high-field $C^\beta$ (Pro) resonances concomitantly with high-field $C^\alpha$ (Pro) resonances indicate eclipsing of the $C^\beta$ and the Pro (C'=O) (eg, a low trans' $\psi$ angle; cf Reference 93). These NMR criteria allow the recognition of specific bends as presently under discussion.≫

for Pro$_{i+2}$ in bends (I), II*, IV, V*, and VI, as discussed below.

## b. (Aaa)$^1$-Pro-Xxx-(Aaa)$^4$

In contrast to X-Pro sequences, interresidue medium-range interactions do not disallow existing single-residue minima. In addition, some new minima are found in Pro-X sequences, especially in regions A, D, and G for (Xxx)$_{i+2}$. It is not without surprise therefore that one finds many minimum conformations corresponding to bends with Pro at the ($i$ + 1)-position, and indeed for eight different sequences, a total of 292 minima were calculated[1] to be probable, among which 114 (39%) belong to bends. This is slightly more than the average frequency of $\beta$ turns (32%) found in proteins.[76] No specific bend types are preferred markedly over the others, and ensembles of types I, IV, V, and VII, and to a lesser extent II and III, represent possible local minima. Slight preferences are predicted for type I (if X = Ser, Val) and type IV (if X = Gly, Leu, Ser); types II and V are well populated if X = Gly. Pro-Gly bending is relatively favored, simply because Gly may occupy enantiomeric regions (eg, bends II* and V*). For the dipeptide model Pro-Gly, rather extended zones for $\psi$ (Pro) seem to be allowed, ranging from 80° to 200°[61] and 0° to −90°.[7] Pro-Leu is special in that it may occupy type III (as does Pro-Val and in contrast to Pro-Phe). Here there seems to be some controversy between PCILO calculations (cf Figure 7-2) and earlier results[100] in that this would correspond to the normal tendency for single-Leu to occupy regions A, A*, and D more than any other X residue (except Ala and Gly). We refer to the original literature (Reference 1) for further exemplification of the subtle changes of preferred conformational populations depending on the nature of Xxx.

Experimental evidence has been advanced[86] for the occurrence of bends V and IV that would exist about Pro-Xxx (Xxx = Gly or D-Ala) in Gly-Pro-Xxx-Gly with ($\phi,\psi$) sets very close to those of calculated stable forms.[1] The high mole fractions of bend forms found in Xxx-Pro-Gly-Yyy (Xxx = Ala, Val, Yyy = Val or Gly)[101] are also in good agreement with the theoretical considerations.

### c. Pro-Pro Sequences

Pro-Pro sequences have garnered much attention because[117] of the high proportion of Pro in numerous fibrous protein structures (eg, collagen) and the existence of poly-Pro [all-cis (PP I) and all-trans (PP II)]. Well-elaborated theoretical studies have been performed on PP[6,7] as well as on Ac-Pro-NMe$_2$[3], which is representative of an isolated section of Pro-Pro. Experimental data for linear cases are also available (PP I and PP II: References 78–80) as well in the solid[119] and in solution.[82,96,103−106] Local conformational features may be discussed in terms of backbone parameters $\omega$ (cis–trans), $\phi$ (and connected pyrrolidine ring behavior), and $\psi$ (cis'–trans'), and they have some noticeable aspects. There is a net tendency for high cis populations in solution, especially in nonpolar solvents.

≪The observation that (all-)trans forms are favored in the solid state does not necessarily indicate that large differences would exist in conformational potentials between solid and solute, as often believed and as found in the case of linear homo L-Pro-oligopeptides.[82] Also, interestingly enough, Pro sequences of enantiomeric composition tend to crystallize in *cis*-peptide forms (Reference 119 and references cited therein). It seems quite general that *cis*-PP has little crystallinic affinity (eg, PP I is an oil), and the bias to the trans state from a multitude of local minimal conformational forms is to a great extent triggered by kinetic aspects of crystallization. Thus the solid state conformations may or may not represent a global energy minimum in the conformational space. It is sufficient for one form to represent a local minimum with an energy profile shallow enough to be trapped in this state during crystallization. We personally believe (cf Reference 151) that a solid state conformation is a good representative for one of the (relatively stable) forms in solution, where eventually a distribution among a multitude of forms is more feasible because of the relatively low energy barriers that separate these forms. Thus although Piv-L-Pro-D-Pro-OMe crystallizes in the cis conformation,[119] *t*.Boc-L-Pro-D-Pro-OMe does it in the trans.[124]»

This tendency is even more pronounced in sequences of different chirality.[119,122,123] This feature is common to all *imino* acid residues, as illustrated in Tables 7-4 and 7-5. Although the amount of cis increases with increasing bulkiness (eg, on $C^\alpha$), there is no simple relation between $K_e$ and the steric parameter $E_S$. Also in Pro-Pro sequences the equilibrium shifts in favor of the cis form, with the bulkiness of the amino acids following this dipeptide sequence [eg, in *t*.Boc-(Pro)$_2$ − X with X = Ala-OH (15%); = Ala-OMe (30%); = Leu-OH (30%); and = 3,5 diBr-Tyr-OH (40%)].[96]

In *t*. Boc-(Pro)$_n$-OBz no cis,cis nor cis,cis,cis forms for $n$ = 2, 3 were observed except for random distributions.[103] The all-trans form is the unique form if $n$ = 5, 6.[106] In the tripeptide Ser-Pro-Pro-OH[105] typically the cis,cis form is the second most stable form in protic solvents, but this depends strongly on the ionization state of the terminal Pro. It was a surprise to find that the population of this isomer (cis,cis) exceeds by far the simple summation of the probabilities observed for the cis,trans and trans,cis forms, indicating that medium-range contributions to the respective relative stabilities are present. The proposed contributions, however, all involve charges, and it remains an interesting point to consider for data from other models that represent nonterminal peptide fragments. The barrier to cis–trans isomerization is relatively high (eg, $\Delta G^\ddagger$ = 18–19 kcal/mol).[115] Distortions about the peptide bond are sometimes observed for L,D sequences and in *cis*-Piv-L-Pro−D-Pro-OMe, $\Delta\omega$ is large as − 19°.[119] An observable barrier about $\psi^3$-rotation has been advanced[82] in *t*.Boc-Aib$^1$-L-Pro$^2$-L-Pro$^3$-NHMe in CDCl$_3$ (for the *cis*-Pro-Pro isomer only). In our opinion the observed phenomenon (shifts of NHMe) may equally well be explained by the conversion about the *t*.Boc-Aib urethane bond, for which a relatively low coalescence temperature resulting from barriers of about 10–14 kcal/mol might also be expected. Model building indeed shows that in

TABLE 7-4. Presence of cis Form in $Ac-N(R_N)CH(R_\alpha)CONMe_2$ Peptide Models[a]

| | cis Form% | | | | | | | | | |
|---|---|---|---|---|---|---|---|---|---|---|
| $R_N$: | Me | Et | n-Pr | Bz | i-Pr | H | Me | Et | n-Pr | n-Bu |
| $R_\alpha$: | H | H | H | H | H | Me | Me | Me | Me | Me |
| Peptide | | | | | | | | | | |
| $C_6D_6$ | 20 | 19 | 28 | 27 | 40 | | | | | |
| $CD_2Cl_2$ | 28 | 26 | 29 | 36 | 36 | 0 | 18 | 56 | 74 | 71 |
| log (trans/cis) | 0.602 | 0.630 | 0.410 | 0.432 | 0.176 | ∞ | 0.659 | −0.105 | −0.466 | −0.404 |
| $E_S$ | 0 | −0.07 | −0.36 | −0.38 | −0.5 | +1.24 | 0 | −0.07 | −0.36 | −0.39 |

[a]Reference 115.

TABLE 7-5. Presence of cis Form in $\phi CO -$ N(Me)CH(R$_\alpha$)COOH in CDCl$_3$ at Room Temperature[a]

| R$_\alpha$: | H | Me | n-Pr | i-Bu | i-Pr | s-Bu |
|---|---|---|---|---|---|---|
| cis (%): | 40 | 45 | 50 | 48 | 60 | 55 |

[a]Reference 116.

the cis isomer the urethane bond may approach the carboxamide bond and thus may exercise a magnetic anisotopic effect especially on NHMe. Nevertheless, relatively high activation barriers [$\Delta G^{\ddagger}$ ($\psi$) $\frown$ 14 kcal/mol] are possible.[120,121]

### d. Pro[1]-Pro[2] Dipeptide Models

Taking Ac-Pro-NMe$_2$ as a model compound for Pro[1] in the Pro dimer, Madison has found from a Consistent Force Field (CFF) treatment[3] that all combinations of cis, trans, cis', and trans' are allowed. For a *cis*-peptide, the cis,cis' ($\psi = -50°$) would be somewhat favored over the cis,trans' ($\psi = 130°$) form but only for a $\gamma^-$-pyrrolidine ring. In the *trans*-peptide, the trans,trans' ($100° < \psi < 150°$) form represents the global minimum in comparison to trans, cis' ($\psi = -60°$). The CFF calculations predict a high amount of cis, as experimentally found (60% in nonpolar solvents), and a relatively high barrier for interconversion between cis' and trans' [$\Delta H^{\ddagger}_{cis',trans'}$ $\frown$ 10 kcal/mol). We will see that from ECEPP treatment of the genuine Pro dimer, the assumption that cis,cis' would be relatively favored for a Pro-Pro sequence, too, may be somewhat of an oversimplification, since repulsive interresidue contributions between (CH$_2^\delta$)$_1$ and (CH$_2^\delta$)$_2$ are absent in the dimethylamide model. Other calculations[6] too indicate that the cis,cis' form would be disallowed because of (CH$^\alpha$)$_2$−(CH$_2^\delta$)$_1$ repulsive interactions.

The most extensive calculations on Pro-Pro fragments were performed more than a decade ago by Gō and Scheraga, but results were reported for the *trans*-Pro-Pro case only.[7] However with the use of a naive algorithm for coordination transformation (see Section below), predictions may also be made from this ECEPP treatment for the diastereomeric peptide fragments (eg, for the cis isomer) using the results obtained for the *trans*-L,L form. The ECEPP calculations predict that cis,cis' is disallowed in favor of the conformers cis,trans' ($\psi = 160°$), trans,trans' ($\psi = 160°$), and trans,cis' ($\psi = -40°$). In this approach—as in all others—

≪In the light of increasing evidence that the rather flexible pyrrolidine ring of Pro often takes (more strained) forms other than the simple $\gamma^+/\gamma^-$ (eg, with an increasing pucker about $\chi^1$, $\chi^2$ and even $\chi^3$; ie, with $\phi < -75°$), it may be worthwhile in the future also to allow in calculations a pseudorotation of the pyrrolidine ring further away from the prefixed phase angle—

the more so because L-Pro is predisposed to occur in bends, in which espe-cially types I ($i + 2$), II* ($i + 2$), IV, (V*), $\gamma$, and especially VI [both in the ($i + 1$)- and ($i + 2$)-positions] are to be considered (cf, Table 7-2).≫

only pyrrolidine ring conformations $\gamma^+$ and $\gamma^-$ were considered (fixing $\phi$ at about $-60°$), and the energies were computed with these two possibilities only and as a function of the variable $\psi$ (cis' → trans' isomerization).

We will briefly discuss the results obtained by Gō and Scheraga, further summarized in Table 7-6. Two regions of minima are found for all $^\gamma E$ ($\gamma^+$) and $_\gamma E$ ($\gamma^-$) combinations: a local one for the cis' form ($\psi \simeq -40°$) and a broad global one for trans' ($90° < \psi < 190°$) (entries 1, 3, 6, 8, 9, and 11). In addition, for $\gamma^-/\gamma^+$ combinations the cis' region is split into two shallow minima ($\psi \simeq 0°$ and $\psi \simeq -80°$) with in-between energy barriers (entries 7 and 10), mainly resulting from repulsions between $(CH_2^\delta)_1$ and $(CH_2^\delta)_2$. The most important repulsive effects affecting the main disallowed ranges for $\psi$ arise almost entirely from steric congestions involving the $\delta$-position of res-idue 2 $(CH_2^\delta)_2$ and the atomic fragments of preceding residues [eg, with $(C_0^\alpha,$ $C=O)_0$, $C_0'$, $N_1$, $(CH_2^\beta)_1$, $(CH_2^\gamma)_1$, $(CH_2^\delta)_1$] next to $O_0 >-< C_2^\gamma$ and $O_0 >-<$ $N_2$.

*Algorithm Predicting Stability Profiles for Proline Dimers Other Than trans-L-Pro-L-Pro*

The following simple rules of thumb allow predictions of stable forms for Pro-Pro combinations with opposite chirality and/or with *cis*-peptide bonding.

1. *cis*-L-Pro-D-Pro Sequences.
If the chirality is altered in the second residue of an L,L-Pro¹-Pro² sequence, the following coordination-transformations should be envisaged, allowing one to adapt conclusions reached for *trans*-L,L-sequences as just discussed: (a) $C^\alpha$, $C^\beta$, $C^\gamma$, and $C^\delta$ in the second residue must be replaced by, respectively, $C^\delta$, $C^\gamma$, $C^\beta$, and $C^\alpha$, and (b) concomitantly, the *trans*-peptide bond becomes a *cis*-peptide bond.

This coordination-transformation can be rationalized from molecular models and inspection of Figure 7-5. In general, a simple coordination exchange of $C^\alpha$, $C^\beta$, $C^\gamma$, $C^\delta$ with its reverse order also changes the nature of the preceding peptide bond (eg, trans → cis). [Also $\phi$ and $\psi$ have changed in sign ($\phi \rightarrow \phi^*$ and $\psi \rightarrow \psi^*$), but the original (opposite) signs have to be used in applying Table 7-6]. Therefore, after transformation, the newly interact-ing atom pairs, as reported in Table 7-6, may easily be identified. One addi-tional aspect must, however, be looked for since in this new combination, interactions between $C_2'(C=O)_2$, $N_2$, and the preceding $C_1^\alpha, C_1'$ and $(C=O)_1$ also have been altered. None of these pairwise combinations, however, was important in the L,L-Pro dimer model.[7] Nevertheless, some may engender new repulsive terms and, in particular, interactions of $C_2'(C=O)_2$ and $N_2$

TABLE 7-6. Important Steric Interactions Between Pairs of Atom Groupings for the Four Characteristic $\psi$ Regions in $\gamma^\pm/\gamma^\pm$ L-Pro Dimers[a]

| | | | Pairs of atom groupings | | | | | | | |
|---|---|---|---|---|---|---|---|---|---|---|
| | | | Medium-range interactions ($E_0$–$E_2$) | | | | Interresidue interactions ($E_1$–$E_2$) | | | |
| Entry | $\psi$ Zone | $\gamma_1/\gamma_2$ Combination | $O_0$–$N_2$ | $O_0$–$(CH_2^\delta)_2$ | $C_0'$–$(CH_2^\delta)_2$ | $C_0^\gamma$–$(CH_2^\delta)_2$ | $N_1$–$(CH_2^\delta)_2$ | $(CH_2^\delta)_1$–$(CH_2^\delta)_2$ | $(CH_3^+)_1$–$(CH_2^\delta)_2$ | $(CH_2^\delta)_1$–$(CH_2^\delta)_2$ |
| 1 | +90°/+200° (global min.) | $\gamma^\pm/\gamma^\pm$ (all) | — | — | — | — | — | — | — | — |
| 2 | 0°/+90° (disallowed) | $\gamma^\pm/\gamma^\pm$ (all)[b] | + | + | + | + | + | — | — | — |
| 3 | 0°/–40° (local min.) | $\gamma^+/\gamma^\pm$ | — | — | (+) | — | + | — | — | + |
| 4 | –60°/–170° (disallowed) | $\gamma^+/\gamma^\pm$ | — | — | — | — | — | + | + | + |
| 5 | –90°/–160° (disallowed) | $\gamma^-/\gamma^\pm$ | — | — | — | — | — | + | — | — |
| 6 | 0°/–10° (local min.) | $\gamma^-/\gamma^-$ | — | — | (+) | — | (+) | — | — | (+) |
| 7 | –20°c | $\gamma^-/\gamma^-$ | — | — | — | — | (+) | — | — | + |
| 8 | –30°/–80° (local min.) | $\gamma^-/\gamma^-$ | — | — | — | — | (+) | (+) | — | (+) |
| 9 | –10°/–50° (local min.) | $\gamma^-/\gamma^+$ | — | — | (+) | — | (+) | — | — | + |
| 10 | –60°c | $\gamma^-/\gamma^+$ | — | — | — | — | (+) | — | — | + |
| 11 | –70°/–90° (local min.) | $\gamma^-/\gamma^+$ | — | — | — | — | — | + | — | (+) |

[a] Adapted from Reference 7, Figure 2.4
[b] In the ($\gamma^+/\gamma^\pm$) combinations, there is no $N_1$ >–< $H_2^\delta$ and in the ($\gamma^\pm/\gamma^-$) combinations no $C_0^\gamma$ >–< $H_2^\delta$.
[c] Disallowed zone between two local minima: entries 6 and 8 for $\gamma^-/\gamma^-$ and 9 and 11 for $\gamma^-/\gamma^+$.

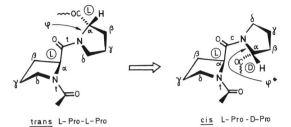

**Figure 7-5.** Illustration of the interrelationship between chirality and *cis-trans* isomerism in the prediction of stable proline dimer sequences.

with the preceding $(CH^\alpha)_1$ must be considered for some of the $\psi_{1,2}$ regions.

2. *cis-* and *trans*-D-Pro-L-Pro Sequences.

If the chirality of the first residue is changed in an L,L-$Pro^1$-$Pro^2$ sequence, the transformations in rule 1 are to be followed with respect to this residue. In addition, a series of new interactions is created [eg, between $C_2^\delta$ and $C_0^\alpha$, $(C=O)_0$] that at first has minor importance but must be considered if $0 < \psi < +120°$ ($\psi$ signs before transposition). This will narrow the global minimum for the trans, trans′ conformation. If, however, the epimeric substitution of the first residue is not combined with concomitant *preceding* trans–cis isomerization, these interactions can be neglected.

3. *cis*-L-Pro-L-Pro Sequences.

In a *cis*-L-Pro-L-Pro dimer the labels $C^\alpha,C^\delta$ and $C^\beta,C^\gamma$ are interchanged in the second Pro residue when referring to the corresponding trans isomer, and only interactions occurring in the trans′ forms are allowed when $120° < \psi < 180°$. This culminates in a well-defined bend (reverse open chain, or bend VI; cf Table 7-2) with backbone coordinates in Xxx-Pro-Pro-Yyy as follows: trans, $-60°$, $>120°$; cis, $-110°$, $+90°$, reinforced in a nonpolar medium by a $(CO)_i \leftarrow (HN)_{i+3'}$ hydrogen bond bridge. The known disallowance of a cis,cis′ form is therefore predicted by the present algorithm, because all the original unfavorable interactions involving the $(CH_2^\delta)_2$ fragment and destabilizing the cis′ region for the trans,cis′ form of the *trans*-L,L isomer (see step 2) now concern the carboxamide-bearing $(C_2^\alpha)_2$ fragment, which is even more hindering. This is also why this cis,cis′ form remains allowed in the Ac-Pro-NMe₂ model, where the $C^\alpha$ is a relatively noncongesting methyl.

### e. Poly-Pro

The main differences between the two possible L-Pro homopolymers (poly-*cis*-PP I, and poly-*trans*-PP II) are: (a) the apolar nature of PP I ($[\alpha]_D^{25} = +50°$ in HOAc) compared to PP II ($[\alpha]_D^{25} = -540°$ in HOAc) with aligned peptide dipoles in the latter, and (b) the more compact nature of the former (molecular volume ratio 1.65) with decreased solvent accessibility. Therefore the major form in protic solvents (H₂O, phenol, . . . ) is PP II,[81,107] and this trend has also been observed in similar smaller fragments such as Ser-Pro-Pro.[105] The isomerization PP I → PP II follows zero-order kinetics

($\Delta H^{\ddagger}$ = 23 kcal/mol), with the remarkable feature that the cis → trans isomerization starts at the carboxyl terminus and gradually and sequentially wobbles along the chain towards the amino terminus (the "zipper-mechanism" of Reference 78). It is an all-or-none transition. Calculations confirm[5] that unfavorable trans,cis sequences

≪In concentrated salt solutions, however trans-cis-trans (and trans-cis-cis) sequences have been observed,[79] whereby metal chelation may stabilize otherwise disallowed trans,cis distributions.≫

are at the origin of this mechanism.

The typical $(\phi, \psi)$ sets

≪Compare the case of collagen, where for the 1-bonded and 2-bonded models respectively, $\omega$, $\phi$, and $\psi$, are 180°, −52.5°, 148.5° and 180°, −77°, 157° (cf Reference 145).≫

for PP I and PP II are reported in Table 7-1. Both PP peptides have $\phi$-values that are at the outer negative range of the allowed region, $\phi = -60° \pm 20°$. It is therefore in accordance with expectation that the puckering zone of the pyrrolidine ring will be displaced from $\chi^2/\chi^3$ toward $\chi^5/\chi^1$ in PP I.[82] It is noticeable that the usually allowed conformational regions for Pro in peptides (eg, A, C, and F, Figure 7-4) become limited to F alone. Because restrictions on X are imposed by Pro in *trans*-X-Pro sequences, resulting in a disfavoring of the A region for X [originating from $(NH)_1 - (CH_2^\delta)_2$ repulsive interactions], it follows (Table 7-6) that in PP II only $\psi$ values greater than about 90° need to be taken into consideration, culminating in the observed value of +149°. Also in PP I the $N_1 - (CH_2^\delta)_2$ unfavorable interactions in the trans model are now replaced by severe $N_1 - (CH^\alpha)_2$ repulsions (cf stability algorithm above), confining conformational freedom even more.

# 3. STATISTICAL TREATMENT OF SOLID STATE DATA OF LINEAR PRO-PEPTIDES

## A. General Aspects

Delos de Tar, and Luthra have amply commented[14] some 40 X-ray data sets of proline derivatives that were available up to 1977. Their main conclusions were: (a) the endocyclic intercarbon lengths do not vary appreciably (1.51–1.54 Å), nor do the nitrogen–carbon bonds [$d(C_0' - N) = 1.33_2 \pm 0.02$ Å; $d(C^\alpha - N) = 1.47_6 \pm 0.03$ Å, and $d(C^\delta - N) = 1.48_7 \pm 0.03$ Å]; (b) valence angles involving the Pro-amide nitrogen have characteristic values ($\overline{C^\alpha N C'}$ = 122.4 ± 3.4°; $\overline{C^\delta N C'}$ = 124.1 ± 3.0°, and $\overline{C^\delta N C^\alpha}$ = 112.7 ± 3.0° with a mean $\bar{\alpha} = 120.0°$); (c) values of $\chi^5 - \phi$ vary greatly (eg, from 54° to 80°), whereas idealized trigonal projections would predict a value of 60°. In a search for an underlying reason, Delos de Tar and Luthra stated that molecular mechanics calculation demonstrated that ring conformations had little control over $\phi$ (or $\psi$).

In our present survey of published X-ray material on proline, we restrict ourselves to linear peptides (bends are considered, but not covalent cyclic peptides and nonrepresentative cases)

≪Omitting the nonrepresentative cases such as Pro–metal complexes and others, we extract from the authors' listing $\chi^5 - \phi = 62.5 \pm 7.4°$.≫

with a substituent on the Pro-nitrogen, which can be either a protecting group or another amino acid.

Since the articles we consulted exhibit no uniformity in the measurements of certain angles, we had to recalculate the values of interest from the published internal coordinates. The results are gathered in Tables 7-7 and 7-8. The peptides are ordered in the following way: if proline is the first amino acid of the chain, the peptides are classified alphabetically based on the amino acid *following* proline; if Pro is not the first, we looked to the *preceding* amino acid(s). The protecting group is always counted number one. When we calculated the pseudorotational parameter $P$ of a D-residue, we added 180° to this value, since a D-Pro fragment with, for example, $P = 198°$, is in fact a $\gamma^-$ form because the carboxyl function has the opposite configuration ($P = 198°$ corresponds to the $\gamma^+$ isomer for L-Pro; see Section 4). The other values for these molecules have not been reversed in sign because they are measurable angles.

In our data treatment we ignored values of Boc-Pro-Aib-Ala-Aib-OBzl,[175] because it is not possible to form a Pro ring using the reported coordinates (bond length = 6 Å); those of Boc-Pro-OH,[128] where the $\phi$ value is not consistent (also reported by Delos de Tar and Luthra[14]); and those of Se-Bzl-Cys-Pro-Leu-Gly-NH$_2$,[183] where some reported torsion angles are inconsistent with the given atomic coordinates.

Entries 55 and 56 of Tables 7-7 and 7-8, which are studies of the same molecule, published simultaneously by different authors, can give us an idea of the very small scatter of X-ray studies. Entries 85 and 86 give a recalculation by the author of his measurements published in 1958 (entries 83 and 84).

Some authors reported the occurrence of two forms of the same molecule in the crystal [entries 4 and 5, 57 and 58, 67 and 68, 70 and 71, 74(D), 75 (L) in Tables 7-7 and 7-8]. In two cases (entries 19 and 20 and 83 and 84, recalculated from 85 and 86), a strong delocalization of the $\gamma$ carbon of proline is observed, so that the existence of two "half" atoms, at both sides of the ring, had to be considered in the author's calculations.

TABLE 7-7. Internal Coordinates (°) Obtained from Solid State Data as Explained in the Text

| Entry | Peptide[a] | Ref. | $P$ | $\chi_M$ | $\chi^5$ | $\omega$ | $\phi$ | $\psi^b$ | $\kappa_N$ | $\kappa_C$ |
|---|---|---|---|---|---|---|---|---|---|---|
| 1 | Piv-Pro-Aib-NHMe | 131 | 17 | 42.4 | 0.8 | 175.1 | −57.8 | 139.2 | 176.3 | 117.6 |
| 2 | p.Cl-Bzoyl-Pro-Aib-Ala-Aib-Ala-OMe | 132 | 185 | 39.9 | −9.1 | −178.6 | −55.9 | 145.2 | 165.9 | 119.1 |
| 3 | Boc-Pro-Ala-Gly-NH₂ | 139 | 180 | 20.8 | −6.4 | −12.6 | −54.3 | 148.0 | 167.5 | 119.5 |
| 4 | Poc-Pro-Ala-Gly-OH, mol. A | 141 | 176 | 22.9 | −8.6 | −2.5 | −68.5 | 155.0 | 180.2 | 120.3 |
| 5 | Poc-Pro-Ala-Gly-OH, mol. B | 141 | 180 | 18.1 | −5.5 | 2.7 | −69.9 | 158.4 | 182.4 | 117.9 |
| 6 | iB-Pro-L-Ala-NHiP | 148 | 14 | 6.2 | 0.3 | 177.4 | −59.5 | 136.8 | 183.1 | 123.3 |
| 7 | iB-Pro-D-Ala-NHiP | 148 | 19 | 29.8 | −0.1 | −178.2 | −62.9 | 137.5 | 181.5 | 118.6 |
| 8 | Piv-Pro-Me-Ala-NHiP | 149 | 13 | 38.1 | 2.4 | −178.0 | −62.3 | 135.0 | 184.8 | 120.1 |
| 9 | Piv-Pro-Me-Ala-OMe | 157 | 177 | 32.2 | −11.5 | −179.9 | −69.6 | 153.3 | 176.4 | 118.4 |
| 10 | Piv-Pro-Me-D-Ala-NHMe, anhyd. | 158 | 144 | 7.7 | −6.3 | 179.5 | −57.6 | 135.6 | 172.7 | 121.5 |
| 11 | Piv-Pro-Me-D-Ala-NHMe · H₂O | 158 | 43 | 30.8 | −12.7 | 175.9 | −68.6 | 164.3 | 177.0 | 121.1 |
| 12 | Z-Pro-Aze-OH | 104 | 24 | 37.8 | −4.3 | −3.5 | −60.0 | 151.5 | 174.4 | 118.8 |
| 13 | Piv-Pro-Gly-NMe₂ | 159 | 173 | 37.2 | −16.0 | −178.7 | −71.0 | 157.0 | 172.7 | 117.7 |
| 14 | Piv-Pro-Gly-NHiP | 160 | 25 | 33.0 | −4.4 | 177.7 | −61.2 | 137.2 | 177.1 | 120.3 |
| 15 | Boc-Pro-Gly-OH | 161 | 85 | 10.8 | −10.3 | −12.1 | −61.2 | 147.1 | 167.9 | 117.0 |
| 16 | Ac-Pro-Gly-Phe-OH | 162 | 2 | 36.2 | 10.2 | −176.2 | −58.8 | 128.0 | 188.5 | 119.5 |
| 17 | Tos-Pro-Hyp-OH^c | 134 | 213 | 34.3 | 9.3 | −76.0 | −102.6 | 157.2 | 217.1 | 105.2 |
| 18 | Boc-Pro-Ile-Gly-OH | 163 | 187 | 39.8 | −8.1 | 11.2 | −81.0 | 154.9 | 190.7 | 117.8 |
| 19 | Ac-Pro-L-Lac-NHMe, 50% | 164 | 355 | 45.9 | 18.2 | −171.7 | −62.0 | −22.9 | 191.7 | 111.4 |
| 20 | Ac-Pro-L-Lac-NHMe, 50% | 164 | 190 | 42.8 | −3.9 | −171.7 | −62.0 | −22.9 | 191.7 | 133.5 |
| 21 | Ac-Pro-D-Lac-NHMe | 165 | 190 | 38.4 | −5.3 | −178.1 | −62.7 | 140.5 | 174.5 | 117.1 |
| 22 | Boc-Pro-Leu-OBzl | 166 | 170 | 36.4 | −17.4 | −13.5 | −72.5 | 157.3 | 172.0 | 117.0 |
| 23 | Z-Pro-Leu-OEt | 166 | 177 | 37.2 | −13.5 | −11.2 | −62.6 | 171.2 | 166.0 | 116.9 |
| 24 | Boc-Pro-Leu-Gly-OH | 167 | 0 | 40.7 | 12.3 | −171.4 | −65.1 | −20.6 | 198.8 | 121.4 |
| 25 | Ac-Pro-NH₂ | 129 | 178 | 38.9 | −13.6 | −173.4 | −80.1 | −14.2 | 188.4 | 122.0 |
| 26 | Ac-Pro-NHMe | 126 | 185 | 36.3 | −7.7 | −177.0 | −76.3 | −15.9 | 190.0 | 121.4 |
| 27 | Aoc-Pro-OH | 127 | 182 | 31.0 | −8.5 | −2.1 | −69.7 | −9.1 | 180.9 | 119.7 |
| 28 | Prop-Pro-OH | 130 | 167 | 40.4 | −20.8 | −8.1 | −75.3 | 175.5 | 173.1 | 118.6 |
| 29 | HSMeProp-Pro-OH | 125 | 38 | 27.8 | −9.8 | 173.3 | −67.3 | 161.9 | 178.0 | 120.5 |
| 30 | Boc-Pro-Pro-OH, Pro-2 | 137 | 173 | 31.3 | −12.9 | −7.0 | −66.6 | 141.5 | 171.6 | 117.8 |

| | | | | | | | | | | |
|---|---|---|---|---|---|---|---|---|---|---|
| 31 | Boc-Pro-Pro-OH, Pro-3 | 137 | 154 | 39.6 | −27.9 | −175.8 | −95.1 | −168.8 | 189.3 | 122.1 |
| 32 | Z-Pro-Pro-OH, Pro-2 | 104 | 23 | 35.0 | −3.6 | 3.1 | −64.8 | 155.3 | 179.4 | 118.2 |
| 33 | Z-Pro-Pro-OH, Pro-3 | 104 | 357 | 36.1 | 12.4 | −178.9 | −53.3 | 141.7 | 184.1 | 111.1 |
| 34 | Boc-Pro-D-Pro-OH, Pro-2 | 124 | 35 | 29.8 | −9.1 | −3.6 | −69.5 | 140.3 | 176.2 | 115.8 |
| 35 | Boc-Pro-D-Pro-OH, Pro-3 (D) | 124 | 181 | 40.4 | 11.7 | 173.5 | 70.2 | −151.1 | 177.1 | 118.6 |
| 36 | Boc-Pro-D-Pro-OMe, Pro-2 | 124 | 179 | 35.7 | −12.0 | −1.5 | −71.8 | 143.2 | 179.4 | 119.7 |
| 37 | Boc-Pro-D-Pro-OMe, Pro-3 (D) | 124 | 188 | 40.4 | 6.9 | 173.6 | 69.6 | −152.6 | 179.2 | 116.5 |
| 38 | Piv-Pro-D-Pro-OMe, Pro-2 | 119 | 28 | 23.9 | −4.2 | −175.7 | −69.7 | 129.1 | 186.1 | 117.6 |
| 39 | Piv-Pro-D-Pro-OMe, Pro-3 (D) | 119 | 176 | 21.2 | 7.4 | −19.1 | 84.3 | −168.6 | 195.9 | 119.0 |
| 40 | Piv-D-Pro-Pro-Ala-NHMe, Pro-2 (D) | 168 | 8 | 38.3 | −6.8 | 173.8 | 59.0 | −135.6 | 184.8 | 119.0 |
| 41 | Piv-D-Pro-Pro-Ala-NHMe, Pro-3 | 168 | 13 | 36.8 | 3.0 | −178.9 | −58.7 | −23.1 | 182.3 | 120.6 |
| 42 | Boc-Pro-Pro-Gly-NH₂, Pro-2 | 169 | 11 | 33.4 | 4.0 | −2.2 | −60.9 | 156.3 | 182.5 | 117.6 |
| 43 | Boc-Pro-Pro-Gly-NH₂, Pro-3 | 169 | 195 | 39.2 | −1.9 | −178.8 | −64.9 | −23.0 | 185.4 | 122.3 |
| 44 | Aoc-Pro-Pro-Pro-OH, Pro-2 | 136 | 169 | 28.0 | −14.3 | −10.5 | −59.0 | 155.1 | 164.7 | 120.0 |
| 45 | Aoc-Pro-Pro-Pro-OH, Pro-3 | 136 | 170 | 20.0 | −9.7 | 171.9 | −66.0 | 138.9 | 176.0 | 119.7 |
| 46 | Aoc-Pro-Pro-Pro-OH, Pro-4 | 136 | 182 | 37.5 | −10.3 | −179.0 | −73.2 | 166.1 | 182.0 | 119.1 |
| 47 | Boc-D-Pro-D-Pro-Pro-OH, Pro-2 (D) | 170 | 20 | 36.3 | 1.7 | −5.0 | 70.8 | −149.7 | 187.0 | 117.9 |
| 48 | Boc-D-Pro-D-Pro-Pro-OH, Pro-3 (D) | 170 | 9 | 39.6 | −5.7 | −173.7 | 61.5 | −137.7 | 184.5 | 117.3 |
| 49 | Boc-D-Pro-D-Pro-Pro-OH, Pro-4 | 170 | 171 | 36.2 | −16.9 | 11.1 | −78.3 | −175.6 | 182.4 | 121.0 |
| 50 | Boc-Pro-Pro-Pro-OBzl, Pro-2 | 138 | 11 | 15.2 | 1.6 | −1.1 | −67.8 | 151.8 | 186.2 | 116.8 |
| 51 | Boc-Pro-Pro-Pro-OBzl, Pro-3 | 138 | 170 | 31.9 | −15.5 | 168.8 | −69.8 | 164.9 | 173.1 | 118.8 |
| 52 | Boc-Pro-Pro-Pro-OBzl, Pro-4 | 138 | 183 | 28.5 | −7.0 | 176.4 | −61.1 | 152.3 | 174.9 | 120.8 |
| 53 | Boc-Pro-Pro-Pro-OBzl, Pro-5 | 138 | 355 | 16.6 | 6.3 | 179.8 | −57.8 | −38.5 | 182.7 | 118.6 |
| 54 | Boc-Pro-Sar-OBzl | 171 | 160 | 36.1 | −22.6 | −13.0 | −71.3 | 170.2 | 166.0 | 117.3 |
| 55 | Boc-Pro-Sar-OH | 172 | 182 | 29.8 | −7.9 | −7.0 | −67.3 | 151.6 | 176.0 | 116.6 |
| 56 | Boc-Pro-Sar-OH | 173 | 184 | 29.8 | −6.6 | −6.8 | −66.8 | 151.9 | 176.4 | 116.2 |
| 57 | Boc-Pro-Val-Gly-OH, mol. A | 174 | 174 | 29.3 | −11.5 | 8.0 | −81.1 | 163.1 | 187.8 | 118.3 |
| 58 | Boc-Pro-Val-Gly-OH, mol. B | 174 | 177 | 28.7 | −10.4 | 2.1 | −76.1 | 153.9 | 180.5 | 114.8 |
| 59 | Boc-Leu-Aib-Pro-OH | 175 | 174 | 39.9 | −16.7 | 176.8 | −71.3 | 160.9 | 175.0 | 120.3 |
| 60 | Boc-Leu-Aib-Pro-OBzl | 175 | 180 | 38.2 | −11.8 | 169.3 | −74.3 | 161.7 | 181.8 | 119.2 |
| 61 | Z-Aib-Pro-NHMe | 176 | 16 | 30.0 | 0.5 | −174.0 | −64.9 | −25.5 | 185.8 | 120.4 |

TABLE 7-7. Internal Coordinates (°) Obtained from Solid State Data as Explained in the Text (continued)

| Entry | Peptide[a] | Ref. | P | $\chi_M$ | $\chi^5$ | $\omega$ | $\phi$ | $\psi^b$ | $\kappa_N$ | $\kappa_c$ |
|---|---|---|---|---|---|---|---|---|---|---|
| 62 | Z-Aib-Pro-Aib-Pro-OMe, Pro-3 | 177 | 13 | 34.7 | 2.8 | -179.5 | -71.8 | -3.7 | 194.9 | 120.4 |
| 63 | Z-Aib-Pro-Aib-Pro-OMe, Pro-5 | 177 | 190 | 41.2 | -7.1 | -177.1 | -76.9 | 158.1 | 188.4 | 118.6 |
| 64 | H-Ala-Pro-p.NO₂An.HCl | 178 | 190 | 31.3 | -4.4 | -173.5 | -69.3 | 153.8 | 184.1 | 119.2 |
| 65 | Boc-Ala-Pro-OH | 179 | 173 | 33.4 | -14.4 | 170.4 | -71.8 | -21.7 | 177.2 | 119.7 |
| 66 | iB-Ala-Pro-NHiP | 180 | 195 | 31.0 | -1.4 | -178.8 | -67.2 | -22.3 | 188.0 | 122.2 |
| 67 | Piv-D-Ala-D-Pro-NHiP, mol. A (D) | 181 | 354 | 13.3 | -5.3 | -173.9 | 57.3 | -142.0 | 181.0 | 118.4 |
| 68 | Piv-D-Ala-D-Pro-NHiP, mol. B (D) | 181 | 177 | 35.5 | 12.6 | -179.6 | 82.7 | -156.3 | 187.9 | 117.8 |
| 69 | Piv-D-Ala-Pro-NHiP | 182 | 158 | 38.6 | -24.6 | 176.3 | -89.2 | 9.4 | 181.5 | 116.9 |
| 70 | S-Bzl-Cys-Pro-Leu-Gly-NH₂, mol. A | 183 | 174 | 33.3 | -13.8 | -175.5 | -70.2 | -16.0 | 177.5 | 121.1 |
| 71 | S-Bzl-Cys-Pro-Leu-Gly-NH₂, mol. B | 183 | 185 | 35.0 | -7.9 | -178.1 | -63.9 | -27.9 | 177.8 | 121.8 |
| 72 | Boc-Gly-Pro-OH | 144 | 178 | 35.9 | -12.3 | 177.4 | -70.5 | 155.5 | 177.3 | 119.1 |
| 73 | Boc-Gly-Pro-OBzl | 144 | 168 | 34.3 | -17.4 | -179.9 | -75.6 | 158.2 | 177.7 | 119.6 |
| 74 | Z-Gly-D,L-Pro-OH, mol. A (D) | 184 | 40 | 15.3 | 6.1 | -171.4 | 66.7 | -170.1 | 177.6 | 117.0 |
| 75 | Z-Gly-D,L-Pro-OH, mol. B | 184 | 151 | 25.4 | -18.4 | 171.4 | -69.5 | 173.3 | 174.5 | 123.4 |
| 76 | Z-Gly-Pro-OH | 185 | 177 | 34.9 | -12.3 | 179.6 | -69.5 | 150.9 | 175.6 | 118.4 |
| 77 | Z-Gly-Pro-Leu-OH (D) | 186 | 174 | 33.0 | 13.3 | 3.8 | 71.7 | -144.9 | 178.1 | 119.7 |

| No. | Compound | | | | | | | | |
|---|---|---|---|---|---|---|---|---|---|
| 78 | p.Br-Z-Gly-Pro-Leu-Gly-OH | 135 | 0 | 20.9 | 5.8 | −174.4 | −57.6 | −33.4 | 188.7 | 125.3 |
| 79 | o.BrZ-Gly-Pro-Leu-Gly-Pro-OH, Pro-3 | 187 | 351 | 26.9 | 12.7 | −168.7 | −65.0 | −26.7 | 191.5 | 113.9 |
| 80 | o.Br-Z-Gly-Pro-Leu-Gly-Pro-OH, Pro-6 | 187 | 174 | 38.1 | −14.6 | −177.8 | −72.2 | 159.2 | 175.2 | 117.6 |
| 81 | Z-Gly-Pro-Leu-Gly-Pro-OH, Pro-3 | 188 | 23 | 14.9 | −1.5 | −176.5 | −63.2 | −23.0 | 182.7 | 120.9 |
| 82 | Z-Gly-Pro-Leu-Gly-Pro-OH, Pro-6 | 188 | 188 | 33.3 | −5.6 | −178.8 | −67.1 | 135.1 | 179.3 | 117.8 |
| 83 | H-Leu-Pro-Gly-OH, 50% | 133 | 177 | 31.4 | −11.1 | 175.2 | −68.2 | 161.9 | 179.2 | 122.1 |
| 84 | H-Leu-Pro-Gly-OH, 50% | 133 | 45 | 23.0 | −11.1 | 175.2 | −68.2 | 161.9 | 179.2 | 122.1 |
| 85 | H-Leu-Pro-Gly-OH, 59% | 189 | 174 | 37.8 | −15.5 | 175.6 | −66.7 | 160.7 | 178.4 | 127.2 |
| 86 | H-Leu-Pro-Gly-OH, 41% | 189 | 15 | 46.5 | 2.6 | 175.6 | −66.7 | 160.7 | 178.4 | 109.1 |
| 87 | Z-Ala-D-Phe-Pro-OH | 190 | 187 | 28.2 | −5.0 | −174.4 | −66.7 | 157.1 | 181.4 | 119.7 |
| 88 | H-Tyr-Pro-Asp-Gly-OH | 191 | 359 | 32.7 | 10.8 | 176.1 | −53.3 | 141.8 | 185.5 | 121.4 |
| 89 | Boc-Val-Pro-Gly-Gly-OBzl | 192 | 20 | 33.4 | −1.2 | 179.0 | −62.3 | 135.5 | 179.7 | 118.6 |
| 90 | Boc-Val-Pro-Gly-Val-OH | 193 | 4 | 34.3 | 8.6 | 173.2 | −53.1 | 138.7 | 183.0 | 121.4 |
| 91 | Boc-Val-Pro-Gly-Val-OH · H$_2$O | 193 | 359 | 37.6 | 12.1 | −175.2 | −56.6 | 136.8 | 190.2 | 121.4 |
| 92 | Boc-Val-Pro-Gly-Val-Gly-OH | 194 | 172 | 37.8 | −16.8 | −176.5 | −88.2 | 144.8 | 188.2 | 116.8 |

[a]Abbreviations used: Piv = pivaloyl; p.Cl-Bzoyl = p-chlorobenzoyl; Boc = t-butyloxycarbonyl; Z = benzyloxycarbonyl; Lac = lactyl; Bzl = benzyl; Aoc = amyloxycarbonyl; Prop = propionyl; HSMeProp = 1-(D-3-mercapto-2-methylpropionyl)-; Aib = α-aminoisobutyryl; p.NO$_2$-An = p-nitroanilide; (o)p.BrZ = (o)p-bromobenzyloxycarbonyl.

[b]For a terminal carboxyl function, the ψ value is calculated based on the oxygen, reported to bear the H atom.

[c]Not considered in any calculation, since a tosylamide differs too much from a peptide bond.

TABLE 7-8. Internal Coordinates Obtained from Solid State Data as Explained in the Text

| Entry | Bond lengths (Å) | | | $\overline{C^\delta NC^\alpha}$ | $\overline{C^\alpha NC'}$ | $\overline{C'NC^\delta}$ | $\bar{\alpha}^a$ | $\Delta\omega$ | $\theta'_c$ | $\theta_N$ | $\tau$ |
| | $d(C^\delta-N)$ | $d(C^\alpha-N)$ | $d(C'-N)$ | | | | | | | | |
|---|---|---|---|---|---|---|---|---|---|---|---|
| 1 | 1.451 | 1.477 | 1.363 | 111.1° | 132.4° | 116.3° | 120.0° | −4.9° | 0.5° | 4.7° | 177.2° |
| 2 | 1.462 | 1.469 | 1.345 | 111.5 | 127.8 | 118.9 | 119.4 | 1.4 | 0.2 | 16.7 | 189.7 |
| 3 | 1.462 | 1.458 | 1.336 | 113.2 | 122.9 | 122.7 | 119.6 | −12.6 | −2.6 | 13.7 | −4.5 |
| 4 | 1.470 | 1.475 | 1.331 | 113.2 | 122.8 | 124.0 | 120.0 | −2.5 | −2.7 | −0.3 | −1.3 |
| 5 | 1.423 | 1.466 | 1.306 | 113.2 | 122.6 | 124.2 | 120.0 | 2.7 | 3.3 | −2.6 | −0.2 |
| 6 | 1.456 | 1.478 | 1.334 | 111.8 | 129.5 | 118.7 | 120.0 | −2.6 | −5.2 | −3.7 | 178.2 |
| 7 | 1.468 | 1.453 | 1.343 | 111.6 | 128.0 | 120.4 | 120.0 | 1.8 | 1.1 | −1.7 | 180.4 |
| 8 | 1.485 | 1.478 | 1.355 | 111.1 | 131.0 | 117.7 | 119.9 | 2.0 | 1.3 | −5.9 | 178.4 |
| 9 | 1.458 | 1.472 | 1.345 | 111.4 | 131.0 | 117.5 | 120.0 | 0.1 | −0.8 | 4.4 | 182.7 |
| 10 | 1.471 | 1.436 | 1.384 | 111.3 | 129.9 | 118.2 | 119.8 | −0.5 | 2.8 | 8.9 | 182.5 |
| 11 | 1.473 | 1.478 | 1.340 | 108.4 | 132.5 | 118.9 | 120.0 | −4.1 | 2.2 | 3.8 | 176.6 |
| 12 | 1.459 | 1.463 | 1.330 | 112.6 | 123.3 | 123.9 | 119.9 | −3.5 | 0.5 | 6.2 | −0.6 |
| 13 | 1.466 | 1.476 | 1.345 | 112.2 | 130.1 | 117.2 | 119.8 | 1.3 | 0.9 | 8.9 | 185.3 |
| 14 | 1.466 | 1.466 | 1.347 | 110.4 | 132.0 | 117.4 | 120.0 | −2.3 | −1.0 | 3.6 | 180.0 |
| 15 | 1.476 | 1.469 | 1.348 | 114.6 | 120.8 | 123.5 | 119.6 | −12.1 | −1.7 | 12.9 | −4.8 |
| 16 | 1.476 | 1.465 | 1.335 | 113.0 | 125.9 | 120.5 | 119.8 | 3.8 | −0.2 | −9.7 | 179.0 |
| 17[b] | 1.475 | 1.500 | 1.589 | 111.1 | 120.4 | 116.4 | 116.0 | | | | |
| 18 | 1.485 | 1.479 | 1.308 | 110.7 | 121.9 | 126.5 | 119.7 | 11.2 | 1.7 | −11.8 | 4.4 |
| 19 | 1.486 | 1.474 | 1.331 | 112.2 | 125.9 | 120.7 | 119.6 | 8.3 | −1.0 | 13.3 | 182.1 |
| 20 | 1.486 | 1.474 | 1.331 | 112.2 | 125.9 | 120.7 | 119.6 | 8.3 | −1.0 | −13.3 | 182.1 |
| 21 | 1.490 | 1.449 | 1.326 | 114.4 | 126.2 | 119.1 | 119.9 | 1.9 | 0.5 | 6.2 | 185.2 |
| 22 | 1.458 | 1.444 | 1.329 | 112.2 | 123.3 | 124.0 | 119.8 | −13.5 | −5.1 | 8.8 | −6.5 |
| 23 | 1.467 | 1.464 | 1.339 | 112.7 | 122.2 | 123.5 | 119.5 | −11.2 | −1.2 | 15.3 | −2.9 |
| 24 | 1.476 | 1.464 | 1.360 | 111.8 | 124.6 | 120.6 | 119.0 | 8.6 | −0.9 | −21.3 | 178.4 |
| 25 | 1.483 | 1.464 | 1.345 | 112.2 | 126.2 | 121.1 | 119.8 | 6.6 | −0.8 | −9.7 | 182.2 |
| 26 | 1.476 | 1.471 | 1.338 | 112.2 | 125.5 | 121.4 | 119.7 | 3.0 | −1.3 | −11.4 | 178.0 |
| 27 | 1.469 | 1.455 | 1.328 | 112.7 | 121.7 | 125.6 | 120.0 | −2.1 | 0.6 | −1.0 | −2.8 |
| 28 | 1.473 | 1.462 | 1.331 | 111.9 | 121.7 | 126.0 | 119.9 | −8.1 | −2.5 | 7.5 | −3.1 |

| | | | | | | | | | | | |
|---|---|---|---|---|---|---|---|---|---|---|---|
| 29 | 1.487 | 1.482 | 1.309 | 111.2 | 127.0 | 121.8 | 120.0 | −6.7 | −3.9 | 2.4 | 176.5 |
| 30 | 1.456 | 1.453 | 1.351 | 113.3 | 122.0 | 124.1 | 119.8 | −7.0 | 0.0 | 9.2 | −2.4 |
| 31 | 1.480 | 1.456 | 1.323 | 109.5 | 128.6 | 121.1 | 119.7 | 4.2 | −1.0 | −11.3 | 179.1 |
| 32 | 1.474 | 1.454 | 1.328 | 113.0 | 122.1 | 124.9 | 120.0 | 3.1 | 1.3 | 0.6 | 2.7 |
| 33 | 1.465 | 1.481 | 1.342 | 113.0 | 127.8 | 119.1 | 120.0 | 1.1 | −1.2 | −2.8 | 180.3 |
| 34 | 1.457 | 1.470 | 1.340 | 113.0 | 123.0 | 123.8 | 120.0 | −3.6 | 1.1 | 4.2 | −2.1 |
| 35 | 1.489 | 1.465 | 1.332 | 112.8 | 127.6 | 119.6 | 120.0 | −6.5 | −3.5 | −3.3 | 173.5 |
| 36 | 1.462 | 1.452 | 1.321 | 110.5 | 125.1 | 124.4 | 120.0 | −1.5 | 2.2 | 0.6 | −2.3 |
| 37 | 1.497 | 1.444 | 1.355 | 111.9 | 128.8 | 119.2 | 120.0 | −6.4 | −4.0 | −0.9 | 175.1 |
| 38 | 1.465 | 1.464 | 1.354 | 111.8 | 130.6 | 117.5 | 120.0 | 4.3 | 2.8 | −3.8 | 181.0 |
| 39 | 1.459 | 1.471 | 1.342 | 113.0 | 119.2 | 126.0 | 119.4 | −19.1 | −6.4 | 16.8 | −7.5 |
| 40 | 1.478 | 1.472 | 1.349 | 111.2 | 131.0 | 117.6 | 119.9 | −6.2 | −2.6 | 5.9 | 178.0 |
| 41 | 1.472 | 1.464 | 1.342 | 112.0 | 127.9 | 120.1 | 120.0 | 1.1 | 1.3 | −2.7 | 179.1 |
| 42 | 1.463 | 1.468 | 1.341 | 114.1 | 121.5 | 124.3 | 120.0 | −2.2 | −1.3 | −2.6 | −2.8 |
| 43 | 1.475 | 1.470 | 1.353 | 112.2 | 127.5 | 120.1 | 119.9 | 1.2 | −4.2 | −6.3 | 180.2 |
| 44 | 1.465 | 1.446 | 1.337 | 113.8 | 120.9 | 123.5 | 119.4 | −10.5 | −3.1 | 16.3 | −0.8 |
| 45 | 1.471 | 1.455 | 1.328 | 113.1 | 128.3 | 118.4 | 120.0 | −8.1 | −2.7 | 4.7 | 175.6 |
| 46 | 1.481 | 1.456 | 1.332 | 111.8 | 127.1 | 121.1 | 120.0 | 1.0 | −0.7 | −2.3 | 180.2 |
| 47 | 1.463 | 1.463 | 1.349 | 112.8 | 122.8 | 124.1 | 119.9 | −5.0 | −1.5 | 7.7 | −0.4 |
| 48 | 1.482 | 1.478 | 1.336 | 112.7 | 127.3 | 119.8 | 119.9 | 6.3 | 6.0 | 5.2 | 185.9 |
| 49 | 1.468 | 1.466 | 1.346 | 112.9 | 120.5 | 126.5 | 120.0 | 11.1 | 5.1 | −2.6 | 7.3 |
| 50 | 1.463 | 1.467 | 1.334 | 115.3 | 120.9 | 123.6 | 119.9 | −1.1 | −3.5 | −6.5 | −2.6 |
| 51 | 1.471 | 1.431 | 1.357 | 114.7 | 126.4 | 118.5 | 119.9 | −11.2 | −5.0 | 7.8 | 175.2 |
| 52 | 1.502 | 1.443 | 1.311 | 111.9 | 125.7 | 122.2 | 119.9 | −3.6 | 0.3 | 5.9 | 179.1 |
| 53 | 1.505 | 1.460 | 1.314 | 113.2 | 128.1 | 118.7 | 120.0 | 0.2 | −2.9 | −3.1 | 179.7 |
| 54 | 1.461 | 1.452 | 1.347 | 112.7 | 121.6 | 124.2 | 119.5 | −13.0 | −1.4 | 15.2 | −4.7 |
| 55 | 1.455 | 1.446 | 1.347 | 114.2 | 121.7 | 123.9 | 120.0 | −7.0 | −1.0 | 4.3 | −4.4 |
| 56 | 1.450 | 1.465 | 1.339 | 114.0 | 122.5 | 123.4 | 120.0 | −6.8 | −0.9 | 3.9 | −4.4 |
| 57 | 1.483 | 1.465 | 1.342 | 113.3 | 121.9 | 124.4 | 119.8 | 8.0 | 0.6 | −8.5 | 3.4 |
| 58 | 1.482 | 1.472 | 1.323 | 114.2 | 120.8 | 125.0 | 120.0 | 2.1 | 1.0 | −0.6 | 1.3 |
| 59 | 1.480 | 1.482 | 1.292 | 111.6 | 129.5 | 118.7 | 119.9 | −3.2 | −3.1 | 6.1 | 181.4 |
| 60 | 1.463 | 1.464 | 1.346 | 111.0 | 130.6 | 118.4 | 120.0 | −10.7 | −7.5 | −2.2 | 172.0 |

TABLE 7-8. Internal Coordinates Obtained from Solid State Data as Explained in the Text (*Continued*)

| Entry | Bond lengths (Å) | | | $\overline{C^\delta NC^\alpha}$ | $\overline{C^\delta NC'}$ | $\overline{C^\alpha NC'}$ | $\overline{\alpha}^a$ | $\Delta\omega$ | $\theta'_c$ | $\theta_N$ | $\tau$ |
| | $d(C^\delta - N)$ | $d(C^\alpha - N)$ | $d(C' - N)$ | | | | | | | | |
|---|---|---|---|---|---|---|---|---|---|---|---|
| 61 | 1.499 | 1.481 | 1.335 | 110.9 | 129.8 | 119.0 | 119.9 | 6.0 | 2.2 | −7.1 | 181.4 |
| 62 | 1.490 | 1.481 | 1.341 | 111.1 | 130.6 | 116.2 | 119.3 | 0.5 | −3.3 | −18.5 | 172.9 |
| 63 | 1.474 | 1.457 | 1.314 | 109.9 | 130.7 | 118.7 | 119.8 | 2.9 | −4.7 | −10.4 | 180.0 |
| 64 | 1.464 | 1.472 | 1.322 | 112.8 | 128.5 | 118.6 | 119.9 | 6.5 | −0.8 | −4.9 | 184.4 |
| 65 | 1.472 | 1.465 | 1.342 | 112.4 | 127.6 | 120.0 | 120.0 | −9.6 | −3.8 | 3.3 | 173.9 |
| 66 | 1.474 | 1.457 | 1.349 | 112.8 | 126.4 | 120.3 | 119.8 | 1.2 | −1.5 | −9.1 | 177.4 |
| 67 | 1.470 | 1.474 | 1.337 | 112.0 | 127.5 | 120.4 | 120.0 | 6.1 | 4.9 | 1.2 | 184.2 |
| 68 | 1.476 | 1.468 | 1.320 | 112.7 | 127.1 | 119.7 | 119.8 | 0.4 | 4.8 | 9.1 | 182.6 |
| 69 | 1.479 | 1.485 | 1.342 | 111.8 | 127.2 | 121.0 | 120.0 | −3.7 | 0.2 | −1.8 | 175.3 |
| 70 | 1.489 | 1.472 | 1.335 | 114.2 | 126.3 | 119.4 | 120.0 | 4.5 | −0.2 | 2.8 | 186.0 |
| 71 | 1.496 | 1.468 | 1.331 | 113.0 | 126.4 | 120.6 | 120.0 | 1.9 | −0.6 | 2.5 | 183.4 |
| 72 | 1.462 | 1.452 | 1.326 | 112.5 | 126.6 | 120.8 | 120.0 | −2.6 | −0.5 | 3.1 | 179.2 |
| 73 | 1.472 | 1.449 | 1.332 | 112.6 | 126.8 | 120.6 | 120.0 | 0.1 | −1.9 | 2.6 | 182.3 |
| 74 | 1.459 | 1.459 | 1.344 | 112.8 | 127.8 | 119.4 | 120.0 | 8.6 | 1.9 | −2.7 | 186.3 |
| 75 | 1.479 | 1.446 | 1.330 | 112.6 | 126.0 | 121.1 | 119.9 | −8.6 | −1.6 | 6.3 | 175.3 |
| 76 | 1.474 | 1.457 | 1.333 | 112.8 | 126.6 | 120.5 | 119.9 | −0.4 | −0.9 | 5.0 | 182.6 |
| 77 | 1.469 | 1.473 | 1.340 | 112.9 | 121.9 | 125.2 | 120.0 | 3.8 | 0.9 | −2.1 | 2.3 |
| 78 | 1.459 | 1.456 | 1.371 | 113.9 | 123.9 | 121.6 | 119.8 | 5.6 | −3.1 | −9.6 | 182.3 |
| 79 | 1.460 | 1.502 | 1.333 | 113.7 | 125.7 | 119.5 | 119.6 | 11.3 | −1.3 | −13.0 | 185.5 |

| | | | | | | | | | | | |
|---|---|---|---|---|---|---|---|---|---|---|---|
| 80 | 1.548 | 1.464 | 1.320 | 114.0 | 121.0 | 124.8 | 119.9 | −2.2 | 0.9 | 5.1 | 179.9 |
| 81 | 1.477 | 1.472 | 1.335 | 111.9 | 126.7 | 121.3 | 120.0 | 3.5 | −2.1 | −3.1 | 183.0 |
| 82 | 1.476 | 1.451 | 1.329 | 112.9 | 126.4 | 120.8 | 120.0 | 1.2 | 0.5 | 0.8 | 181.4 |
| 83 | 1.459 | 1.452 | 1.339 | 113.3 | 126.1 | 120.6 | 120.0 | −4.8 | −4.2 | 0.9 | 177.8 |
| 84 | 1.459 | 1.452 | 1.339 | 113.3 | 126.1 | 120.6 | 120.0 | −4.8 | −4.2 | 0.9 | 177.8 |
| 85 | 1.479 | 1.465 | 1.329 | 112.1 | 127.2 | 120.7 | 120.0 | −4.4 | −1.7 | 1.8 | 177.3 |
| 86 | 1.479 | 1.465 | 1.329 | 112.1 | 127.2 | 120.7 | 120.0 | −4.4 | −1.7 | 1.8 | 177.3 |
| 87 | 1.468 | 1.458 | 1.325 | 114.7 | 127.0 | 118.2 | 120.0 | 5.6 | 2.8 | −1.6 | 183.5 |
| 88 | 1.480 | 1.470 | 1.339 | 112.0 | 128.5 | 119.3 | 119.9 | −3.9 | −2.8 | −6.5 | 174.2 |
| 89 | 1.465 | 1.458 | 1.344 | 112.0 | 128.3 | 119.7 | 120.0 | −1.0 | −1.0 | 0.4 | 179.7 |
| 90 | 1.472 | 1.479 | 1.328 | 113.9 | 129.0 | 117.1 | 120.0 | −6.8 | −3.9 | −3.6 | 173.4 |
| 91 | 1.474 | 1.463 | 1.343 | 112.6 | 127.7 | 118.7 | 119.7 | 4.8 | −0.6 | −11.9 | 179.2 |
| 92 | 1.472 | 1.471 | 1.337 | 112.7 | 126.9 | 119.8 | 119.8 | 3.5 | −1.0 | −9.5 | 179.3 |
| $\bar{x}$ | 1.473 | 1.464 | 1.337 | 112.5 | 126.1 | 121.0 | 119.8 | | | | |
| Items | (88)[c] | (88)[c] | (88)[c] | (88)[c] | (88)[c] | (88)[c] | | | | | |
| s | 0.015 | 0.012 | 0.014 | 1.27 | 3.29 | 2.66 | | | | | |
| Max. | 1.548 | 1.502 | 1.384 | 115.3 | 132.5 | 126.5 | | | | | |
| Min. | 1.423 | 1.431 | 1.292 | 108.4 | 119.2 | 116.2 | | | | | |

[a] Mean of $\overline{C^\delta NC^\alpha}$, $\overline{C^\delta NC'}$, and $\overline{C^\alpha NC'}$.

[b] Not considered in any calculation, since a tosylamide differs too much from a peptide bond.

[c] Entries 17, 20, 84, and 86 are not considered for the calculation of the mean.

## B. Pyrrolidine Ring Conformational Behavior

Of the 92 proline rings we considered, 36 have a North conformation (mean form near envelope $\gamma^-$), while 56 prefer the South conformational space ($^\gamma_\beta T$) (see Section 4). When a scatter diagram of the Pro-ring conformations (Fig. 7-6) is plotted on the pseudorotational wheel, it is striking that the variations in phase angle $P$ (mean for North = 16.5 (19.11); South = 178.3 (8.3)] and maximum ring pucker $\chi_M$ [mean for North = 30.4 (9.9); South = 33.2 (6.6)] are more pronounced for the North than for the South region. This indicates that the energy minimum is better defined in the South region than in the North, which is not in agreement with calculations on the isolated molecule by Delos de Tar and Luthra[14] (see also Figure 7-18, in Section 4, which is adapted from Reference 14). We would conclude that crystal packing forces seem to be more able to disturb a North conformation than a South one.

In view of this statement, it seemed to be of interest to verify whether these ring shapes would deviate from pseudorotational behavior.

≪As discussed in Section 4, the ideal pseudorotation is the one whereby an equilateral five-membered ring system can be described by the following pseudorotation equation[150,151]:

$$\chi^{j+2} = \chi_M \cos\left(\frac{P + 4\pi j}{5}\right) \quad \text{with } j = 0, 1, \dots, 4$$

where $\chi^6 = \chi^4$ for Pro.≫

It is accepted[151–153] that for Pro systems with nonexaggerated pucker (eg,

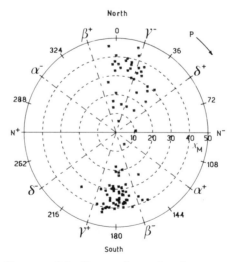

**Figure 7-6.** Scatter diagram of the Pro conformational parameters $P$ and $\chi_M$ on the pseudorotational wheel.

$\chi_M < 50°$), the deviations in $\chi_j$ should not exceed $0.6°$.[154] Our calculations disclose (not shown in the tables of Section 3) that the Pro ring in most of its derivatives can perfectly be described by pseudorotation, corroborating the results of Delos de Tar and Luthra.[14] No discrimination either for the North or the South was disclosed.

## C. Backbone Conformation

When the Ramachandran plot (Figure 7-7) of these proline residues is compared with Figure 7-2, it is clear that only three regions are occupied: F, A, and F*. In fact this is not too surprising because $\phi$ is restricted by the Pro ring, and the carboxyl function takes only the cisoid and transoid positions (cf discussion accompanying Figure 7-2).

Since this is quite a crude classification, we wondered whether it is possible to extract from this experimental material a more refined classification on the basis of the observed $\phi_i$, $\psi_i$, and $\omega_{i-1}$ values of proline. A cluster analysis

≪In a cluster analysis of $m$ items, each characterized by $n$ parameters, $m$ points are placed in a $n$-dimensional space. In this space the distance between each point and the others is calculated. On the basis of these distances, "dense" regions can be detected, which are also called "clusters." The program used was CLUSTAN2, version 2.1.[195]≫

with these values as input disclosed that there were in fact six to seven clusters to be distinguished (Table 7-9).

The bends situated in group 1 are mainly of type II. The members of this group that do not form a bend have two successive Pro units and/or a carboxy terminal Pro unit. Item 88 is an exception.

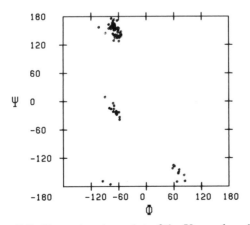

**Figure 7-7.** Ramachandran plot of the X-rayed prolines.

TABLE 7-9. Clustering on the Basis of $\phi_i$, $\psi_i$, and $\omega_{i-1}$ Values: Standard Deviations Given in Parentheses

| Group | Entry | Torsional angles (°) | | |
|-------|-------|------------|----------|---------|
| | | $\omega_{i-1}$ | $\phi_i$ | $\psi_i$ |
| 1 | 1,[a] 2,[a] 6,[a] 7,[a] 8,[a] 10,[a] 14,[a] 16,[a] 21,[a] 33, 38, | 179.5 | −60.0 | 137.2 |
| | 45, 82, 88, 89,[a] 90,[a] 91[a] | (3.8) | (4.9) | (4.7) |
| 2 | 3,[b] 4, 5, 12, 15, 18, 22, 23, 28, 30, 32, 34, 36, | −3.6 | −68.5 | 155.8 |
| | 42, 44, 49, 50, 54, 55, 56, 57, 58 | (7.6) | (7.2) | (10.0) |
| 3 | 9, 11, 13, 29, 31, 46, 51, 52, 59, 60, 63,[b] 64, | 178.2 | −72.1 | 159.3 |
| | 72, 73, 75, 76, 80,[b] 83, 85, 87, 92 | (5.0) | (7.4) | (6.5) |
| 4 | 19,[a] 24,[a] 25, 26, 41,[a] 43,[a] 53, 61,[a] 62,[a] 65, 66, | 182.9 | −67.6 | −20.5 |
| | 69,[a] 70,[a] 71,[a] 78,[a] 79,[a] 81[a] | (4.9) | (8.3) | (10.9) |
| 5 | 27 | −2.1 | −69.7 | −9.1 |
| 6 | 35, 37, 40,[a] 48, 67, 68, 74 | 180.3 | 66.7 | 149.3 |
| | | (6.7) | (8.7) | (12.1) |
| 7 | 39, 47, 77 | −6.8 | 75.6 | 154.4 |
| | | (11.6) | (7.6) | (12.5) |

[a]Reported to form a bend, with Pro part of it.
[b]Reported to form a bend, but with Pro outside the bend.

The members of group 2 all possess a *cis*-amide (or urethane) function.

Although members of group 3 do not differ significantly by $\phi,\psi$ from those belonging to group 1, it is the more remarkable that our clustering discriminates them by the failure of Pro to take part in bend formation.

Group 4 consists of peptides that mainly form a type I bend. Groups 6 and 7 are D-residues, but since there are only 10 cases, they are classified roughly as *cis*- and *trans*-amides (or urethanes). Item 40 of group 6 can be viewed as the mirror image belonging to group 1.

As a rule of thumb, we can state that if proline is built in a peptide and the $\omega_{i-1}$, $\phi_i$, and $\psi_i$ values fall within the limits of group 1 or 4, the propensity for Pro participation in bend formation is high, except when Pro is the terminal amino acid.

In all these bends, proline occupies the $(i + 1)$-position, except for entry 61, where it takes the $(i + 2)$-position, which is quite unusual, and is classified in group 4 (bend I).

## D. Nitrogen: A Link Between the Conformation of the Pro Ring and the Backbone?

The mean values of $\overline{C^\delta NC'}$ and $\overline{C^\alpha NC'}$, as indicated by their exaggerated $\Delta$ values in Table 7-8, cannot be considered to be representative. The histograms of these values (Figures 7-8 and 7-9) make this clear. The distribu-

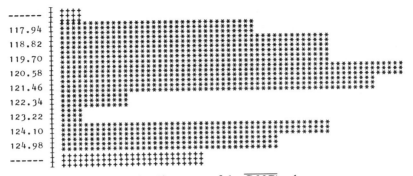

**Figure 7-8.** Histogram of the $\overline{C^nNC'}$ values.

tions cannot be considered to be "normal"; at least two subpopulations must be considered. The distribution of the ring valence angle $\overline{C^\delta NC^\alpha}$, on the contrary, seems to be uniform (not shown), and therefore it is representative for the whole group.

A cluster analysis, with these three valence angles as input, revealed that finally three subgroups are present (Table 7-10):

Cluster 1 groups the *trans*-amide Pro fragments, for which the preceding substituent possesses a bulky tertiary carbon atom, such as pivaloyl or $\alpha$-amino isobutyryl. This explains the increase in $\overline{C^\delta NC'}$ (decrease in $\overline{C^\alpha NC'}$).

Cluster 2 groups the other *trans*-amide Pro fragments, and indeed compared to the first group, $\overline{C^\delta NC'}$ is diminished by 2.6° and $\overline{C^\alpha NC'}$ is 2° larger, due to lack of steric hindrance. These values can be considered to be representative.

Cluster 3 gives the typical values for the *cis*-amide (and urethane) Pro fragments.

**Figure 7-9.** Histogram of the $\overline{C^\delta NC'}$ values.

TABLE 7-10. Clustering Based on the Three Valence Angles (°) Around the Pro-Nitrogen: Standard Deviations Given in Parentheses

| Cluster | Entry | $\overline{C^{\delta}NC^{\alpha}}$ | $\overline{C^{\delta}NC'}$ | $\overline{C^{\alpha}NC'}$ |
|---|---|---|---|---|
| 1 | 1, 8, 9, 10, 11, 13, 14, 31, 38, 40, 59, 60, 61, 62, 63 | 110.9 (1.0) | 130.7 (1.1) | 118.0 (1.2) |
| 2 | 2, 6, 7, 16, 19, 21, 24, 25, 26, 29, 33, 35, 37, 41, 43, 45, 46, 48, 51, 52, 53, 64, 65, 66, 67, 68, 69, 70, 71, 72, 73, 74, 75, 76, 78, 79, 81, 82, 83, 85, 87, 88, 89, 90, 91, 92 | 112.6 (0.8) | 127.1 (1.2) | 120.0 (1.0) |
| 3 | 3, 4, 5, 12, 15, 18, 22, 23, 27, 28, 30, 32, 34, 36, 39, 42, 44, 47, 49, 50, 54, 55, 56, 57, 58, 77, 80 | 113.1 (1.0) | 122.0 (1.1) | 124.5 (1.0) |

In groups 1 and 2, $\overline{C^{\delta}NC'}$ was larger each time than $\overline{C^{\alpha}NC'}$. Typically this situation is reversed for cis fragments, but not to the same extent. It is remarkable that although item 80 is a *trans*-amide junction, it possesses values typical for a cis fragment and was classified in group 3. Some of these numbers deviate appreciably from the typical values reported for acyclic amino residues (Table 7-11).

It has been found[14] that $|\chi^5 - \phi|$ values may vary widely (eg, $65 \pm 15°$). Whereas it can be stated (see Figure 7-10) that:

$$\chi^5 - \phi = \kappa_N - \kappa_c$$

with $\kappa_N = \overline{C^{\delta}NC'_{i-1}}$ and $\kappa_c = \overline{C^{\beta}C^{\alpha}C'_i}$, two mechanisms may lie at the origin of the discrepancies:

1. A bond angle deformation at $C^{\alpha}$; thus $\kappa_c$ deviates in its trigonal projection from 120°.
2. A nonplanarity of the amide bond ($\kappa_N \neq 180°$).

Angles $\theta_N$, $\theta'_c$, and $\tau$ used in the literature[196,197] for the description of nonplanarity of the amide bond

$$\omega(C^{\alpha}_i C'NC^{\alpha}_{i+1}) = \omega_1 = \omega_{i-1} \qquad \theta'_C = \omega_1 - \omega_3 + \pi = -\omega_2 + \omega_4 + \pi$$
$$\omega(OC'NC^{\delta}) = \omega_2 \qquad\qquad\quad \theta_N = \omega_2 - \omega_3 + \pi = -\omega_1 + \omega_4 + \pi \qquad (\text{mod } 2\pi)$$
$$\omega(OC'NC^{\alpha}) = \omega_3 \qquad\qquad\quad \tau = \tfrac{1}{2}(\omega_1 + \omega_2)$$
$$\omega(C^{\alpha}_i C'NC^{\delta}) = \omega_4$$

TABLE 7-11. Standard Values of Peptide Fragments for Acyclic Amino Acid Residues

| | $\overline{HNC^{\alpha}}$ or $\overline{C^{\delta}NC^{\alpha}}$ | $\overline{HNC'}$ or $\overline{C^{\delta}NC'}$ | $\overline{C^{\alpha}NC'}$ | $d(C'_0-N)$ | Ref. |
|---|---|---|---|---|---|
| trans | 113.4° | 121.9° | 119.7° | 1.33 Å | 140 |
| cis | 113° | 121° | 126° | 1.32 Å | 142 |

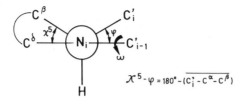

$$\chi^5 - \varphi = 180° - (\overline{C_i' - C^\alpha - C^\beta})$$

**Figure 7-10.** Projection on a Newman circle along the NC$^\alpha$ bond.

are listed in Table 7-8: $\theta_N$ is the nonplanarity of nitrogen; $\theta_C'$ is the nonplanarity of the carbonyl function, and $\tau$ is a measure for the twist of the amide bond. As Table 7-8 indicates, the deformation at nitrogen is in general more appreciable than that at carbonyl, but there are exceptions (entries 4, 5, 6, 29, 32, 36, 37, 48, 49, 58, 60, 67, 83, and 87); $\theta_N$ can take values as high as $-21.8°$ (entry 24).

The Pierson correlation coefficient between $\theta_N$ and $\theta_C'$ is zero when all cases are considered, but when restricted to peptides that form a bend, the correlation coefficient is 0.30. While this is still a small value, the difference between the two classes is remarkable. It means that there exists a certain tendency in bends to receive the tension engendered by the formation of a 10-membered ring in deformations at both nitrogen and carbonyl. We did not find any correlation between $\tau$ or $\theta_N$ and the carbon–nitrogen bond length $d(C'-N)$; hence the deformation of nitrogen or the twist of the amide bond is apparently not reflected in a change of this carbon–nitrogen bond length. In other cases such a correlation has been observed.[143,198] In general we were not able to find any trends for the three bond lengths around nitrogen. Ramachandran and co-workers[21,197] have stated that $\Delta\omega$

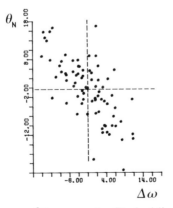

**Figure 7-11.** Scatter diagram of $\theta_N$ versus $\Delta\omega$. Dashed lines are the mean values of the two parameters.

TABLE 7-12. North Conformations

|              | $P$     | $\chi^5$ |
|--------------|---------|----------|
| $\kappa_N$   | -0.7399 | 0.7404   |
| $\theta_N$   | 0.7282  | -0.7379  |

≪For *trans*-amides $\Delta\omega = \omega - 180°$ (mod $2\pi$); for *cis*-amides $\Delta\omega = \omega - 0°$.≫

is correlated to $\theta_N$ by the relation $\theta_N = -a\Delta\omega$, with $a$ having a value between 1.5 and 2.0. From the present proline data, we calculated a correlation coefficient of $-0.70$ for all data; the correlation being larger for the *trans* amides ($-0.86$) than for the *cis* amides ($-0.54$). The scatter diagram (Figure 7-11) shows that a value of $a \approx 1$ is better suited.

To search for any link between the pyrrolidine ring conformation on the one hand and the nonplanarity at nitrogen on the other, we calculated the correlation matrices for $\kappa_N, \theta_N$ with $P$ and $\chi^5$ for the Pro ring in the North and in the South regions (Tables 7-12 and 7-13, respectively).

It is clear from Tables 7-12 and 7-13 that definitive correlations exist between the parameters of interest. Because the sampling area of the phase angle $P$ is larger in the North than in the South region (see subsection B, above, "Pyrrolidine Ring Conformational Behavior"), the North forms show a higher correlation between the parameters for nonplanarity around nitrogen and the Pro-ring conformational parameters. Figures 7-12 through 7-15 are the scatter diagrams. Since the dashed lines represent the mean value of the plotted parameters, it is clear that the populations of the quadrants are different.

We finally also calculated the correlation between $\kappa_C$ and the Pro-ring conformation, but it was found to be negligible.

In conclusion we would state that there exists a possibility that part of $|\chi^5 - \phi|$ is relaxed in changes of the pyrrolidine ring conformation. As a consequence, but only to a small extent, the $\phi$ value can be considered to contribute to the ring shape (or the reverse). This is in accord with earlier theoretical considerations[3] and experimental evidence[8] whereby a transmission effect backbone-ring shape seems to be weak, except perhaps for more exaggerated $\phi$ values.

TABLE 7-13. South Conformations

|              | $P$     | $\chi^5$ |
|--------------|---------|----------|
| $\kappa_N$   | 0.4398  | 0.3652   |
| $\theta_N$   | -0.4347 | -0.3594  |

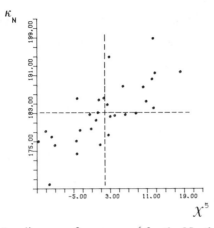

**Figure 7-12.** Scatter diagram of $\kappa_N$ versus $\chi^5$ for the North conformations.

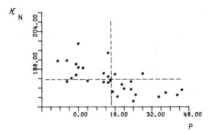

**Figure 7-13.** Scatter diagram of $\kappa_N$ versus $P$ for the North conformations.

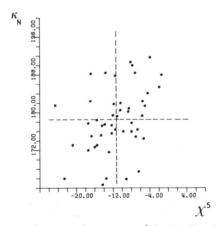

**Figure 7-14.** Scatter diagram of $\kappa_N$ versus $\chi^5$ for the South conformations.

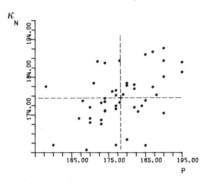

**Figure 7-15.** Scatter diagram of $\kappa_N$ versus $P$ for the South conformations.

# 4. SYMBOLS, NOMENCLATURE, AND CONFORMATIONAL NOTATION

## A. Peptide Chain Conformation

The torsion angles along the peptide backbone, $\phi$, $\psi$, and $\omega$, are described following the 1971 IUPAC conventions,[146] represented in Fig. 7-16; a positive value is assigned to a clockwise rotation. The conformation of the amino acid side chain residue can be described in terms of $\chi_i^j$, where $j = 1$ for the first side chain torsion angle ($NC^\alpha C^\beta C^\gamma$) and $i$ stands for the $i$th amino acid in the peptide chain, starting at the amino terminus.

## B. Pyrrolidine Conformation

Among the many proposed detailed descriptions (eg, References 147 and 199) of the conformation of the proline ring, we refer to the one of Altona, Geise, and co-workers[113,150,151,156] (except for the definition of endo-exo), who rationalize the conformation of the five-membered ring in terms of a phase angle ($P$) and the maximum pucker ($\chi_M$). The parameter $P$ can be under-

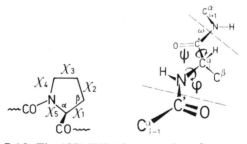

**Figure 7-16.** The 1971 IUPAC conventions for $\phi$, $\psi$, and $\omega$.

stood as describing the shape of the ring, while $\chi_M$ is a measure for the intensity of that pucker.

Given five endocyclic torsion angles (eg, from X-ray data), the ring conformation parameters can be calculated according to the following equations:

$$\tan P = \frac{(\chi^1 + \chi^4) - (\chi^3 + \chi^5)}{3.0777\chi^2}$$

$$\chi_M = \frac{\chi^2}{\cos P}$$

The individual torsion angles $\chi^j$ in function of $P$ and $\chi_M$ may be calculated using the equation:

$$\chi^j = \chi_M \cos [P + (j - 2)\delta]$$

where $j$ = 2, 3, 4, 5, and 6 with $\chi^6$ equivalent to $\chi^1$
$\delta = 144$

The accordance between experimental values and the internal torsion angles calculated as above was better than 0.5°.

The phase angle $P$ varies between 0° and 360°, whereby each multiple of 18° coincides with a characteristic form $E$ (envelope) or $T$ (twist or "half-chair"). Arbitrarily, $P$ was set zero for the North conformation ($^\beta_\gamma T$ or $\beta^+ - \gamma^-$, representing in Figure 7-17); the antimeric skeleton ($P$ = 180°) is then the limiting South conformation.

In this way any conformation of a five-membered ring can be mapped (Figure 7-18), with $P$ being the angle starting at the arbitrary zero and $\chi_M$ being the distance from the middle of the conformational wheel. The upper hemisphere in the wheel ($-90 < P < 90$) thus defines the N-conformational region. In this representation, the center represents a completely flat ring system.

There has been much confusion in the literature about shorthand notations for envelopes and twists. We use[14,155] a *positive* sign in combination with the symbol of a well-defined ring atom to indicate a position of that atom to be *at the same side* as C' (the carboxamide grouping as the reference); hence $\gamma^+ = {}^\gamma E = \gamma_{endo}$ (envelope with C$^\gamma$ endo, ie, with C$^\gamma$ as the flap up on the same side as C'). Similarly, twist forms may be indicated by, for example, $^\beta_\gamma T$ (C$^\beta$ up or endo and C$^\gamma$ down or exo) (Figure 7-17). Alternatively,

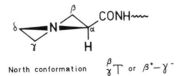

North conformation          $^\beta_\gamma T$ or $\beta^+ - \gamma^-$

**Figure 7-17.** The North conformation $^\beta_\gamma T$ (or $\beta^+ - \gamma^-$), represented for an L-residue.

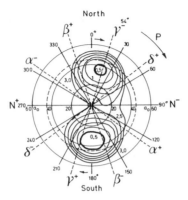

**Figure 7-18.** Conformational wheel representing calculated isoenergetic lines (kcal/mol) for *trans*-Ac-Pro-OMe showing the puckering amplitude $\chi_M$ being radial and the angular phase angle $P$. The limiting N conformation is at $P = 0°$ (north pole), the N hemisphere ranging from $-90°$ (or 270°) at $N^+$ ($^NE$, envelope with nitrogen as the flap up and at the same side of C'), to $+90°$ at $N^-$. (Adapted from Reference 14.)

the notation $\beta^+ - \gamma^-$ may be used. Thus, by the notation $(\gamma^+, \gamma^+ - \beta^-)$ we denote an in-between form located between the $^\gamma E$ and $^\gamma_\beta T$ ($P \simeq 189°$). A symmetrical twist has $C_2$ symmetry, while an envelope has $C_v$. Figure 7-18 represents the conformational wheel for *trans*-Ac-Pro-OMe, showing calculated energy contours, adapted from DeTar and Luthra[14] and consistent with the present symbolization.

It is also of general usage in structural descriptions of proline to distinguish two ring types with notations A and B.[147] According to this definition, A designates the $\gamma^-$ zone, characterized by a negative endocyclic $\chi^1$ value (for an L-residue), and B designates the $\gamma^+$ zone (a positive $\chi^1$ value for L). These forms define in fact regions of (usually) global energy minima in the conformational space of the pyrrolidine ring.[147] In this notation, the endocyclic torsion angle $\chi^5$ is also symbolized as $\theta$. A typical A form therefore has $-10° < \theta < 15°$ (usually slightly positive) together with a negative $\chi^1$ value, while for the B form $-10° < \theta < 0°$ (usually slightly negative) together with a positive $\chi^1$ value. The A and B conformational regions have their transition point at $\chi^1 = 0$ ($P = +54$, $^\delta E$, $\delta^+$, and $P = 234$, $_\delta E$, $\delta^-$), while the north and south hemispheres have their transition at $\chi^2 = 0$ ($P = 90°$, $_N E$, $N^-$, and $P = -90°$, $^N E$, $N^+$). As seen in Figure 7-18, the latter represents a maximum in energy, with $\chi^1$ usually of about 35° according to Reference 14.

In this latter classification, the conformations of an L-Pro ring can be placed into two classes depending on the sign of $\chi^2$, the N type being characterized by a positive $\chi^2$ value ($-90° < P < 90°$) and the S type by a negative value of $\chi^2$ ($90° < P < 270°$). For D-proline fragments the signs are of course reversed, for these become mirror images.

# REFERENCES

1. Zimmerman, S.S.; Scheraga, H.A. *Biopolymers* **1977**, *16*, 811.
2. Ramachandran, G.N.; Lakshminarayan, A.V.; Balasubramanian, R.; Tegoni, G. *Biochim. Biophys. Acta* **1970**, *221*, 165.
3. Madison, V. *Biopolymers* **1977**, *16*, 2671.
4. Torchia, D.A. *Macromolecules* **1971**, *4*, 440.
5. Tanaka, S.; Scheraga, H.A. *Macromolecules* **1975**, *8*, 504.
6. Venkatalacham, C.M.; Price, B.J.; Krimm, S. *Biopolymers* **1975**, *14*, 1121.
7. Gō, N.; Scheraga, H.A. *Macromolecules* **1970**, *3*, 188.
8. Anteunis, M.J.O.; Callens, R.; Asher, V.; Sleeckx, J. *Bull. Soc. Chim. Belg.* **1978**, *87*, 41.
9. Smith, J.A.; Pease, L.G. In "Critical Reviews in Biochemistry," Vol. 8(4), Fasman, G.D., Ed.; CRC Press: Boca Raton Fla., 1980, pp. 315–400.
10. Pogliani, L.; Ellenberger, M.; Valat, J. *Org. Magn. Reson.* **1975**, *7*, 61.
11. Jankowski, K.; Söler, F.; Ellenberger, M. *J. Mol. Struct.* **1978**, *48*, 63.
12. DeLos, F. De Tar; Luthra, N.P. *J. Org. Chem.* **1979**, *44*, 3299.
13. Nishikawa, K.; Ooi, T. *Prog. Theor. Phys.* **1971**, *46*(2), 670.
14. DeLos, F. De Tar; Luthra, N.P. *J. Am. Chem. Soc.* **1977**, *99*, 1232.
15. (a) Pullman, B. In "Proteins, Nucleic Acids and Their Constituents in Quantum Mechanics of Molecular Conformations," Pullman, B., Ed.; Wiley: New York, 1970, Chapter 4, pp. 296–377. (b) Pullman,B.; Maigret, B.; Peraha, D. *Theor. Chim. Acta* **1970**, *18*, 44. (c) Pullman, B. *Int. J. Quantum Chem.* **1971**, *4*, 319.
16. Hodes, Z.I.; Némethy, G.; Scheraga, H.A. *Biopolymers* **1979**, *18*, 1565, 1611.
17. Higashijima T.; Tasumi, M.; Miyazawa, T. *Biopolymers* **1977**, *16*, 1259.
18. London, R.E. *J. Peptide Protein Res.* **1979**, *14*, 377.
19. Winkler, F.K.; Dunitz, J.D. *J. Mol. Biol.* **1971**, *59*, 169.
20. Ramachandran, G.N.; Lakhsminarayan, A.V.; Sasisekharan, A.S. *Biochim. Biophys. Acta* **1973**, *303*, 8.
21. Ramachandran, G.N.; Kolaskar, A.S. *Biochim. Biophys. Acta* **1973**, *303*, 385.
22. Kolashkar, A.S.; Lakshminarayanan, A.V.; Sasisekharan, V. *Biopolymers* **1975**, *14*, 1081.
23. Scheiner, S.; Kern, C.W. *J. Am. Chem. Soc.* **1977**, *99*, 7042.
24. Karle, I.L. *J. Am. Chem. Soc.* **1979**, *101*, 181.
25. (a) Benedetti, E.; Christensen, A.; Gilon, C.; Fuller, W.; Goodman, M. *Biopolymers* **1976**, *15*, 2523. (b) Mabouzami, T.; Iitama, Y. *Acta Crystallogr. Sect. B* **1971**, *27*, 507.
26. Cf Callens, R.E.A.; Anteunis, M.J.O. *Biochim. Biophys. Acta* **1979**, *577*, 337.
27. London, R.E. *J. Am. Chem. Soc.* **1978**, *100*, 2678, and references cited therein.
28. Haasnoot, C.A.G. Ph.D. Thesis Leiden, 1981, Chapter VI, pp. 58–73.
29. Cuypers, P.; Borremans, F.; Sleeckx, J.; Anteunis, M.J.O. To be published.
30. Prange, T.; Garbay-Jaureguiberry, C.; Roques, B.; Anteunis, M. *Biochem. Biophys. Res. Commun.* **1974**, *67*, 104–109.
31. (a) Liquori, A.M. *Q. Rev. Biophys.* **1969**, *2*, 65. (b) Zimmerman, S.S.; Pottle, M.S.; Némethy, G.; Scheraga, H.A. *Macromolecules* **1977**, *10*, 1.
32. Clothia, C. *Nature* **1975**, *254*, 304; Kuntz, I.D. *J. Am. Chem. Soc.* **1972**, *94*, 4009.
33. Lewis, P.N.; Momany, F.A.; Scheraga, H.A. *Biochim. Biophys. Acta* **1973**, *303*, 211.
34. Venkatalacham, C.M. *Biopolymers* **1968**, *6*, 1425.
35. Chou, P.Y.; Fasman, G.D. *Adv. Enzym.* **1978**, *47*, 45.
36. Burgess, A.W.; Ponnuswamy, P.K.; Scheraga, H.A. *Isr. J. Chem.* **1974**, *12*, 239.
37. Matthews, B.W. *Macromolecules* **1972**, *5*, 818.
38. Némethy, G.; Printz, M.P. *Macromolecules* **1972**, *5*, 755.
39. See, eg, Kessler, H. *Angew. Chem. Int. Ed. Engl.* **1982**, *21*, 512, and references cited therein.
40. Eg, Matthews, B.W.; Weaver, L.H.; Kester, W.R. *J. Biol. Chem.* **1974**, *249*, 8030.
41. Chandrasekaran, R.; Lakshminarayanan, A.V.; Pandya, U.V.; Ramachandran, G.N. *Biochim. Biophys. Acta* **1973**, *303*, 14.
42. Lewis, P.N.; Momany, F.A.; Scheraga, H.A. *Proc. Natl. Acad. Sci, U.S.A.* **1971**, *68*, 2293.
43. Crawford, J.L.; Lipscomb, W.N.; Schellman, C.G. *Proc. Natl. Acad. Sci. U.S.A.* **1973**, *70*, 538.
44. Glickson, J.D.; Cunningham, W.D.; Marshall, C.R. *Biochemistry* **1973**, *12*, 3684.

45. Matheson, R.R., Jr.; Scheraga, H.A. *Macromolecules* **1978**, *11*, 819.
46. Robson, B.; Pain, R.H. *J. Mol. Biol.* **1971**, *58*, 237.
47. Finkelstein, A.V.; Ptitsyn, O.B. *J. Mol. Biol.* **1971**, *62*, 613.
48. Wu, T.T.; Kabat, E.A. *J. Mol. Biol.* **1973**, *75*, 13.
49. Chou, P.Y.; Fasman, G.D. *Biochemistry* **1974**, *13*, 222.
50. Chou, P.Y.; Fasman, G.D. *J. Mol. Biol.* **1977**, *115*, 135.
51. Howard, J.C.; Ali, A.; Scheraga, H.A.; Momany, F.A. *Macromolecules* **1975**, *8*, 607.
52. Schellman, J.A.; Schellman, C. In "The Proteins," Vol. II, 2nd ed., Neurath, H., Ed.; Academic Press: New York, 1964.
53. Traub, W.; Schmueli, U. *Nature* **1963**, *198*, 1165.
54. Ramachandran, G.N.; Mitra, A.K. *J. Mol. Biol.* **1976**, *107*, 85.
55. Kaufmann, *W. Adv. Protein Chem.* **1959**, *14*, 1.
56. Clothia,C. *Nature* **1974**, *248*, 338.
57. Cramm, R.D., III. *J. Am. Chem. Soc.* **1977**, *99*, 5408.
58. Poland, D.C.; Scheraga, H.A. *Biopolymers* **1965**, *3*, 321.
59. Némethy, G.; Scheraga, H.A. *J. Phys. Chem.* **1962**, *66*, 1773.
60. Nishikawa, K.; Momany, F.A.; Scheraga, H.A. *Macromolecules* **1974**, *7*, 797.
61. Simon, I.; Némethy, G.; Scheraga, H.A. *Macromolecules* **1978**, *11*, 797.
62. Scheraga, H.A. *Pure Appl. Chem.* **1973**, *36*, 1.
63. Lewis, P.N.; Momany, F.A.; Scheraga, H.A. *Isr. J. Chem.* **1973**, *11*, 121.
64. Lewis, P.N.; Scheraga, H.A. *Arch. Biochem. Biophys.* **1971**, *144*, 576.
65. Anderson, J.S.; Scheraga, H.A. *Macromolecules* **1978**, *11*, 812.
66. Zimmerman, S.S.; Scheraga, H.A. *Biopolymers* **1978**, *17*, 1849.
67. Zimmerman, S.S.; Scheraga, H.A. *Biopolymers* **1978**, *17*, 1871.
68. Manavalan, P.; Ponnuswamy, P.K. *Arch. Biol. Biophys.* **1977**, *184*, 476.
69. Warme, P.K.; Morgan, R.S. *J. Mol. Biol.* **1978**, *118*, 273.
70. Warme, P.K.; Morgan, R.S. *J. Mol. Biol.* **1978**, *118*, 289.
71. Morgan, R.S.; Tatsch, C.E.; Gushard, R.H.; McAdon, J.M.; Warme, P.K. *Int. J. Peptide Protein Res.* **1978**, *11*, 209.
72. Morgan, R.S.; McAdon, J.M. *Int. J. Peptide Protein Res.* **1980**, *15*, 177.
73. Lynn Bodner, B.; Jackman, L.; Morgan, R.S. *Biochem. Biophys. Res. Commun.* **1980**, *94*, 807.
74. Efremov, E.S.; Kostetsy, P.V.; Ivanov, V.T.; Popov, E.M.; Ovchinnikov, Y.A. *Khim. Prir. Soedin (U.S.S.R.)* **1973**, 348.
75. Dickerson, R.E. In "The Chemical Basis of Life." W.H. Freeman: San Francisco, 1973, pp. 83–95.
76. Chou, P.Y.; Fasman, G.D. *Biochemistry* **1974**, *13*, 211.
77. Szent-Györgi, A.G.; Cohen, C. *Science* **1957**, *126*, 3276.
78. Torchia, D.A.; Bovey, F.A. *Macromolecules* **1971**, *4*, 246.
79. Dorman, D.E.; Torchia, D.A.; Bovey, F.A. *Macromolecules* **1973**, *6*, 80.
80. Kricheldorf, H.R. *Makromol. Chem.* **1978**, *179*, 247.
81. Tanaka, S.; Scheraga, H.A. *Macromolecules* **1975**, *8*, 504, 516.
82. Balaram, H.; Prasad, B.V.V.; Balaram, P. *Biochem. Biophys. Res. Commun.* **1982**, *109*, 825.
83. Wüthrich, K.W.; Tun-Kiji, A.; Schwyzer, R. *FEBS Lett.* **1972**, *25*, 104.
84. Lin, L.-N.; Brandts, J.F. *Biochemistry* **1979**, *18*, 43.
85. Garel, J-R. *Proc. Natl. Acad. Sci. U.S.A.* **1980**, *77*, 795.
86. Kopple, K.D.; Go, A.; Pilipauskas, D.R. *J. Am. Chem. Soc.* **1975**, *97*, 6830.
87. Damiani, A.; De Santis, P. *Nature* **1970**, *226*, 542.
88. Maigret, B.; Pullman, B.; Caillet, J. *Biochem. Biophys. Res. Commun.* **1970**, *40*, 808.
89. Schimmel, P.R.; Flory, P.J. *Proc. Natl. Acad. Sci. U.S.A.* **1967**, *58*, 52.
90. Brant, D.A.; Miller, W.G.; Flory, P.J. *J. Mol. Biol.* **1967**, *23*, 47.
91. Kessler, H., Hehlein, M.; Schuck, R. *J. Am. Chem. Soc.* **1982**, *104*, 4534.
92. Kopple, K.D.; Go, A.; Schamper, T.; Wilcox, C. *J. Am. Chem. Soc.* **1973**, *95*, 6090.
93. Pease, L.G.; Niu, C.H.; Zimmerman, G. *J. Am. Chem. Soc.* **1979**, *101*, 184.
94. Grathwohl, C.; Wüthrich, K. *Biopolymers* **1976**, *15*, 2043.
95. Grathwohl, C.; Wüthrich, K. *Biopolymers* **1976**, *15*, 2025.
96. Deslauriers R.; Becker, J.M.; Steinfeld, A.S.; Nader, F. *Biopolymers* **1979**, *18*, 523.

97. Anteunis, M.J.O.; Borremans, F.A. Stewart, J.M.; London, R.E. *J. Am. Chem. Soc.* **1981**, *103*, 2187.
98. Zimmerman, S.S.; Scheraga, H.A. *Macromolecules* **1976**, *11*, 408.
99. Patel, D.S. *Biochemistry* **1973**, *12*, 667.
100. Gō, N.; Scheraga, H.A. *J. Chem. Phys.* **1969**, *51*, 4751.
101. Urry, D.W.; Ohnishi, T. In "Second Rehovot Symposium on Peptides, Polypeptides, and Proteins," Blout, E.R.; Bovey, F.A.; and Goodman, M. Eds.; Wiley: New York, 1974, p. 230.
102. Maxfield, F.R.; Scheraga, H.A. *Macromolecules* **1975**, *8*, 491.
103. Tonelli, A.E. *J. Am. Chem. Soc.* **1970**, *92*, 6187.
104. Blessing, R.H.; Smith, G.D. *Acta Crystallogr. Sect. B* **1982**, *38*, 1203.
105. London, R.E.; Matwiyoff, N.A.; Stewart, J.M.; Cann, J.R. *Biochemistry*, **1978**, *17*, 2277.
106. Deber, C.M.; Bovey, F.A.; Carver, J.P.; Blout, E.R. *J. Am. Chem. Soc.* **1970**, *92*, 6191.
107. Rothe, M.; Rott, H. *Angew. Chem.* **1976**, *88*, 844.
108. Stern, A.; Gibbons, W.A.; Craig, L.C. *Proc. Natl. Acad. Sci. U.S.A.* **1968**, *61*, 734.
109. Urry, D.W.; Onishi, M. In "Spectroscopic Approaches to Biomolecular Conformation," Urry, D.W., Ed.; American Medical Association: Chicago, 1970, Chapter VII.
110. Eg, in tyrothricin: Kuo, M.; Gibbons, W.A. *Biochemistry* **1979**, *18*, 5855, and references cited therein.
111. Gramicidine S.; Ivanov, V.T. *Peptide Proc. 5th Annu. Peptide Symp.* **1977**, 307.
112. Pease, L.G.; Watson, C.J. *J. Am. Chem. Soc.* **1978**, *100*, 1279.
113. de Leeuw, F.A.A.M.; Altona, C.; Kessler, H.; Bernel, W.; Friedrich, A.; Krack, G.; Hull, W.E. *J. Am. Chem. Soc.* **1983**, *105*, 2237; except for definition of endo-exo.
114. Tonelli, A.E. *J. Am. Chem. Soc.* **1973**, *95*, 5946.
115. Sisido, M.; Imanishi, Y.; Higashimura, T. *Biopolymers* **1973**, *12*, 2375.
116. Davies, J.S.; Thomas, W.A. *J. Chem. Soc. Perkin Trans. 2*, **1978**, 1157.
117. Cf Ramachandran, G.N.; Sasisekharan, V. *Adv. Protein Chem.* **1968**, *23*, 283.
118. Levitt, M. *J. Mol. Biol.* **1982**, *145*, 251.
119. Benedetti, E.; Bavoso, A.; Di Blasio, B.; Pavone, V.; Pedone, C.; Toniolo, C.; Bonora, G.M. *Int. J. Peptide Protein Res.* **1982**, *20*, 312, and references cited therein.
120. Nagaray, R.; Venkatachalapathi, Y.V.; Balaram, P. *Int. J. Peptide Protein Res.* **1980**, *16*, 291.
121. Deber, C.M., Fossel, E.T.; Blout, E.R. *J. Am. Chem. Soc.* **1974**, *96*, 4015, for cyclic peptides.
122. Bavoso, A.; Benedetti, E.; Di Blasio, B.; Pavone, V.; Pedone, C.; Toniolo, C.; Bonora, G.M. *Macromolecules* **1981**, *15*, 54.
123. Tanaka, S.; Scheraga, H.A. *Macromolecules* **1974**, *7*, 698.
124. Benedetti, E.; Di Blasio, B.; Pavone, V.; Pedone, C.; Toniolo, C.; Bonora, G.M. *Macromolecules* **1980**, *13*, 1454.
125. Fujinaga, M.; James, M.N.G. *Acta Crystallogr. Sect. B.* **1980**, *36*, 3196.
126. Matsuzami, T.; Iitama, Y. *Acta Crystallogr. Sect. B.* **1971**, *27*, 507.
127. Benedetti, E.; Ciajolo, A.; Di Blasio, B.; Pavone, V.; Pedone, C.; Toniolo, Cl.; Bonora, G.M. *Int. J. Peptide Protein Res.* **1979**, *14*, 130.
128. Benedetti, E.; Ciajolo, M.R.; Maisto, A. *Acta Crystallogr. Sect. B.* **1974**, *30*, 1783.
129. Drück, U.; Littke, W.; Main, P. *Acta Crystallogr. Sect. B.* **1979**, *35*, 253.
130. Kamwaya, M.E.; Oster, O.; Bradaczek, H. *Acta Crystallogr. Sect. B.* **1981**, *37*, 364.
131. Venkataram Prasad, B.V.; Hemalatha, B.; Balaram, P. *Biopolymers* **1982**, *21*, 1261.
132. Cameron, T.S. *Cryst. Struct. Commun.* **1982**, *11*, 321.
133. Leung, Y.C.; Marsh, R.E. *Acta Crystallogr.* **1958**, *11*, 17.
134. Fridrichsons, J.; Mathieson, A.McL. *Acta Crystallogr.* **1962**, *15*, 569.
135. Ueki, T.; Ashida, T.; Kakudo, M.; Sasada, Y.; Katsube, Y. *Acta Crystallogr. Sect. B.* **1969**, *25*, 1840.
136. Kartha, G.; Ashida, T.; Kakudo, M. *Acta Crystallogr. Sect. B.* **1974**, *30*, 1861.
137. Kamwaya, M.E.; Oster, O.; Bradaczek, H. *Acta Crystallogr. Sect. B.* **1981**, *37*, 1564.
138. Matsuzaki, T. *Acta Crystallogr. Sect. B.* **1974**, *30*, 1029.
139. Kojima, T.; Tanaka, I.; Ashida, T. *Acta Crystallogr. Sect. B.* **1982**, *38*, 221.
140. Ramachandran, G.N.; Kolaskar, A.S.; Ramakrishnan, C.; Sasisekharan, V. *Biochim. Biophys. Acta* **1974**, *359*, 298.

141. Yamada, Y.; Tanaka, I.; Ashida, T. *Bull. Chem. Soc. Japan* **1981,** *54,* 69.
142. Ramachandran, G.N.; Sasishekaran, V. *Adv. Protein Chem.* **1968,** *23,* 283.
143. Reed, L.L.; Johnson, P.L. *J. Am. Chem. Soc.* **1973,** *95,* 7523.
144. Marsh, R.E.; Narashima Murthy, M.R.; Venkatesan, K. *J. Am. Chem. Soc.* **1977,** *99,* 1251.
145. Hospital, M.; Courseille, C.; Leroy, F.; Roques, B.P. *Biopolymers* **1979,** *18,* 1141.
146. IUPAC–IUB Commission on Biochemical Nomenclature. *Biochem. J.* **1971,** *121,* 577.
147. Ramachandran, G.N.; Lakshminarayanan, A.V.; Balasubramanian, R.; Tegoni, G. *Biochim. Biophys. Acta* **1970,** *221,* 165.
148. Aubry, A.; Protas, J.; Boussard, G.; Marraud, M. *Acta Crystallogr. Sect. B* **1977,** *33,* 2399.
149. Vitoux, B.; Aubry, A.; Manh Thong, C.; Boussard, G.; Marraud, M. *Int. J. Peptide Protein Res.* **1981,** *17,* 469.
150. Altona, C.; Geise, H.J.; Romers, C. *Tetrahedron* **1968,** *24,* 13.
151. Altona, C.; Sundaralingam, M. *J. Am. Chem. Soc.* **1972,** *94,* 8205.
152. de Leeuw, H.P.M.; Haasnoot, C.A.G.; Altona, C. As quoted in Reference 28.
153. de Leeuw, H.P.M.; Haasnoot, C.A.G.; Altona, C. *Isr. J. Chem.* **1980,** *20,* 108.
154. Koeners, H.J.; Altona, C. As quoted in Reference 28.
155. Anteunis, M.J.O.; Callens, R.; Asher, V.; Sleeckx, J. *Bull. Soc. Chim Belg.* **1978,** *87,* 41.
156. Fuchs, B. In "Topics in Stereochemistry," Vol. 10, Eliel, E.L.; and Allinger, N.L., Eds.; Wiley: New York, 1978, pp. 1–94.
157. Aubry, A.; Vitoux, B.; Marraud, M. *Cryst. Struct. Commun.* **1982,** *11,* 135.
158. Aubry, A.; Vitoux, B.; Boussard, G.; Marraud, M. *Int. J. Peptide Protein Res.* **1981,** *18,* 195.
159. Aubry, A.; Marraud, M.; Protas, J. *C.R. Acad. Sci. Paris, Ser. C* **1975,** *280,* 509.
160. Aubry, A.; Protas, J.; Boussard, G.; Marraud, M. *Acta Crystallogr. Sect. B* **1980,** *36,* 2822.
161. Benedetti, E.; Pavone, V.; Toniolo, C.; Bonora, G.M.; Palumbo, M. *Macromolecules* **1977,** *10,* 1350.
162. Brahmachari, S.K.; Bhat, T.N.; Sudhakar, V.; Vijayan, M.; Rapaka,. R.S.; Bhatganar, R.S.; Ananthanarayanan, V.S. *J. Am. Chem. Soc.* **1981,** *103,* 1703.
163. Yamada, Y.; Tanaka, I.; Ashida, T. *Acta Crystallogr. Sect. B* **1980,** *36,* 331.
164. Lecomte, C.; Aubry, A.; Protas, J.; Boussard, G.; Marraud, M. *Acta Crystallogr. Sect. B* **1974,** *30,* 1992.
165. Lecomte, C.; Aubry, A.; Protas, J.; Boussard, G.; Marraud, M. *Acta Crystallogr. Sect. B* **1974,** *30,* 2343.
166. Sugino, H.; Tanaka, I.; Ashida, T. *Bull. Chem. Soc. Japan* **1978,** *51,* 2855.
167. Ashida, T.; Tanaka, I.; Shimonishi, Y.; Kakudo, M. *Acta Crystallogr. Sect. B* **1977,** *33,* 3054.
168. Nair, C.M.K.; Vijayan, M. *J. Chem. Soc. Perkin Trans. 2* **1980,** 1800.
169. Tanaka, I.; Ashida, T.; Shimonishi, Y.; Kakudo, M. *Acta Crystallogr. Sect. B* **1979,** *35,* 110.
170. Bavoso, A.; Benedetti, E.; Di Blasio, B.; Pavone, V.; Pedone, C.; Toniolo, C.; Bonora, G.M. *Macromolecules* **1982,** *15,* 54.
171. Kojima, T.; Kido, T.; Itoh, H.; Yamane, T.; Ashida, T. *Acta Crystallogr. Sect. B* **1980,** *36,* 326.
172. Itoh, H.; Yamane, T.; Ashida, T. *Acta Crystallogr. Sect. B* **1978,** *34,* 2640.
173. Benedetti, E.; Ciajolo, A.; Di Blasio, B.; Pavone, V.; Pedone, C.; Toniolo, C.; Bonora, G.M. *Macromolecules* **1979,** *12,* 438.
174. Tanaka, I.; Ashida, T. *Acta Crystallogr. Sect. B* **1980,** *36,* 2164.
175. Smith, G.D.; Pletnev, V.Z.; Duax, W.L.; Balasubramanian, T.M.; Bosshard, H.E.; Czerwinski, E.W.; Kendrick, N.E.; Mathews, F.S.; Marshall, G.R. *J. Am. Chem. Soc.* **1981,** *103,* 1493.
176. Venkataram Prasad, B.V.; Shamala, N.; Nagaraj, R.; Chandrasekaran, R.; Balaram, P. *Biopolymers* **1979,** *18,* 1635.
177. Venkatachalipathi, Y.V.; Nair, C.M.K.; Vijayan, M.; Balaram, P. *Biopolymers* **1981,** *20,* 1123.
178. Reck, G.; Barth, A. *Cryst. Struct. Commun.* **1981,** *10,* 1001.
179. Kamwaya, M.E.; Oster, O.; Bradaczek, H.; Ponnuswamy, M.N.; Parthasarathy, S.; Nagaraj, R.; Balaram, P. *Acta Crystallogr. Sect. B* **1982,** *38,* 172.
180. Aubry, A.; Protas, J.; Boussard, G.; Marraud, M. *Acta Crystallogr. Sect. B* **1980,** *36,* 2825.

181. Aubry, A.; Protas, J.; Boussard, G.; Marraud, M. *Acta Crystallogr. Sect. B* **1980,** *36,* 321.
182. Aubry, A.; Protas, J.; Boussard, G.; Marraud, M. *Acta Crystallogr. Sect. B* **1979,** *35,* 694.
183. Rudko, A.D.; Low, B.W. *Acta Crystallogr. Sect. B* **1975,** *31,* 713.
184. Kojima, T.; Yamane, T.; Ashida, T. *Acta Crystallogr. Sect. B* **1978,** *34,* 2896.
185. Tanaka, I.; Kozima, T.; Ashida, T.; Tanaka, N.; Kakudo, M. *Acta Crystallogr. Sect. B* **1977,** *33,* 116.
186. Yamane, T.; Ashida, T.; Shimonishi, K.; Kakudo, M.; Sasada, Y. *Acta Crystallogr. Sect. B* **1976,** *32,* 2971.
187. Ueki, T.; Bando, S.; Ashida, T.; Kakudo, M. *Acta Crystallogr. Sect. B* **1971,** *27,* 2219.
188. Bando, S.; Tanaka, N.; Ashida, T.; Kakudo, M. *Acta Crystallogr. Sect. B* **1978,** *34,* 3447.
189. Marsh, R.E. *Acta Crystallogr. Sect. B* **1980,** *36,* 1265.
190. Nair, C.M.K.; Nagaraj, R.; Ramaprasad, S.; Balaram, P.; Vijayan, M. *Acta Crystallogr. Sect. B* **1981,** *37,* 597.
191. Précigaux, G.; Geoffre, S.; Hospital, M.; Leroy, F. *Acta Crystallogr. Sect. B* **1982,** *38,* 2172.
192. Ayato, H.; Tanaka, I.; Ashida, T. *J. Am. Chem. Soc.* **1981,** *103,* 6869.
193. Yagi, Y.; Tanaki, I.; Yamane, T.; Ashida, T. *J. Am. Chem. Soc.* **1983,** *105,* 1242.
194. Ayato, H.; Tanaka, I.; Ashida,T. *J. Am. Chem. Soc.* **1981,** *103,* 5902.
195. The BS2000 version of the program was run on a SIEMENS 75/51 computer (8 Mb). The program was obtained from the Program Library Unit, Edinburgh University, 18 Buccleuch Place, Edinburgh EH8 9LN, Scotland, U.K.
196. Winkler, F.K.; Dunitz, J.D. *J. Mol. Biol.* **1971,** *59,* 169.
197. Ramachandran, G.N.; Lakshminarayanan, A.V.; Kolaskar, A.S. *Biochim. Biophys. Acta* **1973,** *303,* 8.
198. Gilli, G.; Bertolasi, V. *J. Am. Chem. Soc.* **1979,** *101,* 7704.
199. Ashida, T.; Kakudo, M. *Bull. Chem. Soc. Japan* **1974,** *47,* 1129.

CHAPTER 8

# Theoretical Calculations in the Study of Electronic Substituent Effects

**Ronald D. Topsom**

**La Trobe University, Bundoora, Victoria, Australia**

## CONTENTS



## 1. INTRODUCTION

In recent years theoretical calculations have developed to the extent of providing experimentally meaningful results for certain properties and molecules. Thus, as discussed below, it is now possible to obtain relative gas phase proton affinities for certain series of small molecules to within a few kilocalories per mole of experiment. Reliable values for absolute proton


MOLECULAR STRUCTURE AND ENERGETICS, Vol. 4

affinities can also be obtained for yet smaller molecules (one or two atoms other than hydrogen), but considerably more sophisticated calculations are required.

This generally good agreement between calculation and experiment gives confidence in theoretical results for similar systems where experimental results are not available. This has proved particularly fruitful in the study of both the nature and the magnitudes of the various transmission mechanisms by which a change in structure at some point can influence a property measured elsewhere in the molecule. We limit our discussion here to such substituent effects that are electronic rather than arising fully or partly from energy or entropy effects ensuing from nonbonded repulsions.

This chapter considers first the available theoretical methods and their limitations. In Section 3, we give a variety of examples to illustrate the ability of the theoretical calculations to reproduce experimental data. In section 4, we consider the various types of electronic substituent effect that are possible. Section 5 then looks at the use of theoretical calculations to investigate and help understand the relative importance of the various substituent effects. Finally, in Section 6, we examine the theoretical derivation of scales of various electronic substituent effects for a variety of common substituents.

# 2. CHOICE OF THEORETICAL APPROACH

In principle, we can calculate any property of a molecule or system that we wish to study. This is still not possible for many properties and molecules, however, and most theoretical calculations have been concerned with energies and electron distributions in relatively small molecules.

The results quoted or reported in this chapter have all been obtained by the ab initio molecular orbital approach using the Gaussian series of programs developed[1-3] by Pople, Hehre, and their collaborators. We are not thereby criticizing the use of the simple semiempirical approaches involving varying degrees of neglect of differential overlap such as CNDO/2,[4] MINDO/3,[5] and MNDO.[6] However, the latter methods are necessarily limited in the particular case of substituent effect studies. This follows since, in the semiempirical methods, it is necessary to use experimental data to parameterize each substituent, and this leads to a related substituent response in other sets of circumstances. Thus subtle substituent interactions could be lost. Equally, while other ab initio methods have been used, the STO-$n$G basis set calculations have been extensively tested[2] against experimental data and provide the vast majority of published substituent effect studies.

A combination of present programs and computer availability usually limits the use of minimal basis STO-3G[3] calculations to some 12–14 atoms other than hydrogen (referred to as heavy atoms below). The split valence bases 3–21G[7] and 4–31G[3] are similarly limited to some nine heavy atoms, and the 6–31G* basis, containing a larger basis set and polarization functions, to five heavy atoms. (See Chapter 4 by Gordon in Volume 1 of this series.) Most calculations have also been limited to hydrogen and atoms in the first two rows of the periodic table; the minimal basis STO-3G level is not likely[2] to give generally useful results for second-row elements.

Very few substituent effect calculations have included allowance for electron correlation. Such allowance does not seem to be of great importance in studies of electron distribution in molecules or for energies of equilibria where there is conservation of the number and type of bonds involved. Such reactions are known as isodesmic,[8] where the number and formal type (carbon–oxygen double bond, carbon–carbon single bond, etc), of bond is conserved. Isodesmic processes are particularly useful in substituent effect studies, since results here are generally referred to a standard. Thus reaction (8-1) yields the relative energy of a substituent when joined to benzene as against a methyl group, referred to hydrogen as a standard substituent.

$$\bigcirc \; + \; CH_3X \; \rightleftharpoons \; \underset{X}{\bigcirc} \; + \; CH_4 \qquad (8\text{-}1)$$

A subclassification of such reactions occurs when the numbers of bonds of a particular type ($C_{sp3}-C_{sp2}$, $C^2_{sp-C_{sp2}}$, etc) are conserved. These are known as homodesmotic[9] reactions and are even less affected by correlation energy. They include such proton transfer processes as reaction 8-2, which measures, for example, the relative acidities of a series of parasubstituted benzoic acids.

$$\underset{X}{\overset{CO_2H}{\bigcirc}} \; + \; \overset{CO_2^-}{\bigcirc} \; \rightleftharpoons \; \underset{X}{\overset{CO_2^-}{\bigcirc}} \; + \; \overset{CO_2H}{\bigcirc} \qquad (8\text{-}2)$$

Electron correlation depends[10] on the number and distribution of electron pairs and is thus of much greater significance in processes such as absolute proton affinities (8-3) particularly of highly charge-localized atoms or anions. Obviously theoretical substituent effect studies are restricted here to smaller molecules.

$$A + H^+ \rightleftharpoons AH^+ \qquad \text{or} \qquad A^- + H^+ \rightleftharpoons AH \qquad (8\text{-}3)$$

It has been shown[2,3] that major improvements in the calculations generally occur in going from the minimal basis STO-3G to a split valence basis

(3–21G or 4–31G) and then again on going to a larger basis also including polarization functions (6–31G*). The size of the molecules involved limits many studies to the minimal basis, but we include the results of split valence and polarization basis set calculations where available. Details of the calculations are available in the literature, and their limitations and performance have been extensively reviewed (see, eg, References 2 and 3). Theoretical results are given for geometry-optimized structures where available. Results for standard geometries are shown as STO-3G, 3–21G, and so on, while those for optimized structures as 3G//3G, 4–31G//4–31G, and so on, where the first and second symbols denote the basis sets used for the final calculation and for the optimization, respectively.

## 3. COMPARISON OF THEORETICAL AND EXPERIMENTAL RESULTS

Before going further, it is useful to consider the performance of ab initio molecular orbital calculations versus experimental results. Strictly, the results of theoretical calculations should be compared to experimental results for molecules in the gas phase at absolute zero temperature without any zero-point vibrational energies. However, the temperature effects for the experimental results seem small and susceptible to correction for most substituent effect properties of interest.[11] These are mainly either gas phase isodesmic reactions—particularly proton transfer equilibria, which have been the basis of much physical organic chemistry—or properties dependent on electron distribution. Calculations of equilibrium geometries have proved very successful,[2,3] but such properties are not greatly affected[12] by substituent effects. The calculation of spectroscopic properties such as infrared frequencies and intensities has been an area of recent interest, but here substituent effect studies are limited. Chapter 7 in Volume 2 by Dixon and Lias compares calculational and gas-phase experimental data.

### A. Proton Transfer Equilibria

The general reactions of proton transfer equilibria can be represented as one of the two processes 8-4 and 8-5.

$$X-G-YH^+ + H-G-Y \rightleftharpoons X-G-Y + H-G-YH^+ \qquad (8\text{-}4)$$
$$X-G-YH + H-G-Y^- \rightleftharpoons X-G-Y^- + H-G-YH \qquad (8\text{-}5)$$

Here X is the variable substituent, G is the molecular framework in the system of interest, and Y is the acid or base center. It is both traditional and frequently particularly instructive to take the simplest of all substituents, the H atom, as the standard substituent.

Most ab initio molecular orbital calculations reported for such series with

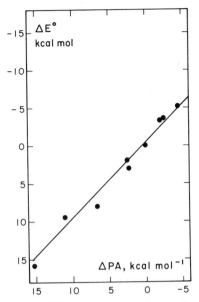

**Figure 8-1.** Plot of $\Delta E°$ values (STO-3G) for proton transfer reactions of substituted methylamines versus experimental $\Delta PA$ values. (Reprinted with permission from *Accounts of Chemical Research,* **1983,** *16,* 292. Copyright 1983, American Chemical Society.)

a representative range of substituents have been at the minimal basis STO-3G level. Calculations at a higher level are precluded in most cases because the total number of heavy atoms in X, G, and Y exceeds program limitations.

The calculations are remarkably successful where G contains at least one atom, that is, where the probe (Y) is not directly joined to the substituent (X). In a simple example,[13] G = $CH_2$, Y = $NH_2$, that is, process 8-6:

$$XCH_2NH_3^+ + CH_3NH_2 \rightleftharpoons XCH_2NH_2 + CH_3NH_3^+ \qquad (8\text{-}6)$$

A plot of the calculated $\Delta E°$ values for process 8-6 at the STO-3G basis versus experimental $\Delta PA$ (proton affinity) values is shown in Figure 8-1. A recent compendium has appeared[13b] listing gas phase basicities and proton affinities for a wide variety of bases.

Another series of nitrogen bases for which both calculated and experimental gas phase results are available are proton transfer reactions of pyridines:

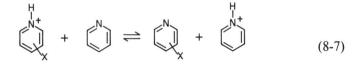

$$(8\text{-}7)$$

**TABLE 8-1.** Ab Initio Orbital Calculations (STO-3G) for Equilibria Involving Substituted Pyridines: Equation 8-7

| Substituent | $\Delta E_{calc}^{\circ}$ (kcal/mol) | $\Delta G_{exp}^{\circ}$ (kcal/mol) | $\Delta E_{int}$ (kcal/mol)[a] (Cation) | (Neutral) | $\Delta q_H$ ($\times 10^4$ electrons)[b] |
|---|---|---|---|---|---|
| 3-NH$_2^c$ | 3.5 | 9.5[d] | 4.4 | 1.0 | −18 |
| 4-NH$_2^c$ | 16.1 | 15.6[d] | 18.0 | 1.8 | −142 |
| 3-OMe | −0.3 | 3.2 | −0.3 | 0.0 | −17 |
| 4-OMe | 9.3 | 7.2 | 9.9 | 0.6 | −76 |
| 3-Me | 2.5 | 2.9 | 2.3 | −0.2 | −27 |
| 4-Me | 5.0 | 4.3 | 5.7 | 0.6 | −49 |
| 3-F | −6.3 | −7.0 | −7.3 | −1.0 | 51 |
| 4-F | −0.2 | −4.1 | −0.9 | −0.7 | −10 |
| 3-CF$_3$ | −6.8 | −8.6 | −8.0 | −1.2 | 37 |
| 4-CF$_3$ | −6.8 | −8.3 | −5.7 | −1.0 | 34 |
| 3-CN | −12.4 | −12.0 | −14.3 | −1.8 | 73 |
| 4-CN | −11.4 | −11.2 | −13.8 | −2.4 | 51 |
| 3-CHO | −3.6 | — | −4.8 | −1.2 | 13 |
| 4-CHO | −2.9 | — | −3.8 | −0.9 | −3 |
| 3-NO$_2$ | −17.4 | — | −19.7 | −2.3 | 115 |
| 4-NO$_2$ | −16.7 | −12.7 | −19.7 | −3.0 | 149 |

[a]See text.
[b]Parent $q_H = 0.6891$.
[c]Planar geometry.
[d]NMe$_2$.

Table 8-1 shows[14] the $\Delta E^{\circ}$ values calculated for process 8-7 compared with corresponding experimental $\Delta G^{\circ}$ values, which are almost identical to $\Delta H^{\circ}$ values.[11] The other columns contain data to be discussed in later sections, included here for convenience. Once again, the agreement between calculated and experimental results is excellent except for the nitro substituent, for which the minimal basis set seems inadequate.[14,15]

An example of oxygen as the probe center is provided by results for process 8-8. The calculations[16] are again at the STO-3G basis, and here experimental gas phase values are available both from ion-cyclotron resonance (ICR) and high-pressure mass spectrometric (HPMS) techniques.

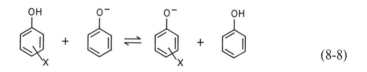

$$(8-8)$$

Table 8-2 shows the excellent agreement obtained between the calculated and experimental values, except for the strongly interacting paracyano and paranitro groups.

**TABLE 8-2.** Ab Initio Molecular Orbital Calculations (STO-3G) for Proton Transfer Reactions Involving Substituted Phenols: Equation 8-8

| | $\Delta G_{exp}^{\circ}$ (kcal/mol) | | |
| Substituent | ICR (298 K) | HPMS (600 K) | $\Delta E_{calc}^{\circ}$ (kcal/mol) |
|---|---|---|---|
| 3-Me | 0.4 | 0.4 | 0.4 |
| 4-Me | 1.1 | 1.3 | 1.0 |
| 3-NH$_2$ | 1.4 | 0.9 | 0.1 |
| 4-NH$_2$ | 3.3 | 4.2 | 4.9 |
| 3-OMe | −1.1 | −1.5 | −2.1 |
| 4-OMe | 1.2 | 0.8 | 3.0 |
| 3-F | −5.3 | −5.8 | −5.4 |
| 4-F | −2.3 | −2.6 | −1.6 |
| 3-NO$_2$ | −14.4 | −15.7 | −18.1 |
| 4-NO$_2$ | −20.9 | | −29.2 |
| 3-CN | −13.0 | −14.3 | −14.5 |
| 4-CN | −16.6 | −17.7 | −21.4 |
| 3-CF$_3$ | −9.6 | | −8.4 |
| 4-CF$_3$ | −11.9 | | −11.5 |
| 3-CHO | −8.5 | | −5.9 |
| 4-CHO | −15.8 | | −13.8 |

Similar good agreement is found between calculated[19] and experimental[18] gas phase data for substituent benzoic acids according to process 8-9.

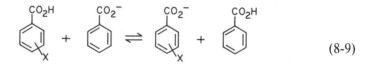

$$(8-9)$$

The results are shown in Table 8-3.

Rather fewer results are available for aliphatic systems, since conformational mobility or synthetic difficulties preclude comparison or make it difficult to obtain experimental results. A series of 4-substituted bicyclo[2.2.2]octyl-1-carboxylic acids are, however, being studied,[20] according to the following process:

$$(8-10)$$

TABLE 8-3. Ab Initio Molecular Orbital
Calculations (STO-3G) for Proton Transfer
Equilibria of 4-Substituted Benzoic Acids:
Equation 8-9

| Substituent | $\Delta E^{\circ}_{calc}$ (kcal/mol) | $\Delta G^{\circ}_{exp}$ (kcal/mol) |
|---|---|---|
| NH$_2$ | 3.1[a] | 2.3 |
| OMe | 0.6[b] | 0.8 |
| Me | 1.1 | 1.1 |
| F | $-1.5$ | $-2.9$ |
| CHO | $-4.5$ | — |
| CF$_3$ | $-5.3$ | — |
| CN | $-9.9$ | $-10.3$ |
| NO$_2$ | $-13.6$ | $-11.1$ |

[a]Pyramidal NH$_2$.
[b]Calculation is for OH.

The results, to date, are shown in Table 8-4, and once again the agreement seems to be very good.

A more rigorous test is provided by the reaction series involving the relative proton affinities of substituted cyanides:

$$XCNH^+ + CH_3CN \rightleftharpoons XCN + CH_3CNH^+ \qquad (8-11)$$

Here[21] the methyl cyanide has been used as a standard partly for reasons relating to the optimization of the geometries in the calculated results. The data shown in Table 8-5 are for optimized geometries at the STO-3G, 3–21G and 6–31G* levels, where possible. The agreement between theory and experiment at the minimal basis level is surprisingly good, both since here

TABLE 8-4. Ab Initio Molecular Orbital
Calculations (STO-3G) for Equilibria of 4-
Substituted Bicyclo[2.2.2]octyl-1-carboxylic
Acids: Equation 8-10

| Substituent | $\Delta E^{\circ}_{calc}$ (kcal/mol) | $\Delta G^{\circ}_{exp}$ (kcal/mol) |
|---|---|---|
| Me | 0.0 | 1.0 |
| NH$_2$ | 0.1 | — |
| OH | 2.0 | 2.7[a] |
| F | 3.2 | 5.6 |
| CHO | 2.7 | 3.2[b] |
| CN | 6.8 | 8.3 |
| NO$_2$ | 8.7 | 8.7 |

[a]Measurement is for OMe.
[b]Measurement is for CO$_2$Et.

TABLE 8-5. Ab Initio Molecular Orbital Calculations for Proton Transfer Equilibria of Cyanides Referred to Methyl Cyanide: Equation 8-11

| Substituent | $\Delta E^\circ_{calc}$ (kcal/mol) | | | $\Delta G^\circ_{exp}$ (kcal/mol) |
| | 3G//3G | 3–21G//3–21G | 6–31G*//6–31G* | |
| --- | --- | --- | --- | --- |
| NMe$_2$ | 17.4 | 19.5 | — | 17.1 |
| C$_6$H$_5$ | 11.1 | 7.5 | — | 7.1 |
| t-Bu | 8.3 | 7.3 | — | 7.0 |
| i-Pr | 6.0 | 5.3 | — | 5.7 |
| Et | 3.3 | 2.9 | 2.8 | 3.0 |
| CHCH$_2$ | 3.6 | 2.2 | 2.8 | 0.6 |
| CH$_2$F | −7.2 | −13.2 | −11.4 | −7.9 |
| COMe | −7.7 | −17.4 | — | −6.9 |
| H | −18.4 | −16.0 | −15.9 | −16.7 |
| CF$_3$ | −17.8 | −37.4 | — | −23.8 |
| CN | −30.1 | −33.2 | −33.3 | −26.7 |

the substituent is directly joined to the probe center and since agreement is much poorer at the split valence level for COMe and CF$_3$.

Results are also available for comparison of the effects on proton transfer equilibria of the attachment of one or more water molecules to the acid and base forms. For example, theoretical[22] and gas phase experimental[23] results are available for process 8-12.

$$(8-12)$$

Here again, agreement is reasonable[11] for the limited number of substituents available.

Rotational barriers also seem to be well reproduced[2] by calculations at the minimal basis level. Thus Table 8-6 gives values[24] for the relative C−O barrier for parasubstituted phenols. Overall then, we can have some confidence in calculated results for isodesmic processes of the type discussed above.

## B. Electron Density Distribution

A small but increasing number of calculations have been made using ab initio molecular orbital theory[24] for overall electron distribution patterns in molecules. These can be compared with experimental determinations made

TABLE 8-6. Ab Initio Molecular Orbital Calculations (STO-3G) for the
Barriers to Rotation Around the CO Bond in Parasubstituted Phenols

| | Values relative to phenol (kcal/mol)[a] | |
| Substituent | Experimental | Calculated |
|---|---|---|
| OH | −0.87 | −0.95 |
| F | −0.60 | −0.53 |
| Me | −0.32 | −0.28 |
| CHO | 0.87 | 0.47 |
| CN | 0.70 | 0.66 |
| NO$_2$ | 0.98 | 1.02 |

[a]These values are the twofold component of the barrier.

from comparisons either of high-and low-incidence X-ray studies (X—X)
or of X-ray and neutron diffraction results (X−N). The agreement made is
generally good[24] at the 6–31G* basis level, particularly if some allowance is
made for electron correlation, and at the split valence level if additional
bond orbitals are included.[25]

The apportioning of overall electron density to atoms, although not of
fundamental meaning, is useful in substituent effect studies, in visualizing
the various contributions and their magnitude. A recent apparently successful approach[26] has been to fit the overall density by spheres centered on
nuclear positions. The size of the sphere for a particular atom varies from
compound to compound in proportion to the atomic electron population.

The most used approach is to obtain atomic electron population by a partitioning scheme proposed by Mulliken.[27] This scheme is mathematically
explicit and readily incorporated in the computer programs. However, the
method of apportioning the electrons between atoms is clearly unreasonable
in some cases.[2,28] Thus the charge on the carbon atom in methane is calculated[29] to be as high as 0.79 electron at the 3–21G//3–21 level and is 0.47 at
the 6–31G**//6–31G** level. The actual charge is likely[28] to be very much
lower and possibly of opposite sign. However, it does appear that for series
of related molecules, Mulliken electron populations follow those obtained
from a topological analysis of the overall density distribution. Thus the calculated Mulliken populations at the hydrogen of methyl derivatives CH$_3$X
follow[29] values obtained from the overall density distribution as shown in
Figure 8-2. This is true for various basis levels, since the Mulliken populations at one level seem[29] to be proportional to those at another. Another
example[30] is found for the $\pi$-electron populations at the oxygen atoms of
simple aldehydes and ketones and, to lesser accuracy, for the total electron
populations at the same oxygen of the aldehydes taken separately.

Atom electron populations certainly seem to mirror expectation. Thus the
$^{13}$C chemical shifts at the para position in monosubstituted benzene might
be expected[31] to follow the local electron density at that atom. Values are
given in Table 8-7, and Figure 8-3 shows the excellent linearity obtained[31]

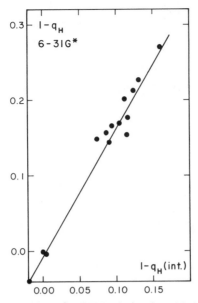

**Figure 8-2.** Plot of $1 - q_H$ values for $CH_3X$ derivatives (6–31G* std) versus $1 - q_H$ values obtained by an integration method. (Reprinted with permission from *J. Mol. Struct.* **1982,** *89,* 83.)

when such $^{13}C$ shifts[32] are plotted against the corresponding changes in $\pi$-electron population for the *p*-carbon atom. A similar plot is obtained[31] if the change in the total electron charge at this carbon is used since the total and $\pi$-charges are linearly related as the para-substituent changes.

A simple example of the significance of the electron population is pro-

**TABLE 8-7.** $^{13}C$ Substituent Chemical Shifts (SCS) and Calculated (STO-3G) $\pi$-Electron Populations for Carbon Atoms in Monosubstituted Benzenes

| Substituent | SCS (ppm)[a] | $10^3\Delta q_\pi(C)$ ($\times$ $10^3$ electrons)[b] | | | |
|---|---|---|---|---|---|
| | | ipso | ortho | meta | para |
| $NMe_2$[c] | −11.16 | 51 | −91 | 28 | −61 |
| $NH_2$[d] | −9.86 | 44 | −70 | 23 | −44 |
| OH | −7.63 | 25 | −68 | 24 | −39 |
| F | −4.49 | −7 | −42 | 16 | −21 |
| Me | −2.89 | 28 | −18 | 6 | −12 |
| H | 0.00 | 0 | 0 | 0 | 0 |
| CN | 3.80 | −56 | 24 | 1 | 28 |
| $CF_3$ | 3.19 | −35 | 14 | 1 | 16 |
| CHO | 5.51 | −20 | 20 | 2 | 18 |
| $NO_2$ | 5.53 | −90 | 42 | −3 | 43 |

[a]Negative is upfield.
[b]As changes from neutrality.
[c]Planar about N.
[d]Pyramidal about N.

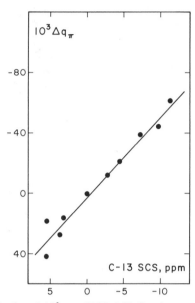

**Figure 8-3.** Plot of calculated $10^3 \Delta q_\pi$ (STO-3G) for the para position in monosubstituted benzene versus $^{13}C$ substituent chemical shifts.

vided by the $\pi$-electron charges given in Table 8-7 for the ipso, ortho, meta, and para carbon atoms in monosubstituted benzenes. These results clearly fit in with simple resonance concepts where, for example, a substituent such as $NH_2$ or OMe, capable of $\pi$-charge transfer to the ring, would be expected to increase the $\pi$-electron density at the ortho and para positions as shown in structure 1.

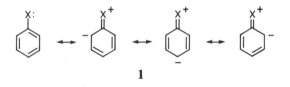

**1**

Ab initio molecular orbital calculations also[29] give dipole moments in reasonable agreement with experimental values as shown in Table 8-8. The average deviation over a variety of simple compounds was found[29] to be 0.47 D on the STO-3G basis decreasing to 0.25 D for calculations at the 6-31G* basis with geometry optimization.

In summary, we can have some confidence in the figures obtained for relative atomic electron populations, particularly in a closely similar series of compounds such as frequently encountered in substituent effect studies.

TABLE 8-8. Electric Dipole Moments (debyes)

| Molecule | Calculated | | | Experimental |
|---|---|---|---|---|
| | STO-3G//<br>STO-3G | 3–21G//<br>3–21G | 6–31G*//<br>6–31G* | |
| HF | 1.25 | 2.17 | 1.97 | 1.82 |
| $H_2O$ | 1.71 | 2.39 | 2.20 | 1.85 |
| $NH_3$ | 1.87 | 1.75 | 1.92 | 1.47 |
| HCN | 2.45 | 3.04 | 3.21 | 2.98 |
| $CH_3F$ | 1.15 | 2.34 | 1.99 | 1.85 |
| $CH_3OH$ | 1.48 | 2.15 | 1.87 | 1.70 |
| $CH_3NH_2$ | 1.62 | 1.44 | 1.62 | 1.31 |
| $CH_3CN$ | 3.10 | 3.88 | 4.04 | 3.92 |
| $H_2CO$ | 1.54 | 2.66 | 2.67 | 2.33 |
| $HCO_2H$ | 0.63 | 1.40 | 1.60 | 1.41 |
| $HCF_3$ | 1.13 | 2.12 | 1.87[a] | 1.65 |

[a]6–31G*//4–31G*.

# 4. ELECTRONIC SUBSTITUENT EFFECTS

There are four basic types of electronic substituent effect[33] that are possible, differing in their origin and mode of transmission.

| | |
|---|---|
| Field effects | $F$ |
| Electronegativity effects | $X$ |
| Polarizability effects | $\alpha$ |
| Resonance or $\pi$-electron delocalization effects | $R$ |

Various secondary effects can occur; those of particular importance are considered under the corresponding primary effect.

## A. Field Effects

The field effect (now[37] designated F) arises from the change in local dipole caused by replacing a hydrogen atom, taken as standard, by a substituent. The transmission occurs via a through-space electrostatic effect on the probe site, that is, the site that involves the reaction or measurement of interest.

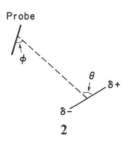

2

Simple electrostatics now allows calculation of the expected effect.[38] For a dipolar substituent Equation 8-13 should apply if the probe reaction involves charge $q$, as in proton transfer reactions where $q$ is unity in the charged species.

$$\Delta E = \frac{q_1\, \mu \cos \theta}{Dr^2} \tag{8-13}$$

Here $\mu$ is the local dipole, $\theta$ the angle it subtends to the probe site as shown in **2**, D the effective dielectric constant, and $r$ the distance from the dipole center to the probe site. For measurements relating to the polarization of a bond or group at the probe site, Equation 8-14 holds.

$$E_{\text{bond}} = \mu\, \frac{(2 \cos \theta \cos \Phi - \sin \theta \sin \Phi)}{Dr^3} \tag{8-14}$$

If the substituent itself has charge $q_2$, then the corresponding equations are 8-15 and 8-16, referring to a charged reaction site or to polarization of a bond at the reaction site, respectively. (Equation 8-16 is for the case in which the substituent is on the axis of the polarized probe bond.) All these equations are based on the assumption of point charges.

$$\Delta E = \frac{q_1 q_2}{Dr} \tag{8-15}$$

$$\Delta E = \frac{2q_2 \cos \phi}{Dr^2} \tag{8-16}$$

Such field effects are now considered[34-36] to provide the predominant mechanism of transmission of electronic substituent effects in molecules where the substituent is not able to conjugate with a $\pi$-electron system. It is possible[34,39] that lines of force may tend to follow the molecular framework in systems where $D$ is relatively high. For the gas phase, however, it would appear that the equations above should hold in the absence of secondary effects. An inherent assumption here is that we can treat a charged probe or substituent as a point charge.

An important secondary effect occurs when the primary field effect polarizes a $\pi$-electron system,[40] such as a benzene ring, only part of which constitutes the measurement site. An example[41] is the $^{13}C$ substituent chemical shift (relative to hydrogen as the reference substituent) of the para positon of compounds of structure **3**.

**3**

Here we refer[37] to the effect by the symbol $F_\pi$, that is, a field effect on a $\pi$ system. Secondary effects on $\sigma$ systems, $F_\sigma$, such as on CH bonds may also modify[42] the interaction of a substituent with a probe.

## B. Electronegativity Effects

Electronegativity effects originate in the difference in electronegativity between a substituent X and the carbon atom to which it is attached. They should thus be a function of the group electronegativity of X. Such effects are transmitted by a progressive but diminishing relay of polar effects along a chain of carbon atoms depicted as follows:

$$\begin{array}{cccc} \delta- & \delta+ & \delta\delta+ & \delta\delta\delta+ \\ X-CH_2 & -CH_2 & -CH_2 & - \end{array}$$

The effect is now[35,37] given the symbol $X$, although it was earlier referred to as a $\sigma$-inductive effect. Its magnitude obviously depends on the number and length of the chains of atoms connecting the substituent to the probe site. It should be otherwise independent of the relative geometry of the substituent and probe.

## C. Polarizability Effects

A substituent may also interact with a probe by virtue of its polarizability, $\alpha$. Thus a charged probe, in particular, may polarize the substituent thereby modifying the overall measured result. The appropriate equation[11] for a charged substituent is:

$$E = -\frac{\alpha q^2}{2Dr^4} \tag{8-17}$$

where $\alpha$ is the polarizability of the substituent and $r$ the distance of the mode to the center of substituent polarizability. For a dipolar substituent the effect follows $1/r^6$. In either case, the effect obviously is very small if $r$ and $D$ are relatively large. Such effects are of significance[11] in gas phase ions with different alkyl groups as substituents but may also[43] be more important than previously appreciated for other substituents both in the gas phase and in solvents of low dielectric constant.

## D. Resonance Effects

Resonance effects arise from $\pi$-charge transfer between the orbitals of a substituent and those of a $\pi$ system to which it is attached. Transmission is via the resulting changes in electron distribution in the $\pi$ system, although secondary effects can result, for example, from a transmission further through space of the change in electron distribution created by the primary resonance effect. A simple example of a resonance effect was seen in structure **1**. Such effects can considerably influence a probe, particularly if it is also directly attached to the $\pi$ system.

Substituents such as methyl groups also have orbitals of suitable symmetry to interact with a $\pi$ system; such effects are sometimes referred to as hyperconjugation.

## 5. THEORETICAL INSIGHTS INTO THE NATURE OF ELECTRONIC SUBSTITUENT EFFECTS

### A. Introduction

As discussed above, the four major primary electronic substituent effects frequently act simultaneously. It is therefore rather difficult to separately examine each effect in a detailed way from direct experimental data. Investigations have often centered on model compounds where only one of the effects might be expected to clearly predominate or to behave in a distinctive manner. Such compounds have frequently been hard to synthesize, particularly since rigid systems are needed to avoid conformational mobility.

Appropriate theoretical calculations thus offer great advantages, provided comparison with experiment has been shown to be good for related systems. For example, experimental comparison is only reasonable for structures **4** and **5**, since structure **6** is just one of the possible conformations (and usually a sparsely populated one) for a 1,4-disubstituted butane. (In these structures, Y is taken to be the probe group, and X the substituent as in Equations 8-4 and 8-5.)

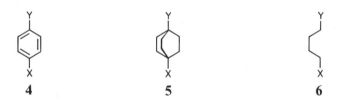

Such structures are interesting in providing similar relative geometry for the probe and substituent. A great advantage of theoretical calculations is that they can be made with fixed conformations,[44] such as structure **6**, with all heavy atoms coplanar, thus facilitating comparison of the effects of different molecular frameworks on the transmission of substituent effects. Indeed one can go further[42,45] and simulate structures by isolated molecules—for example,[42,44] employing **7**, where the groups X and Y have the same relative positions as in **6**.

Similarly, properties in 4-substituted styrene (**8**) can be simulated by calculations[45] on an isolated ethylene molecule having a $CH_3X$ molecule at the corresponding geometry (**9**).

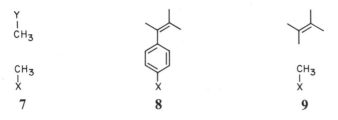

7                                8                                9

In the reactions below, we look at some theoretical studies of the nature and relative importance of the various electronic substituent effects. These are not intended to be inclusive; we concentrate on results from our own and related work.

## B. The Relative Importance of Electronegativity and Field Effects

There has been considerable controversy about the relative importance of these two modes of transmission. It is possible to eliminate resonance effects by studying $\sigma$-bonded systems, and polarizability effects would be expected to be small for significant distances between probe and substituent. Much effort has been concentrated on the investigation of various model compounds.[34,36] There is now considerable experimental evidence[35,46,47] that electronegativity effects are not important after the first atom to which the substituent is attached. Thus [13]C chemical shifts in methyl derivatives[46] and $^1J_{CC}$ coupling constants originating at the point of substitution[47] may reflect electronegativity effects. Nevertheless, there remains[48] a small amount of evidence for the presence of electronegativity effects for more remote probe sites.

Theoretical calculations have proved most useful in this area. First, calculations on molecules such as 1-substituted butanes do not show any significant effect on the electron population past the first carbon atom. Thus STO-3G calculations for 1-fluorobutane give substituent-induced changes in electron populations at carbon atoms, as compared to butane itself, as shown in **10** in electron units.

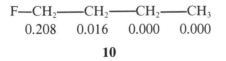

F—CH₂——CH₂——CH₂——CH₃

0.208    0.016    0.000    0.000

**10**

A more sophisticated study looked at the effect of a fluorine substituent on the proton affinity of an amino group in various $\omega$-fluoroalkylamines.[44] Values (STO-3G) of $\Delta E^\circ$ for three conformations of 4-fluorobutylamine (**11–13**) are shown below, according to Equation 8-4.

If the effect of the substituent were transmitted solely via the carbon atoms, identical values of $\Delta E^\circ$ would occur and thus the electronegativity effect clearly does not have major significance. In fact the results for these

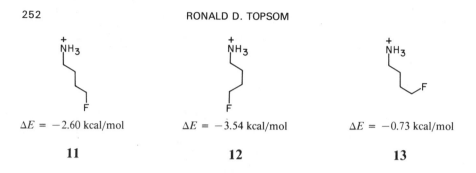

$\Delta E = -2.60$ kcal/mol          $\Delta E = -3.54$ kcal/mol          $\Delta E = -0.73$ kcal/mol

**11**                                      **12**                                      **13**

and other related compounds can be explained[44] in terms of the field effect
as given in Equation 8-13.

It has further been shown that similar results are obtained when the
molecular framework between substituent and probe is removed by using
isolated molecules. Thus the calculated (STO-3G) results[29,44] for structures
**14–16** may be compared with structure **12**.

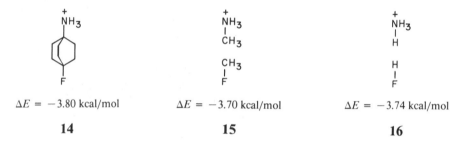

$\Delta E = -3.80$ kcal/mol          $\Delta E = -3.70$ kcal/mol          $\Delta E = -3.74$ kcal/mol

**14**                                      **15**                                      **16**

These results are particularly compelling in indicating that little, if any,
transmission occurs via the electronegativity mechanism shown in section
**4**. Calculations,[44] using isolated molecules, at various geometries give results
in accord with field effects (Subsection A of Section 4, "Field Effects") both
for proton transfer reactions and for bond polarizations in the probe.

Similar results are found where Y is a carboxylic acid or hydroxy group
as shown for the $\Delta E°$ values calculated[29] for the proton transfer equilibria
corresponding to Equation 8-5.

Similar studies have been made for isolated molecules containing a variety of substituents. Typical comparative results are shown in **17–20**. The

$\Delta E = -3.17$ kcal/mol          $\Delta E = -3.31$ kcal/mol

**17**                                      **18**

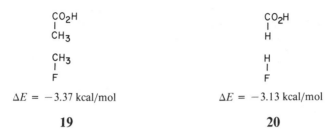

$$\Delta E = -3.37 \text{ kcal/mol} \qquad\qquad \Delta E = -3.13 \text{ kcal/mol}$$

**19**                      **20**

isolated molecule results vary with geometry in accord with the electrostatic equations given in section 3.1 and are discussed in more detail in Subsection B of Section 6, "Theoretical $^\sigma F$ Values." We note further that the results show clearly[38] that for dipolar substituents the variation in effect with distance differs by a power in $r$ compared to results for charged substituents as anticipated from Equations 8-13 to 8-16.

## C. $\pi$-Polarization Effects

The presence of a substituent dipole or pole can lead to the polarization of a $\pi$ system elsewhere in the molecule. This has been the subject of many investigations[40] and of some controversy, although the polarization of an isolated multiple bond by a substituent at a relatively remote position is broadly understood. Here the polarization follows Equation 8-14 for a dipolar substituent and Equation 8-16 for a pole (charge). Calculations, particularly those involving isolated molecules, can be particularly useful here. Thus for structure **21**, a plot of the ln $\Delta q_\pi$ at position $\alpha$ (3-21G) versus distance $r$ for X = F is shown in Figure 8-4.

$$\overset{\alpha}{H_2C} = CH_2 \qquad\qquad H-X$$
$$\underset{r}{\underbrace{\qquad\qquad\qquad}}$$

**21**

The plot is nicely linear, with a slope of 2.90 corresponding to $\Delta q_\pi$ depending on $r^{2.90}$. (A value of 2.85 has been found[48] for the polarization of the carbonyl group formaldehyde by methyl fluoride.) A similar plot for polarization by the monopole X = $NH_3^+$ leads to a dependency on a one-lower power in $r$, $r^{2.01}$, as anticipated. A similar polarization (**22**) occurs for benzene, although here the distance must be taken to carbon atom $\alpha$, at which the primary polarization seems to occur. At close distances there is some evidence for polarization as in **23** in accord with resonance ideas that an increase in $\pi$-electron density at $C_\alpha$ leads to a depletion at the ortho and para positions. In both **22** and **23** the HF molecule is in the plane of the benzene ring.

The polarization of ethylene by various substituents X in **21** follows[40,49]

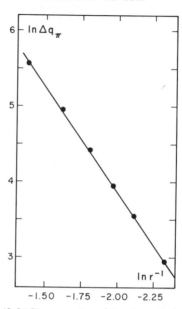

**Figure 8-4.** Plot of ln $q_\pi$ (3-21G) versus ln $r^{-1}$ for the polarization of ethylene by HF. (Reprinted with permission from *J. Chem. Soc. Perkin Trans. 2,* **1984,** 113.)

the field effects of the group X, and this is discussed in greater detail in Subsection B of Section 6. The relationship found[40] was much superior to that[50] between the $^{13}C$ substituent shifts and substituent field effects for the $sp^2$ carbon atom of compounds $CH_2=CH-(CH_2)_nX$ where other factors, such as conformational flexibility, cause problems.

The polarization of conjugated systems such as butadiene is also readily examined by use of isolated molecule calculations. Such studies require that the $\pi$-electron populations of the butadiene be corrected for the polarization of ethylene molecules at positions corresponding to the double bonds as shown in **24**.

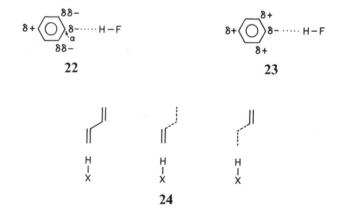

There is then found an underlying additional polarization of the butadiene $\pi$ system that is in accord with resonance concepts (**25**) leading to an increase in $\pi$-electron density at $C_1$ and a decrease at $C_4$ rather than a polarization of the system as a whole.

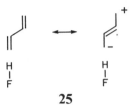

**25**

## D. Resonance Effects

Resonance effects of significance require that substituents be attached to a $\pi$-electron system. The experimental measurements for such systems usually include at least field effects as well. Much effort has been put into the separation of the $F$ and $R$ effects, as well as in studying the alteration of resonance effects for various substituents depending on the $\pi$-electron sufficiency or deficiency of the system to which they are attached. The intensities of certain infrared absorptions on molecules such as substituted benzenes, ethylenes, and acetylenes do seem to follow[51] only resonance interactions; but apart from these, theoretical calculations offer a simple and direct way to investigate resonance effects. (See also Chapter 5 by Greenberg and Stevenson in Volume 3 of this series.)

The primary result of interest is the overall $\pi$-electron density transferred between a substituent and $\pi$ system to which it is attached. Such $\Sigma\Delta q_\pi$ values are taken as negative if the transfer is to the $\pi$ system, positive if it is to the substituent. It has been established[52] that such transfers are colinear between substituted benzenes, ethylenes, and acetylenes. For example,[52] Figure 8-5 plots the $\Sigma\Delta q_\pi$ values for a series of monosubstituted ethylenes versus those of the corresponding monosubstituted benzenes at the STO-3G basis level. A similar result is found at the 4-31G basis level. This means that calculations of the resonance scale can be made for the simpler substituted ethylenes rather than for the benzenes for which the majority of experimental evidence is available. We return to the matter of the determination of resonance interactions through theoretical calculations in Subsection D of Section 6, "Theoretical $\sigma_R{}^X$ Values" We note, however, that calculations[52] in either substituted ethylenes or benzenes at the 4-31G basis are not colinear against those at the minimal STO-3G basis if both $\pi$-electron withdrawing and donating substituents are included.

We now consider some important topics involving resonance interactions where theoretical calculations have proved very useful.

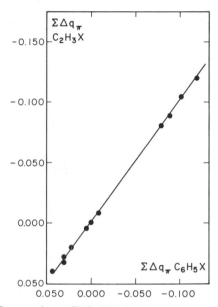

**Figure 8-5.** Plot of $\Sigma \Delta q_\pi$ values (STO-3G) in monosubstituted ethylenes versus those in monosubstituted benzenes. (Reprinted with permission from *J. Mol. Struct.* **1984,** *106,* 277.)

### a. Substituent–Substituent Resonance Interactions

There has been considerable interest in the magnitude of substituent–substituent resonance interactions, particularly in compounds such as paranitroaniline where one $\pi$-electron-donating substituent is conjugated with a $\pi$-electron-withdrawing substituent. The resonance interaction here, sometimes referred to as through conjugation, can be visualized in terms of the canonical structure **26.**

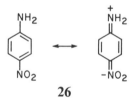

**26**

Evidence for such effects has mainly come from structure determinations suggesting the presence of some contribution from the quinonoid structure above, from dipole moment measurements on disubstituted benzenes, and from the broad interpretation of data on the reactions of such compounds. A more detailed examination[12,53] of the structural evidence, particularly of recent gas phase investigations, does not support any significant bond length changes in such compounds. This is supported by theoretical determinations of optimized structures for benzenes[53] and ethylenes.[54] Such calculations suggest that significant bond length changes occur only in compounds such as the paranitrophenoxide ion, where charge is thereby delocalized (**27**), rather than as in **26** where charge is separated. Structural determinations are underway[55] on a tetraalkylammonium salt of **27**.

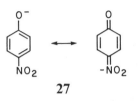

**27**

Theoretical calculations can provide a direct measurement of the energy involved in such substituent–substituent interactions by use of the energy change for the homodesmotic process (8-18) which is only amenable to experimental determination if accurate heats of formation are available for each species.

$$\text{(8-18)}$$

Values (STO-3G) for parasubstitued anilines[14] and phenols[16] are given in Table 8-9; values for parasubstituted pyridines appeared in Table 8-1. They are referred to as interaction energies, $\Delta E_{int}$. It can be seen that the values are relatively small and certainly do not support really significant interaction energies for even species such as paranitroaniline. Indeed, the interac-

TABLE 8-9. Interaction Energies for Parasubstituted Anilines, Anilium Ions, Phenols, and Phenoxides

| | STO-3G Energies (kcal/mol)[a] | | | |
|---|---|---|---|---|
| Substituent | Aniline | Anilium cation | Phenol | Phenoxide |
| NH₂ | −1.6 | 4.5 | −1.5 | −6.4 |
| OMe | −1.5 | 2.8 | −1.3 | −4.3 |
| F | −0.8 | −2.0 | −0.9 | 0.7 |
| Me | −0.5 | 2.0 | −0.4 | −1.4 |
| CN | 1.4 | −9.0 | 0.8 | 22.2 |
| CF₃ | 0.8 | −5.1 | 0.4 | 11.5 |
| CHO | 1.0 | −3.2 | 0.7 | 14.3 |
| NO₂ | 2.2 | −12.6 | 1.3 | 30.4 |

[a]Negative sign denotes stabilization; see Equation 8-18.

tion energy of 2.2 kcal/mol here drops to 1.3 kcal/mol when the structures are optimized.[56a] (Recent $\Delta H_f^\circ(g)$ data has shown that the para isomer[56b] is 1.7 kcal/mol more stable than the meta isomer.) Calculations[54] made for optimized structures at the 4–31G basis level do, however, suggest an interaction energy of some 8.5 kcal/mol in *trans*-2-nitro-1-aminoethylene (28), suggesting much more importance for quinoidal resonance here.

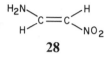

**28**

The interaction energies found for the charged anilinium and phenoxide ions (Table 8-9), and pyridinium ions (Table 8-1) are much greater than for the corresponding neutral species. These include dipole–charge interactions between the substituent and the charge as well as resonance interactions. That the latter can be important here is shown by a calculation (STO-3G) of 22.6 kcal/mol for the interaction energy in the paranitrophenoxide ion in which the plane of the nitro group is held at right angles to that of the benzene ring. This should be approximately the energy of the dipole–charge interaction in the planar form and leaves the difference of 7.8 kcal/mol as arising from the resonance interaction as in **27**.

These results indicate[57] that the proton transfer equilibria of species such as substituted anilines and phenols are mainly controlled by interaction in the charged forms, not in the neutrals. This is not true for molecules such as certain substituted methylamines[58] where direct hyperconjugative interaction is possible between the amino group and the substituent.

### b. Resonance Interactions Under Conditions of Varying Demand

It follows from the discussion above that resonance interactions alter depending on the demand, particularly of a positive or negative charge. Variations in electron demand can be readily followed by theoretical calculations[59] of the $\pi$-electron transfer to or from substituents attached to a variety of $\pi$-electron systems. Thus the $\pi$-electron transfers for a few common substituents are shown in Table 8-10 as calculated at the STO-3G basis with $C-X$ bond length optimization. It is clear that a substituent such as a hydroxy group donates considerably more $\pi$-electron density when attached to the strongly $\pi$-electron-demanding $-CH_2^+$ cation than in phenol and has effectively zero $\pi$-electron transfer when attached to the $CH_2^-$ anion. By contrast, the cyano substituent gains considerably by $\pi$-electron transfer in $XCH_2^-$ and by a small amount in $C_6H_5X$, but is calculated to be a $\pi$-electron donor in $XCH_2^+$. Some experimental evidence[59] is available for the latter type of $\pi$-electron donation.

The results can be more highly quantified if some relative assessment can be made of the varying $\pi$-electron demands of the various systems. This is

TABLE 8-10. Calculated (STO-3G)$^a$ $\pi$-Electron Transfer in Various Molecules

| Substituent | $\Sigma\Delta q_\pi$ (electrons $\times$ $10^3$) | | | | |
|---|---|---|---|---|---|
| | $XCH_2^{+b}$ | $p-XC_6H_4CH_2^{+b}$ | $XC_6H_5$ | $p-XC_6H_4CH_2^{-11}$ | $XCH_2^{-b}$ |
| $NH_2$ | $-566$ | $-284$ | $-115$ | $-53$ | $0$ |
| OH | $-486$ | $-202$ | $-90$ | $-45$ | $0$ |
| F | $-353$ | $-134$ | $-70$ | $-39$ | $0$ |
| $CHCH_2$ | $-427$ | $-148$ | $0$ | $105$ | $430$ |
| $CH_3$ | $-113$ | $-29$ | $-8$ | $8$ | $53$ |
| $CF_3$ | $-29$ | $-4$ | $10$ | $32$ | $120$ |
| CN | $-262$ | $-33$ | $21$ | $104$ | $326$ |
| CHO | $-155$ | $-20$ | $27$ | $142$ | $435$ |
| $NO_2$ | $-76$ | $-10$ | $19$ | $130$ | $398$ |
| $\Delta q_\pi H$ | $+1.00$ | $+0.24$ | $0.00$ | $-0.25$ | $-1.00$ |

$^a$The C$-$X bond length was optimized in each case.
$^b$Planar geometry about C.

provided by the $\pi$-electron density calculated for the position of substituent attachment in the compound with X = H. Such $\Sigma q_\pi H$ values are also listed in Table 8-10. Figure 8-6 $\Sigma\Delta q_\pi$ for some common substituents versus $\Delta q_\pi(H)$. It is seen that the response is monotonic but not linear and also not of the same form from one substituent to another. Substituents such as CN, CHO, and $NO_2$ are calculated to act as $\pi$ donors when attached to $\pi$-electron-deficient systems where $\Delta q_\pi H$ is less than $-0.2$ electron (see also Chapter 9, this volume). The pyridinium ion has a $1 - \Delta q_\pi H$ value of $-0.2$ elec-

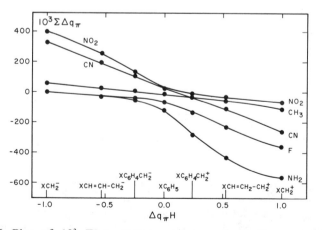

**Figure 8-6.** Plot of $10^3$ $\Sigma\Delta q_\pi$ versus $\pi$-electron demand parameters $\Delta q_\pi(H)$. (Reprinted with permission from *Accounts of Chemical Research*, **1983**, *16*, 292. Copyright 1983, American Chemical Society.)

tons, and this presumably explains why such substituents do not appear to show resonance effects here.[14]

## E. Field-Induced Resonance Effects

Infrared results for systems such as **29** suggested that a relatively remote substituent (CN here) could alter the resonance interaction between a second group (OMe) and a $\pi$ system. This would presumably arise from stabilization of the charged canonical form in **29**. Other results have been similarly interpretated.[60] This is another area in which theoretical calculations using isolated molecules can be usefully employed. A study has been made of the polarization of monosubstituted ethylenes by HF and by $NH_4^+$.

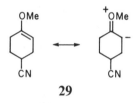

**29**

For the relative geometry shown in **30**, it was found that the additional $\pi$-electron transfer between X and the double bond was very small compared to that in the unperturbed $C_2H_3X$ molecule. For the geometry shown in **31**, corresponding to that in a 4-fluoro-1-X-substituted benzene, there is small additional $\pi$-electron transfer, but only of the order of 0.0005–0.0009 electron in expected direction, for various substituents. The $NH_4^+$ group also causes only a relatively small field-induced resonance effect. Thus this effect is likely to be relatively unimportant in most systems compared to normal resonance interactions.

**30**                                                              **31**

## F. Effects on Energy Versus Effects on Electron Distribution

Substituent effects are sometimes described in terms of energy, such as their effect on proton affinities, and sometimes in terms of electron distribution, as in resonance concepts. It follows from Equations 8-13 to 8-16 for a given system (ie, one having a fixed relative geometry of substituent and probe)

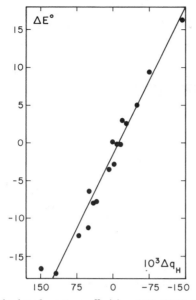

**Figure 8-7.** Plot of calculated proton affinities $\Delta E°$ (STO-3G) of substituted pyridines versus the charge on the proton, $\Delta q_H$, in the corresponding pyridinium ion.

that the field effect of substituents on charge should be directly proportional to their effect on bond polarization. Such a relationship is not so obvious for resonance effects. Nevertheless a series of calculations[58] (STO-3G) on proton transfer reactions of compounds as diverse as methylamines, pyridines, anilines, phenols, and toluenes all show approximately linear relationships between $\Delta E°$ and the corresponding electron populations of hydrogen atoms attached to the acidic center. Thus Figure 8-7 plots calculated $\Delta E°$ values for the proton affinities of substituted pyridines versus the charge on the proton $\Delta q_H$ in the neutral form (Table 8-1). The relationships are not precise but do give general support for the different conceptual ways of describing substituent effects.

# 6. THEORETICAL CALCULATIONS OF SUBSTITUENT EFFECT PARAMETERS

## A. Introduction

The Hammett substituent parameters, $\sigma$, provide a numerical scale of substituent effects. Many different scales have been proposed[62] at various times, but the basic ones must refer to the four major substituent effects and can accordingly be labeled $\sigma_F$, $\sigma_X$, $\sigma_\alpha$, and $\sigma_R$. From the discussion in Subsection D of Section 5, "Resonance Effects," it is clear that the scale of resonance interactions may alter from when the substituent is attached to a relatively neutral system such as a benzene ring ($\sigma_R^o$) to when it is attached to systems that are highly $\pi$-electron deficient ($\sigma_R^+$) or rich ($\sigma_R^-$). Presently, theoretical values are available for $\sigma_F$, $\sigma_X$, and $\sigma_R^o$ constants for many common substituents and work is proceeding on scales of $\sigma_\alpha$, $\sigma_R^+$, and $\sigma_R^-$ values.

Theoretically calculated values have two advantages: they are free from solvent effects on the substituent, and they are obtained directly rather than by separation from a combination of substituent effects. The figures therefore can be referred to as inherent values.[17]

## B. Theoretical $\sigma_F$ Values

Experimentally determined values of $\sigma_F$ often depend quite markedly on the solvent used, and values for either the gas phase or nonpolar solvents are limited. Only for relatively few substituents is there agreement to $\pm 0.02$ for the $\sigma_F$ values among various authors.[49] Clearly, therefore, it is extremely valuable to have theoretically calculated $\sigma_F$ values for a wide variety of substituents.

In early work[13] we showed that calculated $\Delta E^\circ$ values (STO-3G) for the proton transfer of $\beta$-substituent ethylamines (process 8-19)

$$XCH_2CH_2NH_3^+ + CH_3CH_2NH_2 \leftrightharpoons XCH_2CH_2NH_2 + CH_3CH_2NH_3^+$$

$$(8\text{-}19)$$

followed experimental $\sigma_F$ values. However the agreement was not precise, probably since such calculations include[49] the effect of indirect polarization via the CH bonds. By contrast, three simple processes have been found[49] that lead to $\sigma_F$ values. The first two (8-20 and 8-21) involved the polarization of either a hydrogen molecule or of the $\pi$ system of an ethylene molecule by an HX molecule.

$$\overset{\Delta q_H}{H-H\,\text{------}\,H-X} \tag{8-20}$$

$$\overset{\Delta q_\pi}{H_2C=CH_2\,\text{----}\,H-X} \tag{8-21}$$

The values of $\Delta q_H$ (4–31G//4–31G) at the first hydrogen atom in process 8-20 are accurately linear against $\Delta q_\pi$ at the first carbon atom in process 8-21, and therefore the first and simpler system is preferred.

The third system involves calculating (4–31G//4–31G) the $\Delta E^\circ$ values for the proton transfer process 8-22.

$$
\begin{array}{ccccccc}
\overset{+}{N}H_3 & & NH_2 & & NH_2 & & \overset{+}{N}H_3 \\
| & & | & & | & & | \\
H & & H & & H & & H \\
& + & & \rightleftharpoons & & + & \\
H & & H & & H & & H \\
| & & | & & | & & | \\
X & & H & & X & & H
\end{array}
\tag{8-22}
$$

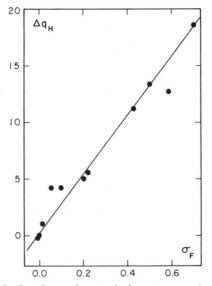

**Figure 8-8.** Plot of calculated atomic populations $\Delta q_H$ on the hydrogen (4–31G) in $H_2$ as polarized by HX molecules versus $\sigma_F$ values for substituents X.

The $\Delta E°$ values obtained here for various substituents show[49] very good linearity versus the $\Delta q_H$ values from process 8-20. It appears that small polarizability effects occur in process 8-22 for some substituents, and thus process 8-20 is preferred overall.

The $\Delta q$ values for process 8-20 are plotted against literature inherent best values[17] for $\sigma_F$ in Figure 8-8. It is seen that the theoretical values show excellent linearity against the literature figures, which are for substituents where $\sigma_F$ values have been moderately well defined. For many other substituents, however, there is considerable uncertainty as to the inherent $\sigma_F$ value. Here the theoretical approach seems preferred. To generate such an identity, we need to define one scale factor. We used the $\sigma_F$ values $CF_3 = 0.42$, $NO_2 = 0.65$, which are well agreed, to obtain Equation 8-23. Values of $\sigma_F$ obtained are given in Table 8-11.

$$\sigma_F = -35.6\Delta q_H \qquad (8-23)$$

One important aspect of this work[49] is that $\sigma_F$ values can be obtained for various conformations of substituents such as $CH_2OH$, $CO_2Me$, and NHOH. Thus for $CO_2Me$, it is found that the $\sigma_F$ values for the two conformations below are as listed.

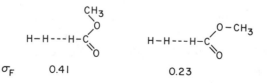

**TABLE 8-11.** Theoretical $\sigma_F$, $\sigma_X$, and $\sigma_R$ Values

| Substituent | $\sigma_F$ | $\sigma_X$ | $\sigma_R^\circ$ |
|---|---|---|---|
| NMe$_2$[a] | 0.15 | 0.34 | −0.58 |
| NH$_2$[a] | 0.15 | 0.33 | −0.47 |
| OMe | 0.29 | 0.55 | −0.41 |
| OH | 0.30 | 0.54 | −0.41 |
| F | 0.47 | 0.70 | −0.30 |
| Me | −0.01 | 0.00[b] | −0.08 |
| t−Bu | −0.01 | −0.02 | — |
| CHCH$_2$ | 0.04 | 0.02 | −0.05 |
| CCH | 0.17 | 0.22 | −0.01 |
| CF$_3$ | 0.42[b] | 0.02 | 0.04 |
| CN | 0.45 | 0.30 | 0.09 |
| CHO | 0.22 | −0.05 | 0.18 |
| COMe | 0.19 | −0.04 | 0.20 |
| CO$_2$Me | 0.23 | 0.04 | 0.17 |
| NO$_2$ | 0.66[b] | 0.46 | 0.19 |
| NO | 0.55 | 0.33 | 0.29 |

[a]Planar
[b]Taken to define scales.

Such important conformational effects have some experimental support[49] but are not generally recognized.

The excellent correspondence[49b] between $\sigma_F$ and the older $\sigma_I$ values clearly show that the latter measured a field effect.

## C. Theoretical $\sigma_X$ Values

Electronegativity values are not well defined experimentally for substituent groups.[47] An analysis[63] of much experimental data led to approximate values for a few common substituents, but even here considerable variation was found in figures from various techniques even for simple substituents such as CH$_3$. The major problem is to obtain a set of experimental data that depends solely on electronegativity differences.

It is thus highly desirable to obtain $\sigma_X$ values theoretically. These can then be used, in conjunction with $\sigma_F$ values, to analyze experimental data. Our approach has been to obtain atomic electron populations for the hydrogen atom in compounds HX. This has been done[47] at the 6–31G* level with geometry optimization. Values for some common substituents are listed in Table 8-12 in terms of 1.00 $q_H$, that is, differences from values in H$_2$(X = H). Such figures were earlier[47] used directly as defining theoretical $\sigma_X$ values.

The $q_H$ values should be free from any hyperconjugative interactions because the hydrogen has no pseudo-$\pi$ orbitals. However, the Mulliken population analysis is known to be particularly poor in partitioning electron density between carbon and hydrogen. This shows up in the calculated val-

TABLE 8-12. Calculated (6–31G*//6–31G*)
Values of $1.00 - q_H$ for HX Versus Radius
of Sphere Fitted to H in HX (3–21G//3–
21G)

| Substituent | $q_H$ | $r_H$ (Å) |
|---|---|---|
| SiH$_3$ | −0.13 | 1.363 |
| SH | 0.12 | 1.235 |
| CHO | 0.14 | 1.206 |
| CH$_2$F | 0.16 | 1.193 |
| CH$_3$ | 0.17 | 1.193 |
| H | 0.00 | 1.19 |
| CF$_3$ | 0.17 | 1.178 |
| CH$_2$CN | 0.23 | 1.176 |
| Cl | 0.28 | 1.177 |
| CHCH$_2$ | 0.17 | 1.169 |
| CN | 0.31 | 1.130 |
| NH$_2$ | 0.33 | 1.129 |
| CCH | 0.28 | 1.121 |
| OH | 0.43 | 1.083 |
| F | 0.52 | 0.010 |

ues, which suggest a remarkably large electronegativity difference between H and Me. Most determinations suggest that Me and H have very similar electronegativities, with that for Me being slightly larger.

Also listed in Table 8-12 are values [26,64] for the radius of H in HX as determined by fitting spheres to the total electron density surface (see Subsection B of Section 3, "Electron Density Distribution"). These should give a more reliable guide to the hydrogen electron populations. They are plotted versus the Mulliken atom populations in Figure 8-9. It is seen that the agreement is very good provided that the values for X = H are omitted.

We thus choose to use the methyl substituent as a standard and take $\sigma_X(Me) = 0.00$. The difference in $q_H$ between HX and HCH$_3$ could be used directly as a measure of $\sigma_X$, but the scale of values would then be much less than for $\sigma_F$ and $\sigma_R^\circ$ for which common substituents have values in the range 0.0–0.7. We therefore define $\sigma_X$ as in Equation 8-24.

$$\sigma_X = 2(q_H HCH_3) - q_H(HX))$$  (8-24)

Values of such theoretically derived $\sigma_X$ values are given in Table 8-11.

## D. Theoretical $\sigma_R^\circ$ Values

The simplest theoretical approach to $\sigma_R$ values is to calculate the $\pi$-electron charge transfer ($\Sigma\Delta q_\pi$) between substituents and the system of interest. Obviously it follows from the discussion in Subsection D of Section 5, "Resonance Effects," that the values will alter depending on the system chosen

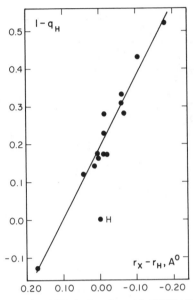

**Figure 8-9.** Plot of $1 - q_H$ for HX derivatives (6–31G*//6–31G) versus the radius difference, $r_X - r_H$, obtained by best fitting of spheres (3–21G//3–21G).

(eg, being quite different in monosubstituted benzenes compared to substituted cyanides XCN). In the latter case the substituents will have a greater tendency to donate $\pi$-electron density than in the benzenes. The $\sigma_R^\circ$ scale applies to $\pi$-electron interactions in substituted benzenes, and calculations of $\Sigma\Delta q_\pi$ gives values showing good linearity[35] against experimental values applicable to nonpolar solvents. However, the calculations are restricted to the minimal basis level for many substituted benzenes. An alternative system is monosubstituted ethylenes, and it has been demonstrated[52] that $\Sigma\Delta q_\pi$ values obtained here are linear against $\Sigma\Delta q_\pi$ values for the corresponding monosubstituted benzenes at either the STO$-$3G or 4–31G basis levels. We therefore choose to use 4–31G calculations on monosubstituted ethylenes as our standard. To fit an equation, we took[65] experimental values for 10 substituents for which $\sigma_R^\circ$ values seem well agreed and obtained Equation 8-25 with a correlation coefficient of 0.991.

$$\sigma_R^\circ = 0.0041\Sigma\Delta q_\pi - 0.044 \qquad (8\text{-}25)$$

The equation applies equally well to substituents in either of the first two major rows of the periodic table. The intercept follows[65] because ethylenes rather than benzenes are used as the defining series; hydrogen always has $\sigma_R^\circ = 0.0$ but cannot be included in such an equation plotting the response in one system against a second of differing $\pi$-electron demand. Values of $\sigma_R^\circ$ have been calculated in this way for a wide range of substituents and show excellent agreement against experimental $\sigma_R^\circ$ values where these are well established. Calculated values for some common substituents are given in Table 8-11. For certain substituents such as NO and NCO having a very variable response in differing systems, the theoretical approach would seem to provide the most reliable inherent values presently available.

For symmetrical substituents (ie, those with local symmetry of $C_3$ or higher), such as F, $CH_3$, $CF_3$, and CN, the $\pi$-electron interaction with the attached system is independent of the conformation. This is not true for other groups and, for example, $\sigma_R^\circ$ is expected to be zero for groups such as $NO_2$ and CHO when they are twisted so that their planes are at right angles to that of the attached $\pi$ systems. Minimum values for substituents on rotation have been symbolized[51] as $\sigma_R^\circ$ (twist) and have been experimentally estimated. Such $\sigma_R^\circ$ (twist) values can also be calculated by comparison of the $\pi$ interactions of a substituent with the two separate $\pi$-electron systems in monosubstituted acetylenes. They show good agreement with the experimental values. One particularly interesting result is that while the nitroso group (NO) is quite a strong $\pi$-electron-withdrawing group when coplanar with an attached benzene or ethylene molecule, it is predicted to be a moderate $\pi$-electron donor ($\sigma_R^\circ$ (twist) $= -0.17$) if the steric strain is sufficient to rotate it 90° out of plane.

## ACKNOWLEDGMENT

I am grateful to Prof. R. W. Taft for helpful discussion and to the Australian Research Grants Scheme for financial assistance.

## REFERENCES

1. For a general discussion, see References 2 and 3. Particular basis sets are given in Reference 7.
2. Hehre, W.J.; Radom, L.; Schleyer, P.v.R.; Pople, J.A. "Ab Initio Molecular Orbital Theory". Wiley: New York, 1986.
3. Hehre, W.J. *Acc. Chem. Res.* **1976,** *9,* 399, and references therein.
4. Pople, J.A.; Beveridge, D.L. "Approximate Molecular Orbital Theory". McGraw-Hill: New York, 1970.
5. Bingham, R.C.; Dewar, M.J.S.; Lo, D.H. *J. Am. Chem. Soc.* **1975,** *97,* 1285, 1294, 1302, 1307.
6. Dewar, M.J.S.; Thiel, W. *J. Am. Chem. Soc.* **1977,** *99* 4899, 4907.
7. Binkley, J.S.; Pople, J.A.; Hehre, W.J. *J. Am. Chem. Soc.* **1980,** *102,* 939; Gordon, M.S.; Binkley, J.S.; Pople, J.A.; Pietro, W.J.; Hehre, W.J. ibid, **1982,** *104,* 2797.
8. Hehre, W.J.; Ditchfield, D.; Radom, L.; Pople, J.A. *J. Am. Chem. Soc.* **1970,** *92,* 4796.
9. George, P.; Trachtman, M.; Brett, A.M.; Bock, C.W. *J. Chem. Soc. Perkin Trans. 2,* **1977,** 1036.
10. Carsky, P.; Urban, M. "Ab Initio Calculations". Springer-Verlag: New York, 1980.
11. Taft, R.W. *Prog. Phys. Org. Chem.* **1983,** *14,* 247.
12. Topsom, R.D. *Prog. Phys. Org. Chem.* In press.
13. (a) Taagepera, M.; Hehre, W.J.; Topsom, R.D.; Taft, R.W. *J. Am. Chem. Soc.* **1976,** *98,* 7438. (b) Lias, S.G.; Liebman, J.F.; Levin, R.D. *J. Chem. Phys. Ref. Data* **1984,** *13,* 695.
14. Taagepera, M.; Summerhays, K.D.; Hehre, W.J.; Topsom, R.D.; Pross, A.; Radom, L.; Taft, R.W. *J. Org. Chem.* **1981,** *46,* 891.

15. Reynolds, W.F.; Mezey, P.G.; Hehre, W.J.; Topsom, R.D.; Taft, R.W. *J. Am. Chem. Soc.* **1977**, *99*, 5821.
16. Pross, A. Radom, L.; Taft, R.W. *J. Org. Chem.* **1980**, *45*, 818.
17. Fujio, M.; McIver, R.T.; Taft, R.W. *J. Am. Chem. Soc.* **1981**, *103*, 4017.
18. McMahon, T.B.; Kebarle, P. *J. Am. Chem. Soc.* **1977**, *99*, 2222.
19. Mezey, P.G.; Reynolds, W.F. *Can. J. Chem.* **1977**, *55*, 1567. See also: Bohm, S.; Kutham, J. *Init. J. Quantum Chem.* **1984**, *26*, 21.
20. Koppel, I.; Mishima, M.; Stock, L.; Taft, R.W.; Topsom, R.D. Unpublished results.
21. Marriott, S.; Lebrilla, C.B.; Koppel, I.; Mishima, M.; Taft, R.W.; Topsom, R.D., *J. Mol. Struct.* **1896**, *137*, 133.
22. Arnett, E.M.; Chawla, B.; Bell, L.; Taagepera, M.; Hehre, W.J.; Taft, R.W. *J. Am. Chem. Soc.* **1977**, *99*, 5729.
23. Davidson, W.R.; Sunner, J.; Kebarle, P. *J. Am. Chem. Soc.* **1979**, *101*, 1675.
24. Coppens, P.; Hall, M.B., Eds. "Electron Distributions and the Chemical Bond". Plenum Press: New York, 1983.
25. Wiberg, K.B. *J. Am. Chem. Soc.* **1980**, *102*, 1229.
26. Francl, M.M.; Hout, R.F.; Hehre, W.J. *J. Am. Chem. Soc.* **1984**, *106*, 563.
27. Mulliken, R.S. *J. Chem. Phys.* **1955**, *23*, 1833, 1841, 2338, 2343.
28. Fliszar, S. "Charge Distributions and Chemical Effects". Springer-Verlag: New York, 1983.
29. Marriott, S.; Topsom, R.D. *J. Mol. Struct.* **1982**, *89*, 83.
30. Grier, D.L.; Streitweiser, A. *J. Am. Chem. Soc.* **1982**, *104*, 3556.
31. Hehre, W.J.; Taft, R.W.; Topsom, R.D. *Prog. Phys. Org. Chem.* **1976**, *12*, 159.
32. Bromilow, J.; Brownlee, R.T.C.; Lopez, V.O.; Taft, R.W. *J. Org. Chem.* **1979**, *44*, 4766. Taft, R.W. Unpublished results.
33. For a general discussion, see References 11, 12, and 34–37.
34. Topsom, R.D. *Prog. Phys. Org. Chem.* **1976**, *12*, 1, and references therein.
35. Topsom, R.D. *Acc. Chem. Res.* **1983**, *16*, 292.
36. Stock, L.M. *J. Chem. Educ.* **1972**, *49*, 400, and references therein.
37. Reynolds, W.F. *Prog. Phys. Org. Chem.* **1983**, *14*, 165.
38. Marriott, S.; Reynolds, W.F.; Topsom, R.D. *J. Org. Chem.* **1985**, *50*, 741.
39. Craik, D.J.; Brownlee, R.T.C.; Sadek, M. *J. Org. Chem.* **1982**, *47*, 657. See also, Kalfus, K.; Friedl, Z.; Exner, O. *Collect. Czech, Chem. Commun.* **1984**, *49*. 179.
40. Marriott, S.; Topsom, R.D. *J. Chem. Soc. Perkin Trans. 2*, **1984**, 113, and references therein.
41. Hamer, G.K.; Peat, I.R.; Reynolds, W.F. *Can. J. Chem.* **1973**, *51*, 897; Reynolds, W.F.; Peat, I.R.; Freedman, M.J.; Lyerla, J.R. ibid, 1973, **51**, 1857; Brownlee, R.T.C.; Butt, G.; Chan, C.H.; Topsom, R.D. *J. Chem. Soc. Perkin Trans. 2*, **1976**, 1486.
42. Topsom, R.D. *Tetrahedron Lett.* **1980**, 403.
43. Taft, R.W. Personal communication.
44. Topsom, R.D. *J. Am. Chem. Soc.* **1981**, *103*, 39.
45. Reynolds, W.F.; Mezey, P.G.; Hamer, G.K. *Can. J. Chem.* 1977, *55*, 522.
46. Reynolds, W.J. *J. Chem. Soc. Perkin Trans. 2* **1980**, 985.
47. Marriott, S.; Reynolds, W.F.; Taft, R.W.; Topsom, R.D. *J. Org. Chem.* **1984**, *49*, 959.
48. Brownlee, R.T.C.; Craik, D.J. *J. Chem. Soc. Perkin Trans. 2* **1981**, 760.
49. (a) Marriott, S.; Topsom, R.D. *J. Am. Chem. Soc.* **1984**, *106*, 7. (b) Marriott, S.; Topsom, R.D. *Tetrahedron Lett.* **1982**, 1485.
50. Hill, E.A.; Guenther, H.E. *Org. Magn. Reson.* **1981**, *16*, 177.
51. Katritzky, A.R.; Topsom, R.D. *Chem. Rev.* **1977**, *77*, 639.
52. Marriott, S.; Topsom, R.D. *J. Mol. Struct.* **1984**, *106*, 277.
53. Van Nagy-Felsobuki, E.; Topsom, R.D.; Pollack, S.; Taft, R.W. *J. Mol. Struct.* **1982**, *88*, 255.
54. Marriott, S.; Topsom, R.D. *J. Mol. Struct.* **1984**, *109*, 305.
55. Butt, G.B.; Mackay, M.; Topsom, R.D. Work in progress.
56. (a) Pollack, S.; Taft, R.W.; Topsom, R.D. Unreported calculations. (b) Nishiyama, K.; Sakiyama, M.; Seki, S. *Bull. Chem. Soc. Japan* **1983**, *56*, 3171.
57. For a general coverage, see Pross, A.; Radom, L. *Prog. Phys. Org. Chem.* **1981**, *13*, 1.
58. Hehre, W.J.; Taagepera, M.; Taft, R.W.; Topsom, R.D. *J. Am. Chem. Soc.* **1981**, *103*, 1344.
59. Gassman, P.G.; Talley, J.J. *J. Am. Chem. Soc.* **1980**, *102*, 1214; Gassman, P.G.; Saito, K. *Tetrahedron Lett.* **1981**, *22*, 1311; Dixon, D.A.; Charlier, P.A.; Gassman, P.G. *J. Am. Chem. Soc.* **1980**, *102*, 3957, 4138; Dixon, D.A.; Eades, R.A.; Frey, R.; Gassman, P.G.; Hendewerk, M.L.; Paddon-Row, M.N.; Houk, K.N. ibid, **1984**, *106*, 3885.

60. See, for example, Reference 14 for substituted anilines, Reference 40 for conjugated polyenes, and Reference 16 for substituted phenols. See also Butt, G.; Topsom, R.D. *Spectrochim. Acta* **1982,** *38A,* 649.
61. Marriott, S.; Topsom, R.D. *J. Chem. Soc. Perkin Trans. 2* **1985,** 697.
62. See, for example, Exner, O. In "Correlation Analysis in Chemistry", Chapman, N.B.; and Shorter, J., Eds.; Plenum Press: New York, 1978.
63. Wells, P.R. *Prog. Phys. Org. Chem.* **1968,** *6,* 111.
64. Hehre, W.J. Personal communication.
65. Marriott, S.; Topsom, R.D.; *J. Chem. Soc. Perkin Trans. 2,* **1985,** 1045.

CHAPTER 9

# Substituent Effects in Carbocations Substituted at Cationic Carbon

## Marvin Charton

**Pratt Institute, Brooklyn, New York**

## CONTENTS

## 1. INTRODUCTION

A linear free energy relationship (LFER) is the simplest and generally the most effective method for quantitatively relating molecular structure to chemical reactivities, chemical and physical properties, and biological activities. Its use requires only a small investment of time and money. The basic

MOLECULAR STRUCTURE AND ENERGETICS, Vol. 4

principle of the method is that a change in some quantity of interest resulting from a change in molecular structure is a linear function of the change
in some reference quantity that results from the same structural change in
some reference chemical species. Thus,

$$\frac{\Delta Q}{\Delta R} = a_1 \frac{\Delta Q^\circ}{\Delta R} + a_0 \tag{9-1}$$

where $Q$ and $Q^\circ$ are the quantity of interest and the reference quantity,
respectively, and $\Delta R$ represents the structural change in both the chemical
species under study and the reference chemical species.

The most commonly encountered structural change is variation of a substituent, X. The quantity $\Delta Q^\circ/\Delta R$ is then taken to be a parameter, often
called a substituent constant, which models the change in $Q$ with change in
X. Equation 9-1 then becomes:

$$Q_X = a_1 \sigma_X + a_0 \tag{9-2}$$

The Hammett and Taft equations are examples of Equation 9-2. The
method is also applicable to the quantification of solvent effects.

## 2. ELECTRICAL EFFECTS OF SUBSTITUENTS

### A. Effects in XGY Systems

Almost invariably, LFERs have been applied to systems that have the form
XGY, where X is a variable substituent, Y is an active site (the atom or
group of atoms responsible for the measurable property $Q$), and G is a skeletal group to which X and Y are bonded. Normally Y and G are held constant throughout the data set, and the variation of $Q$ with X is studied. This
variation in $Q$ may result from some combination of electrical, steric, or
intermolecular force effects exerted by X. The electrical effect of X may be
conveniently separated into two components, the localized (field and/or
inductive) and the delocalized (resonance) effects.[1,2] When X is bonded to a
C atom of G that is $sp^n$ hybridized, only the localized electrical effect is usually observed for $n = 3$, while both localized and delocalized electrical
effects are observed when $n \leq 2$. Thus, in the former case when only electrical effects are involved, the correlation equation is:

$$Q = L\sigma_{lX} + h \tag{9-3}$$

while in the latter case it is

$$Q = L\sigma_{lX} + D\sigma_{DX} + h \tag{9-4}$$

The nomenclature used throughout this chapter is given in Appendix I.

## a. The Localized Electrical Effect

Parameters for the localized electrical effect are generally defined from XGY systems with G chosen such that:

1. G is rigid, excluding conformational equilibria.
2. Steric effects are absent.
3. Intramolecular hydrogen bonding does not occur.
4. The measured properties have small experimental errors.
5. The system is reasonably sensitive to electrical effects ($L > 1$).
6. There are three or more sp$^3$-hybridized C atoms between X and Y; therefore only the localized electrical effect occurs.

The most widely used localized electrical effect parameters are the $\sigma_I$ constants.[2] Considerable evidence has accumulated to suggest that the localized electrical effect in XGY systems is transmitted predominantly through space and is therefore a field effect rather than an inductive (through-bond) effect.[3] The magnitude of the localized effect varies therefore, with the square of the distance $r$ from Y to the midpoint of the X$-$G bond and the cosine of the angle $\theta$ made by $r$ with the XG bond (Figure 9-1). It also varies with the effective dielectric constant $\epsilon_E$ of the space between X and Y, as described by the Kirkwood–Westheimer equation. Usually $\epsilon_{EG} = \epsilon_{EG}^{\circ}$. Then the magnitude of the localized effect is given by the expression:

$$L_G = \left(\frac{\cos \theta_G}{r_G^2}\right) \cdot \left(\frac{r_{G^{\circ}}^2 \, L_{G^{\circ}}}{\cos \theta_{G^{\circ}}}\right) \qquad (9\text{-}5)$$

or

$$L_G = \left(\frac{\cos \theta_G}{r_G^2}\right) \gamma_{G^{\circ}} \qquad (9\text{-}6)$$

where G is some skeletal group of interest, G$^{\circ}$ is a reference skeletal group, and $L$ the localized effect coefficient in a correlation equation.

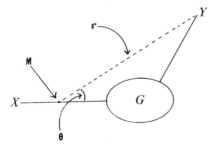

**Figure 9-1.** Definition of $r$ and $\theta$ in the Kirkwood–Westheimer equation: $M = XG/2$, $r = MY$, $\odot = YMG$.

There are some G to which Equation 9-6 is not applicable. Examples are the vinylidene, methylene, 2-substituted pyridinium, 2-substituted styrylene, 2-substituted phenylene, and 2-substituted phenylethynylene groups, for which $L$ is much larger than the value predicted from Equation 9-6. It seems likely that in these cases there is an important inductive contribution to the localized effect.

A basic assumption of electrical effect LFER is that the localized effect is independent of the hybridization state of the atom in G to which X is bonded. It is also independent of interaction of p, $\pi$, or $n_\pi$, orbitals on Y with orbitals of these types on G. This is inherent in the definition of $\sigma_D$ constants.[1,2]

### b. The Delocalized Electrical Effect

Four different types of $\sigma_D$ constant are available.[1,2] The choice of which to use depends on both the nature of G and the electronic requirements of Y. When Y is bonded to an sp³-hybridized C atom of G, the $\sigma_R°$ constants are required. When Y has an empty p or $\pi$ orbital that can interact with a $\pi$ orbital on G, the $\sigma_R{}^+$ constants are needed; when it has partially empty (low electron density) p or $\pi$ orbitals capable of interacting with a $G_\pi$ orbital, the $\sigma_R$ constants give best results. If Y has a full p or $\pi$ orbital that can interact with a G orbital, the $\sigma_R{}^-$ constants must be used. These statements are summed up in Table 9-1. The $\sigma_D$ constants, no matter what type, represent the perturbation of $\pi$, $\pi^*$, and $n_\pi$ orbitals on G by the variable substituent X. Thus, in an XGY system there is no direct interaction of X with Y. Consider for example, PhX (X = $NH_2$, OMe, F, H, $NO_2$, O=CH, or CN). The benzene $\pi$ orbitals are shown in Figure 9-2. When G is vinylene, ethynylene, or arylene and X is substituted carbonyl, sulfonyl, or phosphonyl, trihaloalkyl, cyano, or nitro, relative to H, X is a delocalized effect acceptor. The XG group will stabilize negative and destabilize positive charges. When X is ZO, SZ, $NZ^1Z^2$, or alkyl (Ak), it is a donor by the delocalized effect.

TABLE 9-1. Electrical Effect Parameters in XGY Systems[a]

| Hybridization, sp$^n$ | | | Electrical effect parameters required | |
|---|---|---|---|---|
| $n$ of $C^X$ | $n$ of $C^Y$ | Y | Parameters | Typical system |
| 3 | 3 | All | $\sigma_I$ | 1,3-X-[1.1.1.]bcPeCO$_2$H |
| $1 \leq n \leq 2$ | $1 \leq n \leq 2$ | All | $\sigma_I, \sigma_R°$ | 4-X-PnCH$_2$NH$_3^+$ |
| $1 \leq n \leq 2$ | $1 \leq n \leq 2$ | M$^{\delta+}$ | $\sigma_I, \sigma_R$ | E-2-VnCO$_2$H |
| $1 \leq n \leq 2$ | $1 \leq n \leq 2$ | M$^+$ | $\sigma_I, \sigma_R{}^+$ | 4-XPnCHClPh |
| $1 \leq n \leq 2$ | $1 \leq n \leq 2$ | M$^-$ | $\sigma_I, \sigma_R-$ | 4-X-PnNH$_2$ |

[a]$C^X$ is the C atom in G to which X is bonded; $C^Y$ is that to which Y is bonded; M is the atom of Y that is bonded to G; bc = bicyclo, Pe = Pentane, Pn = phenylene, Vn = vinylene. For a list of abbreviations used, see Appendix II.

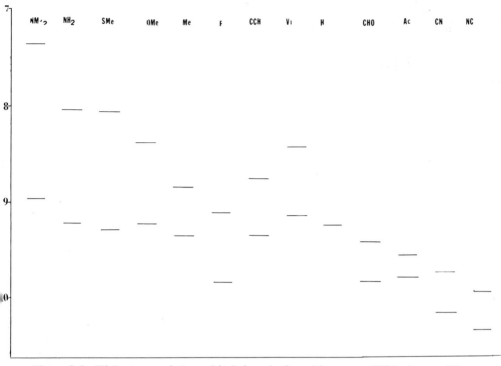

**Figure 9-2.** Highest occupied $\pi$ orbitals in substituted benzenes. With the possible exception of NO$_2$, the lower energy orbital is $b_1$, the higher is $a_2$.

## B. Electrical Effects in XY Systems

Electrical effects in XY systems are clearly different in some ways from those in XGY systems. Both localized and delocalized electrical effects will be observed in all XY systems. Transmission of the localized effect is predominantly, if not entirely, in the through-bond (inductive) mode. The $\sigma_I$ constants represent successfully the localized electrical effect in both XGY systems for which the mode of transmission is the field effect and in those for which the inductive effect seems to make an important contribution. It seems reasonable then to conclude that the $\sigma_{IX}$ values are independent of the mode of transmission of the localized electrical effect. It should therefore be possible to apply them, in the absence of medium effects, to modeling structural effects on quantities in XY systems.

It is in the case of the delocalized effect of substituents with $\pi$ orbitals capable of interacting strongly with p or $\pi$ orbitals on Y that the most dramatic differences between XY and XGY systems are observed. Consider the interaction of the X group, $-M^1 = M^2$ with three particular types of Y: the carbenium ion $^+CH_2$, the radical $^{\cdot}CH_2$, and the carbanion $\ddot{C}H_2^-$; $M^1$ and $M^2$

are any two atoms capable of $\pi$ bonding with each other. From simple valence bond (resonance) theory we have:

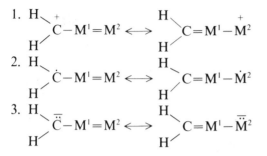

1. $\overset{H}{\underset{H}{>}}\overset{+}{C}-M^1=M^2 \longleftrightarrow \overset{H}{\underset{H}{>}}C=M^1-\overset{+}{M^2}$

2. $\overset{H}{\underset{H}{>}}\overset{\cdot}{C}-M^1=M^2 \longleftrightarrow \overset{H}{\underset{H}{>}}C=M^1-\overset{\cdot}{M^2}$

3. $\overset{H}{\underset{H}{>}}\overset{\ddot{\ \ }}{C}-M^1=M^2 \longleftrightarrow \overset{H}{\underset{H}{>}}C=M^1-\overset{\ddot{\ \ }}{M^2}$

From *qualitative* molecular orbital (MO) theory we have the orbitals and correlation diagrams given in Figure 9-3.

Obviously, a group of the $M^1 = M^2$ type should be capable of stabilizing carbenium ions, radicals, and carbanions by the delocalized effect. Such

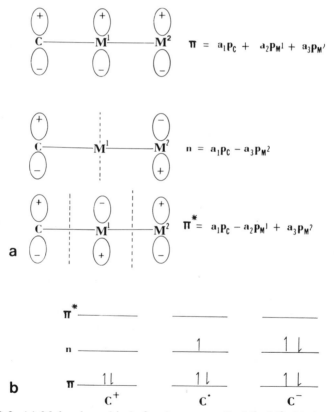

$\pi = a_1 P_C + a_2 P_{M^1} + a_3 P_{M^1}$

$n = a_1 P_C - a_3 P_{M^2}$

$\overset{*}{\pi} = a_1 P_C - a_2 P_{M^1} + a_3 P_{M^2}$

**Figure 9-3.** (a) Molecular orbitals for the system $C-M^1=M^2$. (b) Correlation diagrams for carbenium ions, radicals, and carbon ions. *They are qualitative and idealized.*

groups include Ph, Vi, $-C\equiv CH$, $NO_2$, CN, 2-furyl, $-C=O$, and $-C=S$. This is in sharp constrast to the behavior of such groups in XGY systems, where Ph, Vi, HCC$-$ and 2$-$furyl groups are clearly delocalized electrical effect donors and will not stabilize a carbanion, while $NO_2$, CN, and ZCO groups such as acetyl and carbomethoxy are delocalized electrical effect acceptors and will not stabilize a carbenium ion.

## C. Evidence for Carbenium, Radical, and Carbanion Stabilization by X$_\pi$ Groups Directly Bonded to C$^+$, C$^-$, and C$^\bullet$

There are three common types of delocalized effect donor and acceptor groups, $X_\pi$, $X_p$, and $X_h$. The $X_\pi$ groups have a $\pi$ orbital that can overlap with p or $\pi$ orbitals on G or Y; $X_p$ groups have a full (donor) or empty (acceptor) p orbital that can do the same. The $X_h$ groups act by hyperconjugation. The $X_d$ groups in which a full (donor) or empty (acceptor) d orbital on X can overlap with a p or $\pi$ orbital on Y are a much smaller category. While X groups of other types can be visualized, they are rare. Some groups are capable of interacting with G or Y in more than one way; they may be called amphielectronic. Fairly common examples are the ZSO group; and the NO and $-N=NZ$ groups; for example, MeSO possesses a lone pair and probably some degree of dp$\pi$ bonding in the SO bond. The MeSO group has a negative $\sigma_R^+$ and a positive $\sigma_R^-$, value, indicating that it may function either as a donor or as an acceptor when bonded to a skeletal group with which it can interact. Examples of $X_\pi$, $X_p$, $X_d$, and $X_h$ (hyperconjugative) donors and acceptors are given in Table 9-2.

The $X_\pi$ ($M^1=M^2$) groups may be doubly bonded, triply bonded, part of an allenyl system, aryl groups, or heteroaryl groups. These groups may be delocalized ($D$) effect XY donors or XGY acceptors. If all their $\sigma_D$ values ($\sigma_R^+$, $\sigma_R$, $\sigma_R^\circ$, $\sigma_R^-$) are negative, they are classified as $D_G$ donors. If all their $\sigma_D$ values are positive, they are classified as $D_G$ acceptors (where $D_G$ refers to the delocalized effect in an XGY system). From our previous discussion, an $X_\pi$ group that is a $D_G$ donor should be capable of stabilizing the negative

TABLE 9-2. Delocalized Effect Substituent Types

| Type[a] | Examples | | Donor |
|---|---|---|---|
| | Acceptor | | |
| $X_\pi$ | $COCH_3$, $NO_2$, CN, $POMe_2$, $SO_2Me$, | | Ph, HCC$-$, 2-furyl |
| $X_p$ | $BMe_2$, $BF_2$ | | OMe, NMeH, $NMe_2$ |
| $X_d$ | $SiMe_3$, $GeMe_3$ | | HgMe |
| $X_h$ | $CF_3$, $CCl_3$ | | Me, Et, $i$-Pr, $t$-Bu, Pr |
| $X_{\pi,(p)}$ | | SOMe, $-NO$, | $-N_2Ph$ |

[a]The $X_\pi$ groups include both pp-$\pi$ bonding as in Ph and HCC$-$ and pd bonding as in the PO and SO bonds. The $X_{\pi,(p)}$ groups have either a pp- or a pd-$\pi$ bond and a full p orbital available for interaction. The $X_{(p)}$ and $X_{(d)}$ groups have an empty or full p or d orbital, respectively, that is available for interaction. The $X_h$ groups interact by hyperconjugation.

charge in the system $X-\ddot{Y}^-$, while an X acceptor should be capable of stabilizing the positive charge in $XY^+$. The $X_\pi$ groups are much more capable of delocalizing either charge or an unpaired electron in a p orbital on C than are the $X_r$, $X_h$, or $X_d$ groups.

It follows, then, that the order of electron donor or acceptor strength observed for a set of groups of all types in an XGY system should differ from that of the same set of groups in an XY system. If and when it is observed, this difference in behavior constitutes a major piece of evidence in favor of the discussion above. A second type of evidence is the observation that an $X_\pi$ donor in an $XY^+$ system will be more effective than it is in an $XGY^+$ system. Similarly, an $X_\pi$ acceptor in an $XY^-$ system will be more effective than it is in an $XGY^-$ system.

### a. Stabilization of Carbenium Ions

The stabilization of carbenium ions, radicals, and carbanions by triply bonded $X_\pi$ groups has been reviewed.[4] The evidence for stabilization may be divided into three categories:

Chemical reactivities.

Physical properties.

Quantum chemical calculations.

The chemical reacticity data generally have involved comparisons of rates of solvolysis of the systems **1a, 1b,** and **2.**[5]

$$
\begin{array}{c}
Ak^1 \\
| \\
X-\!\!\!\overset{}{\underset{|}{C}}\!\!\!-Lg \\
Ak^2
\end{array}
\qquad\qquad
\begin{array}{c}
Ak^1 \\
| \\
X-CH_2-\!\!\!\overset{}{\underset{|}{C}}\!\!\!-Lg \\
Ak^2
\end{array}
$$

1a. X = H
1b. X ≠ H
Ak = alkyl, Lg = leaving group

**1**                                                      **2**

We define $K_{rel}$ as the ratio $k_H/k_X$. Values of $k_{rel}$ for X groups are generally much smaller than would have been predicted from the localized effect of $X_\pi$ acting alone. They are even smaller than the value of $k_{rel}$ for **2** in which only the localized effect of X makes a major contribution. Values of log $k_{rel}$, reported in Table 9-3, can be accounted for only if $X_\pi$ groups are capable of stabilizing carbenium ions and therefore can act as delocalized effect donors in $XY^+$ systems. In a related study, Creary[6] has compared the solvolysis values of **3** and **4.** We define $k^*_{rel} \equiv k_3/k_4$. This term differs from $k_{rel}$ because is represents $k_{CH_2}/k_{CO}$ rather than $k_H/k_X$.

**3**                                                      **4**

**TABLE 9-3.** Values of Log $k_{rel}$ for 1

| X | Ak$^1$, Ak$^2$ | Lg | Medium | T (°C) | Log $k_{rel}$ | Ref. |
|---|---|---|---|---|---|---|
| CN | Me, Me | OTs | CF$_3$CH$_2$OH | 25 | 3.54 | 5a |
| CN | $-(CH_2)_7-$ | OTs | CF$_3$CH$_2$OH | 25 | 3.28 | 5a |
| CN | | OTs | CF$_3$CH$_2$OH | 25 | 2.30 | 5b |
| CF$_3$ | Me, Me | OTs | CF$_3$CO$_2$H | 25 | 5.04 | a |
| H | Me, Me | OTs | CF$_3$CO$_2$H | 25 | 0 | By definition |
| HC≡C | Me, Me | Cl | 80% aq. EtOH | 25 | -2.09 | b |
| HC≡C | Me, Me | Cl | HCO$_2$H | 15 | -1.72 | b |
| Vi | Me, Me | Cl | 80% aq. Et.OH | 25 | -6.02 | b |
| MeC≡C | Me, Me | Cl | 80% aq. Et.OH | 25 | -5.43 | b |
| MeC≡C | Me, Me | Cl | HCO$_2$H | 15 | -3.73 | b |
| Bz | Me, Me | OMs | AcOH | 25 | -0.079 | c |
| Bz | Me, Me | OMs | HCO$_2$H | 25 | -0.078 | c |
| Bz | Me, Me | OMs | CF$_3$CO$_2$H | 25 | -0.017 | c |
| COtBu | Me, Me | OMs | AcOH | 25 | 0.043 | c |
| COtBu | Me, Me | OMs | HCO$_2$H | 25 | 0.046 | c |
| COtBu | Me, Me | OMs | CF$_3$CO$_2$H | 25 | 1.29 | c |
| CO$_2$Me | Me, Me | OMs | AcOH | 25 | -1.71 | c |
| CO$_2$Me | Me, Me | OMs | HCO$_2$H | 25 | -1.70 | c |
| CO$_2$Me | Me, Me | OMs | CF$_3$CO$_2$H | 25 | -2.25 | c |
| PO(OEt)$_2$ | Me, Me | OMs | AcOH | 25 | 1.91 | 10b |
| PO(OEt)$_2$ | Me, Me | OMs | HCO$_2$H | 25 | 2.74 | 10b |
| PO(OEt)$_2$ | Me, Me | OMs | CF$_3$CO$_2$H | 25 | 2.39 | 10b |

[a]Koshy, K.M.; Tidwell, T.T. *J. Am. Chem. Soc.* **1980,** *102,* 1216.
[b]Richey, H.G., Jr.; Richey, J.M. In "Carbonium Ions", Vol II, Olah, G.A.; and Schleyer, P.v.R., Eds.; Wiley: New York, 1970.
[c]Creary, X. *J. Am. Chem. Soc.* **1984,** *106,* 5568.

The evidence based on physical properties for carbenium ion stabilization is rather limited. The reaction of Ph$_2$C(OH)CN with FSO$_3$H in SO$_2$ClF at $-78$°C produces the stable carbenium ion Ph$_2$$\overset{+}{C}$CN. The NMR spectrum of this ion indicates delocalization of the positive charge by CN as well as by Ph.[7a]

The solvolysis of substrates bearing a diethoxyphosphonyl group on the C atom to which the leaving group is attached is more rapid than would be expected from the $\sigma^+$ value of PO(OEt)$_2$.[7b]

Values of $\Delta A_X$, the difference in appearance potential between XCH$_2^+$ and CH$_3^+$ in the gas phase reaction:

$$XCH_3 + e \rightarrow X\overset{+}{C}H_2 + 2e + H^+ \qquad (a)$$

have been determined.[8] They are reported to be a measure of the effect of X in stabilizing the carbenium ion. The nitro group stabilizes the carbenium ion. The cyano group shows overall destabilization, but to a much lesser extent than would have been the case in an XGY system.

Quantum chemical calculations supply some additional evidence for carbenium ion stabilization. Thus, calculations have been carried out for

MeCN and for the carbenium ions **5** and **6**. Values of the bond order and for some ions of the localized molecular orbital (LMO) density have been reported.[9]

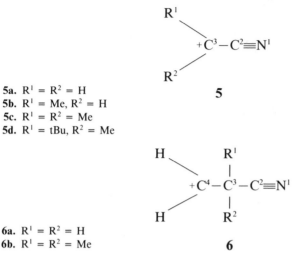

**5a.** $R^1 = R^2 = H$
**5b.** $R^1 = Me, R^2 = H$
**5c.** $R^1 = R^2 = Me$
**5d.** $R^1 = tBu, R^2 = Me$

**5**

**6a.** $R^1 = R^2 = H$
**6b.** $R^1 = R^2 = Me$

**6**

The CN and $C^2C^3$ bond orders are about the same in **6a** and **6b** as in MeCN. In **5a–5d** the CN bond order is significantly less than that in MeCN, while the $C^2C^3$ bond order is significantly greater. As H is replaced by Me on the $C^+$ atom of **5** ($C^3$), the CN bond order increases and the $C^2C^3$ bond order decreases. the LMO electron densities increase on N and decrease on $C^3$ as H is replaced by Me.

Ab initio calculations using a 4–31G basis set have been carried out on $XCH_2^+$ carbenium ions with X equal to CN, CHO, $CF_3$, and $NH_3^+$. The authors give values of $\Delta E$ of $-9.9$, $-6.1$, $-37.3$, and $-156.9$ kcal/mol, respectively, based on the isodesmic reaction of $XCH_2^+$ with $CH_4$. Thus, the order of carbenium ion stabilization is CHO > CN ≫ $CF_3$.[10a]

Partial retention of diatomic differential overlap (PRDDO) and ab initio calculations on cyano- and formyl-substituted cations support the conclusion that these groups are σ-electron acceptors and π-electron donors.[10b]

### b. Stabilization of Radicals

Chemical evidence for radical stabilization by $X_\pi$ groups has been derived from radical substitution and dissociation reactions.

The Hammett $\rho$ value for the reaction:

$$\underset{7}{X-Pn-\overset{\displaystyle Z^1}{\underset{\displaystyle Z^2}{C}}-H} + \cdot CCl_3 \rightarrow \underset{8}{X-Pn-\overset{\displaystyle Z^1}{\underset{\displaystyle Z^2}{C}}\cdot} + HCCl_3 \qquad \text{(b)}$$

(Pn = phenylene)

when $Z^1$ = CN and $Z^2$ = H is 0.55.[11a] It was reported elsewhere that values for this reaction can be calculated from the following equation[11b]:

$$\rho_{Z^1Z^2} = -0.606 \; \Sigma\sigma_p^+ Z + 0.195 \; \Sigma \; E_{SZ} - 1.063 \qquad (9\text{-}7)$$

Substitution of the $\sigma_p^+$ and Taft steric parameter values for CN and H into Equation 9-11 gives calculated $\rho$ values of $-0.98$ to $-1.10$ depending on the choice of the $E_S$ value for CN. The difference between calculated and observed values has been ascribed to stabilization of **8** by the CN group. The replacement of Cl or H by CN in the reaction of $ClCH_2CN$ or $Cl_2CHCN$ with cyclohexyl radicals decreases $E_{Cl}$, the chlorine transfer activation energy, by about 3.70 and 7.60 kcal/mol, respectively. It has been suggested that this is due to transition state stabilization by the CN group.[12]

Values of log $k_{rel}$ for the thermolysis of **9** show that the CN group is effective at stabilizing radicals, although somewhat less so than the ethynyl, propynyl, and phenyl groups,[13] The cyano group is said to stabilize the biradical intermediate resulting from thermolysis of cyano-, $E$-, or $Z$-1,2-dicyano- and 1-cyano-3-methylene-cyclobutanes by 6–10 kcal/mol.[114]

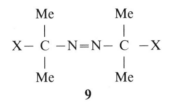

**9**

Values of $\Delta H°D$, the enthalpy of dissociation of the C—H bond in $XCH_3$ according to the equation:

$$XCH_3 \rightarrow XCH_2\cdot + H^\cdot \qquad (c)$$

at 25°C in the gas phase show clearly that the Ac and CN groups are capable of radical stabilization.[15]

Evidence based on physical properties has come from electron spin resonance (ESR) spectroscopy studies. The splitting $a_H^{Me}$ resulting from the $\beta$ protons in **10** is proportional to the spin density $\rho_\alpha$ at the atom $\alpha$ to the radical C atom.

**10**

Thus,

$$a_H^{Me} = Q_H^{Me}\rho_\alpha \qquad (9\text{-}8)$$

where $Q_H^{Me}$ is the proportionality constant.[16] It has been shown that $\rho_\alpha$ can be calculated from the equation:

$$\rho_\alpha = \prod_{i=1}^{3} (1 - \Delta_{Xi}) \qquad (9\text{-}9)$$

where $\Delta_X$ is a constant characteristic of the spin-withdrawing (electron-delo-calizing) effect of X. Then $\Delta_X$ should be a measure of the ability of X to stabilize a radical. The $\Delta_X$ values available show that CN, Ac, and $CO_2Et$ are all capable of stabilizing a radical.

### c. Stabilization of Carbanions

To demonstrate the special nature of $X_\pi$ delocalized electrical effects on the stabilization of carbanions, we must show that $X_{\pi d}$ ($\pi$ donor) groups can function as electron acceptors. Chemical evidence for this can be found in the $pK_a$ values of $XCH_2Z$ in $Me_2SO$ measured by Bordwell and his group.[18] Their results show that the phenyl, vinyl, naphthyl, and butadiynyl groups are all capable of delocalized effect stabilization of $X\overset{..}{C}HZ^-$.

These results show clearly that $X_{\pi a}$ ($\pi$ acceptor) groups such as $NO_2$, Ac, and CN can stabilize carbenium ions of the type $XY^+$ while $X_{\pi d}$ groups such as Ph, Vi, and NH can stabilize carbanions of the type $XY:^-$, reversing their behavior observed in $XGY^+$ or $XGY:^-$ systems. When we look at the behavior of $X_h$ and $X_p$ groups in $XY^+$ and $XY:^-$ systems, they show no such dramatic change.

## 3. THE DEFINITION OF $\sigma_R^\oplus$ CONSTANTS

We have established that certainly $X_\pi$ groups exert in XY systems electrical effects very different from those they exhibit in XGY systems. It is quite possible that other substituent types ($X_h$, $X_{(p)}$, $X_{(d)}$) will also show some differences in electrical effects in the two systems, although any discrepancy will be much smaller than that found for $X_\pi$ groups. Then if correlation analysis is to be extended to XY systems, it is necessary to define two new types of $\sigma_D$ constant, one for $XY^+$ and another for $XY:^-$ systems. A brief description of the definition of a set of $\sigma_R^\ominus$ constants for use in $XY:^-$ systems has appeared.[4] We now consider the problem of the definition of a set of $\sigma_R^\oplus$ constants for $XY^+$ systems.

### A. Requirements for the Definition of Substituent Constants

To obtain the best possible set of parameters, it is necessary to meet certain requirements. These include:

1. High sensitivity to the parameterized effect.
2. Small experimental errors for the data.
3. A large amount of available data with a complete range of substituent type.

4. Absence of steric effects and conformational equilibria.
5. New experimental determinations that are inexpensive and simple.
6. A measured phenomenon that is well understood and clearly shows its dependence on structural effects.
7. When a parameter must be defined from a multiparameter equation, it must be shown that the known parameters used are valid under the experimental conditions of the chosen reference data set. Thus, if a $\sigma_D$ parameter is to be defined from a data set dependent on both $\sigma_I$ and $\sigma_D$ using the equation:

$$\sigma_{DX} = \frac{Q_X^\circ - L\sigma_{IX} - h}{D} \tag{9-10}$$

7. It must be shown that the $\sigma_I$ values are valid under the experimental conditions of the chosen reference set.
8. The parameters must be scaled such that correlations with different electrical effect parameters are comparable. This permits conclusions to be drawn concerning relative sensitivity to substituent effects.

In our opinion, the data set that best meets these requirements is that of the ionization potentials of the highest occupied $\pi$ molecular orbital ($\pi$ HOMO) in PhX. If only spectroscopic (S), photoelectron spectroscopic (PES), and photoionization (PI) data are used, the experimental values are generally good to $\pm 0.05$ eV or less. The ionization process may be written as follows:

$$PhX \rightarrow \overset{+}{PhX} + e^- \tag{d}$$

Thus Ph$^+$ is a special case of Y$^+$.

## B. The Validity of the $\sigma_I$ Constants in the Gas Phase

Evidence has been presented that supports the validity of the $\sigma_I$ constants in aprotic media including the gas phase. Gas phase data are particularly pertinent to this work because ionization potentials determined by the PES, PI, and S methods are all measured in this medium. Results of correlations of proton transfer data and ionization potentials for suitable model systems (Table 9-4) with the equations:

$$Q_X = L\sigma_{IX} + A\alpha_X + h \tag{9-11}$$
$$Q_X = L\sigma_{IX} + A\alpha_X + Sv_X + h \tag{9-12}$$

are given in Tables 9-5 and 9-6. The $\alpha$ parameter is defined as follows:

$$\alpha_X \equiv \frac{MR_X - MR_H}{100} \tag{9-13}$$

where MR is the group molar refractivity. It is a measure of polarizability. The factor $\frac{1}{100}$ is introduced to convert the molar refractivities to a scale

TABLE **9-4.** Data in the Gas Phase[a]

| N5 | $\Delta G°_{600}$ (kcal/mol[1]): $XCH_2CO_2^- + H_2O \rightleftharpoons XCH_2CO_2H + OH^{-b}$ H, 45.2; Me, 46.4; Et, 47.3; $ClCH_2CH_2$, 48.1; MeCHCl, 51.4; $ClCH_2$, 52.5; Cl, 58; Br, 59.1; $F^X$, 56 |
|---|---|
| N9 | Vertical, ionization potentials, (eV), 4-substituted quinuclidines PES[c] <br> H, 8.05; Me, 8.06; Et, 8.05; $i$-Pr, 7.99; $t$-Bu, 7.97; $C_2H$, 8.30; Ph, 8.13; $CH_2OH$, 8.17; OAc, 8.42; Cl, 8.55; Br, 8.46; CN, 8.71; $NO_2$, 8.81; $OH^X$, 8.48; I, 8.35; $NO_2$, 8.81 |
| N12 | Vertical ionization potentials, (eV), $Ph_2PCH_2X$, PES[d] <br> $CO_2H$, 8.22; $CO_2Et$, 8.16; $CONH_2$, 8.18; $CONMe_2$, 8.05; CN, 8.44; H, 8.05; Me, 7.98 |
| N14, N15 | $\delta\Delta G°$, (kcal/mol): $XCH_2OH_2^+ + MeOH \rightleftharpoons XCH_2OH + MeOH_2^{+e}$ <br> $XCH_2O^- + MeOH \rightleftharpoons XCH_2OH + MeO^-$ <br> $t$-Bu, 9.1, 7.3; $i$-Pr, 8.0, 6.1; Et, 6.5, 4.8; Me, 4.5, 3.1; H, 0, 0; Ph, 4.8, 10.3; $MeOCH_2$, 1.0, 7.0; $CHF_2$, −4.3, 13.4; $CF_3$, −10.0, 16.6 |
| N16, N17 | $\delta\Delta G°$ (kcal/mol[1]): $X(CH_2)_nNH_2 + NH_4^+ \rightleftharpoons X(CH_2)_nNH_3^+ + NH_3$; 25°C[f], $n = 1$ for N16, $n = 2$ for N17 <br> H, 9.1, 11.8; Me, 11.8, 13.0; Et, 13.0, 13.6; Pr, 13.6, −; $i$-Pr, 16.1, −; $CF_3$, − 1.4, 6.7; $CF_2H^X$, 4.0, −; $F^X$, −, 8.0; $CF_3CH_2$, 6.7; 10.1; CN, −, 3.0; $HC \equiv C^X$, 6.7, −; Vi, 11.3, −; $NCCH_2$, 3.0, − |
| N21 | $\Delta\nu_{CO}$ (cm$^{-1}$), 4-substituted camphors[g] <br> Et, −0.7; Me, 0.1; H, 0; Vi, 0.2; Ph, −0.3; I, 4.6; Cl, 6.1; Br, 5.5; CN, 6.6; $NO_2$, 7.8; OMe, 2.6; $C \equiv CH$, 2.4; $NMe_2$, −0.5; $NH_2$, 1.6 |

[a]Superscript X indicates excluded from the correlation. Groups that are strongly medium dependent such as OH, or groups that exhibit intramolecular hydrogen bond formation such as F, are in this category.
[b]Kebarle, P. In "Environmental Effects on Molecular Structure and Properties", Reidel: Dordrecht, 1976, p. 81.
[c]Bieri, G.; Heilbronner, E. *Helv. Chim. Acta,* **1974,** *57,* 546.
[d]Dahl, O.; Henrikson, L. *Acta Chem. Scand.* **1977,** *B31,* 427.
[e]Taft, R.W. In "Kinetics of Ion–Molecule Reactions", Ausloos, P., Ed.; Plenum Press: New York, 1979, p. 271.

$$\delta\Delta° = \Delta G°_{XCH_2OH} - \Delta G°_{CH_3OH}$$

[f]Arnett, E.M. In "Proton Transfer Reactions", Caldin, E.; and Gold, V., Eds.; Chapman & Hall: London, 1975, p. 79.

$$\delta\Delta G° = \Delta G°_{XNH_2} - \Delta G_{NH_3}$$

[g]Laurence, C.; Bertholet, M.; Lucon, M.; Helbert, M., Morris, D.G.; Gal, J.F. *J. Chem. Soc. Perkin Trans. 2,* **1984,** 705.

$$\Delta\nu_{CO,X} \equiv \nu_{CO,X} - \nu_{CO,H}$$

comparable in magnitude to those of the electrical and steric parameters. The results obtained support the validity of the use of $\sigma_I$ values in the gas phase (Tables 9-5 and 9-6).

The observation that good results for the sets studied were obtained with Equations 9-11 and 9-12 suggested the possibility that ionization potentials (IP) of the highest occupied molecular orbital (HOMO) in PhX would be well correlated by the *LDA* equation:

$$\text{IP}_{\pi\text{HOMO,X}} = L\sigma_{IX} + D\sigma_{RX}+ + A\alpha_X + \text{h} \qquad (9\text{-}14)$$

TABLE 9-5. Results[a] of Correlations with Equation 9-11: Data from Table 9-4

| Set | $L$ | $A$ | $h$ | $100R^2$ | $F$ (see note) | $r_{12}$ | $S_{est}$ | $S_L$ | $S_A$ | $S_h$ | $S°$ | $n$ |
|-----|-----|-----|-----|----------|----------------|----------|-----------|-------|-------|-------|------|-----|
| N5 | 25.7 | 12.2 | 45.8 | 97.92 | 117.8 | 0.008 | 0.898 | 1.69 | 6.92[b] | 0.701 | 0.182 | 8 |
| N9 | 1.07 | −0.464 | 8.07 | 97.29 | 179.3 | 0.244 | 0.0506 | 0.0600 | 0.231[c] | 0.0321 | 0.188 | 13 |
| N12 | 0.807 | −1.23 | 8.05 | 99.81 | 1030. | 0.345 | 0.00820 | 0.0179 | 0.0626 | 0.00609 | 0.058 | 7 |
| N21 | 12.7 | −9.40 | 0.000 | 96.50 | 151.8 | 0.105 | 0.608 | 0.761 | 2.91[d] | 0.374[e] | 0.211 | 14 |

[a]Column heads as follows:
$100R^2$ = percent of the variance of the data accounted for by the regression equation
$F$ = $F$ test for significance
$r_{12}$ = correlation coefficient for correlation of the $i$th independent variable with the $j$th independent variable
$S$ = standard errors of the estimate and the regression coefficients
$S°$ = standard error of the estimate divided by the root mean square of the data
$n$ = number of data points in the set
[b]80% confidence level (CL).
[c]90% CL.
[d]99.0% CL.
[e]0% CL.

NOTE: The $F$ test determines whether the regression coefficients are significantly different from zero. It is a measure of the goodness of fit of the data to the correlation equation. The larger the value of the $F$ statistic, the better the fit.
The statistic $100R^2$ ($100r^2$ for simple regression with only one independent variable) represents the percentage of the variance of the data that is accounted for by the regression equation. The closer its value is to 100, the better the regression equation models the data. A value of 100 for $100R^2$ would mean that the regression equation accounts completely for the variance of the data.
One of the measures of goodness of fit is the standard error of the estimate. It represents the probable error encountered in the prediction of a new value of $Q$ from the regression equation. Clearly the smaller the value of $S_{est}$, the better the regression equation will be as a predictor. The magnitude of $S_{est}$ depends, however, not only on the goodness of fit but on the magnitude of $Q$ as well. It is not possible therefore to make meaningful comparisons of $S_{est}$ values obtained in the correlation of different types of $Q$. Such comparisons can be made by using the statistic $S°$ in which the magnitude of $Q$ is accounted for by dividing $S_{est}$ by the root mean square of the data. Thus, $S°$ values for different types of $Q$ are comparable. The smaller the value of $S°$, the better the fit of the data to the regression equation.
The $r_{ij}$ and $R_i$ determine whether the independent variables are interrelated. If the confidence levels of $r_{ij}$ and/or $R_i$ are $\geq$ 80.0, the variables in question are interrelated. Should that be the case, no conclusion can be drawn as to the validity of the correlation equation *no matter how good the fit of the data*.

The results obtained for a fairly complete range of substituent types indicate that this is not the case. To verify the point, we have examined the correlation of IP$_{\pi HOMO}$ values for substituted ethylenes (ViX) with Equation 9-14. Again the results, while significant, were not acceptable, as shown by consideration of the ratio of the standard error of the estimate to the experimental error in the measured ionization potentials. The maximum experimental error in the ionization potentions is 0.05 eV. The standard errors of the estimate for adiabatic and vertical ionization potentials of PhX (sets 01 and 02, Table 9-7; results of correlations are set forth in Table 9-8) are 0.202 and 0.212, respectively. They are therefore about four times as large as the experimental error. The standard errors of the estimate for adiabatic and vertical ionization potentials of ViX (sets 501 and 521, Table 9-7; results are reported in Table 9-8) are 0.229 and 0.189, respectively; again they are about four times as large as the experimental error. It follows then, that Equation 9-14 does not adequately model the data.

TABLE 9-6. Results[a] of Correlations with Equation 9-12: Data from Table 9-4

| Set | L | A | S | h | $100R^2$ | F | $r_{12}$ | $r_{13}$ | $r_{23}$ | $R_1$ | $R_2$ | $R_3$ | $S_{est}$ | $S_L$ | $S_A$ | $S_S$ | $S_h$ | $S°$ | n |
|---|---|---|---|---|---|---|---|---|---|---|---|---|---|---|---|---|---|---|---|
| N14 | −33.8 | 25.1 | 3.08 | 1.15 | 95.82 | 38.19 | 0.247 | 0.229 | 0.477 | 0.466 | 0.602 | 0.597 | 1.61 | 4.13 | 9.05[b] | 2.16[c] | 1.28[d] | 0.274 | 9 |
| N15 | 30.0 | 16.3 | 3.66 | 0.516 | 98.83 | 140.4 | 0.247 | 0.229 | 0.477 | 0.466 | 0.602 | 0.597 | 0.703 | 1.81 | 3.96[e] | 0.934[f] | 0.559[g] | 0.145 | 9 |
| N16 | −20.4 | 62.5 | −6.98 | 10.63 | 88.97 | 18.83 | 0.277 | 0.491 | 0.527[h] | 0.799[e] | 0.810[e] | 0847[e] | 2.09 | 7.24[b] | 26.1[b] | 5.14[c] | 1.85 | 0.416 | 11 |
| N17 | −17.1 | 3.98 | 1.57 | 11.8 | 99.62 | 174.5[e] | 0.006 | 0.320 | 0.582 | 0.397 | 0.616 | 0.666 | 0.401 | 0.791[e] | 6.74[i] | 0.793[c] | 0.375[e] | 0.107 | 6 |

[a]Column heads as follows:

$100R^2$ = percent of the variance of the data accounted for by the regression equation

F = F test for significance; see Note, Table 9-5.

r = correlation coefficients for correlation of the $i$th independent variable with the $j$th independent variable

R = multiple correlation coefficients for the correlation of the $i$th independent variable with all the remaining independent variables; the confidence levels (CL) of F and the standard errors of the regression coefficients are 99.9% unless otherwise noted; the CLs of r and R are less than 90.0% unless otherwise noted

S = standard errors of the estimate and the regression coefficients; $S°$ = standard error of the estimate divided by the root mean square of the data

n = number of data points in the set

[b]95.0% confidence level (CL).
[c]70.0% CL.
[d]50.0% CL.
[e]99.0% CL.
[f]98.0% CL.
[g]60.0% CL.
[h]90.0% CL.
[i]30.0% CL.

TABLE 9-7. Ionization Potentials$^a$ of $\pi$-$_{HOMO}$ Orbitals (eV)

| | |
|---|---|
| 01 | PhX, adiabatic |
| | H, 9.241$^b$; Me, 8.821; C$_2$H, 8.815; Vi, 8.43; Et, 8.77; Pr, 8.72; *i*-Pr, 8.69; Bu, 8.69; *s*-Bu,. 8.68; *i*-Bu, 8.69$^b$; *t*-Bu, 8.68; Ph, 8.23; NH$_2$, 7.69; CN, 9.705$^b$; NMe$_2$, 7.14$^b$; OMe, 8.20$^b$; CH$_2$OMe, 8.85; OEt, 8.13$^b$; NO$_2$, 9.8$^b$; F, 9.200$^b$; CF$_3$, 9.685$^b$; Cl, 9.08$^b$; CH$_2$Cl, 9.10; Br, 9.03; I, 8.78; $-$CMe=CH$_2$, 8.35; MeNH, 7.34; NHAc, 8.30; CH$_2$SiMe$_3$, 8.35; SO$_2$Pr, 9.21; SO$_2$Ph, 9.16; PhCH$_2$, 8.55; PhC$_2$, 7.90; MeS, 7.96 |
| 02 | PhX, vertical |
| | NH$_2$, 8.05; H, 9.24$^c$; MeS, 8.07$^a$; OMe, 8.41$^c$; Ac, 9.55; Cl, 9.05$^c$; Br, 9.06$^c$; NO$_2$, 9.99$^c$; CN, 9.73$^c$; NMe$_2$, 7.37$^c$; Me, 8.85$^c$; I, 8.801$^c$; *t*-Bu, 8.82$^c$; F, 9.11$^c$; NHAc, 8.46$^c$; SO$_2$Me, 9.74; CH$_2$Cl, 9.30$^c$; CH$_2$Br, 9.23; CF$_3$, 9.90$^c$; $-$CMe=CH$_2$, 8.50$^c$; c-Pr, 8.66; OEt, 8.36$^c$ |
| 501 | ViX, adiabatic |
| | H, 10.507; Me, 9.727; Et, 9.58; Pr, 9.50; CN, 10.91; OMe, 8.93; OAc, 9.19; F, 10.363; Cl, 10.00; Br, 9.80; CH$_2$Cl, 10.05; CH$_2$Br, 10.06 |
| 521 | ViX, vertical |
| | H, 10.68; Me, 9.86; Et, 9.77; Pr, 9.68; CN, 10.92; OMe, 9.05; F, 10.56; Cl, 10.15; Br, 9.87; CH$_2$Cl, 10.20; CH$_2$Br, 10.18 |

$^a$Unless otherwise noted, ionization potentials are from: Rosenstock, H.M.; Draxl, K.; Steiner, B.W.; Herron, J.T. *J. Phys. Chem. Ref. Data* **1977**, *6* suppl. 1. or Levin, R.D.; Lias, S.G. "Ionization Potential and Appearance Potential Measurements, 1971–1981." NSRDS–NBS71, National Bureau of Standards: Washington, D.C., 1982. (A revised edition will appear shortly).
$^b$Member of set 01C. Set 01B includes all members of set 01C except H. Outliers in set 01A are the groups other than H that are not in 01C.
$^c$Member of set 02A.

# C. The Definition of $\sigma_R^{\oplus}$ Constants

Previous attempts to correlate ionization potentials suggest that the first IPs of benzene derivatives are best correlated by $\sigma_R^+$. Our own correlations support this conclusion.

We have therefore taken the best available IP values (set 01; Table 9-7) for substituted benzenes (only those obtained by PES, UV spectroscopy, and PI were considered) and correlated them with the *LD* equation. On the basis of the result obtained (set 01A; Table 9-9), we excluded all the points that were serious outliers. The remaining IP values were again correlated with the *LD* equation using $\sigma_R^+$ as $\sigma_D$. An excellent correlation was obtained (set 01B); however, the value for X = H did not lie on the correlation line. We had found the same problem in the definition of $\sigma_R^-$ and used the same technique to resolve it.[2] We define from the correlation equation for set 01B, a set of $\sigma_D$ constants represented by $\sigma_{Rb}^{\oplus}$. Then from Equation 138 of Reference 2:

$$\sigma_{RbH}^{\oplus} = \frac{h - Q_H}{D} \qquad (9\text{-}15)$$

**TABLE 9-8.** Results[a] of Correlations with Equation 9-14: Data from Table 9-7

| Set | L | D | A | h | 100R² | F | r₁₂ | r₁₃ | r₂₁ | R₁ | R₂ | R₃ | S_est | S_L | S_D | S_A | S_h | S° | n |
|---|---|---|---|---|---|---|---|---|---|---|---|---|---|---|---|---|---|---|---|
| 01 | 0.616 | 1.53 | −2.50 | 9.24 | 90.17 | 91.71 | 0.156 | 0.177 | 0.147 | 0.255 | 0.236 | 0.250 | 0.202 | 0.170 | 0.105 | 0.423 | 0.0909 | 0.334 | 34 |
| 02 | 0.777 | 1.42 | −2.70 | 9.32 | 91.15 | 61.83 | 0.239 | 0.263 | 0.129 | 0.335 | 0.249 | 0.272 | 0.212 | 0.238[b] | 0.127 | 0.976[c] | 0.139 | 0.329 | 22 |
| 501 | 0.742 | 2.27 | −6.01 | 10.54 | 87.65 | 18.92 | 0.199 | 0.247 | 0.047 | 0.310 | 0.199 | 0.247 | 0.229 | 0.318[d] | 0.369 | 1.59[b] | 0.176 | 0.430 | 12 |
| 521 | 0.746 | 2.09 | −5.06 | 10.62 | 89.67 | 26.05 | 0.008 | 0.088 | 0.174 | 0.088 | 0.174 | 0.194 | 0.189 | 0.230[c] | 0.276 | 1.22[b] | 0.142 | 0.386 | 13 |

[a]Column heads as follows:

$100R^2$ = percent of the variance of the data accounted for by the regression equation

$F$ = $F$ test for significance; see Note, Table 9-5

$r$ = correlation coefficients for correlation of the $i$th independent variable with the $j$th independent variable

$R$ = multiple correlation coefficients for the correlation of the $i$th independent variable with all the remaining independent variables; the confidence levels (CL) of $F$ and the standard errors of the regression coefficients are 99.9% unless otherwise noted; the CLs of $r$ and $R$ are less than 90.0% unless otherwise noted

$S$ = standard errors of the estimate and the regression coefficients; $S°$ = standard error of estimate divided by the root mean square of the data

$n$ = number of data points in the set

[b]99.0% confidence level (CL).

[c]98.0% CL.

[d]95.0% CL.

where $h$ and $D$ are from set 01B, Table 9-8, and $Q_H$ is the first IP for benzene. This results in a value of $-0.18$ for $\sigma^{\oplus}_{RbH}$. Then we may write[2]:

$$\sigma^{\oplus}_{RcX} = \sigma^{\oplus}_{RbX} - 0.18 \tag{9-16}$$

These constants were used to correlate adiabatic $\pi_{HOMO}$ IP values of PhX in set 01 (Table 9-7) to give set 01C (Table 9-9). This correlation equation was then used to calculate $\sigma^{\oplus}_{RX}$ constants. Thus,

$$\sigma^{\oplus}_{RX} = \frac{Ip_X - 1.10\,\sigma_{IX} - 9.243}{1.66} \tag{9-17}$$

The $\sigma^{\oplus}_{RX}$ constants, together when possible with their estimated errors, are given in Table 9-10. Vertical IP values (set 02) were also correlated with the $LD$ equation to give:

$$\sigma^{\oplus}_R = \frac{(IP_X - 1.03\sigma_{IX} - 9.32)}{1.56} \tag{9-18}$$

which was used to calculate additional $\sigma^{\oplus}_R$ values. The results of the correlations of set 02A are given in Table 9-8. Values of $\sigma^{\oplus}_R$ calculated from set 02A are reported in Table 9-10. The $\sigma_I$ values used in the calculation of $\sigma^{\oplus}_R$ were taken from our compilation or estimated as described therein.

Reliable $\sigma^{\oplus}_R$ values are given in Table 9-10 with their estimated errors. The calculation of the error has been described elsewhere. Values of $\sigma^{\oplus}_R$ that are uncertain because the $\sigma_I$ value used to determine them was estimated or uncertain, or because no error was reported for the ionization potential used, are designated in Table 9-10 as U (uncertain). Values of $\sigma^{\oplus}_R$ that have been obtained by the estimation methods described in Subsection B of Section 5 are designated in Table 9-10 as E (estimated).

# 4. APPLICATIONS OF THE $\sigma^{\oplus}_R$ CONSTANTS

In Section 3 we developed a set of $\sigma_D$ values, the $\sigma^{\oplus}_R$ constants, which are specifically intended for use in $XY^+$ systems. To show that these constants are indeed both useful and necessary in modeling substituent effects in these systems, we must establish that:

1. The use of $\sigma^{\oplus}_R$ gives consistently good results in $XY^+$ systems.
2. The use of $\sigma_R$ and $\sigma^+_R$ in these same systems gives results inferior to those obtained with $\sigma^{\oplus}_R$.

Since the most commonly encountered $XY^+$ systems in organic chemistry are those involving carbocations, we have concentrated on these as test cases.

TABLE 9-9. Results[a] of Correlations of Ionization Potentials with Equation 9-14: Data from Table 9-11

| Set[b] | L | D | h | $100R^2$ | F | $r_{12}$ | $S_{est}$ | $S_L$ | $S_D$ | $S_h$ | $S°$ | n |
|---|---|---|---|---|---|---|---|---|---|---|---|---|
| 01A | 0.821 | 1.42 | 8.83 | 78.72 | 57.35 | 0.156 | 0.293 | 0.241 | 0.150 | 0.0860 | 0.483 | 34 |
| 01B | 1.11 | 1.66 | 8.94 | 99.86 | 2444. | 0.318 | 0.0361 | 0.0518 | 0.0294 | 0.0223 | 0.045 | 10 |
| 01C | 1.10 | 1.66 | 9.24 | 99.99 | 69395. | 0.171 | 0.00688 | 0.00891 | 0.00510 | 0.00400 | 0.009 | 11 |
| 02A | 1.02 | 1.56 | 9.32 | 98.29 | 401.8 | 0.143 | 0.0937 | 0.113 | 0.0615 | 0.0530 | 0.144 | 17 |
| 2R | 0.655 | 1.33 | 8.07 | 94.10 | 39.90 | 0.259 | 0.153 | 0.300[c] | 0.171 | 0.123 | 0.307 | 8 |
| 2P | 0.559 | 1.06 | 8.10 | 93.55 | 36.24[d] | 0.299 | 0.160 | 0.318[e] | 0.143 | 0.131 | 0.321 | 8 |
| 2C | 0.996 | 1.17 | 8.18 | 98.71 | 190.9 | 0.113 | 0.0715 | 0.137 | 0.0683 | 0.0604 | 0.144 | 8 |
| 3R | 0.640 | 1.45 | 8.60 | 96.79 | 75.40 | 0.315 | 0.122 | 0.193[f] | 0.142 | 0.0848 | 0.227 | 8 |
| 3P | 0.643 | 1.13 | 8.58 | 95.74 | 56.25 | 0.316 | 0.140 | 0.223[f] | 0.129 | 0.0964 | 0.261 | 8 |
| 3C | 0.942 | 1.25 | 8.68 | 99.20 | 308.3 | 0.165 | 0.0610 | 0.093[f] | 0.0604 | 0.0445 | 0.113 | 8 |
| 4R | 1.34 | 1.54 | 8.57 | 71.81 | 7.641[g] | 0.270 | 0.304 | 0.531[f] | 0.434[h] | 0.197 | 0.650 | 9 |
| 4P | 0.953 | 1.22 | 8.61 | 82.51 | 14.15[f] | 0.053 | 0.240 | 0.403[c] | 0.249[h] | 0.157 | 0.412 | 9 |
| 4C | 1.15 | 1.48 | 8.90 | 97.96 | 144.2 | 0.147 | 0.0818 | 0.139 | 0.0931 | 0.0628 | 0.175 | 9 |
| 5R | 1.37 | 1.39 | 8.51 | 61.96 | 4.072[c] | 0.274 | 0.370 | 0.647[c] | 0.575[c] | 0.240 | 0.780 | 8 |
| 5P | 1.01 | 1.11 | 8.56 | 73.42 | 6.905[f] | 0.038 | 0.309 | 0.520[e] | 0.344[f] | 0.202 | 0.652 | 8 |
| 5C | 1.23 | 1.48 | 8.87 | 98.76 | 199.8 | 0.139 | 0.0667 | 0.113 | 0.0819 | 0.0516 | 0.141 | 8 |
| 6R | 0.431 | -0.191 | 8.37 | 13.07 | 0.301[i] | 0.787[j] | 0.400 | 1.11[j] | 1.77[k] | 0.269 | 1.23 | 7 |
| 6P | 0.988 | 1.41 | 8.50 | 21.84 | 0.559[j] | 0.724[c] | 0.379 | 0.940[l] | 2.07[m] | 0.294 | 1.17 | 7 |
| 6C | 1.15 | 1.59 | 8.83 | 96.78 | 60.14[d] | 0.420 | 0.0769 | 0.145[h] | 0.156 | 0.0644 | 0.237 | 7 |
| 13R | 0.953 | 1.41 | 9.18 | 92.97 | 59.69 | 0.093 | 0.164 | 0.244[h] | 0.185 | 0.106 | 0.325 | 9 |
| 13P | 0.765 | 1.14 | 9.20 | 91.43 | 32.02 | 0.194 | 0.181 | 0.273[f] | 0.167 | 0.118 | 0.358 | 9 |
| 13C | 1.01 | 1.35 | 9.31 | 96.89 | 93.33 | 0.063 | 0.109 | 0.162 | 0.115 | 0.0750 | 0.216 | 9 |
| 15R | 0.588 | 1.49 | 9.12 | 94.49 | 34.28[d] | 0.036 | 0.115 | 0.202[f] | 0.195[h] | 0.0831 | 0.311 | 7 |

| | | | | | | | | | | | |
|---|---|---|---|---|---|---|---|---|---|---|---|
| 15P | 0.513 | 1.32 | 9.10 | 91.35 | 21.12[h] | 0.086 | 0.144 | 0.254[e] | 0.220[h] | 0.103 | 0.389 | 7 |
| 15C | 0.889 | 1.23 | 9.16 | 95.87 | 46.41[d] | 0.156 | 0.0996 | 0.177[h] | 0.138 | 0.0735 | 0.269 | 7 |
| 301R | 0.477 | 0.924 | 8.08 | 86.95 | 13.33[g] | 0.146 | 0.174 | 0.355[n] | 0.195[n] | 0.120 | 0.478 | 7 |
| 301P | 0.539 | 0.778 | 8.11 | 92.20 | 23.66[h] | 0.106 | 0.134 | 0.273[e] | 0.123[h] | 0.0940 | 0.369 | 7 |
| 301C | 0.822 | 0.771 | 8.15 | 97.10 | 66.87 | 0.055 | 0.0819 | 0.166[h] | 0.0720 | 0.0585 | 0.225 | 7 |
| 302R | 0.602 | 0.992 | 8.10 | 88.26 | 18.79[d] | 0.187 | 0.191 | 0.390[e] | 0.179[h] | 0.131 | 0.433 | 8 |
| 302P | 0.660 | 0.802 | 8.13 | 93.10 | 33.72[d] | 0.155 | 0.146 | 0.297[c] | 0.107 | 0.102 | 0.332 | 8 |
| 302C | 0.944 | 0.788 | 8.16 | 97.61 | 102.0 | 0.027 | 0.0861 | 0.173[h] | 0.0604 | 0.0611 | 0.196 | 8 |
| 312R | 1.02 | 1.73 | 8.88 | 86.02 | 30.76 | 0.158 | 0.220 | 0.278[h] | 0.277 | 0.106 | 0.426 | 13 |
| 312P | 1.04 | 1.69 | 8.90 | 92.46 | 61.33 | 0.142 | 0.161 | 0.203 | 0.188 | 0.0785 | 0.313 | 13 |
| 312C | 1.31 | 1.43 | 9.05 | 96.53 | 139.0 | 0.007 | 0.109 | 0.137 | 0.104 | 0.0580 | 0.212 | 13 |

[a]Column heads as follows:

$100R^2$ = percent of the variance of the data accounted for by the regression equation

$F$ = $F$ test for significance; see Note, Table 9-5

$r$ = correlation coefficients for correlation of the $i$th independent variable with the $j$th independent variable

$R$ = multiple correlation coefficients for the correlation of the $i$th independent variable with all the remaining independent variables; the confidence levels (CL) of $F$ and the standard errors of the regression coefficients are 99.9% unless otherwise noted; the CLs of $r$ and $R$ are less than 90.0% unless otherwise noted

$S$ = standard errors of the estimate and the regression coefficients; $S$ = standard error of the estimate divided by the root mean square of the data

$n$ = number of data points in the set

[b]Sets designated $-R$, $-P$, and $-C$ were correlated with $\sigma_R$, $\sigma_R^{\ddagger}$, and $\sigma_R^{\oplus}$, respectively.

[c]90.0% confidence level (CL).

[d]99.0% confidence level (CL).

[e]80.0% CL.

[f]95.0% CL.

[g]97.5% CL.

[h]99.0% CL.

[i]<90.00% CL.

[j]20% CL.

[k]0.% CL.

[l]60.0% CL.

[m]40.0% CL.

[n]70.0% CL.

**TABLE 9-10.** Values of $\sigma_R^\oplus$

| X | $\sigma_R^\oplus$ | Error | IP (eV) | Error | Set |
|---|---|---|---|---|---|
| Alkyl, cycloalkyl | | | | | |
| Me | −0.25 | 0.015 | 8.821 | 0.01 | 01C |
| Et | −0.28 | 0.015 | 8.77 | 0.01 | 01C |
| c-Pr | −0.43 | U | 8.66 | — | 02A |
| Pr | −0.31 | 0.007 | 8.72 | 0.01 | 01C |
| i-Pr | −0.34 | 0.007 | 8.69 | 0.01 | 01C |
| c-Bu | −0.35 | U | 8.77 | — | 02A |
| Bu | −0.33 | 0.007 | 8.69 | 0.01 | 01C |
| i-Bu | −0.33 | 0.007 | 8.69 | 0.01 | 01C |
| s-Bu | −0.33 | 0.007 | 8.68 | 0.01 | 01C |
| t-Bu | −0.33 | 0.009 | 8.68 | 0.01 | 01C |
| c-Am | −0.32 | U | 8.81 | — | 02A |
| $CH_2t$-Bu | −0.35 | U | 8.77 | — | 02A |
| Ethynyl, vinyl, aryl, ethynyl-, vinyl-, aralkyl | | | | | |
| $HC_2$ | −0.45 | 0.011 | 8.815 | 0.005 | 01C |
| Vi | −0.56 | 0.007 | 8.43 | 0.01 | 01C |
| $MeC_2$ | −0.78 | 0.053 | 8.41 | 0.02 | 02A |
| E-2-MeVn | −0.67 | 0.014 | 8.20 | 0.02 | 01C |
| 1-MeVn | −0.60 | 0.008 | 8.35 | 0.01 | 01C |
| $ViCH_2$ | −0.40 | U | 8.60 | — | 01C |
| E-2-ViVn | −0.86 | U | 7.95 | U | 01C |
| $Vi(Vn)_2$ | −1.05 | E | — | — | E31 |
| Ph | −0.69 | 0.007 | 8.23 | 0.01 | 01C |
| $PhCH_2$ | −0.44 | 0.018 | 8.55 | 0.03 | 01C |
| E-2-PhVn | −1.01 | U | 7.70 | 0.02 | 01C |
| $PhC_2$ | −1.03 | 0.016 | 7.90 | 0.02 | 01C |
| $Ph(CH=CH_2)$ | −1.23 | U | 7.54 | 0.03 | 02A |
| 1-Nh | −1.02 | E | — | — | E31 |
| 2-Nh | −0.93 | E | — | — | E31 |
| $Ph_2CH$ | −0.57 | U | 8.34 | 0.03 | 1C |
| Haloalkyl | | | | | |
| $CH_2F$ | −0.02 | U | 9.55 | — | 02A |
| $CH_2Cl$ | −0.17 | 0.007 | 9.14 | 0.01 | 01C |
| $CH_2Br$ | −0.19 | U | 9.23 | — | 02A |
| $CH_2I$ | −0.38 | U | 8.91 | — | 02A |
| $CF_3$ | 0.00 | 0.015 | 9.685 | 0.005 | 01C |
| Other substituted alkyl | | | | | |
| $CH_2OH$ | −0.21 | U | 9.11 | — | 02A |
| $CH_2OMe$ | −0.31 | 0.020 | 8.85 | 0.03 | 01C |
| $CH_2CHO$ | −0.37 | U | 8.80 | — | 01C |
| $CH_2SiMe_3$ | −0.52 | U | 8.35 | — | 01C |
| Carbonyl | | | | | |
| HCO | −0.04 | U | 9.40 | — | 01C |
| $H_2NCO$ | −0.10 | U | 9.45 | — | 02A |
| Ac | −0.05 | U | 9.55 | — | 02A |
| Bz | −0.13 | U | 9.42 | — | 02A |
| $CO_2Me$ | −0.12 | 0.018 | 9.40 | 0.025 | 01C |

**TABLE 9-10.** (continued)

| X | $\sigma_R^{\oplus}$ | Error | IP (eV) | Error | Set |
|---|---|---|---|---|---|
| Aza | | | | | |
| NH$_2$ | −1.05 | 0.014 | 7.69 | 0.02 | 01C |
| NHMe | −1.23 | 0.014 | 7.34 | 0.02 | 01C |
| NMe$_2$ | −1.38 | 0.020 | 7.14 | 0.03 | 01C |
| NHAc | −0.75 | 0.060 | 8.30 | 0.10 | 01C |
| NHPh | −1.40 | U | 7.25 | 0.03 | 01C |
| Oxa | | | | | |
| OH | −0.71 | E | — | — | E11, E13 |
| OMe | −0.83 | 0.020 | 8.20 | 0.02 | 01C |
| OEt | −0.86 | 0.013 | 8.13 | 0.02 | 01C |
| OPh | −0.96 | 0.020 | 8.09 | 0.03 | 01C |
| Phospha | | | | | |
| PH$_2$ | −1.15 | E | — | — | E11, E13 |
| PHMe | −1.31 | E | — | — | E11 |
| PMe$_2$ | −1.47 | E | — | — | E11 |
| PHPh | −1.41 | E | — | — | E12 |
| Thia | | | | | |
| SH | −0.81 | E | — | — | E11, E13 |
| SMe | −0.97 | 0.008 | 7.96 | 0.01 | 01C |
| SEt | −0.99 | 0.015 | 7.88 | 0.02 | 01C |
| SPh | −1.00 | 0.010 | 7.92 | 0.01 | 01C |
| SCH$_2$Ph | −1.02 | 0.065 | 8.00 | 0.05 | 02A |
| Sulfinyl, sulfonyl | | | | | |
| SOMe | −0.70 | U | 8.79 | — | 02A |
| SOPh | −0.81 | U | 8.58 | U | 02A |
| SO$_2$Me | −0.12 | U | 9.74 | U | 02A |
| SO$_2$Pr | −0.40 | U | 9.21 | 0.03 | 01C |
| SO$_2$Ph | −0.42 | U | 9.16 | 0.03 | 01C |
| Selena | | | | | |
| SeH | −0.80 | E | — | — | E11, E13 |
| SeMe | −0.96 | E | — | — | E11, E13 |
| SeEt | −0.97 | E | — | — | E13 |
| SePh | −1.05 | E | — | — | E12 |
| Other | | | | | |
| F | −0.38 | 0.007 | 9.200 | 0.005 | 01C |
| Cl | −0.41 | 0.020 | 9.08 | 0.01 | 01C |
| Br | −0.44 | 0.016 | 9.03 | 0.01 | 01C |
| I | −0.57 | 0.018 | 8.73 | 0.01 | 01C |
| CN | −0.10 | 0.009 | 9.705 | 0.01 | 01C |
| NO$_2$ | −0.08 | 0.024 | 9.85 | 0.03 | 01C |
| N$_3$ | −0.67 | U | 8.72 | 0.02 | 02A |
| NC | −0.30 | U | 9.50 | U | 02A |
| SiMe$_3$ | −0.10 | U | 9.05 | — | 02A |
| H | 0 | — | 9.241 | 0.001 | [a] |

[a]By definition.

## A. The Formation of Carbocations

There are three types of process by which a carbocation or a significantly cationic transition state can be formed.

Dissociation.

Ionization by electron loss.

Association.

Examples of dissociation reactions are:

$$R^1R^2R^3C - Lg \rightleftharpoons R^1R^2R^3\overset{\sigma\ +}{C} \cdots \overset{b}{Lg}^{\sigma -} \rightarrow R^1R^2R^3C^+ + Lg^- \qquad \text{(e)}$$

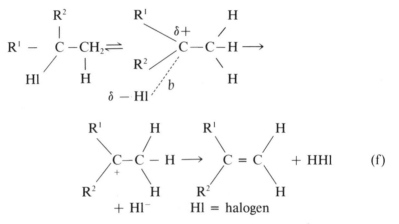

$$\text{Ionization involving loss of an electron from a } \pi \text{ orbital produces a radical cation, thus:}$$

Loss of an electron from a radical leads to a carbocation:

Examples of the formation of carbocations by association are:

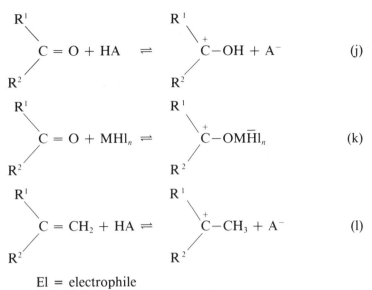

$$\begin{array}{c} R^1 \\ \diagdown \\ \diagup \\ R^2 \end{array} C = O + HA \quad \rightleftharpoons \quad \begin{array}{c} R^1 \\ \diagdown \\ \diagup \\ R^2 \end{array} \overset{+}{C} - OH + A^- \tag{j}$$

$$\begin{array}{c} R^1 \\ \diagdown \\ \diagup \\ R^2 \end{array} C = O + MHl_n \quad \rightleftharpoons \quad \begin{array}{c} R^1 \\ \diagdown \\ \diagup \\ R^2 \end{array} \overset{+}{C} - O\overline{MHl}_n \tag{k}$$

$$\begin{array}{c} R^1 \\ \diagdown \\ \diagup \\ R^2 \end{array} C = CH_2 + HA \quad \rightleftharpoons \quad \begin{array}{c} R^1 \\ \diagdown \\ \diagup \\ R^2 \end{array} \overset{+}{C} - CH_3 + A^- \tag{l}$$

$$El = \text{electrophile}$$

## B. Formation of Carbocations by Electron Loss

### a. Ionization Potentials of Molecules: Radical Cation Formation

A very large number of ionization potentials of $\pi$ HOMOs in substituted benzenes, ethylenes, polycyclic arenes, and heteroarenes has accumulated in the literature. We have chosen from the available data several sets of 4-XPnZ, where Z is constant (sets 2–6, 13, 15). We have also examined data for 1- and 2-substituted naphthalenes (sets 301, 302), 2-substituted thiophenes (set 312), and substituted ethylenes (sets 501, 521). The ionization potential correlations with the *LD* equation are reported in Table 9-9. The data used are given in Table 9-11.

Substituted benzene sets (Z = 4-MeO; 4-I; 2-, 3-, 4-Me; 4-Cl; and 4-Br) gave very much better results with $\sigma_R^\oplus$ than with $\sigma_R^+$ or $\sigma_R$. In the correlation of the 1-substituted naphthalenes, it was necessary to exclude the value for the dimethylamino group because this group behaves sterically like a planar $\pi$-bonded group such as $NO_2$ or Ac. In the 1-position of naphthalene the group cannot adopt a suitable conformation for extensive delocalization. After exclusion of the data point for $Me_2N$, the set gave very good results with $\sigma_R^\oplus$. In the 2-position of naphthalene no steric hindrance of resonance occurs, and inclusion of the value for $Me_2N$ gave very good results with $\sigma_R^\oplus$. For both sets, correlation with $\sigma_R^+$ or $\sigma_R$ was much poorer. Finally, for the 2-substituted thiophenes best results were again obtained with the $\sigma_R^\oplus$ values.

The correlation of ionization potentials of substituted ethylenes with the *LD* equation also gave best fit with the $\sigma_R^\oplus$ values, but the results were not acceptable. The standard error of the estimate is too large. Examination of the results of the correlations of ionization potentials for substituted ben-

TABLE **9-11.** Ionization Potentials (eV) Substituted Benzene Derivatives, Naphthalenes, Thiophenes, and Ethylenes

---

Set 2     4-XPnOMe, adiabatic[a] (Pn = phenylene; see Appendix II)

H, 8.24; Br, 8.11; I, 7.97; $NH_2$, 7.08; $NO_2$, 8.79; CHO, 8.43; OMe, 7.54; $CO_2$Me, 8.24

Set 3     4-XPnI, adiabatic[a]: $^X$ = excluded from the correlation

H, 8.70; Br, 8.52; $NH_2$, 7.51; $NO_2$, 9.24; Me, 8.38; OH, 8.06$^x$; OMe, 7.97; $CO_2$Me, 8.73; CN, 9.13

Set 4     2-XPnMe adiabatic[b]

H, 8.821; Me, 8.58; $NH_2$, 7.44; F, 8.915; Cl, 8.83; Br, 8.78; I, 8.62; Ph, 8.10; OMe, 8.03

Set 5     2-XPnMe, adiabatic[b]

H, 8.821; Me, 8.56; $NH_2$, 7.50; F, 8.915; Cl, 8.83; Br, 8.81; I, 8.61; Ph, 7.95

Set 6     4-XPnMe adiabatic[b]

H, 8.821; Me, 8.445; F, 8.785; Cl, 8.69; Br, 8.67; I, 8.50; Ph, 7.80

Set 13     4-XPnCl, vertical[c]: $^X$ = excluded from the correlation

H, 9.31; F, 9.26; Cl, 9.17; Me, 8.90; $CF_3$, 9.80; $NO_2$, 9.99; CHO, 9.59; $NH_2$, 8.18; Br, 9.04; OH, 8.69$^x$

Set 15     4-XPnBr vertical[c]: $^X$ = excluded from the correlation

H, 9.25; Cl, 9.04; Br, 8.97; Me, 8.71; $CF_3$, 9.55; CN, 9.54; OMe, 8.49; OH, 8.52$^x$

Set 301     1-XNh, vertical[b,d]: $^X$ = excluded from the correlation

H, 8.15; $NH_2$, 7.46; OH, 7.78$^x$; $Me_2$N, 7.59$^x$; Ac, 8.23; CN, 8.61; MeS, 7.67; MeO, 7.72; Me, 8.01

Set 302     2-XNh, vertical[b,d]

H, 8.15; $NH_2$, 7.56; OH, 7.90; $Me_2$N, 7.12; Ac, 8.31; CN, 8.64; MeS, 7.71; MeO, 7.87; Me, 8.01

Set 312     2-X-thiophene[e]: $^X$ = excluded from the correlation

H, 9.12; SMe, 8.10; Me, 8.63; Et, 8.67; tBu, 8.54; Br, 8.93; Cl, 9.06; OMe, 8.30; Ac, 9.20; $CO_2$Me, 9.22; CHO, 9.55; CN, 9.83; $NO_2$, 9.77; $COCF_3$, 9.70$^x$

---

[a]Behan, J.M.; Johnstone, R.A.W.; Bentley, T.W. *Org. Mass Spectrom.* **1976,** *11,* 207.

[b]Rosenstock, H.M.; Draxl, K.; Steiner, B.W.; Herron, J.T. *J. Phys. Chem. Ref. Data* **1977,** *6,* Suppl. 1, or Levin, R.D. and Lias, S.G., "Ionization Potential and Appearance Potential Measurements, 1971–1981". NSRDS–NBS71, National Bureau of Standards: Washington, D.C., 1982 (a revised edition will appear shortly).

[c]Baker, A.D.; May, D.P.; Turner, D.W. *J. Chem. Soc. B,* **1968,** 22.

[d]Utsunomiya, C.; Kobayashi, T.; Nagayiura, S. *Bull. Chem. Soc. Japan,* **48,** 1852; Bock, H., Wagner, G; Kroner, J. *Chem. Ber,* **1972,** *105,* 3850.

[e]Linda, P. Marino, G. Pagnataro, S. *J. Chem. Soc. B,* **1971,** 1585.

zenes and substituted ethylenes with the *LDA* equation (9-14; Table 9-8) showed that the substituted benzenes are much less sensitive to polarizability than are the substituted ethylenes. It seemed reasonable, then, to examine the correlation of the ionization potentials of substituted ethylenes with the *LDA* equation. The results, set forth in Table 9-12, show a striking preference for the $\sigma_R^\oplus$ constants. Thus we may conclude that the $\sigma_R^\oplus$ parameter is

TABLE 9-12. Results[a] of Correlations of Ionization Potentials with Equation 9-14: Data From Table 9-7

| Set[b] | L | D | A | h | $100R^2$ | F | $r_{12}$ | $r_{13}$ | $r_{23}$ | $R_1$ | $R_2$ | $R_3$ | $S_{est}$ | $S_L$ | $S_D$ | $S_A$ | $S_h$ | $S^e$ | n |
|---|---|---|---|---|---|---|---|---|---|---|---|---|---|---|---|---|---|---|---|
| 501AR | 0.619 | 1.88 | −5.37 | 10.47 | 69.45 | 8.335[c] | 0.250 | 0.397 | 0.166 | 0.439 | 0.261 | 0.403 | 0.316 | 0.420[d] | 0.518[e] | 1.76[f] | 0.236 | 0.645 | 15 |
| 501AP | 0.542 | 2.09 | −5.10 | 10.50 | 79.17 | 13.93 | 0.162 | 0.397 | 0.090 | 0.416 | 0.164 | 0.398 | 0.261 | 0.343[d] | 0.422 | 1.45[e] | 0.194 | 0.553 | 15 |
| 501AC | 1.20 | 2.26 | −1.93 | 10.48 | 97.45 | 140.0 | 0.304 | 0.397 | 0.169 | 0.547[g] | 0.438 | 0.500[h] | 0.0913 | 0.130 | 0.136 | 0.537[e] | 0.0665 | 0.187 | 15 |
| 521AR | 0.708 | 2.10 | −5.11 | 10.62 | 67.86 | 8.444[c] | 0.032 | 0.215 | 0.275 | 0.217 | 0.276 | 0.344 | 0.351 | 0.411[d] | 0.496[e] | 1.92[g] | 0.259 | 0.655 | 16 |
| 521AP | 0.600 | 2.28 | −4.84 | 10.64 | 77.35 | 13.66 | 0.033 | 0.215 | 0.211 | 0.230 | 0.226 | 0.307 | 0.295 | 0.346[d] | 0.413 | 1.59[f] | 0.215 | 0.550 | 16 |
| 521AC | 1.14 | 2.37 | −1.92 | 10.66 | 95.67 | 88.39 | 0.156 | 0.215 | 0.061 | 0.274 | 0.184 | 0.236 | 0.129 | 0.153 | 0.164 | 0.679[f] | 0.0925 | 0.240 | 16 |

[a]Column heads as follows:

$100R^2$ = percent of the variance of the data accounted for by the regression equation

F = F test for significance; see Note, Table 9-5

r = correlation coefficients for correlation of the ith independent variable with the jth independent variable

R = multiple correlation coefficients for the correlation of the ith independent variable with all the remaining independent variables; the confidence levels (CL) of F and the standard errors of the regression coefficients are 99.9% unless otherwise noted; the CLs of r and R are less than 90.0% unless otherwise noted

S = standard errors of the estimate and the regression coefficients; $S^e$ = standard error of the estimate divided by the root mean square of the data

n = number of data points in the set

[b]Sets designated -R, -P, and -C were correlated with $\sigma_R$, $\sigma_R^+$, and $\sigma_R^\oplus$, respectively.

[c]99.5% confidence level (CL).

[d]80.0% CL.

[e]99.0% CL.

[f]98.0% CL.

[g]95.0% CL.

[h]90.0% CL.

indeed the best choice for modeling substituent effects on the ionization potentials of the highest occupied $\pi$ orbital.

### b. Electron Loss Resulting in Carbocations

We have noted that Taft, Margin, and Lampe[8] have reported appearance potentials for the ions $XCH_2^+$. The data are set forth in Table 9-11. Correlation of their data with the $LD$ equation (9-4) gave acceptable results with the $\sigma_R^\oplus$ constants (set C1, Table 9-13). The results obtained with $\sigma_R$ and $\sigma_R^+$ are very poor. Ionization potentials of $XCH_2 \cdot$ have been reported by Lossing and co-workers[19] (set C5) and by Kebarle[20] (set C10). Again, the results obtained with $\sigma_R^\oplus$ are dramatically superior to those obtained with $\sigma_R$ or $\sigma_R^+$ (Table 9-13).

Electron loss from substituted cyclopentadienyl radicals, **11** is best fit by correlation with $\sigma_R^+$. It is difficult to assess the meaning of this result because the orbital from which the electron was lost is not known.

**11**

## C. Chemical Reactivities

### a. Formation of Carbocations by Dissociation

Monomolecular nucleophilic substitution reactions result in the formation of carbenium ion intermediates by dissociation. Much of the evidence cited above for the ability of $X_\pi$ acceptors to stabilize carbocations is based on this reaction. It provides an excellent test of the ability of $\sigma_R^\oplus$ constants to successfully model chemical reactivity in $XY^+$ systems. Solvolysis rate constants for $XMe_2CBr$ and $XMe_2CCl$ in 80% aqueous ethanol at 25°C (sets C3 and C9, respectively, Table 9-13) gave excellent correlations with the $LDS$ (Table 9-15) and $LD$ (Table 9-14) equations, respectively, using $\sigma_R^\oplus$ as the $\sigma_D$ parameter. Correlation with $\sigma_R$ or $\sigma_R^+$ gave results that were very much inferior. Both these sets incorporated $X_\pi$ donor groups. Solvolysis rate constants for $XCMe_2Lg$ (Lg = leaving group) in water were correlated with the equation:

$$Q_X = L\sigma_{IX} + D\sigma_{DX} + Sv_X + F\Lambda_{Lg} + h \qquad (9\text{-}19)$$

Where $\Lambda$ is the Swain–Lohmann leaving group parameter.[21] The data points (set C2; Table 9-13) do not include any $X_\pi$ groups. It is not surprising, then, that best results are obtained with the $\sigma_R^+$ constants. It is surprising, however, that correlation with the $\sigma_R$ constants gave better results than that with the $\sigma_R^\oplus$ constants (Table 9-16, set C2). Correlation of solvolysis rates of

**TABLE 9-13.** Carbocation Data Sets ($^x$ = Excluded from the Correlation)

| | |
|---|---|
| C1 | Appearance potentials of $XCH_2^+$, $\Delta A$ (eV)$^a$ |
| | CN, 0.4; H, 0; F, $-1.1$; Cl, $-1.4$; Me, $-1.5$; SCN, $-1.8$; Br, $-2.2$; I, $-2.3$; NO$_2$, $-2.35$; Vi, $-2.35$; C$_2$H, $-2.4$; Ph, $-2.4$; OH, $-2.6x$; SH, $-2.8^x$; OMe, $-3.0$; SMe, $-3.2$; PMe$_2$, $-3.4^x$; NH$_2$, $-4.1$; NHMe, $-4.3$; NMe$_2$, $-4.6$ |
| C2 | Solvolysis rate constants, $XMe_2CLg$, $H_2O$, 25°C$^{b,c,d}$ |
| | H, Cl 0.2115; Me, Cl 297.5; Et, Cl 767.0; Cl, Cl 0.455; Br, Br 0.231; Cl, Br 1.77; MeOCH$_2$, Cl 0.969; H, Br 7.82; H, I 2.79; CF$_3$, OTs $1.89 \cdot 10^{-3}$; CF$_3$, OMs $9.22 \cdot 10^{-11}$; H, OMs 196.5; H, OTs 410.6; Me, Br 759,000; MeF 0.296 |
| C3 | Solvolysis rate constants, $XMe_2CBr$, 80% aq. EtOH, 25°C: values reported are $\log K^d$ |
| | H, $-7.10$; Me, $-3.45$; Et, $-3.20$; $i$-Pr, $-3.36$; $t$-Bu, $-3.32$; C$_2$H, $-5.34$; CH$_2$Br, $-7.24$ |
| C4 | $\Delta G_{253}^{\ddagger}$, kcal/mol, 4-MePh-$\overset{+}{C}$(OH)X$^e$ |
| | H, 67.4; Me, 54.2; Et, 49.4; $i$-Pr, 48.1; $t$-Bu, 42.8; CH$_2$Cl, 53.4; CHCl$_2$, 54.3; CCl$_3$, 50.3; CH$_2$F, 53.8; CHF$_2$, 59.1; CF$_3$, 63.2; CH$_2$Br, 53.1 |
| C5 | Ionization potentials, $XCH_2 \cdot$ (kcal/mol)$^f$ |
| | H, 229.4; Me, 202.5; Et, 200.4; $i$-Pr, 192.6; $t$-Bu, 192.1; 1-MeVn, 185.2; 2-MeVn, 177.8; HC≡C, 190.2; c-Pr, 185.6; c-Am, 201.1; c-Hx, 176.6; Ph, 178.9; Vi, 188.2 |
| C6 | Elimination rate constants, XCHClMe, 360°C$^g$ |
| | H, 0.0047; Me, 0.49; Et, 1.43; $i$-Pr, 2.03; $t$-Bu, 4.54; Vi, 4.10; Ph, 13.62; MeO, 9500; EtO, 12,700; Ac, 0.066; Cl, 0.047; F, 0.016; CH$_2$Cl, 0.058; ViCH$_2$, 41.85; |
| C7 | Elimination rate constants, XCHClMe, 377°C$^h$ |
| | H, 0.00243; Me, 0.141; Vi, 1.06; Ph, 3.39; MeO, 18.1; EtO, 21.2; Cl, 0.0269; 1-MeVn, 3.19 |
| C8 | Elimination rate constants, XCHBrMe, 377°C$^h$ |
| | H, 0.0208; Me, 1.80; Ph, 135; Br, 0.182; |
| C9 | Solvolysis rate constants, $xMe_2Cl$, 80% aq. EtOH, 25°C$^i$ |
| | H, 0.0019; HC$_2$, 0.236; Me, 9.2; Et, 15.3; EtC$_2$, 490; MeC$_2$, 512; Vi, [2000] |
| C10 | Ionization potentials, $XCH_2 \cdot$ (eV)$^j$ |
| | H, 9.85; NC$-$, 10.81; NCCH$_2$-, 9.85$^x$; Ph, 7.76; 1-Nh, 7.35$^x$; 2-Nh, 7.56$^x$; Cl, 9.32; F, 9.37; Me, 8.38; Et, 8.13; Pr, 8.01 |
| C11 | -$\delta\Delta G_i^{\circ}$, HCOX, MeCOX, protonation$^k$ kcal/mol |
| | MeCOX; OMe, $-5.40$; OEt, $-2.80$; Me, $-6.50$; NMe$_2$, 11.70; H, $-15.4$ HCOX; OMe, $-12.80$; Me, $-15.4$; NMe$_2$, 7.60; NHMe, 1.70; Et, $-12.5$, Pr, $-10.7$ |
| C12 | Solvolysis, rate constants, $XMe_2CLg$, $CF_3CO_2H$, 25°C$^{c,d,l,m}$ |
| | CF$_3$, OTs 0.0000268; Bz, OMs 38; CO$t$-Bu, OMs 1100; CO$_2$Me, OMs 0.32; PO(OEt)$_2$, OMs 0.229; H, OMs 56.0; H, OTs 25.1; Me, Cl 603 |
| C13 | Elimination rate constants, $XMe_2COAc^n$ |
| | H, 0.58; Me, 51.9; Et, 90.4; Pr, 168.8; $i$-Pr, 253.4; Bu, 196.3; $t$-Bu, 588.8; Vi, 117.5; Ph, 131.8; c-Pr, 245.5; Ac, 9.54; AcCH$_2$, 45.7; CO$_2$Me, 3.02; CN, 2.19 |
| C14 | Rate constants, acid hydrolysis, $(EtO)_2CRX$, 49.6% aq. dioxane, 25°C$^o$ |
| | Me, H, 0.248; $i$-Pr, H, 0.164; $t$-Bu, H, 0.188; $i$-Bu, H, 0.167; HOCH$_2$, H, $8.47 \cdot 10^{-4}$; |

**TABLE 9-13.** (continued)

---

Et, H, 0.267; PhCH$_2$, H, 8.70 · 10$^{-3}$; ClCH$_2$, H, 1.03 · 10$^{-5}$; Et$_2$CH,H, 0.312$^x$; PhCH$_2$CH$_2$, H, 2.88 · 10$^{-2x}$; H, H, 4.13 · 10$^{-5}$; Me, Me, 752; Et, Me, 720; ClCH$_2$, Me, 0.0860; EtOCH$_2$, H, 8.62 · 10$^{-4x}$; PhCH$_2$, Me, 84.5; PhOCH$_2$, Me, 0.350$^x$; HOCH$_2$, Me, 11.6$^x$; AcOCH$_2$, Me, 0.228$^x$; BrCH$_2$, Me, 0.138; Ph, H, 7.07; MeVn, H, 298

C15   Solvolysis rate constants, XMe$_2$CLg, AcOH, 25°C values reported are log $k$ $^{c,d,l,m}$:
Me, Cl, −6.67; Vi, Cl, −4.80; CF$_3$, OTs, −12.133; H, OMs, −7.1; PO(OEt)$_2$, OMs, −9.01$^x$; Bz, OMs, −7.89; CO$_2$Me, OMs, −8.80; COt-Bu, OMs, −6.670$^x$; H, OTs, −7.11$^x$

C16   p$K_{BH}$+, XCONH$_2$, 33.5°C$^p$
Me, −0.93; Et, −0.99; Pr, −1.03; ClCH$_2$, −.2.80; PhCH$_2$, −1.68; Cl$_2$CH, −4.21; i-Pr, −1.26; t-Bu, −1.43

C17   p$K_{BH}$+, XCOCl, H$_2$SO$_4$, 33.5°C$^q$
Me, 10.90; Et, 10.82; Pr, 10.78; i-Pr, 10.97; Ph, 11.15; CH$_2$Cl, 11.13

C21   Ionization potentials, substituted cyclopentadienyl radicals (eV)$^r$
CN, 9.44; Br, 8.85; Cl, 8.78; F, 8.82; H, 8.69; Me, 8.54; Vi, 8.44; NH$_2$, 7.55

C22   $E_{0.5}$, substituted tropylium ions, MeCN$^s$
MeS, −0.23; MeO, −0.15; t-Bu, −0.11; H, O; Ph, 0.01; Br, 0.23; Cl, 0.26; CO$_2$H, 0.34$^x$; I, 0.44

C31   Protodetritiation rate constants, 9-substituted-10-tritiophenanthracenes, CF$_3$CO$_2$H, 100°C$^t$
H, 1605; Me, 239,500; F, 5420; Cl, 337; Br, 205; I, 234; CN, 6.5

C32   Protodetritiation rate constants, 2-substituted-1-tritionaphthalenes, CF$_3$CO$_2$H, 700°C$^t$
H, 110; MeO, 2,169,000; Me, 31,640; F, 188; Cl, 29.6; Br, 18.2; I, 25.3

C33   Proton affinities of substituted benzenes, 600°C$^u$ kcal/mol
NH$_2$, 209.3; OMe, 199.4; OH, 195.0$x$; Et, 191.0; Me, 190.0; H, 183.7; F, 182.9; Cl, 182.7

C35   $\Delta G°_{600}$, PhH$_2^+$ + PhX $\rightleftharpoons$ XPnH$_2^+$ + PhH$^u$ kcal/mol
NH$_2$, 25.6; OMe, 15.7; OH, 11.3$x$; Et, 7.3; Me, 6.3; H, 0; F, −0.8; Cl, −1.0

C36   Protodetritiation rate constants, 2-substituted-1-tritiobenzenes, 70°C$^t$
H, 0.095; Me, 20.8; t-Bu, 23.0; MeO, 6945; PhCH$_2$, 4.4; PhO, 654; PhS, 316; Ph, 9.31

C37   Protodetritiation rate constants, 4-substituted-1-tritiobenzenes, 70°C$^t$
H, 0.095; Me, 42.7; t-Bu, 50.9; PhCH$_2$, 11.05; PhO, 2930; PhS, 934; Ph, 13.45

---

$^a$Reference 8
$^b$Robertson, R.E., *Prog. Phys. Org. Chem.* **1967**, *4*, 213.
$^c$Koshy, K.M.; Tidwell, T.T. *J. Am. Chem. Soc.* **1980**, *102*, 1216.
$^d$Palm, V., Ed. "Tables of Rate and Equilibrium Constants of Heterolytic Organic Reactions", Vol. III, Pt. I. Viniti: Moscow, 1977.
$^e$Barthelemy, J.F.; Jost, R.; Sommer, J. *Org. Magn. Reson.* **1978**, *11*, 443.
$^f$Reference 19.
$^g$Chuchani, G.; Martin, I.; Alonso, M.E. *Int. J. Chem. Kinet.* **1977**, *9*, 819.
$^h$Benson, S.W.; O'Neal, H.E. "Kinetic Data on Gas Phase Reactions", National Bureau of Standards, NSRDS-NSB21. Government Printing Office: Washington, D.C., 1970.

XCMe$_2$Lg in trifluoroacetic acid and in acetic (sets C12, C15) with the equation:

$$Q_X = L\sigma_{IX} + D\sigma_{DX} + F\Lambda_{Lg} + h \qquad (9-20)$$

gave best results with $\sigma_R^\oplus$. Again, results with $\sigma_R$ or $\sigma_R^+$ are far inferior to those with $\sigma_R^\oplus$. We believe that this result supports rate-determining carbenium ion formation in the solvolysis of all members of these data sets.

The results obtained for correlation of rate constants for gas phase eliminations of XCHHlMe (Hl = Cl or Br) with the $LD$ equation (sets C6, C7, C8; Table 9-13) also indicate rate-determining carbenium ion formation. There is a definite preference for correlation with $\sigma_R^\oplus$ as compared with $\sigma_R$ or $\sigma_R^+$. This does not seem to be the case for gas phase elimination in XCMe$_2$OAc (set C13, Table 9-13), for which somewhat better correlations with the $LDS$ equation are obtained using $\sigma_R^+$ constants.

It must be noted, however, that the correlation obtained with $\sigma_R^+$ does not have a significant value of $L$, and furthermore that $L$ and $D$ differ in sign. It is quite possible that the correlation obtained with $\sigma_R^+$ is an artifact and that the correlation with $\sigma_R^\oplus$ is the best model for structural effects on pyrolytic elimination in alkyl acetates.

Correlation of the rate constants for the acid hydrolysis of XRC(OEt)$_2$ with the equation:

$$Q_X = L\sigma_{IX} + D\sigma_{DX} + S\upsilon_X + Bn_{Me} + h \qquad (9-21)$$

where $n_{Me} = 1$ when R = Me and 0 when R = H give best results with the $\sigma_R$ constants (Table 9-16, set C14). It is possible that carbenium ion formation is not rate determining in this reaction.

The mechanism of acetal hydrolysis has recently been reviewed.[22] When the alkoxycarbenium ion is well stabilized, proton transfer may be rate determining.

Equilibrium constants for the ionization of XCOCl in 100% H$_2$SO$_4$ (set C17; Table 9-13) are best fit by correlation with the $LD$ equation using the

TABLE 9-13. (continued)

[i]Richey, H.G.; Richey, J.M. In "Carbonium Ions", Vol 2; Olah, G.A.; and Schleyer, P.v.R., Eds., Wiley: New York, 1970, p. 899.
[j]Reference 20.
[k]Taft, R.W. In "Proton Transfer Reactions," Caldin, E.; and Gold, V., Eds.; Wiley: New York, 1975. p. 31.
[l]Creary, X. J. Am. Chem. Soc. 1984, 106, 5568.
[m]Reference 7b.
[n]Martin, I.; Chuchani, G.; Avila, I.; Rotinov, A.; Olmos, R., J. Phys. Chem. 1980, 84, 9.
[o]Kreevoy, M.M.; Taft, R.W. J. Am. Chem. Soc. 1955, 77, 5590; ibid, 1957, 79, 4011.
[p]Liler, M. J. Chem. Soc. B. 1969, 385.
[q]Liler, M. J. Chem. Soc. 1966, 205.
[r]Franklin, J.L. In "Carbonium Ions", Vol. I; Olah, G.A.; and Schleyer, P.v.R., Eds.; Wiley: New York, 1968, p. 1579.
[s]Harmon, K.M. In "Carbonium Ions", Vol IV; Olah, G.A.; and Schleyer, P.v.R., Eds.; Wiley: New York, 1973.
[t]Taylor, R. In "Comprehensive Chemical Kinetics", Vol. 13; Bamford, C.H.; and Tipper, C.F., Eds.; Vol. 13; Elsevier: Amsterdam, 1972, p. 244.
[u]Lau, Y.K.; Kebarle, P. J. Am. Chem. Soc. 1976, 98, 7452.

TABLE 9-14. Results[a] of Correlations of Carbocationic Data Sets with the *D*, *LD*, and *DS* Equations: Data from Table 9-13

| Set[b] | L or S[c] | D | h | $100R^2$ | F | $r_{12}$ | $S_{est}$ | $S_{l/S}$ | $S_D$ | $S_h$ | $S°$ | n |
|---|---|---|---|---|---|---|---|---|---|---|---|---|
| C1R | — | 3.39 | -1.28 | 57.85[d] | 19.21 | — | 0.940 | — | 0.773 | 0.331[e] | 0.694 | 16 |
| C1P | — | 2.64 | -1.30 | 67.77[d] | 29.44 | — | 0.822 | — | 0.487 | 0.276 | 0.607 | 16 |
| C1C | — | 3.06 | -0.502 | 81.31[d] | 60.90 | — | 0.626 | — | 0.393 | 0.279 | 0.462 | 16 |
| C4R | -11.4 | 58.9 | 66.6 | 95.53 | 53.44 | 0.229 | 2.00 | 2.20[e] | 7.83 | 1.61 | 0.267 | 8 |
| C4P | -14.0 | 53.5 | 66.0 | 95.16 | 49.13 | 0.071 | 2.08 | 2.23[e] | 7.43 | 1.69 | 0.278 | 8 |
| C4C | -6.25 | 48.6 | 67.5 | 96.15 | 62.36 | 0.482 | 1.86 | 2.27[f] | 5.96 | 1.52 | 0.248 | 8 |
| C5R | -120. | 185. | 223. | 69.12 | 10.07[e] | 0.454 | 8.50 | 32.0[e] | 47.5[e] | 7.24 | 0.642 | 12 |
| C5P | -71.1 | 133. | 216. | 56.52 | 5.851[g] | 0.071 | 10.1 | 34.0[h] | 46.5[i] | 7.38 | 0.761 | 12 |
| C5C | 16.4 | 68.9 | 221. | 88.46 | 34.49 | 0.523[f] | 5.20 | 20.5[j] | 9.23 | 3.66 | 0.392 | 12 |
| C6R | -4.43 | -7.25 | -0.412 | 50.90 | 5.702[g] | 0.452 | 1.49 | 2.48[k] | 2.16[e] | 0.592[l] | 0.790 | 14 |
| C6P | -5.08 | -8.82 | -0.701 | 75.31 | 16.77 | 0.439 | 1.05 | 1.74[i] | 1.53 | 0.428[k] | 0.561 | 14 |
| C6C | -3.31 | -7.01 | -1.92 | 82.04 | 25.12 | 0.270 | 0.900 | 1.39[f] | 0.992 | 0.465[e] | 0.478 | 14 |
| C7R | -2.35 | -5.02 | -0.946 | 43.91 | 1.957[m] | 0.634[f] | 1.24 | 3.66[l] | 2.71[e] | 0.694[n] | 0.947 | 8 |
| C7P | -1.95 | -4.45 | -0.980 | 50.07 | 2.507[m] | 0.568 | 1.17 | 3.25[l] | 2.10[h] | 0.655[k] | 0.894 | 8 |
| C7C | -2.57 | -5.24 | -2.49 | 98.17 | 134.2 | 0.486 | 0.224 | 0.585[e] | 0.329 | 0.171 | 0.171 | 8 |
| C8R | -3.91 | -9.35 | -0.658 | 15.51 | 0.092[m] | 0.765 | 2.60 | 10.4[o] | 22.3[o] | 2.39[p] | 1.84 | 4 |
| C8P | -4.48 | -18.0 | -1.70 | 65.20 | 0.937 | 0.554 | 1.67 | 5.15[i] | 13.2[i] | 1.67[j] | 1.18 | 4 |
| C8C | -4.14 | -6.17 | -1.54 | 99.07 | 53.50[h] | 0.442 | 0.272 | 0.779[k] | 0.599[h] | 0231[h] | 0.192 | 4 |
| C9R | 0.306 | -19.7 | -1.73 | 73.96 | 4.261[m] | 0.282 | 1.46 | 4.60[q] | 7.09[h] | 1.08[n] | 0.722 | 6 |
| C9P | -3.84 | -24.8 | -2.24 | 74.17 | 4.307[m] | 0.533 | 1.45 | 5.19[i] | 8.88[h] | 1.21[k] | 0.719 | 6 |
| C9C | -14.3 | -12.6 | -2.46 | 98.39 | 91.72[r] | 0.787[f] | 0.363 | 1.78[e] | 0.968 | 0.290[e] | 0.179 | 6 |
| C10R | 3.18 | 3.15 | 8.77 | 67.78 | 5.258[h] | 0.271 | 0.713 | 1.04[f] | 1.67[k] | 0.382 | 0.718 | 8 |
| C10P | 2.95 | 4.51 | 8.98 | 77.33 | 8.525[g] | 0.130 | 0.598 | 0.850[i] | 0169[f] | 0.349 | 0.602 | 8 |
| C10C | 2.82 | 3.51 | 9.43 | 92.12 | 29.22 | 0.062 | 0.352 | 0.498[e] | 0.641[e] | 0.248 | 0.355 | 8 |
| C17R | 3.39 | -4.03 | 10.26 | 88.93 | 12.04[f] | 0.956[e] | 0.0667 | 1.28[h] | 3.05[n] | 0.468 | 0.471 | 6 |
| C17P | 2.23 | -1.22 | 10.69 | 89.50 | 12.78[f] | 0.663 | 0.0650 | 0.489[i] | 0.860[n] | 0.139 | 0.458 | 6 |
| C17C | 1.69 | -0.193 | 10.82 | 87.36 | 10.37[f] | 0.207 | 0.0713 | 0.410[f] | 0.180[s] | 0.0668 | 0.503 | 6 |
| C21R | 1.26 | 1.52 | 8.65 | 94.73 | 44.93 | 0.034 | 0.144 | 0.221[e] | 0.196 | 0.0944 | 0.290 | 8 |
| C21P | 1.03 | 1.19 | 8.65 | 96.94 | 79.11 | 0.103 | 0.110 | 0.169[e] | 0.115 | 0.0717 | 0.221 | 8 |
| C21C | 1.13 | 1.31 | 8.84 | 94.45 | 42.51 | 0.046 | 0.148 | 0.227[e] | 0.175 | 0.111 | 0.298 | 8 |

| Set | | | | $F$ | | | | | | $100R^2$ | $n$ |
|---|---|---|---|---|---|---|---|---|---|---|---|
| C22R | 1.07 | 1.02 | 0.0252 | 80.26 | 10.16[g] | 0.446 | 0.122 | 0.262[e] | 0.291[i] | 0.0837[o] | 0.562 | 8 |
| C22P | 0.971 | 0.809 | 0.0168 | 86.21 | 15.63[e] | 0.335 | 0.102 | 0.208[e] | 0.182[e] | 0.0683[p] | 0.470 | 8 |
| C22C | 0.940 | 0.471 | 0.0647 | 64.40 | 4.523[h] | 0.381 | 0.164 | 0.340[f] | 0.219[h] | 0.129[r] | 0.755 | 8 |
| C33R | −30.0 | −37.4 | 184.7 | 95.48 | 42.25[f] | 0.455 | 2.59 | 5.08[e] | 4.18 | 1.66 | 0.281 | 7 |
| C33P | −19.6 | −25.7 | 185.7 | 96.87 | 61.98 | 0.243 | 2.15 | 3.88[e] | 2.37 | 1.33 | 0.234 | 7 |
| C33C | −22.4 | −28.1 | 183.1 | 99.27 | 273.2 | 0.302 | 1.04 | 1.90 | 1.23 | 0.703 | 0.113 | 7 |
| C35R | −30.0 | −37.4 | 0.969 | 95.49 | 42.30[f] | 0.455 | 2.59 | 5.08[e] | 4.17 | 1.66[l] | 0.281 | 7 |
| C35P | −19.6 | −25.7 | 1.95 | 96.88 | 62.02 | 0.243 | 2.15 | 3.88[e] | 2.37 | 1.33[k] | 0.234 | 7 |
| C35C | −22.4 | −28.1 | −0.577 | 99.28 | 275.0 | 0.302 | 1.04 | 1.90 | 1.23 | 0.701 | 0.112 | 7 |
| C37R | 1.07 | −15.1 | −0.901 | 98.73 | 117.1[r] | 0.550 | 0.190 | 0.807[n] | 1.26[e] | 0.167[i] | 0.159 | 6 |
| C37P | 1.22 | −16.1 | −0.969 | 87.05 | 10.08[f] | 0.580 | 0.607 | 2.65[f] | 4.78[f] | 0.595[n] | 0.509 | 6 |
| C37C | −6.19 | −4.95 | −0.541 | 68.36 | 3.241[m] | 0.911[i] | 0.949 | 8.15[f] | 2.92[k] | 0.901[l] | 0.796 | 6 |

Column heads as follows:

$100R^2$ = percent of the variance of the data accounted for by by the regression equation

$F$ = $F$ test for significance; see Note, Table 9-5

$r$ = correlation coefficients for correlation of the $i$th independent variable with the $j$th independent variable

$R$ = multiple correlation coefficients for the correlation of the $i$th independent variable with all the remaining independent variables; the confidence levels (CL) of $F$ and the standard errors of the regression coefficients are 99.9% unless otherwise noted; the CLs of $r$ and $R$ are less than 90.0% unless otherwise noted

$S$ = standard errors of the estimate and the regression coefficients; $S^\circledast$ = standard error of estimate divided by the root mean square of the data

$n$ = number of data points in the set

[b]Sets designated -R, -P, and -C were correlated with $\sigma_R$, $\sigma_R^\dagger$, and $\sigma_R^\circledast$, respectively.

[c]Depending on the correlation equation used.

[d]$100r^2$, where $r$ is the simple correlation coefficients.

[e]99.0% confidence level (CL).

[f]95.0% CL.      [n]70.0% CL.

[g]97.5% CL.      [o]20.0% CL.

[h]90.0% CL.      [p]10.0% CL.

[i]98.0% CL.      [q]0.% CL.

[j]50.0% CL.      [r]99.5% CL.

[k]80.0% CL.      [s]60.0% CL.

[l]40.0% CL.      [t]30.0% CL.

[m]<90.0% CL.

TABLE 9-15. Results[a] of Carbocationic Data Sets with the *LDB*, *LDS*, and *LDF* Equations: Data from Table 9-13

| Set[b] | L | D | h | B/F/S[c] | 100R² | F | $r_{12}$ | $r_{13}$ | $r_{23}$ | $R_1$ | $R_2$ | $R_3$ | $S_{est}$ | $S_L$ | $S_D$ | $S_F$ | $S_h$ | S' | n |
|---|---|---|---|---|---|---|---|---|---|---|---|---|---|---|---|---|---|---|---|
| C3R | −3.17 | −17.7 | −6.69 | 0.378 | 68.59 | 2.183[d] | 0.471 | 0.030 | 0.765[e] | 0.696 | 0.886[h] | 0.851[e] | 1.46 | 6.68[g] | 1.89[h] | 3.11[i] | 1.44[f] | 0.856 | 7 |
| C3P | −6.79 | −24.0 | −7.27 | 0.463 | 89.63 | 8.647[j] | 0.087 | 0.030 | 0.653 | 0.093 | 0.656 | 0.653 | 0.840 | 2.77[j] | 8.12[j] | 1.24[k] | 0.821[l] | 0.492 | 7 |
| C3C | −13.3 | −12.8 | −6.87 | −0.495 | 98.13 | 52.48[d] | 0.376 | 0.030 | 0.655 | 0.530 | 0.774[j] | 0.730[j] | 0.357 | 1.38[l] | 1.62[l] | 0.582[h] | 0.314 | 0.209 | 7 |
| C11R | −46.6 | −38.5 | −19.8 | 6.14 | 96.78 | 70.07 | 0.735[l] | 0.208 | 0.014 | 0.767[l] | 0.754[l] | 0.322 | 1.98 | 7.32 | 2.93 | 1.27[l] | 1.21 | 0.225 | 11 |
| C11P | −28.7 | −22.9 | −17.6 | 5.89 | 93.68 | 34.59 | 0.636[e] | 0.208 | 0.050 | 0.680[e] | 0.663[e] | 0.315 | 2.77 | 8.98[j] | 2.48 | 1.77[j] | 1.57 | 0.315 | 11 |
| C11C | −37.5 | −23.5 | −20.2 | 6.49 | 97.66 | 97.25 | 0.680[e] | 0.208 | 0.052 | 0.722[j] | 0.708[o] | 0.335 | 1.69 | 5.80 | 1.52 | 1.09 | 1.05 | 0.192 | 11 |
| C12R | −66.1 | 180. | 10.83 | −12.3 | 82.40 | 3.121[m] | 0.835[e] | 0.457 | 0.868[e] | 0.993[o] | 0.998[o] | 0.994[o] | 1.77 | 34.6[p] | 111.[s] | 7.55[s] | 5.40[p] | 0.727 | 6 |
| C12P | 41.1 | −140. | −4.80 | 8.98 | 81.18 | 2.875[m] | 0.868[e] | 0.457 | 0.833[e] | 0.992 | 0.997[e] | 0.970[l] | 1.83 | 34.0[q] | 91.7[s] | 6.03[s] | 4.70[h] | 0.751 | 6 |
| C12C | −15.0 | −27.0 | −0.190 | 2.43 | 96.75 | 19.86[e] | 0.095 | 0.457 | 0.755[j] | 0.595 | 0.806[j] | 0.848[j] | 0.758 | 2.21[e] | 5.60[e] | 0.661[j] | 0.658[i] | 0.312 | 6 |
| C13R | −0.326 | −3.45 | 0.265 | 1.97 | 88.15 | 22.32 | 0.791[l] | 0.287 | 0.497[j] | 0.801[l] | 0.840[p] | 0.526[j] | 0.375 | 0.996[k] | 1.53[j] | 0.471[l] | 0.301[h] | 0.414 | 13 |
| C13P | 0.330 | −5.26 | −0.00418 | 1.90 | 96.09 | 73.82 | 0.777[l] | 0.287 | 0.445 | 0.780[l] | 0.811[l] | 0.454 | 0.216 | 0.548[l] | 0.906 | 0.258 | 0.184 | 0.238 | 13 |
| C13C | −1.54 | −1.89 | 0.124 | 1.95 | 93.02 | 39.97 | 0.367 | 0.287 | 0.441 | 0.393 | 0.507[j] | 0.461 | 0.288 | 0.498[l] | 0.489[e] | 0.347 | 0.238[g] | 0.318 | 13 |
| C15R | −3.77 | −13.3 | −7.17 | 0.330 | 64.28 | 1.800[m] | 0.753[l] | 0.428 | 0.912[l] | 0.983[o] | 0.996[o] | 0.993[o] | 1.92 | 24.0[l] | 78.4[l] | 5.86[l] | 4.11[p] | 0.913 | 7 |
| C15P | 24.8 | −92.1 | −11.42 | 6.11 | 91.53 | 10.81[d] | 0.784[e] | 0.428 | 0.888[e] | 0.995[o] | 0.989[o] | 0.989[o] | 0.936 | 10.7[p] | 29.4[l] | 21.9[l] | 1.66[l] | 0.444 | 7 |
| C15C | −11.4 | −12.9 | −8.02 | 1.47 | 96.63 | 28.71[n] | 0.155 | 0.428 | 0.837[f] | 0.566 | 0.867[e] | 0.889[f] | 0.590 | 1.66[e] | 2.38[l] | 0.457[e] | 0.429 | 0.280 | 7 |
| C16R | −7.09 | −7.23 | −1.62 | −0.967 | 99.12 | 112.5[d] | 0.916[l] | 0.232 | 0.508 | 0.955[l] | 0.965[l] | 0.773[l] | 0.0869 | 1.82[e] | 4.16[p] | 0.230[e] | 0.529[l] | 0.143 | 7 |
| C16P | −9.04 | −1.82 | −1.03 | −0.607 | 98.33 | 59.03[d] | 0.936[l] | 0.232 | 0.034 | 0.958[l] | 0.955[l] | 0.570 | 0.119 | 2.56[e] | 4.22[g] | 0.244[j] | 0.732[s] | 0.197 | 7 |
| C16C | −10.75 | 1.07 | −0.444 | −0.575 | 99.48 | 189.7 | 0.534 | 0.232 | 0.351 | 0.536 | 0.583 | 0.355 | 0.0670 | 0.491 | 0.403[j] | 0.121[l] | 0.138[e] | 0.111 | 7 |
| C31R | −5.27 | −6.33 | 3.686 | 0.281 | 90.17 | 9.174[j] | 0.316 | 0.363 | 0.187 | 0.442 | 0.325 | 0.371 | 0.629 | 1.16[l] | 1.48[l] | 1.07[l] | 0.562[l] | 0.479 | 7 |
| C31P | −4.82 | −6.60 | −0.117 | −0.117 | 90.33 | 9.341[e] | 0.241 | 0.363 | 0.253 | 0.395 | 0.613 | 0.536 | 0.623 | 1.12[e] | 1.82[e] | 0.107[l] | 0.557[l] | 0.475 | 7 |
| C31C | −5.02 | −5.72 | 4.139 | −2.74 | 67.89 | 2.114[m] | 0.461 | 0.363 | 0.817[e] | 0.461 | 0.835[e] | 0.817[e] | 1.14 | 2.11[j] | 4.19[s] | 3.11[b] | 0.997[e] | 0.866 | 7 |
| C32R | −9.47 | −10.9 | 1.949 | 1.80 | 98.68 | 74.59[d] | 0.543 | 0.434 | 0.019 | 0.689 | 0.594 | 0.505 | 0.314 | 0.777[l] | 0.800[l] | 0.570[l] | 0.302[l] | 0.176 | 7 |
| C32P | −6.95 | −9.11 | 2.205 | 0.955 | 99.02 | 101.5[d] | 0.367 | 0.434 | 0.497 | 0.567 | 0.405 | 0.464 | 0.269 | 0.587[l] | 0.571 | 0.477[p] | 0.252[l] | 0.151[l] | 7 |
| C32C | −6.56 | −7.76 | 2.920 | −2.95 | 83.89 | 5.207[m] | 0.530 | 0.434 | 0.032 | 0.566 | 0.607 | 0.536 | 1.09 | 2.38[l] | 2.19[g] | 2.03[s] | 0.967[i] | 0.613 | 7 |
| C36R | 1.01 | −6.31 | −0.719 | 1.04 | 93.80 | 20.16[l] | 0.804[l] | 0.035 | 0.095 | 0.804[e] | 0.036 | 0.804[e] | 0.492 | 1.85[g] | 1.59[l] | 0.486[l] | 0.411[p] | 0.352 | 8 |
| C37P | 0.986 | −5.42 | −0.845 | 1.45 | 93.09 | 17.96[l] | 0.801[l] | 0.035 | 0.095 | 0.809[e] | 0.811[e] | 0.209 | 0.519 | 1.98[g] | 1.47[l] | 0.524[l] | 0.450[p] | 0.372 | 8 |
| C37C | 4.23 | −1.37 | −0.244 | 0.678 | 70.37 | 3.167[m] | 0.916[l] | 0.035 | 0.319 | 0.955[o] | 0.959[o] | 0.712[l] | 1.08 | 8.09[g] | 3.97[k] | 1.51[s] | 0.947[l] | 0.770 | 8 |

[a] Column heads as follows:

100R² = percent of the variance of the data accounted for by the regression equation

F = F test for significance; see Note, Table 9-5

r = correlation coefficients for correlation of the ith independent variable with the jth independent variable

R = multiple correlation coefficients for the correlation of the ith independent variable with all the remaining independent variables; the confidence levels (CL) of F and the standard errors of the regression coefficients are 99.9% unless otherwise noted; the CLs of r and R are less than 90.0% unless otherwise noted

S = standard errors of estimate and the regression coefficients; S' = standard error of the estimate divided by the root mean square of the data

n = number of data points in the set

[b] Sets designated −R, −P, and C were correlated with $\sigma_R$, $\sigma_R^+$ and $\sigma_R^{\oplus}$, respectively.

[c] For B or S, depending on the correlation equation used.

[d] 99.5% confidence level (CL).

| | | | |
|---|---|---|---|
| [e] 95.0% CL. | [g] 30.0% CL. | [i] 10.0% CL. | [k] 20.0% CL. |
| [f] 98.0% CL. | [h] 50.0% CL. | [j] 90.0% CL. | [l] 99.0% CL. |

| | | | |
|---|---|---|---|
| [m] <90.0% CL. | [o] 99.9% CL. | [q] 60.0% CL. | [s] 70.0% CL. |
| [n] 97.5% CL. | [p] 80.0% CL. | [r] 40.0% CL. | [t] 0.0% CL. |

TABLE 9-16. Results[a] of Correlations of Carbocationic Data Sets with the $LDSF$ and $LDSB$ Equations

| Set[b] | L | D | S | F or B[c] | h | $100R^2$ | F | $r_{12}$ | $r_{13}$ | $r_{14}$ | $r_{23}$ | $r_{24}$ | n |
|---|---|---|---|---|---|---|---|---|---|---|---|---|---|
| C2R | -12.7 | -16.1 | 2.20 | 1.37 | 1.136 | 88.78 | 15.83 | 0.386 | 0.567[d] | 0.278 | 0.374 | 0.337 | 13 |
| C2P | -11.2 | -16.5 | 2.23 | 1.40 | 1.131 | 89.92 | 17.83 | 0.215 | 0.567[d] | 0.278 | 0.287 | 0.421 | 13 |
| C2C | -12.9 | -12.1 | 0.272 | 1.35 | 1.196 | 82.81 | 9.635[h] | 0.524 | 0.567[d] | 0.278 | 0.620[d] | 0.316 | 13 |
| C14R | -11.8 | -26.9 | -0.974 | 3.80 | -4.358 | 99.01 | 251.1 | 0.432 | 0.083 | 0.403 | 0.712[e] | 0.007 | 15 |
| C14P | -13.4 | -17.5 | 0.755 | 3.80 | -4.122 | 98.18 | 135.1 | 0.420 | 0.083 | 0.403 | 0.475[g] | 0.124 | 15 |
| C14C | -18.9 | -23.9 | 1.81 | 4.15 | -3.313 | 93.64 | 36.82 | 0.001 | 0.083 | 0.403 | 0.323 | 0.331 | 15 |

| Set[b] | $R_1$ | $R_2$ | $R_3$ | $R_4$ | $S_{est}$ | $S_L$ | $S_D$ | $S_S$ | $S_{F/B}$[c] | $S_h$ | $S^\infty$ |
|---|---|---|---|---|---|---|---|---|---|---|---|
| C2R | 0.741[e] | 0.602[d] | 0.644[d] | 0.615[d] | 0.895 | 1.77 | 2.75 | 1.11 | 0.265 | 0.453 | 0.427 |
| C2P | 0.705[f] | 0.545[g] | 0.644[d] | 0.622[d] | 0.848 | 1.59 | 2.65 | 1.05[g] | 0.253 | 0.429 | 0.405 |
| C2C | 0.754[e] | 0.748[e] | 0.691[f] | 0.622[d] | 1.11 | 2.25 | 2.74[i] | 1.45[j] | 0.330[e] | 0.559[g] | 0.529 |
| C14R | 0.719[e] | 0.849[k] | 0.806[k] | 0.591[d] | 0.276 | 1.38 | 3.01 | 0.476; | 0.177 | 0.259 | 0.122 |
| C14P | 0.670[f] | 0.718[e] | 0.586[d] | 0.605[d] | 0.375 | 1.75 | 2.81 | 0.472[o] | 0.244 | 0.338 | 0.165 |
| C14C | 0.429 | 0.519[g] | 0.423 | 0.573[d] | 0.701 | 2.69 | 1.20[g] | 0.790[d] | 0.443 | 0.562 | 0.309 |

[a]Column heads as follows:

$100R^2$ = percent of the variance of the data accounted for by the regression equation

$F$ = $F$ test for significance; see Note, Table 9-5

$r$ = correlation coefficients for correlation of the $i$th independent variable with the $j$th independent variable

$R$ = multiple correlation coefficients for the correlation of the $i$th independent variable with all the remaining independent variables; the confidence levels (CL) of $F$ and the standard errors of the regression coefficients are 99.9% unless otherwise noted; the CLs of $r$ and $R$ are less than 90.0% unless otherwise noted

$S$ = standard errors of the estimate and the regression coefficients; $S^\infty$ = standard error of the estimate divided by the root mean square of the data

$n$ = number of data points in the set

[b]Sets designated $-$R, $-$P, and $-$C were correlated $\sigma_R$, $\sigma_R^+$, and $\sigma_R^\oplus$, respectively.

[c]$F$ or $B$, depending on the correlation equation used.

$\sigma_R^\oplus$ constants. Since the set consists of only six data points, none of which is for an $X_\pi$ substituent, no conclusion may be drawn. It would be interesting to examine well-characterized data sets for acylium ion reactivities and properties.

### b. Formation of Carbocations by Association

We have correlated gas phase $\delta\Delta G°$ values (set C-11; Table 9-13) for the protonation of RCOX (R = H or Me) with the equation:

$$Q_X = L\sigma_{IX} + D\sigma_{DX} + Bn_{Me} + h \qquad (9\text{-}22)$$

where, as in the case of Equation 9-21, $n_{Me} = 1$ when R is Me and 0 when R is H. Somewhat better results are obtained with $\sigma_R^\oplus$ than with $\sigma_R$; $\sigma_R^+$ gives the worst fit (Table 9-14). As set 11 does not include any $X_\pi$ groups this result is not surprising. Values of $pK_{BH}+$ for $XCONH_2$ (set C16; Table 9-13) were correlated with the LDS equation. The results show best fit with the $\sigma_R^\oplus$ constants.

Values of $\Delta G^\ddagger$ for rotation in the ions 4-MePn(OH)X (set C4, Table 9-13) are best correlated by the DS equation. Somewhat better results are obtained for $\sigma_R^\oplus$ than with $\sigma_R^+$ or $\sigma_R$. Again, there are no $X_\pi$ groups in the data set.

We have correlated several sets of rate constants for proton transfer (sets C31, C32, C36, C37; Table 9-13) that presumably involve the formation of Wheland intermediates such as:

**12**

with the LDS and LD equations. In every case the $\sigma_R^\oplus$ constants gave worst fit. Gas phase proton affinities for PhX in the reaction:

$$\text{(structure)} + PhX \rightleftharpoons \text{(structure)} + PhH \qquad (m)$$

when correlated with the LD equation, shows a very slight preference for the $\sigma_R^\oplus$ constants.

### c. Destruction of Carbocations

We have examined the correlation of $\Delta E_{1/2}$ values for substituted tropylium ions, **13**, (set C22, Table 9-13), with the LD equation. Best results were obtained with the $\sigma_R^+$ constants. Perhaps this is due to a transition state that resembles the product much more than the reactant.

**13**

## D. The Applicability of the $\sigma_R^{\oplus}$ Constants to Carbocations

Our results suggest that when a substituent is bonded to an active site carbon atom that has a significant amount of carbenium ion character, its effect is best represented by the $\sigma_R^{\oplus}$ constants. Thus, all the data sets involving solvolysis of $XMe_2CLg$ that included $X_\pi$ groups are best fit by $\sigma_R^{\oplus}$. Gas phase eliminations with Cl or Br as the leaving group also give best results with $\sigma_R^{\oplus}$, as do ionization potentials of substituted methyl radicals and appearance potentials of substituted methyl carbenium ions. Ions generated by protonation of a carbonyl group are also best modeled by the $\sigma_R^{\oplus}$ values. Ionization potentials of HOMO orbitals in substituted ethylenes, naphthalenes, thiophenes, and benzene derivatives give much better correlation with $\sigma_R^{\oplus}$ constants than with either $\sigma_R$ or $\sigma_R^+$ values. By contrast, detritiation rates of substituted arenes are not well represented by the $\sigma_R^{\oplus}$ constants.

## 5. THE NATURE AND ESTIMATION OF THE $\sigma_R^{\oplus}$ CONSTANTS

### A. The Nature of the $\sigma_R^{\oplus}$ Constants

#### a. $\sigma_R^{\oplus}$ as a Function of $\sigma_R^+$ and $\alpha$

We have shown that polarizability is an important factor in gas phase chemical reactivities and in the ionization potentials of $\pi_{HOMO}$ orbitals in substituted ethylenes. The $\sigma_R^{\oplus}$ constants were themselves defined from a correlation obtained with a selected basis set of $\sigma_R^+$ constants. It is of interest then to consider the relationship between $\sigma_R^{\oplus}$ and the $\sigma_R^+$ and $\alpha$ parameters. For this purpose we have correlated $\sigma_R^{\oplus}$ constants with the $DA$ equation in the form:

$$\sigma_{RX}^{\oplus} = D\sigma_{RX}^+ + A\alpha_X + h \qquad (9\text{-}23)$$

Results of this correlation are presented in Table 9-18. They show that Equation 9-23 is a valid approximation that can be used to make rough estimates of values of $\sigma_R^{\oplus}$ when $\sigma_R^+$ and $\alpha$ are known.

#### b. The Brown $\sigma^{c+}$ Constants

Brown and his co-workers[23] have defined a set of composite substituent constants denoted $\sigma^{c+}$ using an XGY system to model data sets with strongly electron-deficient active sites. They have reported the following $\sigma_P^{c+}$ values.

OMe, $-2.02$; Me, $-0.67$; F, $-0.40$; Cl, $-0.24$; Br, $-0.19$; CF$_3$, 0.79; H, 0. We have correlated these values with the equation:

$$\sigma_{pX}^{C+} = \lambda\sigma_{IX} + \delta\sigma_{DX} + h \qquad (9\text{-}24)$$

Best results were obtained with the $\sigma_R^+$ constants. The statistics are: $\lambda = 1.29$, $\delta = 3.29$, $h = -0.113$, $100R^2 = 97.71$, $F = 85.50$, $r_{12} = 0.219$, $S_{est} = 0.158$, $S_\lambda = 0.290$, $S_\delta = 0.254$, $S_h = 0.111$, $S° = 0.200$, $n = 7$. We may express the composition of any composite substituent by means of the percent delocalized effect, defined by the relationship[2]:

$$P_D = \frac{(\delta \cdot 100)}{(\lambda + \delta)} \qquad (9\text{-}25)$$

For the $\sigma_p^+$ constants $P_D$ is 60.7, while for the $\sigma_p^{C+}$ constants it is 71.8. It follows then that the Brown $\sigma_p^{C+}$ constants are simply composite constants with a greater $\sigma_R^+$ contribution than is found in the $\sigma_p^+$ constants.

## B. The Estimation of $\sigma_R^\oplus$ Values

Our experience has been that frequently substituent constants for a group of particular interest are not available in the literature. It is therefore necessary to have available methods of estimating substituent constants. Examples of such methods are presented here.

### a. Values of $\sigma_R^\oplus$ for CH$_2$Z Groups

We have shown[2] that $\sigma_R$ values for CH$_2$Z groups can be estimated from the equation:

$$\sigma_{D,CH_2Z} = L\sigma_{IZ} + a_0 \qquad (9\text{-}26)$$

Extension of Equation 9-26 to the $\sigma_R^\oplus$ constants was not successful. Data used in the correlation are given in Table 9-17. We then considered the equation:

$$\sigma_{R\ CH_2Z}^\oplus = L\sigma_{IZ} + D\sigma_{DZ} + a_0 \qquad (9\text{-}27)$$

On exclusion of the values for Z = CHO we obtained the relationship (set E1, Table 9-18)

$$\sigma_{R\ CH_2Z}^\oplus = 0.407\sigma_{IZ} + 0.297\sigma_R^\oplus - 0.239 \qquad (9\text{-}28)$$

Equation 9-28 may be used to obtain rough estimates of $\sigma_{R\ CH_2Z}^\oplus$. The equation:

$$\sigma_{DZ}1_z2_z3 = L\sum_{i=1}^{3} \sigma_{IZ_i} + D\sum_{j=1}^{3} \sigma_{DZ_i} + a_0 \qquad (9\text{-}29)$$

can be used for estimation. The data available for testing the applicability of this equation to $\sigma_R^\oplus$ values are very limited.

**TABLE 9-17.** Estimation Equation $\sigma_R^\oplus$ Data Sets

---

E1　　X = CH$_2$Z (Z, $\sigma_R^\oplus$)$^a$
H, $-0.25$; Me, $-0.28$; Et, $-0.31$; Pr, $-0.33$; $i$-Pr, $-0.33$; Ph, $-0.44$; Vi, $-0.40$;
OMe, $-0.31$; F$_1$ $-0.02^x$; Cl, $-0.17$; Br, $-0.19$; CHO, $-0.37^x$; SiMe$_3$, $-0.52^x$; $t$-Bu,
$-0.35$; I, $-0.38$

E11　X = MH$_i$Me$_j$ (X, $\sigma_R^\oplus$)
F, $-0.38$; Cl, $-0.41$; Br, $-0.44$; I, $-0.57$; OMe, $-0.83$; SMe, $-0.97$; NH$_2$, $-1.05$;
NHMe, $-1.23$; NMe$_2$, $-1.38$

E12　X = MH$_i$Ph$_j$ (X, $\sigma_R^\oplus$)
F, $-0.38$; Cl, $-0.41$; Br, $-0.44$; I, $-0.57$; OPh, $-0.96$; SPh, $-1.00$; NH$_2$, $-1.05$;
NHPh, $-1.40$

E13　X = MH$_i$Et$_j$ (X, $\sigma_R^\oplus$)
F, $-0.38$; Cl, $-0.41$; Br, $-0.44$; I, $-0.57$; OEt, $-0.86$; SEt, $-0.99$; NH$_2$, $-1.05$

E21　X = WH, X° = WPh (W, $\sigma_{RWH}^\oplus$, $\sigma_{RWPh}^\oplus$)
CH$_2$, $-0.25$, $-0.44$; 2-Vn, $-0.56$, $-1.01$; C$_2$, $-0.45$, $-1.03$; CO, $-0.04$, $-0.13$;
NH, $-1.05$, $-1.40$

E22　X = WMe, X° = WPh ($W$, $\sigma_{RWH}^\oplus$, $\sigma_{RWPh}^\oplus$)
O, $-0.83$, $-0.96$; S, $-0.97$, $-1.00$; SO, $-0.70$, $-0.81$; SO$_2$, $-0.12$, $-0.42$; CH$_2$,
$-0.28$, $-0.44$; 2-Vn, $-0.67$, $-1.01$; C$_2$, $-0.78$, $-1.03$; CO, $-0.05$, $-0.13$; NH,
$-1.05$, $-1.40$

E31　X$_\pi$ (X, $\sigma_R^\oplus$, $q$, $n_C$)
Vi, $-0.56$, $-0.500$, 2; E-2-ViVn, $-0.86$, 0.333, 4; Ph, $-0.69$, 0.571, 6; E-2-PnVn,
$-1.01$, 0.364, 8; PhVn$_2$, $-1.23$, 0.211, 10

E36　X$_\pi$ = M$^1$M$^2$ (X, $\sigma_R^\oplus$, X$^1$, X$^2$)
Vi, $-0.56$, 2.75, 2.75; C$_2$H, $-0.45$, 3.29, 3.29; HCO, $-0.04$, 2.75, 5.54; $-$CN,
$-0.10$, 3.29, 5.07; $-$NC, $-0.30$, 5.07, 3.29

---

$^a$Excluded from the correlation.

## b. The Estimation of $\sigma_R^\oplus$ from Electronegativity

We have reported[3,4,24] the estimation of substituent constants by means of
the relationships:

$$\sigma_{(IMZ^i)_n} = a_M \chi_M + a_i \sum_{i=1}^{q} n_Z i + a_0 \tag{9-30}$$

and

$$\sigma_{(DMZ^i)_n} = a_m\, m + a_i \sum_{i=1}^{q} n_Z i + a_0 \tag{9-31}$$

where $n_Z i$ is the number of Z$^i$ groups bonded to M and is the Allred–
Rochow[25] electronegativity of M. Sufficient data are available to permit us

TABLE 9-18. Results[a] of Correlations with Equations 9-27, 9-38, 9-41, and 9-43: Data from Table 9-17

| Set[b] | $L/b_1$, $L/a_1$, or $L/a_{12}$ [c] | $D/a_2$ or $D/a_{12}$ [c] | $a_0/b_0$ | $100R^2$ | $F$ | $r_{12}$ | $S_{est}$ | $S_{L/b}$ or $S_{L/a}$ [c] | $S_{D/a}$ | $S_{a0/b0}$ | $S°$ | $n$ |
|---|---|---|---|---|---|---|---|---|---|---|---|---|
| E1 | 0.407 | 0.297 | −0.239 | 87.09 | 26.98 | 0.468 | 0.0326 | 0.0605 | 0.0516 | 0.0209 | 0.421 | 11 |
| E21 | 0.705 | — | 0.0951 | 88.91 | 24.05[d] | — | 0.146 | 0.144[e] | — | 0.132[f] | 0.430 | 5 |
| E22 | 0.973 | — | 0.153 | 93.34 | 98.14 | — | 0.110 | 0.0982 | — | 0.0867[g] | 0.293 | 9 |
| E31 | 1.04 | −0.0461 | −1.00 | 99.61 | 256.4[d] | 0.609 | 0.0232 | 0.103[h] | 0.00463[h] | 0.0627[h] | 0.099 | 5 |
| E36 | 0.0709 | 0.188 | −1.28 | 99.72 | 356.1[d] | 0.299 | 0.0166 | 0.00911[e] | 0.00706[h] | 0.0485[h] | 0.084 | 5 |
| E101 | −0.614 | 0.829 | −0.183 | 85.66 | 95.60 | 0.109 | 0.130 | 0.278[i] | 0.0600 | 0.0453 | 0.396 | 35 |

[a]Column heads as follows:

$100R^2$ = percent of the variance of the data accounted for by the regression equation

$F$ = $F$ test for significance; see Note, Table 9-5

$r$ = correlation coefficients for correlation of the $i$th independent variable with the $j$th independent variable

$R$ = multiple correlation coefficients for the correlation of the $i$th independent variable with all the remaining independent variables; the confidence levels (CL) of $F$ and the standard errors of the regression coefficients are 99.9% unless otherwise noted; the CLs of $r$ and $R$ are less than 90.0% unless otherwise noted

$S$ = standard errors of the estimate and the regression coefficients; $S$ = standard error of the estimate divided by the root mean square of the data

$n$ = number of data points in the set

[b]The sets are designated E to indicate their use in establishing estimation equations for $\sigma_R^\oplus$.

[c]The coefficient depends on the correlation equation used.

[d]99.5% confidence CL.

[e]98.0% CL.

[f]40.0% CL.

[g]80.0% CL.

[h]99.0% CL.

[i]95.0% CL.

to apply Equation 9-31 to groups with Z = H and/or Me, and to groups with Z = H and/or Ph. We have obtained the relationships:

$$\sigma_R^{\oplus}, MH_mMe_n = 0.0945\,\chi_M - 0.310n_H - 0.468n_{Me} = 0.727 \qquad (9\text{-}32)$$
$$\sigma_R^{\oplus}, MH_mPh_n = 0.0724\,\chi_M - 0.325n_H - 0.573n_{Ph} - 0.656 \qquad (9\text{-}33)$$
$$\sigma_R^{\oplus}, MH_mEt_n = 0.0929\,\chi_M - 0.305n_H - 0.475n_{Et} - 0.726 \qquad (9\text{-}34)$$

These equations have been used to estimate $\sigma_R^{\oplus}$ values for a number of groups including OH, SH, PH$_2$, SeH, PMe$_2$, SeMe, SePh, and PHPh.

Values for these groups are given in Table 9-10, labeled E. The results of the correlations are reported in Table 9-19.

## c. The Estimation of $\sigma_R^{\oplus}$ by Other Methods

From Equation 9-31 we may write for some set of groups $MZ_n^1$ with constants $Z^1$:

$$\sigma_{DMZ^1_n} = a_M^1\chi_M + a_i^1 n_Z + a_0^1 \qquad (9\text{-}35)$$

while for some reference set of groups $MZ_n^{\circ}$ with constant $Z^{\circ}$,

$$\sigma_{DMZ^{\circ}_n} = a_M^{\circ}\chi_M + a_i^{\circ} n_Z + a_0^{\circ} \qquad (9\text{-}36)$$

We have shown that values of $a_M$ and $a_0$ are constant. Then,

$$a_M^1 = a_M^{\circ} = a_M \qquad \text{and} \qquad a_0^1 = a_0^{\circ} = a_0$$

Rearranging, we obtain:

$$\sigma_{DM1_n} - a_i^1 n_Z = a_M\chi_M + a_0 = \sigma_{DMZ^{\circ}_n} - a^{\circ}_{\,1} n_Z \qquad (9\text{-}37)$$

Then,

$$\sigma_{DMZ1_n} = \sigma_{DMZ^{\circ}_n} + a_i^1 n_Z - a_i^{\circ} n_Z \qquad (9\text{-}38)$$
$$= \sigma_{DMZ^{\circ}_n} + (a_i^1 - a_i^{\circ})n_Z \qquad (9\text{-}39)$$

If we now choose our data sets such that $n_Z$ is constant, as $a_i^1$ and $a_i^{\circ}$ are constants, it follows that:

$$\sigma_{DM1_n} = b_1\sigma_{DMZ^{\circ}_n} + b_0 \qquad (9\text{-}40)$$

where $b_1 = 1$.

We have correlated $\sigma_R^{\oplus}$ values of MH and MMe with those for MPh using Equation 9-40. The results are presented in Table 9-18, the $\sigma_R^{\oplus}$ values used were taken from Table 9-10. The equations obtained can be used to estimate $\sigma_R^{\oplus}$ values. They are:

$$\sigma_{RMH}^{\oplus} = 0.705\,\sigma_{R\,MPh}^{\oplus} + 0.0951 \qquad (9\text{-}41)$$
$$\sigma_{RMMe}^{\oplus} = 0.973\,\sigma_{R\,MPh}^{\oplus} + 0.153 \qquad (9\text{-}42)$$

We have also examined the correlation of $\sigma_R^{\oplus}$ for aryl, vinyl, butadienyl, and arylvinylene groups with the equation:

$$\sigma_R^{\oplus} = a_1 q + a_2 n_C + a_0 \qquad (9\text{-}43)$$

TABLE 9-19. Results[a] of Correlations with Equation 9-29: Data from Table 9-17

| Set[b] | $a_M$ | $a_1$ | $a_2$ | $a_0$ | $100R^2$ | $F$ | $r_{12}$ | $r_{13}$ | $r_{23}$ | $R_1$ | $R_2$ | $R_3$ | $S_{est}$ | $S_{am}$ | $S_{a1}$ | $S_{a2}$ | $S_{a0}$ | $S°$ | $n$ |
|---|---|---|---|---|---|---|---|---|---|---|---|---|---|---|---|---|---|---|---|
| E11 | 0.0945 | −0.310 | −0.468 | −0.727 | 99.19 | 204.3 | 0.063 | 0.041 | 0.162 | 0.082 | 0.177 | 0.170 | 0.0427 | 0.0270 | 0.0217 | 0.0211 | 0.0826 | 0.121 | 9 |
| E12 | 0.0724 | −0.325 | −0.573 | −0.657 | 98.32 | 78.11 | 0.072 | 0.012 | 0.046 | 0.074 | 0.086 | 0.049 | 0.0646 | 0.0409[c] | 0.0329 | 0.0472 | 0.126[d] | 0.183[d] | 8 |
| E13 | 0.0929 | −0.305 | −0.475 | −0.726 | 98.40 | 61.55[e] | 0.059 | 0.015 | 0.258 | 0.059 | 0.264 | 0.258 | 0.0515 | 0.0326[f] | 0.0288[d] | 0.0446[d] | 0.100[d] | 0.193[d] | 8 |

[a]Column heads as follows:

$100R^2$ = percent of the variance of the data accounted for by the regression equation

$F$ = $F$ test for significance; see Note, Table 9-5

$r$ = correlation coefficients for correlation of the $i$th independent variable with the $j$th independent variable

$R$ = multiple correlation coefficients for the correlation of the $i$th independent variable with all the remaining independent variables; the confidence levels (CL) of $F$ and the standard errors of the regression coefficients are 99.9% unless otherwise noted; the CLs of $r$ and $R$ are less than 90.0% unless otherwise noted

$S$ = standard errors of the estimate and the regression coefficients; $S°$ = standard error of the estimate divided by the root mean square of the data

$n$ = number of data points in the set

[b]The sets are designated E to indicate their use in establishing estimation equations for $\sigma_R^{\oplus}$.

[c]99.9% confidence level (CL).

[d]80.0% confidence level (CL).

[e]99.5% CL.

[f]90.0% CL.

where $q$ is the charge on the carbenium carbon atom in $XCH_2^+$ calculated by the method of Longuet-Higgins and Dewar[26] and $n_C$, the number of carbon atoms in the group, is a measure of its polarizability. Though the data set is small, the results are excellent. The correlation equation obtained is:

$$\sigma_{RX}^{\oplus} = 1.04q - 0.0461n_C - 1.00 \qquad (9\text{-}44)$$

It has been used to estimate the values of $\sigma_R^{\oplus}$ for 1- and 2-naphthyl reported in Table 9-10.

We have studied the correlation of $\sigma_R^{\oplus}$ for two-atom X groups with the equation:

$$\sigma_R^{\oplus} = a_{11}\chi_1 + a_{12}\chi_2 + a_0 \qquad (9\text{-}45)$$

where $\chi_1$ and $\chi_2$ are the hybridization state electronegativities[27] of the first and second atoms, respectively, of the group (set E36, Table 9-17). Again, although the set was small, excellent results were obtained. The correlation equation is:

$$\sigma_R^{\oplus} = 0.0709\chi_1 + 0.188\chi_2 - 1.28 \qquad (9\text{-}46)$$

Finally, if we exclude the values for Me and Et, the mean value of $\sigma_R^{\oplus}$ for the remaining nine alkyl and cycloalkyl groups in Table 9-10 is $-0.332 \pm 0.013$. It seems reasonable then to assign a value of $-0.33$ for $\sigma_R^{\oplus}$ to those alkyl and cycloaklyl groups for which no experimentally based value is available.

## C. The Sensitivity of Substituents to Electronic Demand

We have shown elsewhere[23] that the variation in $\sigma_D$ as a function of Y and G may be quantitatively related to the electron demand of Y by means of the equation:

$$\sigma_{DX} = \xi_X \eta_Y + \xi^0 \qquad (9\text{-}47)$$

where the parameter $\eta$ is a measure of the electronic demand of the active site Y. We have *assigned* values of $\eta$ to each system used to define $\sigma_D$ values. The assignments are shown in Table 9-20. The sensitivity of the group X to

TABLE 9-20. Values of $\eta$

| Y Types | Defining system | $\sigma_D$ Type | $\eta^a$ |
|---------|-----------------|-----------------|----------|
| $Y^{\oplus}$ | $XY^+$ | $R^{\oplus}$ | 3 |
| $Y^+$ | $XGY^+$ | $R^+$ | 2 |
| $Y^{\delta+}$ | $XGY^{\delta+}$ | R | 1 |
| $Y^{\circ}$ | $XG^1G^2Y^b$ | $R^{\circ}$ | 0 |
| $Y^{\delta-}$ | $XGY^{\delta-}$ | None | $-1$ |
| $Y^-$ | $XGY^-$ | $R^-$ | $-2$ |
| $Y^{\oplus}$ | $XY^-$ | $R^{\theta}$ | $-3$ |

$^a$By definition.
$^b$G$^2$ has one or more sp3-hybridized C atoms between G$^1$ and Y.

**TABLE 9-21.** Values of $\zeta$, the Sensitivity to the Electronic Demand of Y

| X | $\zeta$ | X | $\zeta$ | X | $\zeta$ |
|---|---|---|---|---|---|
| $X_h$ Groups | | | | | |
| Me | −0.030 | Et | −0.036 | $i$-Pr | −0.040 |
| $CH_2Ph$ | −0.057 | $CH_2Cl$ | −0.024 | $CH_2Br$ | −0.026 |
| $CF_3$ | −0.026 | $CH_2OMe$ | 0.043 | $t$-Bu | −0.036 |
| $X_p$ Groups | | | | | |
| OMe | −0.062 | OEt | −0.070 | OPh | −0.080 |
| $NH_2$ | −0.13 | $NHA_c$ | −0.162 | F | 0.041 |
| $X_{\pi(p)}$ Groups | | | | | |
| Ph | −0.12 | Vi | −0.12 | C≡CH | −0.10 |
| Bz | −0.105 | Ac | −0.092 | HCO | −0.100 |
| $CONH_2$ | −0.055 | $CO_2Me$ | −0.070 | | |
| $NO_2$ | −0.070 | CN | −0.050 | | |
| $X_{\pi(d)}$ Groups | | | | | |
| $MeSO_2$ | −0.047 | $PhSO_2$ | −0.097 | | |
| $X_{p,d}$ Groups[a] | | | | | |
| SMe | −0.129 | SPh | −0.165 | | |
| Cl | ≈0 | Br | ≈0 | | |
| $X_{\pi(d),p,d}$ Group[a] | | | | | |
| SOMe | −0.096 | | | | |

[a]Amphielectronic.

the electronic demand of Y is represented by $\xi$, and $\xi^0$ is equivalent to the calculated value of $\sigma_R^0$. Equation 9-47 is obeyed fairly well by most substituents. Values of $\xi$ are given in Table 9-21 for many common groups. On the basis of these values, we may draw some tentative conclusions:

1. $X_h$ (hyperconjugative) groups show the smallest sensitivity to the electronic demand of Y.
2. The average sensitivity to electronic demand of $X_p$ groups, which may be represented as $MZ_n$, is a linear function of the electronegativity of M.

$$\bar{\xi}_{MZ_n} = -0.185\chi_M + 0.719 \qquad (9\text{-}48)$$

3. $X_{(p)}$ groups (those in which the atom bonded to G or Y is using a p orbital to form a $\pi$ bond) also seem to show a dependence on $\chi$. As $\xi$ decreases, $\bar{\chi}$ the average electronegativity of the $\pi$-bonded atoms, increases.
4. Amphielectronic groups in which S is the atom bonded to G or Y have a high sensitivity to electronic demand.

## 6. CONCLUSION

We have shown that in $XY^+$ systems where Y is carbocationic, a new type of delocalized electrical effect parameter is required to model the effect of X on chemical reactivities and on some ionization potentials. A set of values of this $\sigma_R^{\oplus}$ parameter has been calculated. The 79 groups for which values are available include most of those commonly encountered. The basis set for the definition of the parameter is the ionization potentials of the highest occupied $\pi$ molecular orbital in substituted benzenes. Equations for the estimation of additional $\sigma_R^{\oplus}$ values have been developed. We believe that correlations with this parameter are potentially useful in mechanistic studies. They can help to identify directly substituted carbenium ion intermediates and transition states with considerable carbenium ion character. Such correlations should also be helpful in assigning ionization potentials to molecular orbitals, since they are useful in modeling substituent effects on the highest occupied $\pi$ molecular orbitals in aromatic, heteroaromatic, and ethylenic systems. Finally, we have presented arguments suggesting that the delocalized electrical effect for most groups is a linear function of the electronic demand of the active site.

## APPENDIX I: EQUATION NOMENCLATURE

Correlation equations are named after the symbols used for the regression coefficients of the independent variables. These symbols are given together with the corresponding parameters in Table I-1. Thus, the equation:

$$Q_X = L\sigma_{IX} + D\sigma_{DX} + A\alpha_X + h$$

is referred to as the *LDA* equation, while the relationship:

$$Q_X = D\sigma_{DX} + Sv_X + h$$

is known as the *DS* equation.

TABLE I-1. Parameters and Their Coefficients

| Independent variable | Parameter | Coefficient |
|---|---|---|
| Localized electrical effect | $\sigma_I$ | $L$ |
| Delocalized electrical effect | $\sigma_D$ | $D$ |
| Steric effect | $v$ | $S$ |
| Polarizability | $\alpha$ | $A$ |
| Leaving group activity | $\lambda$ | $F$ |
| Number of methyl groups | $n_{Me}$ | $B$ |

## APPENDIX II: ABBREVIATIONS

The following abbreviations other than those used by *Chemical Abstracts* are found in this chapter.

| Ak | Alkyl |
|---|---|
| El | Electrophile |
| G | Skeletal group |
| Hl | Halogen |
| Lg | Leaving group |
| 1-Nh | 1-Naphthyl |
| 2-Nh | 2-Naphthyl |
| 2-Pn | 1,2-Phenylene |
| 3-Pn | 1,3-Phenylene |
| 4-Pn | 1,4-Phenylene |
| R | H or Ak or Ar or Vi |
| Vi | Vinyl |
| 1-Vn | Vinylidene |
| $E$-2-Vn | *trans*-Vinylene |
| $Z$-2-Vn | *cis*-vinylene |
| X | Substituent |
| Y | Active site |
| Z | Any atom or group of atoms |

# REFERENCES

1. Ehrenson, S.; Brownlee, R.T.C., Taft, R.W. *Prog. Phys. Org. Chem.* **1973**, *10*, 1.
2. Charton, M. *Prog. Phys. Org. Chem.* **1981**, *13*, 119.
3. Charton, M. *J. Org. Chem.* **1984**, *49*, 1997.
4. Charton, M. In "Chemistry of the Functional Groups," Supplement C, "The Chemistry of Triply Bonded Functional Groups", Patai, S., Ed.; Wiley: New York, 1983, p. 269.
5. (a) Gassman, P.G.; Talley, J.J. *J. Am. Chem. Soc.* **1980**, *102*, 1214. (b) Gassman, P.G.; Talley, J.J. ibid, **1980**, *102*, 4138.
6. Creary, X. *J. Org. Chem.* **1979**, *44*, 3938.
7. (a) Olah, G.A.; Surya Prakash, G.K.; Arvanaghi, M. *J. Am. Chem. Soc.* **1980**, *102*, 6640. (b) Creary, X.; Geiger, C.C.; Hilton, K. ibid, **1983**, *105*, 2851.
8. Taft, R.W.; Martin, R.H.; Lampe, F.W. *J. Am. Chem. Soc.* **1965**, *87*, 2490.
9. Dixon, O.A.; Charlier, P.A.; Gassman, P.G. *J. Am. Chem. Soc.* **1980**, *102*, 3957.
10. (a) Paddon-Row, M.N.; Santiago, C.; Houk, K.N. *J. Am. Chem. Soc.* **1980**, *102*, 6561. (b) Dixon, D.A.; Eades, R.A.; Frey, R.; Gassman, P.G.; Hendewerk, M.L.; Paddon-Row, M.N.; Houk, K.N. ibid, **1984**, *106*, 3885.
11. (a) Chess, E.K.; Schatz, B.S.; Gleicher, G.J. *J. Org. Chem.* **(1977)**, *42*, 752. (b) Gleicher, G.J. *Tetrahedron* **1974**, *30*, 935.
12. Gonen, Y.; Horowitz, A.; Rajbenbach, L.A. *J. Chem. Soc. Faraday Trans. 1* **1977**, *73*, 866.
13. Timberlake, J.W.; Garner, A.W.; Hodges, M.L. *Tetrahedron Lett.* **1978**, 309.
14. Sarner, S.F.; Coale, D.M.; Hall, H.K., Jr., Richmond, A.B. *J. Phys. Chem.* **1972**, *76*, 2817.
15. (a) Egger, K.W.; Cocks, T.A. *Helv. Chim. Acta* **1973**, *56*, 1516, 1537. (b) Kerr, J.A. *Chem. Rev.* **1966**, *66*, 645.
16. Norman, R.O.C.; Gilbert, B.C. *Adv. Phys. Org. Chem.* **1967**, *5*, 53.
17. (a) Fischer, H. *Z. Naturforsch. A* **1964**, *19*, 866. (b) Fischer, H. ibid, **1965**, *20*, 428.

18. (a) Bordwell, F.G. *Faraday Symp.* **1975,** *10,* 100. (b) Bordwell, F.G. *Pure Appl. Chem.* **1977,** *49,* 963. (c) Matthews, W.S.; Bares, J.E.; Bartmess, J.E.; Bordwell, F.G.; Cornforth, F.J.; Drucker, G.E.; Margolin, Z.; McCallum, R.J.; McCollum, G.J.; Vanier, N.R. *J. Am. Chem. Soc.* **1975,** *97,* 7006. (d) Bordwell, F.G.; Van der Puy, M.; Vanier, N.R. *J. Org. Chem.* **1976,** *41,* 1883, 1885. (e) Bordwell, F.G.; Bartmess, J.E.; Hautala, J.A. ibid, **1978,** *43,* 3095. (f) Bordwell, F.G.; Bartmess, J.E. ibid, **1978,** *43,* 3101. (g) Bordwell, F.G.; Branca, J.C.; Hughes, D.L.; Olmstead, W.N. ibid, **1980,** *45,* 3305. (h) Bordwell, F.G.; Branca, J.C.; Johnson, C.R.; Vanier, N.R. ibid, **1980,** *45,* 3884.
19. (a) Lossing, F.P. In "Mass Spectrometry", McDowell, C.A., Ed.; McGraw-Hill: New York, 1963, Chapter 11. (b) Lossing, F.P.; Maccoll, A. *Can. J. Chem.* **1976,** *54,* 990.
20. Kebarle, P. In "Ions and Ion Pairs in Organic Reactions", Vol. 1, Szwarcz, M., Ed.: Wiley: New York, 1972, p. 27.
21. Swain, C.G.; Lohmann, K.H. Cited in E. R. Thornton, "Solvolysis Mechanisms". Ronald Press: New York, 1964, p. 164.
22. Bethell, D.; Whittaker, D. In "Reactive Intermediates", Vol. 2, Jones, M., Moss, R.A., Eds.; Wiley: New York, 1981, p. 211.
23. Brown, H.C.; Kelly, D.P.; Periasamy, M. *Proc. Natl. Acad. Sci. U.S.A.* **1980,** *77,* 6956.
24. Charton, M. Abstracts, 18th Middle Atlantic Regional Meeting of the American Chemical Society, Newark, N.J., 1984, p. 109.
25. Allred, A.L.; Rochow, E.G. *J. Inorg. Nucl. Chem.* **1958,** *5,* 264.
26. Dewar, M.J.S. *J. Am. Chem. Soc.* **1952,** *74,* 3341.
27. (a) Hinze, J.; Jaffé, H.H. *J. Phys. Chem.* **1963,** *67,* 1501. (b) Hinze, J.; Jaffé, H.H. *J. Am. Chem. Soc.* **1962,** *84,* 540. (c) Hinze, J.; Whitehead, M.A.; Jaffé, H.H. ibid, **1963,** *85,* 148.

# CHAPTER 10

# Three-Dimensional Molecular Modeling by Computer

## Tamara Gund

New Jersey Institute of Technology, Newark, New Jersey

## Peter Gund

Merck Sharp & Dohme Research Laboratories, Rahway, New Jersey

## CONTENTS

MOLECULAR STRUCTURE AND ENERGETICS, Vol. 4

# 1. BACKGROUND

For more than 100 years, since the time of the tetrahedral carbon atom of van't Hoff and Le Bel, chemists have known that three-dimensional molecular shape influences physical and chemical properties; somewhat later, it was realized that shape also influences biological properties. Last century's chemists were fascinated with the macroscopic symmetry properties of beautiful crystals and succeeded in associating them with a molecule's microscopic properties of local symmetry. With the advent of X-rays and crystallographic structure determination, it was possible to relate macroscopic and microscopic properties directly. Still later, conformational arguments by Robinson and other organic chemists, especially steroid chemists, made it possible to understand chemical reactions in mechanistic detail.

Mechanical models of various types have long been standard implements of the organic chemist, for supplementing the two-dimensional structural diagrams sketched on paper and blackboard, and for better understanding of geometric and steric properties of molecules. Dreiding-type "stick" models give information about bond lengths, bond angles, and dihedral angles, but allow too many conformations of molecules. Corey–Pauling–Koltun-type "space-filling" models give a better sense of the overall shape of molecules, at the expense of seeing the relative geometry of the component atoms; actually too few conformations may be represented with these non-compressible models, and many strained molecules cannot be built at all. Ball-and-stick models are a compromise, showing some spatial information but not allowing easy geometric measurement.

The computer has an already long history of application to problems of chemical conformation. The molecular mechanics approach of describing a molecule as a collection of particles held together by elastic forces was described by Westheimer.[1] An early application of the digital computer to ring conformation was published by Hendrickson,[2] while the first generally applicable program for computing conformations of hydrocarbons appears to be that of Wiberg.[3] Other pioneering studies in this field included Levinthal's use of computer graphics for molecular conformation,[4] and Scheraga's calculations of polypeptide conformation.[5]

Two fairly recent trends have revolutionized the application of computers to conformational analysis. The first of these is the incredible reduction in cost and size of high-performance computers. The second trend is the spreading of computers out of the domain of the specialists (such as chemist-theoreticians) into the laboratories, offices, and homes of the nonspecialists (such as organic chemists). This spread of personal computers was made possible by the development of new, easier-to-use computer programs (software). It is now possible to perform simple molecular modeling calculations and display on personal computers (Apple II, IBM PC)[6] and even on hand-held calculators.[7] The remainder of this chapter reviews some of the molecular modeling functions that may be performed and some of the systems (software and hardware) available for performing them. We note here and in Table 10-5 that products are listed for informational purposes only. Other suitable products are available in addition to those listed. Neither the authors nor VCH Publishers endorse the purchase of specific listed products, nor can they be held responsible for errors.

## 2. WHAT IS MOLECULAR MODELING?

Any simulation of natural processes is called modeling, and the manifestation of such simulation is called a model. Molecules are submicroscopic collections of atoms that, based on observations of macroscopic phenomena, are believed to possess certain properties. A molecular model, then, is an attempt to simulate some aspect of molecular properties in a macroscopic form, which is more easily comprehended and handled. Consider a CPK (space-filling) model of a molecule; it treats atoms as almost spherical, with connectors to other atoms. A Dreiding (stick) model, on the other hand, represents atoms as small nodes and focuses on the characteristic distance between bonded atoms, represented as bonds, and the directionality in space of such bonds. It is important to remember that such models are not true representations of molecules, which are neither composed of spherical atoms nor connected by stick bonds—but may be conceived as dynamic collections of nuclei, vibrating about a characteristic optimal geometry, surrounded by a cloud of valence electrons that (according to Heisenberg's uncertainly principle) may not be located exactly, and constantly being distorted by interactions with neighboring molecules and other forces.

A computer representation of a molecule may take several forms. The molecular structure is represented in the computer as a connection table or directed graph, with further specification of the graph nodes with atomic labels and of the graph connections with bond order. Thus, the connection table for a thiophene derivative (Table 10-1) may be represented as in Figure 10-1. The molecular geometry is specified by giving positions of the component atoms in Cartesian (three-dimensional) space, usually in Ångstrom units (Table 10-2). Alternatively, the structure may be expressed in terms of internal coordinates (bond lengths, bond angles, and dihedral angles: Table 10-3), or as a distance matrix (Table 10-4). The distance matrix does not, however, give the chirality (handedness) of the structure, since distance is a scalar (not a vector) quantity.

Just as a mechanical model of a molecule allows useful inferences about geometric, steric, and even chemical properties (such as reactivity) of that molecule, a computer-based model may also offer useful insights. Furthermore, in terms of the range of properties that may be inferred or calculated,

**TABLE 10-1.** Connection Table for 2-(2-Thiophenyl)-2-
Hydroxybutyric Acid

| Atom | Symbol | Bonded atoms (bond type) |
|------|--------|--------------------------|
| 1 | C | 2 (2), 5 (1), 11 (1) |
| 2 | C | 1 (2), 3 (1), 12 (1) |
| 3 | S | 2 (1), 4 (1) |
| 4 | C | 3 (1), 5 (2), 13 (1) |
| 5 | C | 4 (2), 1 (1). 6 (1) |
| 6 | C | 5 (1), 7 (1), 10 (1), 14 (1) |
| 7 | C | 6 (1), 8 (2), 9 (1) |
| 8 | O | . 7 (2) |
| 9 | O | 7 (1), 15 (1) |
| 10 | O | 6 (1), 16 (1) |
| 11 | H | 1 (1) |
| 12 | H | 2 (1) |
| 13 | H | 4 (1) |
| 14 | H | 6 (1) |
| 15 | H | 9 (1) |
| 16 | H | 10 (1) |

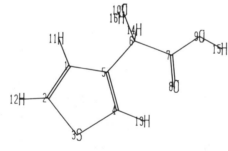

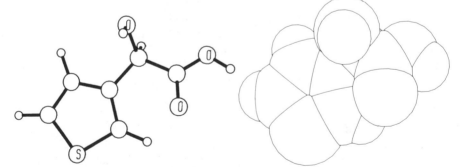

**Figure 10-1.** Model conformation of 2-(2-thiophenyl)-2-hydroxybutyric acid. (*a*)
Stick figure with arbitrary atom numbers. (*b*) Ball-and-stick representations, showing
spatial orientation. (*c*) Space-filling (CPK-type) representation.

TABLE 10-2. Cartesian Coordinates for the Models of 2-(2-Thiophenyl)-2-Hydroxybutyric Acid in Figure 10-1

| Atom | Symbol | Coordinates, (Å) | | |
|------|--------|--------|--------|--------|
| | | X | Y | Z |
| 1 | C | −1.8052 | 0.2564 | −0.5469 |
| 2 | C | −2.5848 | −0.9119 | −0.5335 |
| 3 | S | −1.4839 | −2.2105 | −0.2707 |
| 4 | C | −0.0301 | −1.2922 | −0.1752 |
| 5 | C | −0.4250 | 0.0433 | −0.3526 |
| 6 | C | 0.5672 | 1.1815 | −0.3365 |
| 7 | C | 1.9044 | 0.6772 | 0.1508 |
| 8 | O | 1.9807 | −0.4218 | 0.7416 |
| 9 | O | 2.9352 | 1.3589 | −0.0368 |
| 10 | O | 0.0982 | 2.2105 | 0.5388 |
| 11 | H | −2.2301 | 1.2416 | −0.6946 |
| 12 | H | −3.6578 | −0.9819 | −0.6621 |
| 13 | H | 0.9737 | −1.6651 | −0.0131 |
| 14 | H | 0.6584 | 1.6336 | −1.3285 |
| 15 | H | 3.7169 | 0.9143 | 0.3269 |
| 16 | H | −0.0065 | 1.8764 | 1.4434 |

and in terms of the convenience of computing such properties (once a molecule is in the computer), computer-aided molecular models may offer substantial advantages over the mechanical ones. It is important to remember, however, that the computer models still do not represent reality, and conclusions from such models must still be validated by comparison with experiment.

The basis of computer-aided molecular modeling, as for mechanical molecular models, is that atoms in molecules prefer to orient themselves in well-defined ways. And as the flexible and vibrating aspects of molecular structure may be represented by a mechanical model of balls connected by springs, the computer model may use restraining forces to represent the same aspects more quantitatively.

The discipline of molecular modeling is based on classical physics, which treated the universe as a collection of bodies subject to the classical electrical, magnetic, and gravitational forces. By molecular mechanics, one normally means the application of classical mechanical principles to molecular structure. However, as is well established, the classical mechanical view of matter breaks down at the microscopic level. Therefore quantum mechanical principles are normally utilized as necessary to extend the conclusions derived by molecular mechanics.

Molecular modeling studies may be performed for a variety of reasons. For example, crystallographers perform such studies to find structures to fit their crystallographically determined electron density maps, to study molecular packing, and to compare structures in different crystalline environments. Spectroscopists perform such studies to confirm their spectral assign-

TABLE 10-3. Internal Coordinates (Excluding Hydrogens) for the Models of 2-(2-Thiophenyl)-2-Hydroxybutyric Acid in Figure 10-1

**Bond lengths**

| i | j | Distance (Å) |
|---|---|---|
| 1 | 2 | 1.405 |
| 1 | 5 | 1.410 |
| 2 | 3 | 1.723 |
| 3 | 4 | 1.722 |
| 4 | 5 | 1.40 |
| 4 | 6 | 1.510 |
| 5 | 7 | 1.510 |
| 6 | 10 | 1.430 |
| 7 | 8 | 1.250 |
| 7 | 9 | 1.250 |

**Bond angles**

| i | j | k | Angles (°) |
|---|---|---|---|
| 2 | 1 | 5 | 114.6 |
| 1 | 2 | 3 | 105.9 |
| 2 | 3 | 4 | 98.4 |
| 3 | 4 | 5 | 105.2 |
| 4 | 5 | 1 | 115.9 |
| 4 | 5 | 6 | 122.1 |
| 1 | 5 | 6 | 122.1 |
| 5 | 6 | 7 | 109.5 |
| 5 | 6 | 10 | 109.5 |
| 7 | 6 | 10 | 109.5 |
| 6 | 7 | 8 | 120.0 |
| 6 | 7 | 9 | 120.0 |
| 8 | 7 | 9 | 120.0 |

**Dihedral angles**

| i | j | k | l | Angle (°) |
|---|---|---|---|---|
| 1 | 2 | 3 | 4 | 0.0 |
| 1 | 5 | 4 | 3 | 0.0 |
| 1 | 5 | 6 | 7 | 169.2 |
| 1 | 5 | 6 | 10 | 49.2 |
| 2 | 1 | 5 | 4 | 0.0 |
| 2 | 1 | 5 | 6 | 180.0 |
| 2 | 3 | 4 | 5 | 0.0 |
| 3 | 2 | 1 | 5 | 0.0 |
| 3 | 4 | 5 | 1 | 0.0 |
| 3 | 4 | 5 | 6 | −180.0 |
| 4 | 3 | 2 | 1 | 0.0 |
| 4 | 5 | 1 | 2 | 0.0 |
| 4 | 5 | 6 | 7 | −10.8 |
| 4 | 5 | 6 | 10 | −130.8 |
| 5 | 4 | 3 | 2 | 0.0 |
| 5 | 1 | 2 | 3 | 0.0 |
| 5 | 6 | 7 | 8 | −16.1 |
| 5 | 6 | 7 | 9 | 163.9 |
| 6 | 5 | 4 | 3 | −180.0 |
| 6 | 5 | 1 | 2 | 180.0 |
| 7 | 6 | 5 | 4 | −10.8 |
| 7 | 6 | 5 | 1 | 169.2 |
| 8 | 7 | 6 | 5 | −16.1 |
| 8 | 7 | 6 | 10 | 103.9 |
| 9 | 7 | 6 | 5 | 163.9 |
| 9 | 7 | 6 | 10 | −76.1 |
| 10 | 6 | 5 | 4 | −130.8 |
| 10 | 6 | 5 | 1 | 49.2 |
| 10 | 6 | 7 | 8 | 103.9 |
| 10 | 6 | 7 | 9 | −76.1 |

**Out-of-plane angles, i–j (l)–k**

| i | j | k | l | Angle (°) |
|---|---|---|---|---|
| 4 | 5 | 1 | 6 | 0.0 |
| 4 | 5 | 6 | 1 | 0.0 |
| 1 | 5 | 6 | 4 | 0.0 |
| 6 | 7 | 8 | 9 | 0.0 |
| 6 | 7 | 9 | 8 | 0.0 |
| 8 | 7 | 9 | 6 | 0.0 |

TABLE 10-4. Distance Geometry Description (Excluding Hydrogens) for the Models of 2-(2-Thiophenyl)-2-Hydroxybutyric Acid in Figure 10-1

|    | 1 | 2 | 3 | 4 | 5 | 6 | 7 | 8 | 9 | 10 |
|----|-----|-----|-----|-----|-----|-----|-----|-----|-----|-----|
| 1  | 0.000 | 1.405 | 2.503 | 2.385 | 1.410 | 2.555 | 3.798 | 4.056 | 4.894 | 2.936 |
| 2  | 1.405 | 0.000 | 1.723 | 2.608 | 2.369 | 3.789 | 4.811 | 4.765 | 5.989 | 4.254 |
| 3  | 2.503 | 1.723 | 0.000 | 1.722 | 2.492 | 3.964 | 4.472 | 4.028 | 5.685 | 4.765 |
| 4  | 2.385 | 2.608 | 1.722 | 0.000 | 1.404 | 2.550 | 2.780 | 2.375 | 3.980 | 3.577 |
| 5  | 1.410 | 2.369 | 2.492 | 1.404 | 0.000 | 1.510 | 2.466 | 2.683 | 3.622 | 2.401 |
| 6  | 2.555 | 3.789 | 3.964 | 2.550 | 1.510 | 0.000 | 1.510 | 2.394 | 2.393 | 1.430 |
| 7  | 3.798 | 4.811 | 4.472 | 2.780 | 2.466 | 1.510 | 0.000 | 1.250 | 1.250 | 2.401 |
| 8  | 4.056 | 4.765 | 4.028 | 2.375 | 2.683 | 2.394 | 1.250 | 0.000 | 2.165 | 3.243 |
| 9  | 4.894 | 5.989 | 5.685 | 3.980 | 3.622 | 2.393 | 1.250 | 2.165 | 0.000 | 3.017 |
| 10 | 2.936 | 4.254 | 4.765 | 3.577 | 2.401 | 1.430 | 2.401 | 3.243 | 3.017 | 0.000 |

ments, to extrapolate results in related systems, and to correlate spectral data with structure. Theoreticians perform such studies to test ideas, to develop new or refined theoretical techniques, to visualize their theoretical conclusions, and to find useful generalizations. Inorganic chemists look for relationships between modeled structures and desirable physical or chemical properties. Organic chemists study reactivity, stereoselectivity of reactions, stereoisomerism, reaction mechanisms, and other physical and chemical properties from such models. Polymer chemists attempt to model crystallinity and other polymer properties. Pharmaceutical chemists derive models to explain and predict biological activity of drugs. Biochemists relate such models to biological processes, and enzymologists generate models for enzyme action. In Chapter 11, Hansch makes some use of molecular modeling to promote an understanding of the results of correlation analysis.

Different modeling techniques are used for these different types of study. In general, the larger the molecular structure, the more approximate the modeling methodology that can be applied. The simplest molecule, hydrogen ($H_2$), has been subject to nearly exact theoretical (quantum mechanical) calculations; however, for even as simple a molecule as water ($H_2O$), calculations have been performed many times, with many different levels or sets of assumptions, at various stages approaching and even going beyond the so-called Hartree–Fock limit of the Schrödinger equation. "Ab initio" calculations may be performed (at increasing levels of approximation) for molecules of up to 50 atoms or so. Semiempirical calculations (described below) may handle structures of perhaps 100 atoms, if a large enough computer is available. Classical mechanical computations with a fairly extensive force field can be performed on appropriate structures of up to 100 or 200 atoms, while more approximate force fields have been applied to molecules containing 1000 atoms or more. Macromolecules (polypeptides, nucleic acids, polysaccharides, etc) are usually treated by making additional simplifying assumptions and by moving the "building block" fragments (residues) as units.

# 3. METHODS OF GENERATING MOLECULAR STRUCTURES AND CONFORMATIONS

## A. From Crystal Structures

Several thousand crystal structure determinations have been reported in the literature and are collected in the Cambridge Crystallographic Data File.[8] Where a crystal structure is available, this source is usually the best starting point for a molecular modeling study. However, it is often useful to refine the crystal coordinates before drawing final conclusions; crystal structures are subject to systematic and random errors, distortion by crystal packing forces, and conformational changes.[9] Using standard techniques and programs, the molecular coordinates may be generated from the fractional atomic coordinates of the molecule in the crystallographic unit cell.[10]

## B. From Standard Fragments

Study of many crystal structures has led to the observation that the geometries of certain groups of atoms are more or less constant from one molecular environment to the next. Thus, a nitro group has in many structures about the same molecular dimensions as carboxyl groups, benzene and thiophene rings, and so on. Where a crystal structure of the molecule of interest is not available, fragment geometries may be assembled from related crystal structures and combined to give the desired molecule. Several molecular modeling programs contain files of such fragments, to expedite building complex structures this way. For protein modeling, all the structures are built from standard geometries of amino acid fragment residues.

## C. By Computational Procedures

A number of computer programs have been created to build a reasonable three-dimensional molecular structure from a two-dimensional diagram. Several strategies have been used. One approach uses standard bond lengths, bond angles, and dihedral angles, plus stereochemical information, and a geomentry optimization procedure to alter the initial two-dimensional coordinates into reasonable three-dimensional coordinates; this works best for (poly)cyclic and strained molecules.[11] Another strategy uses the molecular connectivity to assemble the molecule from standard fragments; this works best for unstrained and alicyclic systems,[12] although at least one program uses the approach to systematically generate conformations of cyclic structures also.[13]

## D. By Modifying Molecular Structures

Molecular coordinates may be modified in various ways to create new molecules or new conformations. Atom types may be changed, as may bond lengths, bond angles, dihedral angles, ring puckering, bond order, and charge; a program to perform these operations has been called a molecule editor.[14] Molecules may also be merged, fused, additional bonds created, and so on. After several such changes, an energy calculation may be required to regularize the structure.

## E. By Changing Conformations

Acyclic molecules normally contain one or more rotatable bonds. However, such bonds show preferred dihedral angles. Tetrahedral atoms are typically joined by a single bond showing a threefold barrier to rotation; a tetrahedral atom joined to a trigonal atom would show a sixfold barrier, while two trigonal atoms would be joined with a twofold barrier to rotation. The combinatorics are such that it is prohibitive to systematically survey conformations of molecules with many more than 10 rotatable bonds.

Not all combinations of allowed dihedral angles are reasonable, due to steric repulsion between too close nonbonded atoms. Efficient conformation surveying programs use such information to increase search speed.

When only one or two bonds are allowed to rotate, energy calculations at points along the bond rotations give an indication of the conformational potential energy surface. Contouring lines of isoenergy potentials can give a potential energy contour map, on which minimum energy conformations may be located.

Another technique uses an artifically high torsion barrier to "drive" an energy calculation from a starting conformation to another one. Potential problems with this technique are discussed by Burkert and Allinger.[15]

Cyclic structures are more difficult to survey, because dihedral angles cannot vary independently. For small rings, conformations may be selected from a library or generated by pseudorotation procedures. For large rings, the best approach appears to be to generate all allowed values of dihedral angles around the ring, and throw out all combinations that do not close the ring.[16] In cases where spectral data (eg, NMR coupling constants) can reduce the number of allowed values of dihedral angles, the search can be efficient even for large rings.

# 4. MORE SOPHISTICATED THEORETICAL CALCULATIONS

## A. Semiempirical Quantum Mechanical Methods

While molecular mechanics programs treat molecules as collections of "sticky," soft balls connected by springs, quantum mechanics programs are based on various approximations to the Schrödinger equation, which relates energy of such a system to behavior of a characteristic wave function. The process nominally consists of collecting the component atoms, fixing the nuclei in space, stripping off the valence electrons (or all electrons, depending on the rigor of the procedure), allowing the component atomic orbitals to "mix" according to the variation principle, and filling the resulting molecular orbitals with available electrons. Analysis of the resulting orbitals, preferably after a further round of allowing the original molecular orbitals to split each other (configuration interaction), gives rise to such interesting quantities as partial charges, frontier orbital coefficients, and electrostatic potential energy contours.

The first popular molecular orbital procedure was the Hückel theory, which treated the $\sigma$ framework as fixed and focused only on the $\pi$-electron system. For unsaturated systems, the occupied orbitals highest in energy (HOMO) and unoccupied orbitals lowest in energy (LUMO) tend to be dominated by $\pi$-character; these "frontier orbitals" tend to control chemical reactivity, and the simple theory performs remarkably well for such systems.[17] It is now well established, however, that $\sigma$ and $\pi$ systems interact appreciably, via hyperconjugation and other mechanisms, and simple Hückel theory is an unacceptable simplification for describing many important chemical phenomena. The other amusing aspect of simple Hückel calculations is that no geometric information is necessary; the planarity of the system and the connectivity of the atoms are sufficient for setting up the computations. This is not true of any of the other energy calculations discussed in this chapter.

The next popular program, which became available as significant computer resources appeared on university campuses, was Hoffmann's extended Hückel theory (EHT), which treated $\sigma$ and $\pi$ electrons, but treated the core integrals in too simplistic a way.[18] While EHT has continued to be used for many applications, it is least succesful for molecules containing strong charge alternation.

Perhaps the most popular program for organic chemists, at least until recently, has been Pople's complete neglect of differential overlap (CNDO) theory.[19] Besides accounting in a more realistic way for core integral effects, it was parameterized to give good dipole moments, leading to its extensive use for systems containing multiple heteroatoms. Although limitations of the theory are well known,[20] experience with it is so extensive that the deficiencies can usually be taken into account in drawing conclusions from the calculated results.

A different approach was taken by Dewar's group, with a series of semiempirical programs (MINDO, MNDO) that were parameterized to give reasonable thermodynamic results (heats of formation), were easy to set up and run, and optimized molecular geometries with symmetry constraints.[21] Although these programs also have limitations, such as poor bond angles for certain pathological cases and failure to account for the hydrogen bond attractive energy,[22] development of these formalisms is continuing.

A method from the Pullman group, perturbative configuration interaction using localized orbitals (PCILO), is quite fast and convenient but seems to suffer from many of the deficiencies of CNDO, including the tendency to overemphasize attractive interactions between touching nonbonded atoms.[23]

Of the many other semiempirical methods that have been developed, we will only mention the partial retention of diatomic differential overlap (PRDDO) methodology of Halgren and Lipscomb,[24] which was carefully parameterized to reproduce ab initio minimum basis set results (see next section) with computer times comparable to CNDO; unfortunately this procedure has not been as widely available as some of the other methods.

## B. Ab Initio Computations

When the multicenter integrals are evaluated explicitly rather than estimated by parameterization to experimental data or ignored, the technique is called "ab initio," although the result is usually still far from the Hartree–Fock limit of correctness. The simplest common procedure uses a linear combination of the more tractable Gaussian curves to represent the more correct Slater-type atomic orbitals. The STO-3G procedure uses three Gaussians to represent the basis atomic Slater-type orbitals; more Gaussians are used in 4-31G, 6-31G*, . . . formalisms. Inclusion of electron correlation can now be done using many of the same programs. One of the most popular families of programs for such calculations is the GAUSSIAN series of programs from Pople's group.[25]

## C. Classical Mechanical Calculations

While quantum mechanical programs are considered to be more rigorous, the classical mechanical approach is actually theoretically quite sound, being based on the Born–Oppenheimer assumption that the motion of the heavy nuclei may be decoupled from the much faster (relativistic) motion of the electrons about the ensemble of nuclei. In practice, a trial set of coordinates is taken, and appropriate parameters are assigned for each bond stretch, bond angle bend, dihedral angle twist, nonbonded attractive/repulsive interaction, Coulombic interaction, out-of-plane motion, and possible cross-terms.[15] Common programs include MM2 and MMP2 from Allinger's group,[15] the BIGSTRN programs from Mislow's group,[26] and MODBLD2 from Boyd.[27] For polypeptides, the best known program is from Scheraga's group.[28]

A number of problems affect the results from such programs. The most immediate problem is that of obtaining appropriate parameters for a particular class of molecules, which requires optimizing parameters to reproduce geometries and/or energies for that class of molecules. A complete set of parameters for all types of molecule of possible interest does not appear to be available. Nor is it obvious that it will be possible to develop one homogeneous set of parameters to apply to most classes of molecules.

Another problem relates to the effect of solvation. All calculations discussed above are essentially in vacuo, isolated molecule computations. (Not quite—the parameters for force fields and even semiempirical quantum mechanical calculations are derived from experiments largely carried out in solvent—or from dimensions obtained from crystal structures!) Yet the chemical behavior that makes molecules interesting largely takes place in solvent—and chemical behavior varies with varying solvent! Although accounting for solvation is not a fully solved problem, there appear to be two major approaches. The bulk effect of solvent is handled by assigning a "dielectric constant" to the medium, effectively screening a portion of the partial charges on the component atoms. The specific, directed effects of solvent may often be handled by attaching explicit solvent molecules to strongly solvated groups, such as carbonyl oxygens. A number of groups have recently been trying to account for both classes of effects by performing Monte Carlo simulations, representing the solvent by large numbers of randomly placed solvent molecules.[29] A difficulty with the dielectric constant is that it unfairly treats charge interaction through the medium the same as charge interaction through the interior of the molecule. A number of schemes to surmount this problem allow the dielectric to vary with distance between the interacting groups, or set the dielectric constant to 1.

More complex nonbonded interactions, like charge transfer and polarizability interactions, are also difficult to accomodate in the classical mechanical framework.

## D. Dynamics Calculations

Molecules are not static, despite the appearance of mechanical models and the coordinates derived from crystal structures. Molecules vibrate, rotate, bump into other molecules, and change conformation. One result is that chemistry depends on Gibbs free energy (which includes an entropy term to account for the dynamics), while force field calculations normally report potential energy (enthalpy). Newer programs allow solving the equations of motion for the component atoms in a molecule, to give a dynamic view of molecular motions and (if sufficient time is allowed) to calculate entropy and thus free energy of the system. Unfortunately such calculations are quite expensive in terms of computer time.[30]

## E. Reaction Pathways, Transition State Calculations

Force fields are largely derived for ground state properties. Similarly, semiempirical calculations appear to be most valid for ground state structures. Yet understanding chemical mechanisms requires modeling transition states, unstable intermediates, and reaction pathways. Some work along these lines has been done, but much of it is not of high quality: for example, CNDO calculations are incapable of properly accounting for the rearrangement of electrons in making and breaking bonds (discontinuities result). One technique for forcing PRDDO and ab initio calculations along a postulated reaction pathway has been labeled the synchronous transit method.[31]

# 5. CALCULATING STRUCTURE-RELATED MOLECULAR PROPERTIES

## A. Geometric Properties

A number of properties are calculable directly from the molecular coordinates. Bond lengths, bond angles, dihedral angles, interatomic distances, and ring pucker characteristics are examples. Less straightforward measures include angle between least-squares planes through rings, distance of an atom from a least-squares ring plane, and intermolecular relations involving interatomic distances, angles, or solid angles. Other properties relatable to chemical reactivity include calculated congestion,[32] surface area,[33] molecular volume,[33] and solvent-accessible surface.[34]

## B. Charge Densities

One of the more important yet difficult properties to calculate is electrostatic behavior. The assignment of partial charges to atoms is artificial, but quite useful. In reality the variation of charge is a smooth function of charged nuclei shielded by an average field of electrons in occupied orbitals, and for reasonably small molecules the electrostatic field may be calculated this way (see next section). More typically, electrons are placed in the occupied orbitals, and electrons in the region of an atom are summed, then subtracted from the nuclear charge; electrons in localized bonds are split between the two atoms participating in the bond (Mullikan population analysis). The CNDO technique was parameterized to give partial charges that gave reasonable dipole moments for many classes of molecules. Ab initio calculations presumably give more realistic partial charges, but minimal basis set calculations appear to overemphasize the separation of charges on atoms. Extended Hückel and MNDO-type procedures were not calibrated to give reasonable partial charges and apparently are less reliable. It is possible to assign partial charges in other ways. Allinger uses bond dipoles, summing contributions on individual atoms.[15] Del Re uses a simple Hückel technique over $\sigma$ orbitals.[35] Other schemes, such as one by Gasteiger and Marsili,[36] assign charges based on molecular environment.

Once partial charges have been assigned, they may be used in a number of ways. It used to be common to perform calculations on a series of related compounds and compare the electron density at a particular atom—for example, a basic nitrogen. Although this method occasionally gave correlations with chemical behavior (basicity, reactivity), as often as not it failed. There are several reasons for this failure; beside the artificiality of the concept of partial charge, solvation may determine basicity or reactivity—charge density correlates better with gas phase than with solution basicity. Furthermore, electron density is a ground state property; if the transition state were product-like or intermediate-like, correlation with charge density would not be expected. Finally, it should be remembered than an incoming reagent senses an overall electrostatic field, not the charge on a particular atom; electrostatic maps give a better measure of such averaged fields (see next section).

Another method of calculating basicity that often is useful compares the energy of free base with the protonated form for a series of structures. Placing a standard proton on the free base without altering the geometry would give a least accurate measure of basicity; reoptimizing the geometry of the protonated structure would give a better measure of the protonation energy. Again, solvation effects would not be taken into account. Placing solvent molecules around the free base and the protonated form might account for some of the solvation energy. The reader is encouraged to read Chapters 3 and 4 in this volume concerning experimental and theoretical aspects of proton binding.

## C. Electrostatic Maps

Electrostatic maps have become increasingly common as measures of reactivity of molecules.[37] Again it should be remembered that they give a view of the initial interaction of the molecule with a reactant; as the reaction proceeds, electron transfer between the two systems will alter the electrostatic picture substantially.

## D. Frontier Orbitals

Arguments based on frontier orbitals are common for understanding chemical reactivity.[38] The same point may be made, namely, that they describe the beginning of the reaction coordinate but give no insight into the last part of the pathway. Given molecular coordinates, quantum calculations at varying levels of rigor may be carried out and useful graphic representations of the resulting orbitals created.[39]

# 6. DISPLAY AND MANIPULATION OF MOLECULAR STRUCTURES

## A. Structure Display

The major difference between current molecular modeling and the preceding generation of programs for calculating molecule structure is probably the use of interactive computer graphics for increasing the ease of use of such systems. This difference was made possible by the reduction in cost of appreciable computer power, by the appearance of inexpensive display terminals having sufficient picture resolution, by the presence of graphic languages to simplify the programming for such terminals, and by the use of graphic input devices (tablet, light pen, cursors, mouse, etc) to enhance chemist–machine communication.

Many options for molecular display exist, depending partly on the sophistication of the display device. Table 10-5 lists some display terminals in different price ranges that have been used for molecular modeling. To display molecules clearly, low-cost display terminals need a picture resolution of at least $300 \times 500$ pixels (addressable points in the picture); double that resolution is better. The stick model in Figure 10-1$a$ the ball-and-stick view of Figure 10-1$b$ are typical of molecule display on such terminals. Medium-cost terminals generally offer higher resolution, color, graphics-drawing primitives, and the ability to update part of the picture without erasing the whole screen. High-performance, three-dimensional displays (actually two-dimensional displays using various technical tricks to display three-dimensional data) generally allow dynamic updating of the picture to simulate motion, or very high resolution with multiple colors.

Graphics terminals are of several types. The storage tube display, typified by the Tektronix 4010, is relatively high resolution and gives a stable (non-flickering) picture, which is nevertheless somewhat dim; moreover, it is

TABLE 10-5. Some Graphics Display Terminals Used for Molecular Modeling

NOTE: *Products are listed for informational purposes only. Other suitable products are available in addition to those listed. Neither the authors nor VCH Publishers endorse the purchase of specific listed products, nor can they be held responsible for errors.*

| Make/model | Screen resolution | Number of colors | Input device |
|---|---|---|---|
| **Low-Cost,[a]/Low-Performance Graphics Terminals** | | | |
| VT100, Retrographics (raster) | $400 \times 600$ | monochrome | Light pen |
| Tektronix | | | |
| 4010 (storage tube) | $2000 \times 2000$ | monochrome | Cursors, tablet |
| 4105 | $480 \times 360$ | 8 | Joydisk |
| **Medium-Cost,[b]/Medium-Performance Terminals (All-Raster Displays)** | | | |
| Tektronix 4107, 4109 | $640 \times 480$ | 16 | Joydisk, tablet |
| Envision[c] | $640 \times 480$ | 16 | Mouse |
| Sigmex/Lundy | | | |
| T4688T 8/14 | $768 \times 512$ | 256 | Joystick, tablet |
| T5688S 8/14 | $768 \times 512$ | 256 | Joystick, tablet |
| AED 767 | $767 \times 575$ | 256 | Joystick, tablet |
| **High-Cost,[d]/High-Performance Displays** | | | |
| E & S PS300 (calligraphic) | $8196 \times 8196$ | >1800 | Tablet, joystick |
| Spectrographics (raster) | $1024 \times 1024$ | 256 | Tablet, joystick |

[a]Prices range between $2000 and $5000 (mid-1985).
[b]Prices range between $6500 and $12000 (mid-1985).
[c]Lear-Siegler markets this terminal presently.
[d]A basic system costs (mid-1985) about $50,000; peripherals sharply increase this amount.

impossible to erase part of a picture—the whole screen must be erased. Raster displays use essentially television technology to paint the picture line by line; the higher the picture resolution, the more memory required to describe the picture and associated colors. The vector or calligraphic displays move the electron beam from point to point rather than line by line, giving relatively fast drawing of stick pictures. Until recently, only calligraphic displays could have their views transformed quickly enough to give smooth motion of large images. Molecular display functions are typically performed using different algorithms on the different types of display terminal.

Molecules may be displayed as stick (Figure 10-1*a*), ball-and-stick (10-1*b*), or space-filling figures (Figure 10-1*c*). The views may be reoriented by global rotation about the *x*, *y*, or *z* axis. Internal (bond) rotation around a rotatable bond creates a different molecular conformation. Changing the structure should generate a view of the modified structure; superimposing matching atoms of related structures should give a good sense of the similarity of the molecules.

## B. Advanced Molecular Display

For large or complex structures of molecular surfaces, simple display is inadequate. Three-dimensional information may be imparted in several ways. Side-by-side stereo pairs of images may be viewed by a stereopticon to give a three-dimensional view. Simple fading of the bonds further away from the viewer allows stereochemistry to be perceived. Ball-and-stick or space-filling views, with hidden lines removed, show the relative closeness of atoms. For high-performance displays the molecule may be rotated or "rocked" in real time, giving the illusion of three-dimensionality by the kinetic depth effect.

For crystal-packing views or large molecules like biopolymers, selective display is necessary to pick out significant features of a very complex display. Color and "clipping" (removing portions of the picture outside a selected three-dimensional viewing window) help; so does simplifying the picture—for example, by leaving off hydrogen atoms, showing only the $\alpha$-carbon trace of a polypeptide, or "cartooning" such features as $\alpha$ helix and $\beta$ sheets.

## C. Animation of Reaction Pathways

An increasingly prevalent technique shows movement of molecules or mechanisms by simulation. For example, motion picture films of vibrating structures give a feeling for the dynamic nature of molecules (as opposed to the static structures viewed from crystal structure determinations), simulations of the $S_N2$ reaction give a feeling for the Walden inversion, and simulations of enzyme cleavage of substrate give insights into enzyme mechanism. Some computer graphics languages have become available to expedite such animations. One with which we are familiar is the Graphics for the Multi Picture System (GRAMPS) package of Olson and O'Donnel.[40]

# 7. SOME SYSTEMS FOR MOLECULAR MODELING

## A. Introduction

Until recently, the cost of entry into the molecular modeling field was developing one's own programs, or obtaining academically developed programs as part of a research collaboration. This began to change when a number of molecular modeling packages were made available without fee from academic sources (Prof. Clark Still, Columbia University: MODEL; Prof. Herschel Weintraub, formerly Purdue University: CAMSEQ-M; Prof. Anton J. Hopfinger, formerly Case Western Reserve University: CHEMLAB) and from industry (Dr. David Pensak, DuPont De Nemours & Co.: TRIBBLE). Programs with some modeling capabilities, but not complete molecular modeling systems, are available from Quantum Chemistry Program Exchange.[40]

More recently, commercial systems with increased functionality have become available; many of these are briefly reviewed later in this section.

## B. Hardware and Software Considerations

All the popular systems are interactive, graphics-based programs, partly because molecular modeling is complex enough to make the immediate feedback of real-time processing an advantage, and partly because the chemist's ability to perceive three-dimensional information is vastly improved by such systems. Although early systems were developed for typical laboratory computers (PDP-11's, Modcomp), the introduction of the VAX computers in many chemistry departments resulted in sufficient local computer power to handle most molecular modeling functions. Because of the strong scientific computation component of modeling, many of the available systems are written in FORTRAN, although programs in Unix/C, Pascal, and even BASIC are available. Early systems tended to be written for particular graphics hardware (eg, DEC GT40, Tektronix 4010 terminals); but emergence of Tektronix PLOT-10 graphics software as an industry standard allowed programs to become somewhat independent of dramatic improvements occurring in the performance of graphics terminals.

In general, molecular modeling studies may expand to utilize all computer resources available. For small molecules, computational studies range from empirical and semiempirical calculations on small computers, up to very large extended basis set ab initio computations with geometry optimization on large computers. For macromolecules, simple selective display is

**CAMSEQ-M**

| | |
|---|---|
| Author: | Herschel Weintraub, Abbott Research Laboratories, North Chicago, Ill. |
| Distributor: | Author |
| Capabilities: | Stand-alone molecule creation/display |
| Hardware: | Tektronix microprocessor |

**CHEMGRAF**

| | |
|---|---|
| Author: | Keith Davies, Chemical Design Ltd. |
| Distributor: | Chemical Design Ltd., Unit 12, 7 West Way, Oxford, England OX2OJB20NN |
| Capabilities: | Small-molecule modeling, macromolecular display/manipulation, conformational energy calculations, molecule creation/modification, surface creation/display/manipulation, enzyme/ligand interaction, etc |
| Hardware: | VAX mainframe; Lundy/Sigma, Spectragraphics, Envison, VT100 retrographics, AED, Tektronix, and other terminals |

**CHEMLAB**

| | |
|---|---|
| Author: | Anton J. Hopfinger, University of Illinois, Chicago Circle, Chicago, Ill. |
| Distributor: | Molecular Design Ltd., San Leandro, Calif. (*Note:* MDL also distributes DRAWMOL and PROXBUILDER software for modeling.) |
| Capabilities: | Small-molecule modeling, conformational energy calculations, molecule creation/modification, quantitative structure–activity relationships (QSAR), etc |
| Hardware: | VAX mainframe: Tektronix, VT100 retrographics terminals, Envision, and other terminals. |

**MOLEDIT**

| | |
|---|---|
| Author: | Eugene Fluder, Graham M. Smith, Robert B. Nachbar, Joseph D. Andose, Thomas A. Halgren, R. Blevins, J. Shpungin, Peter Gund, and co-workers, Merck Sharp & Dohme Research Laboratories, Rahway, N.J. |
| Distributor: | Not distributed |
| Capabilities: | Molecule/macromolecule display/manipulation, surface display/manipulation, structure modification, enzyme/ligand interaction, conformation searching, etc |
| Hardware: | VAX mainframe, energy calculations on IBM mainframe; E & S MPS, PS300 graphics processors, VT100 retrofit, Envision, Tektronix, etc, terminals (DI3000 device independent graphics library); IBM personal computers emulating graphics terminals |

**SYBYL**

| | |
|---|---|
| Author: | Garland Marshall and collaborators, Washington University and Tripos Associates, St. Louis |

TABLE **10-6.** (continued)

| | |
|---|---|
| Distributor: | Tripos Associates, 6500 Clayton Rd., St. Louis, MO 63117 |
| Capabilities: | Small-molecule creation/modification/display/manipulation/comparison, systematic conformation search/energy calculation, surface creation/ manipulation/display, receptor site fitting, pharmacophore identification and optimization |
| Hardware: | VAX mainframe; E & S MPS and PS300 graphics processors, Tektronix, Envision, VT100 retrofit, etc, terminals, NEC APC work station |

TRIBBLE

| | |
|---|---|
| Author: | David M. Pensak and various collaborators |
| Distributor: | E.I. DuPont de Nemours Inc., Wilmington |
| Capabilities: | Molecule creation/manipulation/display, energy calculations |
| Hardware: | PDP-10, VAX mainframes; DEC GT-40, Envision, VT100 retrographics, Tektronix, AED, and other terminals |

possible on small systems, but energy calculations normally take very large computers—even supercomputers and/or array processors.

## C. Some Small-Molecule Modeling Systems

Some systems for modeling small molecules and their capabilities are listed in Table 10-6. Since such systems are constantly developing, this list was probably out of date at its time of compilation (April 1985). The list is not complete. The authors apologize for any errors or omissions.

## D. Large-Molecule Display Systems

Systems oriented toward large molecules tended to arise from the environment of protein crystallography; usually they treat macromolecules as collections of residues (amino acid, nucleic acid, saccharide fragments). Some of them have difficulty handling atom- and bond-centered information. Some current systems are collected in Table 10-7.

# 8. CONCLUSIONS

Just as mechanical molecular models are available now from a number of sources, computer-assisted molecular modeling systems are becoming readily available and easy to use. They promise to give a more quantitative feeling of the relationship between three-dimensional structure and physical/ chemical/biological properties. While there is a tendency to become

**TABLE 10-7.** Some Macromolecular Modeling Systems (Listed Alphabetically)
*NOTE:    Products are listed for informational purposes only. Other suitable products are*
*available in addition to those listed. Neither the authors nor VCH Publishers*
*endorses the purchase of specific listed products, nor can they be held responsible for*
*errors.*

BIOGRAPH

| | |
|---|---|
| Authors: | B. Olafson and W. Goddard, University of California at Berkeley and Biological Design Inc., Berkeley |
| Distributors: | Biological Design Inc., Berkeley |
| Capabilities: | Protein-selective display/manipulation, future interface to energy calculations |
| Hardware: | VAX mainframe, PS300 terminal. |

FRODO

| | |
|---|---|
| Author: | T. Alwyn Jones, University of Uppsala, Sweden, and various collaborators |
| Distributor: | Author. PS300 version: Fiorante Quiocho, Department of Biochemistry, Rice University, P.O. Box 1892, Houston, TX 77251 |
| Capabilities: | Protein electron density display, polypeptide model fitting to density, structure manipulation/selectice display, sequence modification, enzyme–ligand "docking," interface to structure refinement and energy calculations, surface creation/display/manipulation |
| Hardware: | PDP-11, VAX, IBM mainframes; Vector General, E & S MPS and PS300 graphics processors |

MMS-X

| | |
|---|---|
| Authors: | G. Marshall, D. Barry, and co-workers, Washington University, St. Louis |
| Distributors: | No longer distributed |
| Capabilities: | Protein electron density display, polypeptide model fitting to density, structure manipulation/selective display, surface creation/display/ manipulation. |
| Hardware: | Custom-designed graphics display, TI microcomputer. |

MIDAS

| | |
|---|---|
| Authors: | R. Langridge, T. Ferrin, and others, Department of Pharmaceutical Chemistry, University of California, San Francisco |
| Distributor: | R. Langridge |
| Capabilities: | Macromolecule display and manipulation, molecular surface display, drug–enzyme docking, etc |
| Hardware: | E & S PS-2, PS330, and other terminals; PDP-11 and VAX mainframes |

MOGLI

| | |
|---|---|
| Author: | Andrew Dearing, Sittingbourne Research Center Shell Research Ltd., Sittingbourne, England ME98AG |

TABLE 10-7. (continued)

| | |
|---|---|
| Distributor: | Evans & Sutherland Computer Corporation, P.O. Box 8700, Salt Lake City, UT 84108 |
| Capabilities: | Molecule/macromolecule display/manipulation, surface display/ manipulation, structure modification |
| Hardware: | VAX mainframe; E & S PS300 display terminals |

entranced with the beautiful pictures and models created with modern molecular modeling systems, one must keep in mind that these too are models of reality, no more useful than the insights they provide to guide further experiments. But they do indeed provide such insights, and they will become even more useful as the methodology continues to develop.

# REFERENCES

1. Westheimer, F.H. *J. Chem. Phys.* **1947,** *15,* 252.
2. Hendrickson, J.B. *J. Am. Chem. Soc.* **1961,** *83,* 4537.
3. Wiberg, K.B. *J. Am. Chem. Soc.,* **1965,** *87,* 1070.
4. Levinthal, C. *Sci. Am.* **1966,** *214* (6), 42.
5. Scheraga, H.A. *Adv. Phys. Org. Chem.* **1968,** *6* 103.
6. For example, the Molecular Animator, an Apple IIe program by J.J. Howbert, from COM-Press, Van Nostrand Reinhold Co., available from the American Chemical Society, Education Division, 1155 Sixteenth St., Washington, D.C. 20036.
7. Clarke, F.H. "Calculator Programming for Chemistry and the Life Sciences". Academic Press: New York, 1981.
8. Kennard, O.; Watson, D.; Allen, F.; Motherwell, W.; Town, W.; Rodgers, J. *Chem. Brit.* **1975,** *11,* 213.
9. Allman, R. In "Chemistry of the Hydrazo, Azo and Azoxy Groups", Patai, S., Ed.; Wiley: London, 1975, p. 23.
10. Dunitz, J.D. "X-Ray Analysis and the Structure of Organic Molecules". Cornell University Press: Ithaca, N.Y., 1979.
11. Computer program supplied by Molecular Design Ltd., San Leandro, Calif.
12. Approach used in PROPHET, SYBIL, and ChemGraf programs, for example.
13. Cohen, N.C.; Colin, P.; Lemoine, G. *Tetrahedron* **1981,** *37,* 1711.
14. Gund, P. In "X-Ray Crystallography and Drug Design", Horn, A.S.; and DeRanter, C.J., Eds.; Clarendon Press: Oxford, 1984, p. 495.
15. Burkert, U.; Allinger, N.L. "Molecular Mechanics". American Chemical Society, Monograph No. 177. ACS: Washington, D.C., 1982.
16. Still, W.C., Galynker, I. *Tetrahedron* **1981,** *37,* 3981.
17. Hückel, E. *Z. Phys.* **1931,** *70,* 204.
18. Hoffmann, R. *J. Chem. Phys.* **1963,** *39,* 1397.
19. Pople, J.A.; Beveridge, D.L. "Approximate Molecular Orbital Theory". McGraw-Hill: New York, 1971.
20. Fernandez-Alonso, J.I. In "Quantum Mechanics of Molecular Conformations", Pullman, B., Ed.; Wiley: London, 1976, p. 117.
21. (a) Bingham, R.C.; Dewar, M.J.S.; Lo, D.H. *J. Am. Chem. Soc.* **1972,** *97,* 1285. (b) Dewar, M.J.S.; Ford, G.P. ibid, **1979,** *101,* 5558.
22. (a) Pople, J.A. *J. Am. Chem. Soc.* **1975,** *97,* 5306. (b) Bielinski, T.J.; Breen, D.L.; Rein, R. ibid, **1978,** *100,* 6266. (c) Klopman, G.; Andreozzi, P.; Hopfinger, A.J.; Kilucki, O.; Dewar, M.J.S. ibid, *100,* 6267.

23. Malrieu, J.P.; Claverie, P.; Diner, S.; Gilbert, M. *Theor. Chim. Acta* **1969**, *15*, 100.
24. Halgren, T.A.; Kleier, D.A.; Hall, J.H., Jr.; Brown, L.D.; Lipscomb, W.N. *J. Am. Chem. Soc.* **1978**, *100*, 6595.
25. GAUSSIAN82 is available from Prof. J.A. Pople, Department of Chemistry, Carnegie–Mellon University, 4400 Fifth Ave., Pittsburgh, PA 15213.
26. BIGSTRN-3: Nachbar, R.N.; Mislow, K. Submitted to Quantum Chemistry Program Exchange, Indiana University, Bloomington.
27. Boyd, R.H.; Sanwal, S.N.; Sharz-Tehrany, S.; McNally, P. *J. Phys. Chem.* **1971**, *75*, 1264.
28. ECEPP: Browman, M.J.; Carruthers, L.M.; Kashuba, K.L.; Momany, F.A.; Pottle, M.S.; Rosen, S.F.; and Rumsey, S.M.; Endres, G.S.; Scheraga, H.A. Quantum Chemistry Program Exchange Program No. 286, Indiana University, Bloomington.
29. For example: Hagler, A.T.; Osguthorpe, D.J.; Robson, B. *Science,* **1980**, *208*, 599.
30. Hagler, A.T.; Osguthorpe, D.J.; Dauber-Osguthorpe, P.; Hempel, J.C. *Science,* **1985**, *227*, 1309.
31. Halgren, T.A.; Lipscomb, W.N. *Chem. Phys. Lett.* **1977**, *49*, 225.
32. Wipke, W.T.; Gund, P. *J. Am. Chem. Soc.* **1976**, *98*, 8107.
33. Pearlman, R.S. In "Physical Chemical Properties of Drugs", Yalkowsky, S.H.; Sinkula, A.A.; and Valvani, S.C., Eds.; Dekker: New York, 1980.
34. Connoly, M.J. *Science,* **1983**, *221*, 709.
35. Del Re, G. *J. Chem. Soc.* **1958**, 4031.
36. Gasteiger, J., Marsili, M. *Tetrahedron* **1980**, *36*, 3219.
37. Politzer, P.; Truhlar, D.G., Eds. "Chemical Applications of Atomic and Molecular Electrostatic Potentials". Plenum Press: New York, 1981.
38. Fleming, I. "Frontier Orbitals and Organic Chemical Reactions". Wiley-Interscience: London, 1976.
39. Quarendon, P.; Naylor, C.B.; Richards, W.G. *J. Mo. Graphics* **1984**, *2*, 4.
40. Available from Quantum Chemistry Program Exchange, Indiana University, Bloomington.

CHAPTER 11

# Structure–Biological Activity Relationships

## Corwin Hansch

### Pomona College, Claremont, California

## CONTENTS

## 1. INTRODUCTION

The computer-based correlation of chemical structure with biological activity has become such a large and diversified endeavor that it is almost impossible for one person to properly review the field. This chapter is limited to a consideration of bridging the gap from the well-established reactions of physical-organic chemistry to those of biochemistry and whole animals using the linear combination of physiochemical parameters.

The interaction of ligands with enzymes is our main focus of interest because an understanding of this process will provide the techniques for understanding the reactions of small organic compounds with biomacromolecules of all kinds.

MOLECULAR STRUCTURE AND ENERGETICS, Vol. 4

For more general reviews the reader is referred to References 1–5. The relationship between chemical structure and reactivity, both chemical and biological, has been of major concern since the very beginnings of science. The Roman poet Lucretius (99–55 B.C.) in his epic poem, "On the Nature of Things," speculated that it is the shapes of atomic particles impinging on our senses that give rise to the variety of our sensations. Modern chemical structure–activity work can be dated from 1869 with publication of Mendeleev's periodic table and, a century later, Crum-Brown and Frazer's postulate[6] that:

$$\phi = f(C) \qquad (11\text{-}1)$$

where $\phi$ is a measure of the biological activity of a compound and $C$ characterizes its chemical structure. Crum-Brown and Frazer[6] envisioned the development of a calculus of biological structure–activity relationships[7] (SARs) arising from the study of the effect of small changes in $C$ on $\phi$.

Although there was great interest in biological SAR in the last third of the nineteenth century,[7] the first important advance was made by two independent researchers, Meyer and Overton, in about 1900. They observed that the narcotic potency of simple neutral organic compounds (alcohols, ketones, esters, etc.) paralleled their oil/water partition coefficients. However, their work was not cast in mathematical terms until the 1960s. Equation 11-2, based on Overton's data for the narcosis of tadpoles, was formulated many years after Overton's publication.[7]

$$\log \frac{1}{C} = 0.94 \log P + 0.87$$
$$n = 51, \qquad r = 0.971, \qquad s = 0.280 \qquad (11\text{-}2)$$

In Equation 11-2, $C$ is the molar concentration of chemical that stopped the movement of tadpoles and $P$ is the octanol/water partition coefficient of the substance. The number of chemicals studied is represented by $n$; $r$ is the correlation coefficient; and $s$ is the standard deviation from the regression. Generalized approaches to biological SAR stalled after Meyer and Overton's seminal discovery because the partitioning effect was only one of a number of factors determining biological activity.

Early in the 1900s the so-called English school of organic reaction mechanisms began to grow. Out of these studies rather clear qualitative ideas about electronic effects of substituents on organic reactions began to emerge. Ingold, who played a major role in the development of these ideas, has reviewed this train of thought.[8]

While these qualitative ideas about inductive and resonance effects and polarizability could be used to rationalize all sorts of after-the-fact observations about organic reactions, their predictive value was low because one had no way of knowing the relative importance of the several general effects. The major breakthrough in this empirical approach to understanding the effect of structural changes on reaction rates and equilibria was made in about 1935 by L.P. Hammett,[9] who used the ionization constants from the model system of the ionization of benzoic acids in water to devise a numerical scale $\sigma$ for the electronic effect of substituents on a reaction center. In the subsequent half-century, thousands of correlation equations defining the effects of structural changes of all kinds on almost every variety of organic reaction were developed using Hammett constants.[10,11]

The next major advance in generalizations was that of Robert Taft,[12] who defined the steric parameter $E_s$:

$$E_s = \log k_{X-CH_2COOEt} - \log k_{H-CH_2COOEt} \qquad (11\text{-}3)$$

where $k$ is the rate of acid hydrolysis of esters of acetic acid. In this definition it is assumed, with good reason, that X has no electronic effect on $k$ and thus it is the steric properties of X that influence the value of $k$. Taft went on to show that a linear combination of steric and electronic parameters could be used to separate and delineate the role of the two independent factors on the reactions of organic compounds.

The stage was now set to bring Meyer and Overton's contributions to the solution of biological SAR.[13–16] A hydrophobic parameter $\pi$ was defined analogously to $\sigma$:[13]

$$\pi = \log P_x - \log P_H \qquad (11\text{-}4)$$

where $P_H$ is the octanol/water partition coefficient of a parent compound (say benzene) and $P_x$ is that of a derivative (eg, chlorobenzene); hence $\pi_{Cl} = \log P_{C_6H_5Cl} - \log P_{C_6H_6} = 2.84 - 2.13 = 0.71$. Positive values of $\pi$ mean that a substituent increases the affinity of a compound for a fatty phase. As the Hammett type of thinking has spread to systems of incredible complexity—as, for example, 600 antimalarial drugs curing mice of an infection—any kind of rigorous theory is exceedingly difficult to support. The expression "correlation analysis" is becoming generally used.[11] One frankly builds empirical mathematical models, by the linear combination of what seem to be appropriate independent variables, in the hope that eventually sound theory will follow to justify the models. A good discussion of this approach to problem solving has been outlined by Daniel and Wood.[17] Attempts to develop some theoretical underpinnings for biological QSAR can be found in References 1 and 15–18.

Although the most fundamental approach to the understanding of chemical reactions, quantum chemistry, began its development in the 1920s and its very rapid expansion with the postwar use of high-speed computers, advances in biochemical SAR continue to depend heavily on the basic ideas of Hammett and Taft of using numerical constants from model systems. In the past two decades what has now come to be called QSAR (quantitative structure–activity relationships) has been applied to an astonishing number of problems where organic chemicals interact with living systems or parts thereof. Almost every type of drug SAR has been treated in terms of QSAR,[18,19] and similar problems in the design and mechanism of action of herbicides,[20-26] insecticides,[27-34] and fungicides[35-37] have also been treated via QSAR. Environmental toxicology is a field in which QSAR finds application in the study of bioconcentration[38,39] and the toxicity of chemicals.[40-45] An article by Geyer and co-workers[42] contains many references to the use of QSAR on environmental problems. In agriculture the binding of chemicals to soils, hence their movement through soils, is being investigated using QSAR.[46] Starting from an initial demonstration that hydrophobicity is an important factor in the sweet taste of chemicals,[47] QSAR has provided insight into the SARs of taste and olefaction.[47-54] Drug metabolism and distribution[55,56] and anesthesiology[57] are fields for QSAR application. In forensic toxicology QSAR finds use in rationalizing the relative lethality of certain classes of drugs.[58,59] Countless studies have been made on the anti-bacterial action[60-71] of compounds of various types. The interaction of chemicals with various subcellular fragments has also been analyzed,[72,73] One of the most widely studied areas is that of enzyme–ligand interactions.[74-83] The first clue[84] that QSAR might be used to correlate protein binding started many studies in this field.[85-93] DNA binding[94] has also been studied, as has the interaction of organic chemicals with antibodies.[95]

## 2. SUBSTITUENT CONSTANTS

### a. Electronic Parameters

Although one could use quantum chemical indices or chemical shift from NMR to account for the electronic effects of structural change on biological activity of the members of a set of congeners, the vast majority of the published work has been based on $\sigma$ constants or $pK_a$ values. A large compilation of substituent constants has been published,[96] and the number of constants continues to grow rapidly. A current data bank contains constants ($\sigma$, $\pi$, $E_s$, etc) for about 3000 different substituents.[97] Sigma constants can be used directly in deriving correlation equations, or they might be used to calculate $pK_a$ values[98] for correlation purposes.

During the first two decades of work with $\sigma$ constants, four types of parameters evolved[96]: $\sigma_{BA}$, $\sigma°$, $\sigma^-$, and $\sigma^+$. The $\sigma_{BA}$ constants, which are the most widely used, have been based on the ionization constants of benzoic acids. However, since there is resonance interaction in benzoic acids with strong electron-releasing groups (eg, OH, $NH_2$, $OCH_3$) in the position para to the carboxyl group, the parameter $\sigma°$ was derived from the ionization constants of phenylacetic acids (where COOH is insulated from the benzene ring by a $CH_2$) to obtain a set of parameters free of thorough resonance. The $\sigma^-$ and $\sigma^+$ constants are employed whenever there is through resonance between substituents and reaction centers in which negative or positive charges are delocalized.[96] In the past two decades efforts have been made to factor $\sigma$ into inductive $\sigma_I$ and resonance $\sigma_R$ components and then to use two parameters to account for the electronic effects of substituents.[99–104] For example, Equations 11-5 and 11-6 show how the linear combination of inductive $F$ and resonance $R$ parameters for meta- and parasubstituted benzoic acids can be used to correlate ionization constants of a set of benzoic acids:

$$pK_a = -1.64\sigma_I^m - 0.34\sigma_R^m - 1.76\sigma_I^p - 1.15\sigma_R^p + 5.76 \qquad (11\text{-}5)$$
$$n = 27, \qquad r = 0.992, \qquad s = 0.083$$

$$pK_a = -1.60F_m - 0.34R_m - 1.57F_p - 1.22R_p + 5.75 \qquad (11\text{-}6)$$
$$n = 27, \qquad r = 0.992, \qquad s = 0.082$$

Equation 11-5 is based on Taft's substituent constants and Equation 11-6 is based on Swain and Lupton's constants.[102,104] The small coefficients with $\sigma_R^m$ and $R_m$ show that resonance effects from metasubstituents are unimportant compared to resonant effects from the para position. The quality of fit is the same for the two equations. There has been much criticism[101] of Swain and Lupton's parameters because the reference point for their system is the substituent $^+N(CH_3)_3$; for which it was assumed there is zero resonance effect.[104] However, this assumption does not affect the relative values of $F$ and $R$, and one can see that essentially the same correlation is obtained using Swain and Lupton's constants as using Taft's values. It must be noted that $F$ and $R$ used in the derivation of Equation 11-6 are not those originally published by Swain and Lupton, but are values that have been more appropriately scaled.[102] Many more values are available from the Swain and Lupton system than from Taft's.[97] Recently, Charton[103] has devised a new set of factored $\sigma$ values: $\sigma_L$ and $\sigma_D$. The $L$ connotates localized electronic effect (inductive and field effects) and $D$ refers to delocalized electronic effects (reasonance). Charton has extensively and critically reviewed the problems associated with the various efforts that have been made to separate resonance and inductive effects.[103]

At the time they published their new parameters, Swain and Lupton thought that a single resonance parameter would suffice to deal with all resonance effects.[104] There is no convincing evidence to support this contention, and thus the best present evidence[105] suggests that four resonance parameters are necessary to account for the various kinds of through resonance. One way of obtaining these is by subtracting the inductive effect from parasubstituent constants as follows:[96]

$$R_{BA} = \sigma_p^{BA} - F \qquad R^- = \sigma_p^- - F$$
$$R^+ = \sigma_p^+ - F \qquad R^0 = \sigma_p^0 - F$$

Since we are primarily concerned with biological SAR, a problem of paramount importance involves the suitability of substituent constants derived from ionization constants of organic compounds in aqueous solution for use in the biochemical systems of "blood and guts." Reactions in biochemical systems could be occurring in phases much less polar than pure water. Equation 11-7, derived from results of Kolthoff and Chantooni, correlates the ionization of benzoic acids in 100% DMSO:[106]

$$pK_a = -1.56(\pm0.36)\sigma + 10.58(\pm0.38)$$ (11-7)
$$n = 17, \quad r = 0.922, \quad s = 0.548$$

This correlation accounts for only 85% of the variance in $pK_a$, and the standard deviation is much larger than one normally expects with the Hammett equation. Equation 11-7 is a rather drastic test of the limits of the Hammett equation, since not only is water absent but the solvent, being aprotic, differs considerably in character from water. The substituents used in this study were predominantly of the type that are "well behaved" (halogen, alkyl, $NO_2$) in Hammett equations rather than those that tend to react with water via hydrogen bonding ($NH_2$, OH, $SO_2NH_2$). Using a wider selection of substituents would very likely reduce the quality of fit.

As poor as Equation 11-7 is from the point of physical-organic chemistry, the quality of correlation is still high enough to be of great help in biochemical problems.

However, a rather good correlation has been obtained when nonpolar benzene is used as the solvent:[107]

$$\log k = 1.80(\pm0.18)\sigma + 5.38(\pm0.07)$$ (11-8)
$$n = 22, \quad r = 0.977, \quad s = 0.148$$

To obtain dissociation constants in benzene, it was necessary to use 1,3-diphenylguanidine as a proton acceptor. There is fair variation among the substituents ($NH_2$, halogen, CN, $OCH_3$, $NO_2$, $N(CH_3)_2$, OH, $CH_3$) used for Equation 11-8. This is probably the worst possible condition for the Hammett equation, and it suggests that if substituents were bound in a truly hydrocarbonlike portion of an enzyme, the Hammett equation might be expected to hold.

Johnson[108] has pointed out that $\sigma$ constants are not highly solvent dependent and in fact show that even in vapor phase reactions, in the absence of solvent, good correlations can be obtained. However, his examples are based on studies with the "well-behaved" substituents. In the example of the vapor phase decomposition of ethylphenyl carbonates, the single strong hydrogen bonding substituent, 4-$NH_2$, is poorly fit.[108]

If one uses solvents with some water, good correlations can be obtained from the ionization of a very wide selection of substituted benzoic acids in, for example, 80% methylcellosolve–20% water or 50–50 ethanol water.[109] This study also showed that there is good additivity of $\sigma$ for 3,4- and 3,5-disubstituted benzoic acids.

Since we do not know how nonpolar the environment is in general for biochemical systems, we have little assurance from studies in nonaqueous solvents about how well behaved $\sigma$ constants will be for such systems.

Except for the study by Kalfus and co-workers,[109] there is little work in which the wide variety of substituents used by medicinal chemists in drug design have been carefully studied in nonaqueous solvents.

Physical organic chemists have in general used small sets of the well-behaved substituents in their studies. A recent paper by Johnels and associates shows that 28 commonly used substituents fall into four rather narrowly spaced groups.[110] Hence if one is not careful, all of a set of substituents could easily be selected from one group and would not be truly representative of the universe of substituents. In fact, it is only recently that researchers have begun to study the properties of substituents with a view to avoiding the collinearity problem.[1,111,112]

This problem arises when a set of substituents is selected so that one or more properties are highly correlated. For example, in the case of the following set of substituents: $CH_3$, $NO_2$, $COCH_3$, $C{\equiv}CH$, $SCH_3$, $NHC_6H_5$, $OCH_2CH_2CH_3$, $SO_2CH_2CH_2CH_3$, I, and $CH_2Cl$, there is extremely high correlation between $\pi$ and the resonance parameter $R (r^2 = 0.94)$. Hence, any data that showed a dependence on $\pi$ would show essentially the same dependence on $R$, and the result would be completely ambivalent.

Thus it seems that the only thing to do is to test the utility of $\sigma$ constants on a wide variety of biochemical problems. During the past two decades, many Hammett equations have been derived for sets of substrates or inhibitors reacting with enzymes. The object of most of these studies has been to find the value of $\rho$ from which to make inferences about the reaction mechanisms. The majority of these studies are seriously flawed in that usually six to eight (or fewer) substituents were studied, and it has been assumed that one can neglect all other substituent effects, such as steric or hydrophobic. Unless one can establish by good statistical procedures that such effects are absent, however, they cannot be neglected. In a number of instances multiple-parameter approaches to enzyme–ligand interactions have been made using a reasonable number and variety of substituents.[74–83,113–118] These examples offer convincing evidence that when steric and hydrophobic effects of substituents can be satisfactorily accounted for, $\sigma$ constants can be shown to play an important independent role in rationalizing enzymic processes. These results are not too surprising, since substituents causing electronic effects may not even be in contact with the enzyme itself. Even if the substituents are in contact, enzymes are rather polar. In fact, it is difficult to imagine placing substituents on an enzyme surface in such a way that they would not be near highly polar amide linkages.

The next step up in complexity in biochemical correlations involves living cells or organisms, and one important question has received very little study, namely: Will a QSAR obtained on isolated bioreceptors (eg, enzymes) prevail in the living cell or organism?

The first workers to investigate this question were Fukuto and Metcalf, who were well ahead of their time. Equation 11-9 correlates their data for the inhibition of fly head cholinesterase by a set of phenyl phosphates X—

$C_6H_4OPO(OC_2H_5)_2$, while Equation 11-10 correlates $LD_{50}$ concentrations of the same class of compounds acting on houseflies.[116]

$$\log \frac{1}{C} = 2.59\sigma^- + 4.29$$

$$n = 12, \qquad r = 0.906, \qquad s = 0.620 \tag{11-9}$$

$$\log \frac{1}{C} = 2.42\sigma^- + 0.26\pi - 0.60$$

$$n = 8, \qquad r = 0.987, \qquad s = 0.228 \tag{11-10}$$

In these equations $C$ is the molar concentration causing 50% enzyme inhibition or killing 50% of the flies. In the living insect, substituents show a small hydrophobic effect that is accounted for by the $\pi$ term in Equation 11-10. Although the SAR of the phosphates is more complex than the equations above imply,[117] the electronic effect of the substituents is quite similar in vitro and in vivo.

Another example showing that the Hammett equation holds in living systems comes from studies on alcohol dehydrogenase.[118] Equation 11-11 correlates the inhibition of purified rat liver alcohol dehydrogenase by 4-substituted pyrazoles, and Equation 11-12 correlates inhibition of the enzyme in living rat liver cells.

$$\log \frac{1}{K_i} = 1.22 \log P - 1.80\sigma_m + 4.87$$

$$n = 14, \qquad r = 0.985, \qquad s = 0.316 \tag{11-11}$$

$$\log \frac{1}{K_i} = 1.27 \log P - 0.20(\log P)^2 - 1.80\sigma_m + 4.75$$

$$n = 14, \qquad r = 0.971, \qquad s = 0.320 \tag{11-12}$$

In these equations, $P$ is the octanol/water partition coefficient of the pyrazoles and $\sigma_m$ is the Hammett constant for meta substituents. There is evidence that a pyrazole nitrogen binds with a positively charged zinc atom in the enzyme; hence one expects and finds a negative coefficient with $\sigma$, indicating that electron-releasing substituents increase inhibition (ie, binding). Since there is a wide range in $\log P$, Equation 11-12 shows a nonlinear dependence on $\log P$ often found in living systems.[15,116,119] However, when proper corrections for the hydrophobic effects of substituents are made in each equation, the electronic effects are, as one would hope, identical. Another example of the applicability of the Hammett equation in vivo involving the enzyme dihydrofolate reductase is discussed below.

## B. Hydrophobic Parameters

Medicinal chemists have been concerned with the lipophilicity of drugs since the work of Meyer and Overton at the turn of the century. Biochemists and chemists in general were slow to appreciate the very great importance of the hydrophobic "bond" until Kauzmann's[120] discussion of its importance in biochemical processes. Starting with the classic work of Frank and Evans,[121] Kauzmann coined the term "hydrophobic" and pointed out certain consequences of hydrophobic interactions for protein chemistry. It was assumed that when an apolar molecule (eg, hydrocarbon) is placed in water, it is solvated by a loosely held layer of water molecules. When molecules partition into a hydrophobic phase, the loosely held water is removed. It was shown that the free energy change (primarily entropic) in this desolvation process is the driving force for partitioning out of the aqueous phase into the hydrophobic phase. The role of the hydrophobic solvent was thought to be minimal. This seemed reasonable when one was discussing the partitioning of a substance like argon or methane into a hydrocarbon phase. However, such simple systems are not of much interest to researchers in biochemical QSAR. Proteins and DNA are highly polar substances. Even hydrophobic regions are closely associated with the polar amide backbone. Attempts to understand the nature of the hydrophobic effect[122–136] validate the assumption that it is not the water phase alone that determines hydrophobicity.[130–135] Indeed, except for academic work on inert compounds partitioning into inert solvents, the term "hydrophobicity" is rather meaningless. In the real world of biochemistry, it is solvents like octanol and butanol that do the best job of modeling partitioning to proteins or nucleic acids. So many interactions are involved to various degrees (hydrogen bonding, dipole–dipole, steric effects, etc, not to mention "hydrophobic" desolvation) that it is difficult to make any rigorous assault on the mechanisms of hydrophobic binding of drugs and proteins. But, using the extrathermodynamic approach with octanol/water as the reference system, one can gain considerable insight into this complex process. Octanol (or a similar solvent) with its dissolved water is a complex reference system that seems to be necessary to rationalize the binding of complex organic compounds to proteins whose structures are in many instances unknown. The fact that such models work suggests that nature utilizes partitioning in a similar way to build cellular units and to sequester critical chemicals for biochemical processes.

Although there were many attempts to use oil/water partition coefficients ($P$) as operational means for defining lipophilicity and its role in the movement of organic compounds through membranes[122–125] and to rationalize degrees of drug potency,[126] success was severely limited because a single parameter will not go far in rationalizing such complex processes. Thus neither the early work of Meyer and Overton nor Kauzmann's review pointing out some of the general principles of the concept of hydrophobicity was enough to sustain much interest in the subject of partition coefficients. The importance of the partition coefficient was so little appreciated that no general review of it was published until 1971.[126] Today there is enormous theoretical[127] as well as practical interest in the hydrophobic effect.

Collander,[122] in developing Meyer and Overton's use of partition coefficients as a model system to help understand the movement of organic compounds in and out of plant cells, realized that Equation 11-13 would have to hold for such a project to succeed:

$$\log P_2 = a \log P_1 + b \tag{11-13}$$

where $P_1$ is the partition coefficient of a reference solvent system (eg, octanol/water) and $P_2$ is a biochemical partitioning process or a second solvent/water system. The Collander Equation does hold for sets of partitioning solutes, provided the two partitioning systems are not too different.[126] Hence the secret for success in selecting a suitable solvent system is to pick one that is similar in properties to the average composition of biological material. Although a variety of solvents have been used to represent the nonaqueous phase since the Meyer–Overton suggestion of olive oil, only the octanol/water system has received truly widespread use. While it does have certain advantageous characteristics,[137] whether it is the "best" solvent remains an open question.

Possibly the best simple test of Equation 11-13 is the study of the binding of organic compounds to proteins. Equation 11-14 correlates the 1:1 binding of a miscellaneous group of organic compounds to bovine serum albumin[85]:

$$\log \frac{1}{C} = 0.75 \log P + 2.30$$
$$n = 42, \qquad r = 0.960, \qquad s = 0.159 \tag{11-14}$$

Unless otherwise noted, $P$ represents the octanol/water system and $C$ the molar concentration of ligand necessary to produce a 1:1 molar complex of ligand–protein. Since the slope of Equation 11-14 is not 1, the average character of the reference solvent (octanol saturated with water) is not exactly like that of the protein. Equation 11-14 is based on a set of 42 simple neutral compounds, including 15 phenols, 6 anilines, nitrobenzene, indole, hydroxyadamantane, naphthalene, azobenzene, 2,4-dichloronitrobenzene, and neopentanol. Equation 11-14 has been reexamined for the possibility of substituting connectivity constants for log $P$. Apparently the correlation between log $P$ and the graphic parameters is much lower than was presumed, because connectivity constants fail to correlate with binding.[139] In a more extensive study, Scholtan[138] used $P$ values from the isobutanol phosphate buffer system to obtain Equation 11-5, correlating binding to human albumin.

$$\log K = 0.91 \log P_{\text{isobutanol}} + \log K_p \tag{11-15}$$

As can be seen from Table 11-1, the intercept of this expression varies considerably depending on the class of drugs studied, so that one must conclude that in addition to "simple" partitioning, each class of compounds may contain a functional group or structure that confers a certain degree of stereoe-

TABLE 11-1

| Compound class | log $K_p$ |
|---|---|
| Sulfonamides | 1.2 |
| Tetracyclines | 0.9 |
| Penicillins | 0.4 |
| Steroid bisguanyl hydrazones | 0.05 |
| Cardenolides | −1.0 |
| Steroid hormones | −1.8 |
| Acridines | −2.6 |

lectronic specificity. It has been shown that log $P_{\text{octanol}}$ and log $P_{\text{isobutanol}}$ are related as follows[97]:

$$\log P_{\text{isobutanol}} = 0.72 \log P_{\text{octanol}} + 0.37 \tag{11-16}$$
$$n = 67, \quad r = 0.992, \quad s = 0.136.$$

Substituting Equation 11-16 into 11-15 yields:

$$\log K = 0.65 \log P_{\text{octanol}} + \log K_p + 0.33 \tag{11-17}$$

Considering that different albumins were employed, different definitions of binding, and that different aqueous phases were used, the agreement in slope between Equations 11-17 and 11-14 is quite good. Scholtan[138] also found similar results for the binding of drugs to yeast ribonucleic acids. Scores of other publications show that partition coefficients model binding of small molecules to macromolecules very well.

Such an example comes from the study by Bird and Marshall on the binding of a large and diverse set of penicillins to human serum[88]:

$$\log \frac{B}{F} = 0.49\pi - 0.63 \tag{11-18}$$
$$n = 79, \quad r = 0.924, \quad s = 0.134$$

In equation 11-18, $B$ is the percentage of bound penicillin, $F$ is the percentage free drug, and $\pi$ is the hydrophobic substituent constant.

The equations above correlate processes in which the overall lipophilicity (as defined by log $P$) of rather large molecules appears to account for protein binding. Albumin is known to have a large hydrophobic pocket that seems to be big enough to engulf the ligand. It is easy to imagine only part of a ligand being able to contact hydrophobic surface on a protein, so that log $P$ for the whole molecule is inappropriate.

This can be illustrated with data on the enzymatic hydrolysis by emulsin of the following phenyl glucosides $(C_6H_{11}O_5-O-C_6H_4-X)$[74]:

Parasubstituted glucosides:

$$\log K = 0.33\pi + 0.62\sigma^- + 1.80 \qquad (11\text{-}19)$$
$$n = 8, \qquad r = 0.921, \qquad s = 0.189$$

Metasubstituted glucosides:

$$\log K = 0.95\sigma + 1.62 \qquad (11\text{-}20)$$
$$n = 6, \qquad r = 0.901, \qquad s = 0.120$$

Without the $\pi$ term in Equation 11-19, one obtains a very poor correlation ($r = 0.75$); hence parasubstituents appear to contact "hydrophobic" space. However, since the coefficient with $\pi$ is far from 1, either the hydrophobic character is not like that of octanol or the substituent is binding to surface, rather than undergoing complete desolvation in a hydrophobic pocket. Adding a term in $\pi$ to Equation 11-20 does not significantly improve the correlation; thus it seems that metasubstituents do not contact the enzyme or, if they do, it is not in hydrophobic space. In these equations $K$ is the enzyme–substrate association constant for the formation of the ES complex. Using data for the catalytic step, Equations 11-21 and 11-22 have been derived:
Parasubstituted glucosides:

$$\log k_{cat} = -0.47\pi + 0.88\sigma^- - 6.32 \qquad (11\text{-}21)$$
$$n = 8, \qquad r = 0.964, \qquad s = 0.221$$

Metasubstituted glucosides:

$$\log k_{cat} = 1.52\sigma - 6.28 \qquad (11\text{-}22)$$
$$n = 6, \qquad r = 0.922, \qquad s = 0.242$$

Again we find that meta substituents do not make hydrophobic contact with the enzyme. Note that the coefficient with $\pi$ in Equation 11-21 is negative, suggesting that in the hydrolytic step one does not want the phenolate moiety sticking to the enzyme. Hydrophobic X slows the rate of reaction by being poorly desorbed. Setting $\pi$ for meta substituents equal to zero, we can combine meta- and para derivatives into single equations as follow:

$$\log K = 0.36\pi + 0.66\sigma^- + 1.76 \qquad (11\text{-}23)$$
$$n = 13, \qquad r = 0.917, \qquad s = 0.163$$

$$\log k_{cat} = -0.60\pi + 0.93\sigma^- - 6.15 \qquad (11\text{-}24)$$
$$n = 13, \qquad r = 0.949, \qquad s = 0.244$$

It is satisfying that one can account for all but 10–15% of the variance in log $K$ and log $k_{cat}$. In addition, one learns something about the hydrophobic nature of the active site and the role of the electronic effects of substituents in the hydrolysis process. There are, however, caveats to be considered. Most important, only a small set of substituents of small size was considered (ie, the usual "well-behaved" substituents). This tells us nothing about the ultimate size of the hydrophobic area. No 3,4-disubstituted congeners were considered. By a proper choice of substituents, one should be able to make much more active substrates. Also, it is not clear why the coefficients with $\sigma$ in Equations 11-21 and 11-22 are not identical. The difference is not large, and maybe a larger set of congeners would more accurately define $\rho$ and show that these differences are not significant. On the other hand, the hydrophobic environment of the 4-substituents may account for its larger $\rho$ value compared to 3-substituents, which are presumed to be in the aqueous phase.

A more complex problem is in the nature of the $\pi$ constants themselves. The $\pi$ constants for the equations above were taken from phenols. That is:

$$\pi_x = \log P_{x-C_6H_5OH} - \log P_{C_6H_5OH}$$

It has been shown that when two substituents are placed on benzene, their $\pi$ constants are only partly additive. Electronic interaction between the two substituents affects their hydrophobicity.[13,135–136] For example, $\pi_{NO_2}$ from benzene is $-0.28$, while $\pi_{NO_2}$ from phenol is 0.50. The increase in hydrophobicity really results from the interaction of the OH on the $NO_2$ and the $NO_2$ on the OH. Hence it is not entirely proper to say that the hydrophobicity of $NO_2$ in phenol is modeled by $\pi = 0.50$.

Clearly hydrophobic constants are of great value in understanding how organic compounds bind to macromolecules in vitro and in living systems. However, exactly what molecular features determine $\log P$ or $\pi$ values requires much more study.

## C. Steric Parameters

A major advance was made in the study of structure–activity relationships when Taft proposed[12] the use of Equation 11-3 for defining steric constants. An illustrative example of the use of $E_s$ in correlation analysis can be seen in equations 11-25 and 11-26.

$$\log k = 1.67E_s - 1.13 \tag{11-25}$$
$$n = 6, \qquad r = 0.989, \qquad s = 0.156$$

$$\log \frac{k_2}{K_m} = 1.76E_s + 0.79\pi + 2.23$$

$$n = 8, \quad r = 0.981, \quad s = 0.201$$

(11-26)

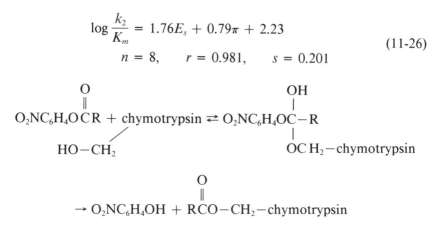

These analyses[140] show that a transesterification reaction with an enzyme has, in this instance, the same kind of steric hindrance as that of a simple organic reaction if a correction is made for the additional hydrophobic effect that occurs with the chymotrypsin accounted by the term in $\pi$. However, in this case we cannot be sure that $\pi$ is modeling a pure hydrophobic effect, since $\pi$ for the set of esters studied is perfectly collinear with molar refractivity (MR). Hence using MR in Equation 11-26 gives the same result as far as the coefficient with $E_s$ is concerned. Further QSAR work needs to be done to better establish the kind of interaction between $R$ and the type of enzymic space with which it binds.[76,141] Equations 11-25 and 11-26 make interesting comparisons because in both instances the transitions states must be similar and the reactions are the type $E_s$ was designed for—intramolecular steric effects. A more important type of steric effect in biochemical problems is that of intermolecular forces. Although $E_s$ was not designed for such problems, it can in some instances contribute to their solution.

An instructive example on the use of $E_s$ for intermolecular steric effects was derived by Kutter and Hansch[142] from the research of Pauling and Pressman on hapten–antibody interactions. Antibodies were prepared by coupling diazotized *p*-aminobenzoic acid to serum proteins and injecting these modified proteins into rabbits. The antibodies were then isolated and, when mixed with antigen, they formed a precipitate. Benzoate ions tend to prevent the union of antigen and antibody and an affinity constant ($k_{rel}$) can be determined for them. Equation 11-27 rationalizes the affinity of the simple benzoate ions (haptens) for the antibody molecules.

$$\log K_{rel} = 0.86E_s^o + 0.08E_s^m - 0.45E_s^p - 0.70$$

$$n = 27, \quad r = 0.934, \quad s = 0.177$$

(11-27)

Bearing in mind that the more sterically hindering the substituent, the more negative its $E_s$ value, the positive coefficient with orthosubstituents ($E_s^o$) shows that orthosubstitution makes less effective compounds for preventing antibody–antigen union. The very small coefficient with $E_s^m$ shows that metasubstituents have essentially no effect. That is, they do not seem to contact the protein. The negative coefficient with $E_s^p$ reveals that the larger the parasubstituent, the more effective it is in blocking antibody–antigen reactions. Electronic and hydrophobic effects of substituents are negligible. Of course the orthosubstituents are ambivalent in their behavior, since they have the potential for both intramolecular and intermolecular effects. The effect seems to be largely intramolecular, since a study of arsonic acids having the rather symmetrical group $-AsO_3H^-$ yield a QSAR with a very small coefficient with $E_s$.

Large parasubstituents on the benzoate ions appear to block the entrance of the antibody cavity for the $-N{=}N-C_6H_4COO^-$ unit of the antigen by a kind of corking action. An interesting aspect of this analysis is that it became clear for the first time that hydrophobic properties of the hapten (benzoate ion) play no role in preventing antibody–antigen union. The union seems to be governed by the negative charge on the carboxylate and the shape of the apolar portion of the benzoates.

Verloop, Hoogenstraaten, and Tipker have developed what would seem to be much more suitable steric parameters for intermolecular steric effects expected with macromolecules.[143] They have calculated five critical dimensions of substituents to obtain five STERIMOL parameters. In making these calculations from known bond angles and distances, some assumptions about the conformations of flexible substituents had to be made. Nevertheless, these parameters show promise in biological QSAR. They have treated the data on which Equation 11-25 is based to obtain Equation 11-28.

$$\log K_{\text{rel}} = -0.41B_1^o + 0.95L^p - 0.07(L^p)^2 - 0.43 \qquad (11\text{-}28)$$
$$n = 36, \qquad r = 0.961, \qquad s = 0.282$$

where $B_1^o$ is the minimum width of orthosubstituents and $L^p$ is length of parasubstituents. The results are similar to those of Equation 11-27, showing that orthosubstitution makes poor inhibitors while large groups in the para position make effective inhibitors. The quadratic term in Equation 11-28 suggests that there is an optimum size for parasubstituents when $L^p = 6.8$. It was possible to include an additional nine data points in Equation 11-28 over 11-27, since $E_s$ values for these substituents are unknown while STERIMOL values could be calculated.

The MR of substituents, which was first used by Pauling and Pressman for correlation of chemical structure and biological activity, has recently received rather wide use.[96] This term is defined as follows:

$$MR = \frac{n^2 - 1}{n^2 + 2} \cdot \frac{MW}{d}$$

where $n$ is the refractive index of the organic compound, $MW$ is the molecular weight, and $d$ is the density. Since there is little variation in $n$ for most organic compounds, MR is actually a kind of corrected molar volume, where $MW/d$ is the volume term. That is, for substances with loosely held electrons, the $(n^2 - 1)/(n^2 + 2)$ term "corrects" molar volume for electron polarizability. MR is a rough measure of the bulk of a substituent.

# 3. PHYSICAL-ORGANIC TO BIOMEDICINAL CHEMISTRY

The foregoing examples provide evidence that the same approach to structure–activity problems so boldly undertaken by Hammett, Taft, and eventually many others can be applied to the seemingly hopelessly complex problems in biomedicinal chemistry. Ever since the work of Meyer and Overton, it has been clear that the lipophilic properties of organic compounds play an enormously important role in all biochemical processes. The so-called nonspecific toxicity of organic compounds is probably heavily dependent on the partitioning of apolar molecules in membranes. Insight can be gained into such processes by the study of model systems. In many such studies the molar concentration of a chemical producing some standard response is taken as the end point. For quasi-equilibrium situations, this can be related to $\log P$ or $\pi$ as follows:

$$P = \frac{\text{conc}_{\text{octanol}}}{\text{conc}_{\text{H}_2\text{O}}}$$

$$P_{\text{bio}} = \frac{\text{conc}_{\text{in biophase}}}{\text{conc}_{\text{in H}_2\text{O}}}$$

The two partition coefficients (equilibrium constants) can be expected to follow a Hammett postulate:

$$\log P_{\text{bio}} = a \log P_{\text{octanol}} + b \qquad (11\text{-}29)$$

If we assume that equivalent numbers of molecules in the biophase produce an equivalent standard response, we can write[60]:

$$(P_{\text{bio}})_{\text{SR}} = \frac{k}{\text{conc}_{\text{H}_2\text{O applied}}} \qquad (11\text{-}30)$$

That is, when the concentration in the aqueous phase is adjusted to give a standard response, the concentration in the biophase $(k)$ is the same for each member of a congeneric series. Taking the logarithm of Equation 11-30 yields:

$$\log \frac{1}{C_{\text{H}_2\text{O applied}}} = \log(P_{\text{bio}})_{\text{SR}} - \log k \qquad (11\text{-}31)$$

and substitution of Equation 11-28 into 11-30 yields:

$$\log \frac{1}{C_{\text{H}_2\text{O applied}}} = a \log P_{\text{octanol}} + \text{constant} \qquad (11\text{-}32)$$

where $C$ is the concentration in moles per liter or moles per kilogram of applied drug.

The use of models and QSARs in membrane study is illustrated by a system of the greatest simplicity—silanized glass beads (0.3 mm diameter). The hydrophobic aggregation of such beads in water can be dispersed by alcohols. The following QSAR for disaggregation has been derived[60]:

$$\log \frac{1}{C} = 0.95 \log P - 0.49$$
$$n = 4, \quad r = 0.989, \quad s = 0.114$$

(11-33)

Apparently the hydrophobic R of ROH partitions into the silane with its OH held at the surface. This produces a partially hydrophobic, partially polar surface on the beads, and the concentration in the "biophase" $(k)$ is the same for each member of a congeneric series. Thus the hydrophobic contact between beads is reduced by the OH groups. This process can be compared with the perturbation of black lipid membranes (BLM)[60] by ROH:

$$\log \frac{1}{C} = 1.16 \log P - 0.51$$
$$n = 7, \quad r = 0.958, \quad s = 0.262$$

(11-34)

Equation 11-34 correlates the concentration of ROH necessary to change the resistance from $10^8$ to $10^6$ $\Omega/cm^2$. The BLMs were prepared by dissolving lipid from sheep red cell ghosts in hydrocarbon and painting the solution over a small orifice to produce a lipid bilayer type of membrane. These results can be compared with living membranes. Equation 11-35 correlates the concentration of alcohol necessary to produce a 10-mV change in the rest potential of a lobster axon by ROH:

$$\log \frac{1}{C} = 0.87 \log P - 0.24$$
$$n = 5, \quad r = 0.993, \quad s = 0.100$$

(11-35)

For practical purposes, the slopes of Equations 11-34 and 11-35 are the same; however, the higher intercept for Equation 11-35 indicates that it takes about half the concentration of isolipophilic alcohol to produce in the lobster nerve changes that are assumed to be caused by the perturbation of the nerve membrane by the lipophilic alcohols.

It is still easier, but parallel in hydrophobic effect, to block the functioning of a frog nerve[60]:

$$\log \frac{1}{C} = 0.88 \log P + 0.63$$
$$n = 25, \quad r = 0.955, \quad s = 0.297$$

(11-36)

The 25 neutral compounds (ROH, phenols, ether, acetanilide, nitrobenzene, pyridine, quinoline, etc) yield an equation very similar to those of the model systems. The difference in the intercepts indicates that it is about 10 times easier to block the frog nerve than to disrupt the synthetic "membranes."

Higher organisms show similar QSARs.[60] The $LD_{100}$ of ROH for cats is given by:

$$\log \frac{1}{C} = 1.06 \log P + 1.37$$

$$n = 8, \quad r = 0.986, \quad s = 0.134$$

(11-37)

Presumably it is disruption of the CNS membranes that kills the cats. It is about five times easier to shut down cats than the frog nerves.

These examples again show how QSAR can be used to compare various living systems with nonliving models. Similar correlations can be done for enzymes. Gitler and Ochoa-Solano[144] studied the hydrolysis of nitrophenyl esters ($RCOOC_6H_4-NO_2$) by $N$-myristoyl-L-histidine in micelles of cetyltrimethylammonium bromide. The highly lipophilic myristoyl moiety ensured that the catalytic entity would be in the micelle phase. The QSAR[61] of Equation 11-38 rationalizes the relative rates of hydrolysis:

$$\log k = 0.62 \log P - 0.28$$

$$n = 5, \quad r = 0.995, \quad s = 0.060$$

(11-38)

The more lipophilic R is, the faster the hydrolysis rate. The same reaction was studied using the polar $N$-acetylhistidine as the catalyst. Equation 11-39 arises from these results:

$$\log k = -0.31 \log P - 0.23$$

$$n = 5, \quad r = 0.948, \quad s = 0.106$$

(11-39)

The negative coefficient with log $P$ shows that the more polar the nitrophenyl ester, the faster the hydrolysis—catalytic hydrolysis now is occurring in the aqueous phase.

A step closer than the system above to a synthetic enzyme is that of the paracyclophane hydrolysis of the same type of nitrophenyl esters. The imidazole moiety attached to the hydrophobic pocket provides the catalytic effect for the ester hydrolysis[145]:

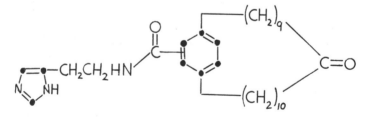

$$\log k = 0.45\pi - 0.53$$

$$n = 11, \quad r = 0.968, \quad s = 0.260$$

(11-40)

In the esters on which Equation 11-40 is based, R ranged in size from methyl to CH$_2$-cyclododecyl and CH$_2$-1-naphthyl. But since Equation 11-40 "explains" 94% of the variance in the rate constants, large specific steric effects do not seem to be involved. Reaction conditions were below critical micelle concentration for paracyclophane. The fact that the coefficient with $\pi$ is only about 0.5 suggests that R in the hydrophobic interactions is only about 50% desolvated and that a surface-to-surface interaction seems to be involved rather than complete engulfing of R in the hydrophobic ring.

# 4. QUANTITATIVE STRUCTURE–ACTIVITY RELATIONSHIPS OF PURIFIED ENZYMES

## A. Papain

Of course the problem of central importance in understanding the effects of organic compounds on living systems is that of their reaction with critical receptors, which are, of course, macromolecules. Since the structures of most of these are still unknown, as far as drug action or toxicity is concerned, the best surrogates for study are enzymes whose X-ray crystallographic structures are known. Although many studies have been made with enzymes whose structures have not yet been worked out, these will not be considered in this discussion.

The hydrolase papain is an instructive example. From the hydrolysis of hippurate esters (X$-$C$_6$H$_4$OCCH$_2$NHCC$_6$H$_4$), the following QSAR was derived for the Michaelis constants[146]:

$$\log \frac{1}{K_m} = 1.03\pi_3' + 0.57\sigma + 0.61MR_4 + 3.80$$

$$n = 25, \quad r = 0.907, \quad s = 0.208$$

(11-41)

In this expression $\pi_3'$ refers to only the more hydrophobic of the two possible metasubstituents. For example, when X = 3$-$OH, the $\pi$ value of O for H was used, since $\pi_3 - $ OH $= -0.67$. Unless such an approach were taken, a good correlation could not have been obtained. Behind this approach were the assumptions that only one of the two metasubstituents could contact hydrophobic space on the enzyme and that the more hydrophilic substituents would project into the aqueous phase.

The positive coefficient with $\sigma$ reveals that electron-withdrawing substituents promote formation of the enzyme–substrate complex (*ES*). However, the most interesting term is MR$_4$, which is for the molar refractivity of X in the para position of the hippurates. Its positive coefficient indicates that the larger X is, the better the binding.

After the development of Equation 11-41, the model shown in View I was constructed via computer graphics.[146] This model was based on the structure for the enzyme–inhibitor complex: benzyloxycarbonyl-L-phenylalanylme-thylene–papain (ZPA–papain) determined by the X-ray crystallographic studies of Drenth and co-workers.[147] To construct this model, the $-COCH_2NHCO-$ segment of the hippurate was superimposed onto the binding positions of the corresponding atoms of ZPA. The remainder of the hippurate was built from standard bond lengths and angles and then fitted into the molecular surface of the papain catalytic site so that the substrate molecule did not approach the enzyme surface at any point closer than the allowed van der Waals distance. The ring containing the Y substituents was placed into the hydrophobic pocket occupied by the phenylalanine portion of ZPA. The red dots color code for hydrophobic surface (carbon), while the blue dots code for polar surface (oxygen and nitrogen). To account for the hydrophobic reaction of metasubstituents, $X_3$ in View I has been placed in a red hydrophobic pocket, which obviously forces the other metasubstituent into the aqueous phase as surmised from correlation equation 11-41.

Substituents $X_4$ project toward a blue bump in the surface caused by the $\epsilon$-NH$_2$ of Gln-142. This surface residue is loosely held and accommodates groups as large 4-SO$_2$NH$_2$. The point of interest, however, is that regardless of whether $X_4$ is highly polar or hydrophobic, correlation is obtained via MR. Residue Gln-142 seems to buttress the substrate in the proper position for ES formation.

The Y substituents bind in a very large hydrophobic pocket, and the nature of this binding has been characterized by the QSAR of Equation 11-42[148] for hippurates of the type: $CH_3OCOCH_2NHCOC_6H_4$-Y.

$$\log \frac{1}{K_m} = 1.01\pi + 1.46$$

$$n = 16, \quad r = 0.981, \quad s = 0.165 \tag{11-42}$$

The equation is based on a good variety of polar and apolar substituents in the 3- and 4-positions of the phenyl ring. When a van der Waals surface is placed over the substrate in View I', a tight fit is observed.[146] The coefficient of 1 with the $\pi$ terms of Equations 11-41 and 11-42 suggests that partitioning of the substrates into the ES complex closely resembles their partitioning into octanol. That is, complete desolvation of the substituents would appear to be the main driving force.

Also in View I the catalytically important SH of cysteine and His-159 are shown in green. The surface of the sulfur is coded red for hydrophobic, but it must be less so than carbon.

In a further study of the hydrolysis of methanesulfonyl glycinates $(X-C_6H_4-OCOCH_2NHSO_2CH_3)$ by papain, the following QSAR was developed[78]:

$$\log \frac{1}{K_m} = 0.61\pi'_3 + 0.55\sigma + 0.46MR_4 + 2.00$$

$$n = 32, \quad r = 0.945, \quad s = 0.178 \tag{11-43}$$

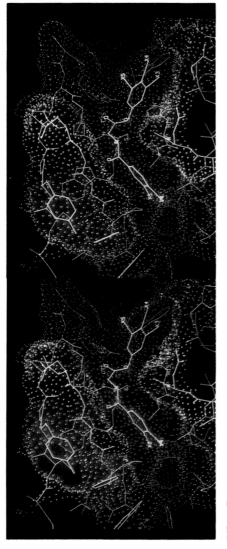

View I

View I'

362

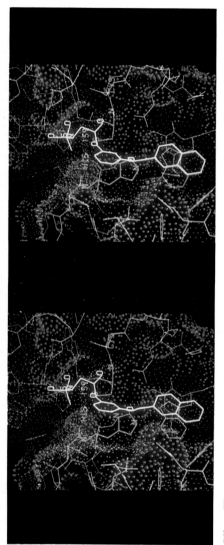

**View II**

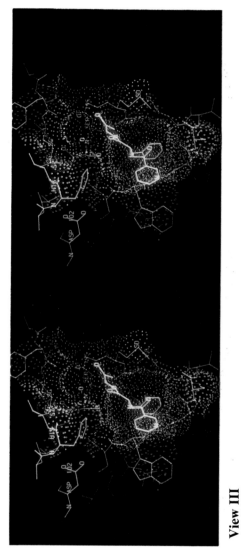

**View III**

The substitution of the $CH_3SO_2NH$ moiety for the $C_6H_5CONH$ gave more soluble substrates, enabling the investigation of larger, more hydrophobic substituents. The surprising difference between Equations 11-43 and 11-41 is that the coefficient with $\pi'_3$ in Equation 11-43 is only about half that in 11-41. The hydrophobicity of X in the methanesulfonyl glycinates was more extensively explored, with substituents as large as $3\text{-}C(CH_3)_3$, $3\text{-}OCH_2C_6H_4\text{-}4'\text{-}Cl$, and $3\text{-}OCH_2\text{-}2'\text{-}naphthyl$.

The results of this QSAR are rationalized in View II. In this stereorepresentation, the phenoxy ring has been rotated sufficiently to place the 3-substituents onto a large hydrophobic surface, much of which is formed by a tryptophane residue. Under the condition of such surface binding, one would expect only about 50% desolvation of X, which would account for the smaller $\pi$ coefficient of Equation 11-43. The question is, Why would the esters having the $CH_3SO_2NH-$ moiety bind differently from those having the $C_6H_5CONH-$ moiety? The respective $\pi$ constants[96] of $-1.18$ and $0.49$ show the much greater hydrophilic character of the sulfonamide group. This group does not bind so well in hydrophobic space, as can be seen from a comparison of the intercepts of Equations 11-41 and 11-43. The hippurates complex about 70 times more strongly. This could mean that the $-NHCOC_6H_5$ moiety of the hippurates produces a conformational change in papain that favors the type of interaction shown in View I; or it is possible that the polar $CH_3SO_2NH-$ moiety does not bind so deeply in the cavity and this leads to the surface binding of 3-X.

QSAR promises to be a powerful means for developing our comprehension of the variable binding potentials of enzymes and other receptors.[149] Via QSARs and graphics, we have noted considerable flexibility in dihydrofolate reductase.[150] Roberts,[151] under the apt title "Flexible Keys and Deformable Locks: Ligand Binding to Dihydrofolate Reductase," has discussed some of the complexities one must face in trying to decipher the binding mechanisms between flexible organic compounds and what more and more appear to be tricky flexible enzymes.

## B. Actinidin

A comparison of the hydrolysis of the phenyl hippurates by papain can be made with results obtained with the similar enzyme actinidin,[152] whose X-ray crystallographic structure is also known:

$$\log \frac{1}{K_m} = 0.50\pi'_3 + 0.74\sigma + 0.24MR_4 + 2.90$$

$$n = 27, \qquad r = 0.927, \qquad s = 0.158$$

(11-44)

The coefficient with $\pi$ in Equation 11-44 resembles that of 11-43 rather than 11-41, as might be expected. This has been assumed to mean that with actinidin, X is binding hydrophobically to the surface rather than with the hydrophobic pocket postulated in View I. As in Equations 11-41 and 11-43, interaction of 4-substituents with the enzyme was not correlated with $\pi$ but with MR, indicating interaction with polar space. Actinidin contains a Lys-145 residue corresponding to Gln-142 in papain. Hence, $X_4$ contacts the hydrated $\epsilon-\overset{+}{N}H_3$ group of the Lys side chain. Why the $MR_4$ term in Equation 11-44 is so much less significant than that in Equation 11-43 is not clear. It might be that the $-CH_2CH_2CH_2CH_2NH_2$ moiety in actinidin is more flexible than the $-CH_2CH_2CONH_2$ moiety in papain, hence is less effective in buttressing the substrate for formation of the ES.

## C. Ficin and Bromelain

The results with papain and actinidin can be compared with those for two other cysteine hydrolases: ficin and bromelain. Equation 11-45 correlates the hydrolysis of X-phenyl hippurates by ficin[153]:

$$\log \frac{1}{K_m} = 0.84\pi_3' + 0.57\sigma + 0.41MR_{4,5} + 3.60$$

$$n = 28, \qquad r = 0.941, \qquad s = 0.147$$

(11-45)

In terms of $\pi_3'$, ficin reacts with the phenyl hippurates more like papain than actinidin and the effect with $\sigma$ is identical to papain. Substituents in the 4-position with ficin behave somewhat differently in that MR now parameterizes both the 4-substituents and the more hydrophilic of the two metasubstituents. This suggests that the metasubstituents ($X_5$) directed away from the hydrophobic surface do contact enzymic polar space.

Hydrolysis of 4-X-phenyl hippurates by bromelain B and bromelain D from the studies of Hawkins and Williams[154] yield Equations 11-46 and 11-47, respectively:

$$\log \frac{k_0}{K_m} = 0.70\sigma + 0.50MR_4 + 2.62$$

$$n = 8, \qquad r = 0.962, \qquad s = 0.137$$

(11-46)

$$\log \frac{k_0}{K_m} = 0.63\sigma + 0.46MR_4 + 2.21$$

$$n = 8, \qquad r = 0.996, \qquad s = 0.041$$

(11-47)

Hawkins and Williams did not report $K_m$ values, but since $k_{cat}(k_0)$ is essentially constant for both the papain and actinidin hydrolysis of the phenyl hippurates, it seems reasonable to compare Equations 11-46 and 11-47 with 11-41. Of course since no metasubstituents are included in the bromelain equations, no $\pi$ terms occur in them. QSARs for the two forms of bromelain are indeed very similar to those of the other cysteine hydrolases. The bromelain B hydrolysis of $4\text{-}X\text{-}C_6H_4COCH_2NHSO_2CH_3$ produces Equation 11-48, which is much like Equation 11-43:

$$\log \frac{k_0}{K_m} = 0.68\sigma + 0.60MR_4 + 1.16$$

$$n = 8, \qquad r = 0.978, \qquad s = 0.107$$

(11-48)

The QSARs of Equations 11-41 to 11-48 show a remarkable consistency for the $\sigma$ term, which is expected to be largely independent of the enzyme. Except for actinidin, the MR term is also rather constant; in fact, one might guess that ficin and bromelain contain a Gln like papain rather than a Lys like actindin.[155]

The role of MR in these equations, suggesting buttressing reactions of polar surface residues whose flexible nature allows them to adjust to accommodate substituents of various sizes, is probably what one might expect to find in a QSAR analysis, since surface residues are mostly polar.

## D. Chymotrypsin

A more complex example of the combined use of QSAR and molecular graphics comes from studies on chymotrypsin (a serine hydrolase) with substrates of the following type:

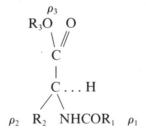

From the hydrolysis of these L-acylamino acid esters, Equation 11-49 was developed[156]:

$$\log \frac{1}{K_m} = 1.09MR_2 + 0.80MR_1 + 0.52MR_3 - 0.63I + 1.26\sigma^*$$

$$- 0.057MR_1 \cdot MR_2 \cdot MR_3 - 1.61$$

$$n = 71, \qquad r = 0.979, \qquad s = 0.332$$

(11-49)

The subscripted MR values refer to the corresponding groups: $R_1$, $R_2$, and $R_3$. The indicator variable[17] $I$ takes the value of 1 when $R_2$-isopropyl. All other examples of $R_2$ are assigned a value of 0 for $I$. The negative coefficient with $I$ indicates that the isopropyl group inhibits formation of the ES complex. The positive sign of $\sigma^*$, the Taft inductive constant for $R_3$, indicates that electron-withdrawing groups promote formation of ES. The cross-product term $MR_1 \cdot MR_2 \cdot MR_3$ shows that when several large R groups are attached, activity is less than expected from a linear combination of $MR_1$, $MR_2$, $MR_3$, $I$, and $\sigma^*$. The effect appears to be more than additive, although groups large enough to firmly establish the validity of this term have not been studied. Almost as good results can be obtained using $(\Sigma MR)^2$ or the bilinear model. One of the difficulties of Equation 11-49 is that for the substituents on which it is based, there is rather high collinearity between $\pi$ and MR, lending to uncertainty as to whether interaction is occurring with polar or hydrophobic space. To clarify this for $R_3$, 14 new compounds were made where $R_2 = CH_3$, $R_1 = 3\text{-}C_5H_4N$ and variation was made in $R_3$ so that collinearity between $\pi$ and MR was broken. Adding these data to those for Equation 11-49 and refitting the data yields:

$$\log \frac{1}{K_m} = 1.13MR_2 + 0.77MR_1 + 0.47MR_3 - 0.56I + 1.35\sigma^*$$

$$- 0.055MR_1 \cdot MR_2 \cdot MR_3 - 1.64 \qquad (11\text{-}50)$$

$$n = 84, \qquad r = 0.977, \qquad s = 0.333$$

The parameters of Equations 11-49 and 11-50 are for practical purposes identical, which illustrates the predictive value of correlation equations. Of course the conclusion is that $MR_3$ is interacting with polar space. View III gives a stereo picture of the situation. In the active site of chymotrypsin, an analogue having the following substituents has been placed[149]: $R_1 = 3\text{-}C_5H_5N$, $R_2 = CH_2C_6H_5$, and $R_3 = -OCH_2COCH_3$. Whereas $R_2$ has been placed in the so-called hydrophobic hole, $R_1$ falls on a surface that is half hydrophobic and half hydrophilic. The N of the pyridine ring has been placed on blue polar space. The yellow O represents water molecules bound so firmly to the enzyme that their position has been established from X-ray crystallographic studies. Thus the whole area behind and around the two O's is very polar, and it is with this surface that $R_3$ participates in a kind of interaction similar to what we saw from Equations 11-41 to 11-48. In fact, the coefficients with these MR terms are all very much the same for the various enzymes except for actinidin, which is smaller.

The overall reaction of chymotrypsin can be visualized as follows:

$$\text{enzyme + substrate} \rightleftarrows ES \xrightarrow{k_2} \text{acylenzyme} \xrightarrow{k_3} \text{enzyme + product}$$

In the step governed by $k_2$, Ser-195 displaces $OR_3$ of the substrate in a transesterification step to produce the acylenzyme. From data on the same kind of esters on which Equation 11-50 is based, the following QSAR for the acylation step has been developed[156]:

$$\log k_2 = -0.52MR_1 + 1.10MR_2 - 1.56I + 0.42 \qquad (11\text{-}51)$$

$$n = 18, \qquad r = 0.971, \qquad s = 0.399$$

The terms carry the same meaning as in Equation 11-50. No term occurs for $R_3$, since substitution in this position was almost constant. The coefficient with $MR_2$ is identical to Equation 11-50, however, the indicator variable shows that the effect of $-CH(CH_3)_2$ is 10 times as great in the acylation step as in the formation of the ES complex. The most interesting aspect of the $k_2$ QSAR is the negative coefficient with $MR_1$, showing that large substituents retard the transesterification step. This could be due to steric inhibition of the formation of the tetrahedral intermediate, or it could be due to the breaking of van der Waals bonding of $R_1$, possibly reinforced by a water matrix of solvation around $R_1$ and the enzyme surface. The latter possibility seems more likely when we consider the following QSAR for $k_3$ for the deacylation step[156]:

$$\log k_3 = 0.75MR_2 - 1.79I_1 - 1.48I_2 - 0.31 \qquad (11\text{-}52)$$
$$n = 33, \qquad r = 0.977, \qquad s = 0.289$$

In this equation $I_1$ is given the value of 1 for D-isomers and 0 for L-isomers. The negative coefficient shows that D-isomers are deacylated about 70 times more slowly on the average. When $R_2$ is isopropyl, $I_2$ is given the value of 1, as in the other equations. The most important variable is $MR_2$, and its positive coefficient seems surprising, since one might expect that in this step the desorption of $R_2$ from the enzyme would hinder the process. Since $MR_2$ plays such as important role in both acylation and deacylation, its main function may be to produce and maintain a conformational change in the enzyme essential for these two reactions. Despite considerable variation in $R_1$, $MR_1$ plays no role in the QSAR for $k_3$. This inactivity suggests that this substituent is no longer in contact with enzyme in the deacylation step, having been pulled away from the enzyme in the preceding acylation reaction.

In the chymotrypsin hydrolysis, MR seems to be playing two roles. The role in $\rho_3$ space appears, from the similarity of the weighting factors with MR, to be much like that of papain and actinidin. The role in $\rho_2$ space, however, must be associated with maintaining of a proper conformation and possibly a hydrophobic partitioning effect. To elucidate the latter, it will be necessary to study a better designed set of $R_2$ groups for which $\pi$ and MR are orthogonal.

## E. Dihydrofolate Reductase

The QSARs for enzymes (bioreceptors) presented above show that we can describe in more or less quantitative terms the properties of a set of congeners that account for their differences in affinity for the active site in a protein. These results, along with similar results for dihydrofolate reductase, provide considerable evidence that two types of parameter for nonspecific binding of ligands to macromolecules are necessary—one for polar and one for hydrophobic space. Other results not yet published on trypsin and subtilisin support these findings. These QSARs show that equations obtained with isolated enzymes hold for enzymes in living cells. However, one does not always obtain the same QSAR in vitro as in vivo, as seen in the following example from dihydrofolate reductase. Equation 11-53 correlates inhibition of highly purified *Lactobacillus casei* dihydrofolate reductase with triazines[157]:

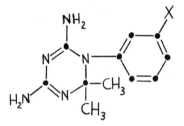

$$\log \frac{1}{K_i} = 0.53\pi_3' - 0.64 \log(\beta \cdot 10^{\pi_3'} + 1) + 1.49I + 0.70\sigma + 2.93 \quad (11\text{-}53)$$

$$n = 44, \qquad r = 0.953, \qquad s = 0.319, \qquad \log \beta = -3.66, \qquad \pi_0 = 4.31$$

In this expression $\pi'$ signifies that for substituents of the type $-CH_2ZC_6H_4-$ Y, $\pi$ for Y was set to zero. Although this was necessary to obtain a good correlation, it was later shown via X-ray crystallography that, in fact, Y does not contact the enzyme.[150] The indicator variable $I$ is assigned the value of 1 for substituents of the types $-CH_2ZC_6H_4-Y$ and $ZCH_2C_6H_4-Y$ and 0 for all other substituents. These bridged phenyl rings increased activity about 30-fold over what one would expect from $\pi$ and $\sigma$ alone.

Equation 11-54 for the action of the same type of inhibitors on *L. casei* cell culture reveals a surprising difference when compared with the pure enzyme equation:

$$\log \frac{1}{C} = 0.82\pi_3' - 1.06 \log(\beta \cdot 10^{\pi_3'} + 1) - 0.94 MR_Y + 0.80I + 4.37 \quad (11\text{-}54)$$

$$n = 34, \qquad r = 0.929, \qquad s = 0.371, \qquad \pi_0 = 2.94, \qquad \log \beta = -2.45$$

A major difference between Equations 11-54 and 11-53 is the negative term of MR for Y in Equation 11-54: Y was not parameterized in any way in Equation 11-53, and we know from X-ray crystallography that there is no reason to do so; hence the negative MR term must be interpreted to mean that in vivo Y encounters some kind of steric hindrance—probably from an adjacent macromolecule or a different conformation of the enzyme in the cell. The negative steric effect of Y might also be accounted for by an unfavorable interaction of Y and the carrier protein in the transport mechanism. However, since the same $MR_Y$ term occurs in a QSAR for cells highly resistant to methotrexate, for which it is believed that their active transport system is impaired, it would seem that the interaction of Y with the transport protein is an unlikely explanation.[157] This steric effect holds regardless of whether Y is hydrophobic or hydrophilic, hence must be related only to the bulk of Y.

There is also a rather pronounced difference between the coefficients with $I$ in the two equations, which points to a possible difference in conformation between isolated enzyme and enzyme in situ. In addition, an electronic term in the form of $\sigma$ cannot be validated for the in vivo equation. Since this is not an important term, it may simply be covered by noise in the data. Thus, there is specific evidence that in some instances there are distinct differences in the behavior of isolated enzyme and the same enzyme in the living cell.

QSAR can be employed to bring out striking differences in the behavior of dihydrofolate reductase (and no doubt other receptors) in resistant and sensitive cells. Equation 11-55 correlates the 50% inhibition of growth of L5178Y leukemia cells sensitive to the antitumor drug methotrexate, and Equation 11-56 correlates the same triazines acting on L5178Y cells highly resistant to methotrexate. The molar concentration of triazine producing 50% inhibition of growth is represented by $C$[158]:

$$\log \frac{1}{C} = 1.40\pi_3 - 1.65 \log(\beta \cdot 10^{\pi_3} + 1) + 0.88\sigma + 0.52I - 0.25OR$$
$$+ 0.63DO + 7.94 \tag{11-55}$$

$$n = 64, \quad r = 0.904, \quad s = 0.298, \quad \pi_0 = 0.89, \quad \log \beta = -0.054$$

$$\log \frac{1}{C} = 0.63\pi_3 - 0.26 \log(\beta \cdot 10^{\pi_3} + 1) - 0.17MR - 0.33OR + 3.11 \tag{11-56}$$

$$n = 61, \quad r = 0.878, \quad s = 0.335, \quad \log \beta = -0.748$$

Where $I$ has the same significance as in Equation 11-54, OR refers to a set of alkoxy-containing congeners that as a class are less active than expected from $\pi$ and MR alone, and DO refers to a set of alkyltriazines that are somewhat more active than expected. In these equations $\pi$ for the complete substituent has been used rather than $\pi'$ as in Equation 11-54. One expects with whole cells to see a hydrophobic interaction of all parts of the compound because of its importance in membrane penetration. Why this is not true for Equation 11-54 is not clear. Actually, since one obtains almost as good correlation in Equation 11-54 with $\pi$ in place or $\pi'$, membrane penetration does not seem to be an overriding factor. The most significant difference between the two equations is that the bilinear QSAR of Equation 11-55 establishes the optimum $\pi$ for X as 0.89, while Equation 11-56 is not a proper bilinear equation. The slope of the left-hand side of the bilinear curve is 0.63 and that of the right-hand side is $0.63 - 0.26 = 0.37$. That is, the slope is still positive and $\pi_0$ cannot be established. However, from an inspection of the data, $\pi_0$ would appear to be around 5. Very hydrophobic triazines are thousands of times more potent than methotrexate against resistant cells. The effect does not hold for other lipophilic antitumor drugs. Apparently in the resistant cells a transport mechanism for the triazines has been turned off, so that the drugs must penetrate through lipophilic membranes, which accounts for the inactivity of hydrophilic triazines and the very hydrophilic methotrexate. The same effect occurs with strains of *L. casei* that are sensitive and resistant to methotrexate.[157]

## 5. ANTIMALARIAL DRUGS IN WHOLE ANIMALS

Of course the most challenging problem for QSAR is the study of large numbers of complex organic compounds acting selectively on pathogenic cells in animals. There are now two examples of QSARs for more than 500 congeners acting in mice having been derived—one for antimalarials[160] and one for antitumor drugs.[163]

Malaria, despite the large number of known antimalarial drugs, is still an extremely serious problem in many parts of the world because of the resistance the parasites develop to drugs after continued use. Hence, research for more effective drugs is of vital importance.

An extensive program of the Walter Reed Army Institute for Medical Research has shown that very effective antimalarials can be developed from phenanthrene carbinols of the type:

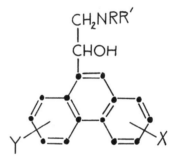

The activity of these compounds was assessed by testing against *Plasmodium berghei* in mice with the cure of 2.5 out of 5 mice being taken as the end point, cure being defined as at least 60 days survival in the Rane mouse test. In a first analysis[159] of about 60 of these compounds in 1969, it was found that substituents X and Y had been selected inadvertently, so that the correlation between $\Sigma\pi$ and $\Sigma\sigma$ was extremely high ($r = 0.919$). With such high collinearity, it was impossible to decide on the roles of these two vectors; in fact, either one would do a good job of correlating the data. This collinearity was broken by the synthesis of about 50 new compounds and the following QSAR was developed[160]:

$$\log\frac{1}{C} = 0.31\pi_{x+y} + 0.78\Sigma\sigma + 0.13\Sigma\pi - 0.015\Sigma\pi^2 + 2.35$$

$$n = 102, \quad r = 0.908, \quad s = 0.263, \quad \pi_0 = 5.05$$

(11-57)

In this expression $C$ is moles of drug per kilogram of mouse and $\Sigma\pi$ refers to $\pi$ for X + Y, and R; $\Sigma\sigma$ is for $\sigma$ of X + Y. Equation 11-57 explains 82% of the variance in log $1/C$ for more than 100 congeners acting in an extremely complex system. Even more surprising is the relatively good correlation obtained by the following simple equation:

$$\log \frac{1}{C} = 0.33\Sigma\pi + 0.83\Sigma\sigma + 2.52$$

$$n = 102, \qquad r = 0.894, \qquad s = 0.278 \tag{11-58}$$

In this simpler expression, parameterization is not made for R and R′. The effect of R groups on activity is almost negligible. Although research with compounds of this type had been conducted for many years, this fact had not become generally appreciated. Later data on about 100 more analogues became available, and from these Equations 11-59 and 11-60 were developed:[161]

$$\log \frac{1}{C} = 0.29\pi_{x+y} + 0.97\sigma_{x+y} + 2.53$$

$$n = 212, \qquad r = 0.849, \qquad s = 0.328 \tag{11-59}$$

$$\log \frac{1}{C} = 0.29\pi_{x+y} + 0.90\sigma_{x+y} + 0.11\Sigma\pi - 0.013\Sigma\pi^2 + 2.41$$

$$n = 212, \qquad r = 0.860, \qquad s = 0.319 \tag{11-60}$$

In Equation 11-60, $\Sigma\pi$ includes $\pi$ for the R groups, which again are seen to contribute little to the correlation. Correlation is not quite as good as with Equation 11-57, partly because of the inclusion of several new aromatic rings for which no special parameterization was included:

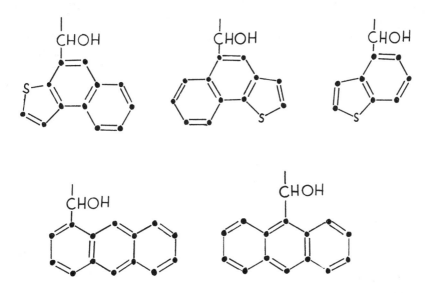

However, the predictability of the earlier equations was quite good. A most important aspect of predictability via correlation equation is:[162] Are predictions being made within explored data space or beyond explored data space? Making correct predictions of compounds of different but similar structure having biological activity within the range of potency already tested is to be more or less expected, but correctly predicting activity higher than any yet found is another quite different problem. Equation 11-57 actually correctly estimates the activity of the new more potent congeners.

The QSAR for the antimalarials has been extended considerably beyond that of Equation 11-60 in the following fashion.[159]

Equation 11-61 has been derived for a larger set of strictly phenanthrene carbinols, omitting the anthracene and thiophene analogues used in Equation 11-60:

$$\log \frac{1}{C} = 0.83\Sigma\sigma + 0.16\ \Sigma\pi + 0.19 \log P - 0.024\ (\log P)^2$$

$$- 0.23(\text{c-side}) + 0.32\text{CNR}_2 - 0.10\text{AB} - 0.64 < 3\text{-cures} + 2.61$$

$$n = 214, \qquad r = 0.925, \qquad s = 0.239, \qquad \log P_0 = 3.9$$

$$(11\text{-}61)$$

where $\pi$ refers to the hydrophobicity of the substituents while $\log P$ refers to the hydrophobic interaction of the whole molecule. It is presumed that $\log P$ relates to the random walk of the drugs in the whole animal, while $\pi$ refers to critical binding at a receptor. The indicator variable c-side takes the value of 1 for congeners having a cycloalkyl group attached to the side-chain nitrogen. The negative sign with this term shows that these congeners are slightly less active than expected. The $\text{CNR}_2$ variable is assigned a value of 1 for congeners containing an extra $\text{CH}_2$ in the side chain, and the AB term is given the value of 1 when substituents are ortho to each other. Because a number of drugs did not produce three cures, their $\log 1/C$ values had to be obtained by extrapolation; hence, these were crudely "corrected" by the use of an indicator variable "< 3-cures." Although Equation 11-61 contains eight variables, these are well supported by the 214 data points, so that this would seem to be a sound equation from which to design new antimalarial agents.

From a set of 216 2-phenylquinolines acting in the same Rane mouse antimalarial test, the following QSAR was formulated:[160]

$$\log \frac{1}{C} = 0.54\Sigma\sigma + 0.16\Sigma\pi + 0.23 \log P - 0.026\ (\log P)^2$$

$$- 0.18(\text{c-side}) - 0.70 < 3\text{-cures} + 0.37(\text{MR-4}'\text{-Q})$$

$$+ 0.23(\text{Me-6,8-Q}) + 0.13(2\text{-Pip}) + 2.49$$

$$n = 216, \qquad r = 0.898, \qquad s = 0.296, \qquad \log P_0 = 4.37$$

$$(11\text{-}62)$$

Several new terms, not of great importance, are necessary for highest correlation. MR-4'-Q is assigned the value of 1 for 4-substituents on the 2-phenyl moiety (these appear to be involved in a positive steric effect not modeled by $\pi$); the Me-6,8-Q indicator takes the value of 1 when methyl groups are in the 6- and 8-positions and the value of 0.5 is used when only one is present. For some obscure reason such methyl groups are more active than expected. The variable 2-Pip is employed with a value of 1 whenever the amino group of the side chain is a piperidinyl group. A third group of 2,6-diphenylpyridines yield QSAR 11-63.

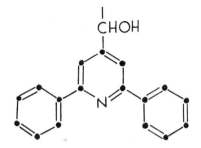

$$\log \frac{1}{C} = 0.95\Sigma\sigma - 0.48 \log P - 0.046 (\log P)^2 - 0.58AB$$

$$+ 0.37\text{NBrPy} - 0.29 < 3\text{-cures} + 1.72 \tag{11-63}$$

$$n = 57, \qquad r = 0.933, \qquad s = 0.212, \qquad \log P_0 = 5.3$$

One new parameter has been introduced for the pyridines (NBrPy), which is assigned the value of 1 for branched alkyl groups attached to the side-chain nitrogen. No $\pi$ term appears in Equation 11-63. In this sense the 2,6-diphenylpyridines resemble the 2-phenylquinolines; however, the MR-4' variable does not improve the QSAR for the pyridines. Unfortunately there is high collinearity between $\Sigma\sigma$, $\Sigma\pi$, and MR-4', which confounds the interpretation of this equation. The $\log P$ and $(\log P)^2$ terms also differ from those of the other two equations, but the confidence limits on these are wide, preventing the inference of a true difference with any degree of certainty. The 491 data points of Equations 11-61 to 11-63 can be combined with 155 miscellaneous compounds of similar structure to formulate Equation 11-64:

$$\log \frac{1}{C} = 0.56\Sigma\sigma + 0.17\Sigma\pi + 0.17 \log P - 0.019 (\log P)^2$$

$$- 0.17(\text{c-side}) + 0.32\text{CNR}_2 - 0.14AB - 0.80 < 3\text{-cures}$$
$$+ 0.28(\text{MR-4'-Q}) + 0.25(\text{Me-6,8-Q}) + 0.081(2\text{-Pip}) \tag{11-64}$$
$$+ 0.17\text{NBrPy} - 0.67\text{Q2P378} + 0.27\text{Py} + 2.69$$
$$n = 646, \qquad r = 0.898, \qquad s = 0.309, \qquad \log P_0 = 4.5$$

One new term (Q2P378) has been added to account for peculiarities of the 155 new congeners. This variable takes the value of 1 for 2-phenylquinolines in which the side chain is attached at positions 3, 7, or 8 rather than the usual 4-position. Attachment at these other positions reduces activity by about a factor of 5 on the average. The indicator variable Py is assigned the value of 1 for the pyridines.

Although Equation 11-64 is a very complex expression containing 14 variables, on the average each variable is supported by 46 data points. The range in activity for the 646 compounds exceeds 1000-fold. With a standard deviation of 0.3, the equation predicts, on the average, activity within a factor of 2. It should be noted that 60 different aryl groups (mostly heterocycles) were used with the $-CHOH-CH_2NRR'$ side chain. In addition to the 646 compounds on which the QSAR is based, an additional 103 similar substances have been reported as inactive or of doubtful activity. These are moderately well accounted for by Equation 11-64. None of them were predicted to be highly active.

Although one would like a simpler equation with fewer indicator variables and more fundamental parameters, the horrendous complexity of the data set as well as uncertainties in the biological assay are formidable barriers to attaining such a relation. The analysis clearly shows that if the QSAR paradigm had been used in the design from the beginning, many of the structural features important for antimalarials for this type of compound could have been defined with far fewer than the 750 compounds actually tested.

The problem of what finally constitutes a good correlation equation for describing a QSAR is too complex for any kind of simple set of guidelines. One must attempt to satisfy two general standards: a qualitative and a quantitative. The quantitative standards fall under the domain of statistical tests. Above all, one wants to avoid the publication of meaningless QSARs. The following criteria should be considered.

1. The problem of chance correlation due to too few data points and variables has been studied by Topliss and co-workers.[164,165] These studies indicate that five or six would be a reasonable minimum standard. However, the number of data points needed per variable to ensure a specific level of confidence depends on the total number of variables screened.[165] Breiman and Freedman[166] have recently presented a highly technical analysis of the problem.

2. One should have a good spread and a wide range in the values for each of the independent variables ($\pi$, $\sigma$ etc). Also, with the usual quality of medicinal chemical data, one normally needs at least a range of one log unit in the dependent variable to obtain a meaningful correlation.

3. QSARs in which the standard deviation from the regression equation is lower in value than what one might expect for the standard deviation on the dependent variable are obviously suspect.

4. As the development of the QSAR proceeds, the collinearity among the variables should be monitored to be sure that collinearity between variables does not arise.

5. The possibility of a variable being closely correlated to one or more other variables is one of the most series problems in the formulation of QSARs. It has been considered by a number of authors.[96,111,166,168] One way to avoid

the problem is to derive new vectors via the method of principal components. A program called SIMCA[169] has been designed for this approach. While working with the principal components solves the problem of having independent variables, it does introduce a new problem, that of trying to deduce the physiocochemical meaning contained in the derived vectors.

6. The most important test of the final QSAR is, Does it make sense? That is, compared to all other structure–activity information about the compounds under consideration, is the equation reasonable? To evaluate the parameters of a QSAR, one must place some kind of confidence limits on them. The coefficients must make sense. For example, in the linear or bilinear models one normally finds a positive slope with the $\pi$ term in the range of 0.3–1.2. This range suggests simple partitioning. A value higher than 1.2 might involve some kind of amplification brought about by the hydrophobic interaction. Ideally one would like to compare coefficients with $\sigma$ with those for comparable reactions with simple organic compounds in homogeneous solutions.

Recent thoughtful general discussions of QSAR have been presented by Wold and Dunn[170] and Lewi.[171]

One must not conclude from the foregoing discussion that the QSAR considered will hold regardless of the structural complexity introduced in the series of congeners. By making more and more gross changes in substituents, sooner or later large enough changes can be made to break the fit. This is in part due to our lack of understanding of how even the best characterized enzymes work. For instance, it is becoming ever more clear that there is a large amount of flexibility in enzymes for which we cannot clearly account. Another shortcoming of QSAR is of course the crudeness of our physicochemical parameters. But, even with these limitations, it is clear that QSAR can be of great help in expanding our understanding of how organic compounds interact with the macromolecules of living systems.

It is hoped that the examples of structure–activity relationships covered in this chapter, which are but a tiny fraction of those published, will encourage chemists to study various aspects of drugs from binding to isolated receptors to action in whole animals with the objective of developing better mathematical models. The stage has been set for the most exciting work, which is yet to come in this field for organic, medicinal, and biochemists.

As Parascandola[172,173] has nicely documented, the controversy over structure–activity relationships during the latter part of the nineteenth century and most of this century has been vigorous. There has been a constant reoccurring dream that we were on the verge of general expressions to relate structure to biological activity. While it is still clear that we are nowhere near a therapeutic millenium brought about by drug design, it is also evident, since the advent of QSAR in the early 1960s, that we are making good and steady progress in understanding how organic compounds react with living systems. Luck will always play a very important role in drug discovery,[174] but good fortune will largely favor those who have well-prepared minds. We believe that QSAR will loom large in the well-prepared mind.

## REFERENCES

1. Martin, Y.C. "Quantitative Drug Design". Marcel Dekker: New York, 1978.
2. Franke, R. "Optimierungsmethoden in der Wirkstufforschung". Academie-Verlag: Berlin, 1980.
3. Hansch, C. In "Correlation Analysis in Chemistry", Chapman, N.B.; and Shorter, J., Eds.; Plenum Press: New York, 1978, p. 397.
4. Cammarata, A.; Rogers, K.S. In "Advances in Linear Free Energy Relationships", Chapman, N.B.; Shorter, J., Eds. Plenum Press: New York, 1972, p. 401.
5. Seydel, J.K.; Schaper, J. "Chemische Struktur und biologische Aktivität von Wirkstoffen". Verlag Chemie: Weinheim, 1979.
6. Crum-Brown, A.; Frazer, T. *Trans. R. Soc. Edinburgh* **1868–1869**, *25*, 151, 693.
7. Hansch, C. In "Drug Design", Vol. 1; Ariens, E.J., Ed.; Academic Press: New York, 1971, p. 271.
8. Ingold, C.K. "Structure and Mechanism in Organic Chemistry", 2nd ed.; Cornell University Press: Ithaca, N.Y., 1969.
9. Hammett, L.P. "Physical Organic Chemistry", 2nd ed.; McGraw-Hill: New York, 1970.
10. Chapman, N.B.; Shorter, J., Eds. "Advances in Linear Free Energy Relationships". Plenum Press: London, 1972.
11. Chapman, N.B.; Shorter, J., Eds. "Correlation Analysis in Chemistry". Plenum Press: London, 1978.
12. Taft, R.W. In "Steric Effects in Organic Chemistry", Newman, M.S., Ed.; Wiley: New York, 1956, p. 556.
13. Fujita, T.; Iwasa, J.; Hansch, C. *J. Am. Chem. Soc.* **1964**, *86*, 5175.
14. Hansch, C.; Maloney, P.P.; Fujita, T.; Muir, R.M. *Nature* **1962**, *194*, 178.
15. Hansch, C. *Acc. Chem. Res.* **1969**, *2*, 232.
16. Hansch, C. In "Structure-Activity Relationships", Vol. 1, Cavallito, C.J., Ed.; Pergamon Press: Oxford, 1973.
17. Daniel, C.; Wood, F.S. "Fitting Equations to Data", 2nd ed.; Wiley-Interscience: New York, 1980.
18. Topliss, J.G., Ed. "Quantitative Structure–Activity Relationships of Drugs". Academic Press: New York, 1983.
19. Dearden, J.C., Ed. "Quantitative Approaches to Drug Design". Elsvier: Amsterdam, 1983.
20. Kakkis, E.; Palmire, V.C.; Strong, C.D.; Bertsch, W.; Hansch, C.; Schirmer, U. *J. Agric. Food Chem.* **1984**, *32*, 133.
21. Briggs, G.G., Jr.; Bromilow, H.; Evans, A.A.; Williams, M. *Pestic. Sci.* **1983**, *14*, 492.
22. Kirino, W.; Hashimoto, S.; Furuzawa, K.; Takayama, C.; Ohshio, H. *J. Pestic. Sci.* **1983**, *8*, 315.
23. Sparatore, F.; Grieco, C.; Silipo, C.; Vittoria, A. *Farmaco Ed. Sci.* **1979**, *34*, 11.
24. Doweyko, A.M.; Bell, A.R.; Minatelli, J.A.; Relyea, D.I. *J. Med. Chem.* **1983**, *26*, 475.
25. Magee, P.S. In Reference 18, p. 393.
26. Fujinami, A.; Mine, A.; Fujita, T. *Agric. Biol. Chem.* **1974**, *38*, 1399.
27. Kirino, O.; Oshita, H.; Oishi, T.; Kato, T. *Agric. Biol. Chem.* **1980**, *44*, 35.
28. Nishimura, K.; Veno, A.; Nakagawa, S.; Fujita, T.; Nakajima, M. *Pestic. Biochem. Physiol.* **1979**, *11*, 83.
29. Magee, P.S. *Chem. Technol.* **1981**, *11*, 378.
30. Nakada, Y.; Ohno, S.; Yoshimoto, M.; Yura, Y. *Agric. Biol. Chem.* **1978**, *42*, 1365.
31. Nishimura, K.; Veno, A.; Nakagawa, S.; Fujita, T.; Nakajima, M. *Pestic. Biochem. Physiol.* **1982**, *17*, 271.
32. Watanabe, K.; Fujita, T.; Takimoto, A. *Plant Cell Physiol.* **1981**, *22*, 1469.
33. Bawden, D.; Gymer, G.E.; Marriott, M.S.; Tute, M.S. *Eur. J. Med. Chem.* **1983**, *18*, 91.
34. Kirino, O.; Yamamoto, S.; Kato, T. *Agric. Biol. Chem.* **1980**, *44*, 2149.
35. Verloop, A.; Tipker, J. *Pestic. Sci.* **1976**, *7*, 379.

36. Takayama, C.; Fujinami, A. *Pestic. Biochem. Physiol.* **1979**, *12,* 163.
37. Hansch, C.; Lein, E.J. *J. Med. Chem.* **1971**, *14,* 653.
38. Neely, W.B.; Branson, D.R.; Blau, G.E. *Environ. Sci. Technol.* **1974**, *8,* 1113.
39. Veith, G.D.; Kosian, P. In "Physical Behavior of PCBs in the Great Lakes", Mackay, D.; Paterson, S.; Eisenreich, S.J.; and Simmons, M.S., Eds.; Ann Arbor Science: Ann Arbor, Mich., 1982, p. 269.
40. Konemann, H. *Toxicology* **1981**, *19,* 209.
41. Saarikoski, J.; Viluksela, M. *Ecotoxicol. Environ. Saf.* **1982**, *6,* 501.
42. Geyer, H.; Sheehan, P.; Kotzias, D.; Freitag, D.; Korte, F. *Chemosphere* **1982**, *11,* 1121.
43. Chiou, C.T.; Porter, P.E.; Schmedding, D.W. *Environ. Sci. Technol.* **1983**, *17,* 227.
44. Kenaga, E.E. *Ecotoxicol. Environ. Saf.* **1979**, *4,* 26.
45. Levitan, H. *Proc. Natl. Acad. Sci. U.S.A.* **1977**, *74,* 2914.
46. Briggs, G.G. *J. Agric. Food Chem.* **1981**, *29,* 1050.
47. Hansch, C. *J. Med. Chem.* **1970**, *13,* 964.
48. Lindley, M. *Chem. Ind.* **1983**, 10.
49. Kaliszan, R.; Pandowski, M.; Szymula, L.; Lamparczyk, H.; Nasal, A.; Tomaszewska, B.; Grzybouski, J. *Pharmazie* **1982**, *37,* 499.
50. Gardner, R.J. *Chem. Senses* **1980**, *5,* 185.
51. Gardner, R.J. *J. Sci. Food Agric.* **1980**, *31,* 23.
52. Greenberg, M.J. *J. Agric. Food Chem.* **1980**, *28,* 562.
53. Iwamura, H. *J. Med. Chem.* **1980**, *23,* 308.
54. Greenberg, M.J. *J. Agric. Food Chem.* **1979**, *27,* 347.
55. Fujita, T. *Adv. Chem.* **1972**, *114,* 6.
56. Hansch, C. *Drug Metab. Rev.* **1973**, *1,* 1.
57. Hansch, C.; Vittoria, A.; Silipo, C.; Jow, P.Y.C. *J. Med. Chem.* **1975**, *18,*. 546.
58. Moffat, A.C.; Sullivan, A.T. *J. Forens. Sci.* **1981**, *21,* 239.
59. King, L.A.; Moffat, A.C. *Med. Sci. Law* **1983**, *23,* 193.
60. Hansch, C.; Dunn, W.J., III. *J. Pharm. Sci.* **1972**, *61,* 1.
61. Dunn, W.J., III; Hansch, C. *Chem. Biol. Interactions* **1974**, *9,* 75.
62. Silipo, C.; Vittoria, A. *Farm Ed. Sci.* **1979**, *34,* 858.
63. Wooldridge, K.R.H. *Eur. J. Med. Chem.* **1980**, *15,* 63.
64. Yasuda, Y.; Tochikubo, K.; Hachisuka, Y.; Tomida, H.; Ikeda, K. *J. Med. Chem.* **1982**, *25,* 315.
65. Coburn, R.A.; Batista, A.J.; Evans, R.T.; Genco, R.J. *J. Med. Chem.* **1981**, *24,* 1245.
66. Mager, P.P. *Acta Histó chem.* **1980**, *66,* 40.
67. Bowden, K.; Dixon, M.S.; Ranson, R.J. *J. Chem. Res. S* **1979**, 8.
68. Ordukhanyan, A.A.; Landau, M.A.; Rudzit, E.A.; Mndzhoyan Sh. L.; Ter-Zakhalyan, Yr. *Z. Farm. Khim. Zh.* **1980**, *14,* 138.
69. Brown, M.R.W.; Tomlinson, E. *J. Pharm. Sci.* **1979**, *68,* 146.
70. Coats, E.A.; Genther, C.S.; Smith, C.C. *Eur. J. Med. Chem.* **1979**, *14,* 261.
71. Pozzo, A.D.; Dansi, A.; Biassoni, M. *Arzneim-Forsch.* **1979**, *29,* 877.
72. Yu, L.; Wilkinson, C.F.; Anders, M.W. *Biochem. Pharmacol.* **1980**, *29,* 1113.
73. Ohyama, T.; Tokahashi, T.; Ogawa, H. *Biochem. Pharmacol.* **1982**, *31,* 397.
74. Hansch, C.; Deutsch, E.W.; Smith, R.N. *J. Am. Chem. Soc.* **1965**, *87,* 2738.
75. Yoshimoto, M.; Hansch, C. *J. Med. Chem.* **1976**, *19,* 71.
76. Grieco, C.; Hansch, C.; Silipo, C.; Smith, R.N.; Vittoria, A.; Yamada, K. *Arch. Biochem. Biophys.* **1979**, *194,* 542.
77. Carotti, A.; Hansch, C.; Mueller, M.M.; Blaney, J.M. *J. Med. Chem.* **1984** *27,* 1401; **1985**, *28,* 261.
78. Carotti, A.; Smith, R.N.; Wong, S.; Hansch, C.; Blaney, J.M.; Langridge, R. *Arch. Biochem. Biophys.* **1984**, *229,* 112.
79. Abdel-Aal, Y.A.I. *Biochem. Pharmacol.* **1977**, *26,* 2187.
80. King, R.W.; Burgen, A.S.V. *Proc. Roy. Soc. London Ser. B* **1976**, *193,* 107.
81. Miller, G.H.; Doukas, P.H.; Seydel, J.K. *J. Med. Chem.* **1972**, *15,* 700.
82. Fujita, T. *J. Med. Chem.* **1973**, *16,* 923.
83. Wu, R.S.; Wolpert-Defilippes, M.K.; Quinn, F.R. *J. Med. Chem.* **1980**, *23,* 256.
84. Hansch, C.; Steward, A.R. *J. Med. Chem.* **1964**, *7,* 691.
85. Helmer, F.; Kiehs, K.; Hansch, C. *Biochemistry* **1968**, *7,* 2858.
86. Scholtan, W. *Arzneim-Forsch.* **1964**, *14,* 1234.
87. Scholtan, W. *Arzneim-Forsch.* **1978**, *28,* 1037.

88. Bird, A.E.; Marshall, A.C. *Biochem. Pharmacol.* **1967,** *16,* 2275.
89. Seydel, J.K.; Trettin, D.; Cordes, H.P.; Wassermann, O.; Malyusz, M. *J. Med. Chem.* **1980,** *23,* 607.
90. Kuchař, M.; Rejholec, V.; Roubal, Z.; Matoušová, O. *Collect. Czech. Chem. Commun.* **1983,** *48,* 1077.
91. Coulson, C.J.; Smith, V.J. *J. Pharm. Sci.* **1980,** *69,* 799.
92. Barbato, F.; Recanatini, M.; Silipo, C.; Vittoria, A. *Eur. J. Med. Chem.* **1982,** *17,* 229.
93. Gota, S.; Yoshitomi, H.; Nakase, M. *Chem. Pharm. Bull.* (Tokyo) **1978,** *26,* 472.
94. Baguley, B.C.; Denny, W.A.; Atwell, G.J.; Cain, B.F. *J. Med. Chem.* **1981,** *24,* 170.
95. Hansch, C.; Moser, P. *Immunochemistry,* **1978,** *15,* 535.
96. Hansch, C.; Leo, A. "Substituent Constants for Correlation Analysis in Chemistry and Biology". Wiley-Interscience: New York, 1979.
97. Pomona College Medicinal Chemistry Project Data Bank, Claremont, Calif.
98. Perrin, D.D.; Dempsey, B.; Serjeant, E.P. "p$K_a$ Prediction for Organic Acids and Bases". Chapman & Hall: London, 1981.
99. Johnson, C.D. "The Hammett Equation". Cambridge University Press: London, 1973.
100. Reference 99 p. 6.
101. Shorter, J. "Correlation Analysis of Organic Reactivity". Research Studies Press: New York, 1982.
102. Hansch, C.; Leo, A.; Unger, S.H.; Kim, K.H.; Nikaitani, D.; Lien, E.J. *J. Med. Chem.* **1973,** *16,* 1207.
103. Charton, M. *Prog. Phys. Org. Chem.* **1981,** *13,* 119.
104. Swain, C. G.; Lupton, E.C., Jr. *J. Am. Chem. Soc.* **1968,** *90,* 4328.
105. Ehrenson, S.; Brownlee, R.T.C.; Taft, R.W. *Prog. Phys. Org. Chem.* **1973,** *10,* 1.
106. Kolthoff, I.M.; Chantooni, M.K., Jr. *J. Am. Chem. Soc.* **1971,** *93,* 3843.
107. Davis, M.M.; Hetzer, H.B. *J. Res. Natl. Bur. Stand.* **1958,** *60,* 569.
108. Reference 99, pp. 20–24.
109. Kalfus, K.; Kroupa, J.; Večeřa, M.; Exner, O. *Collect. Czech. Chem. Commun.* **1975,** *40,* 3009.
110. Johnels, D.; Edlund, U.; Grahn, H.; Hellberg, S.; Sjöström, M.; Wold, S. *J. Chem. Soc. Perkin Trans. 2* **1983,** 863.
111. Hansch, C.; Unger, S.H.; Forsythe, A.B. *J. Med. Chem.* **1973,** *16,* 1217.
112. Austel, V. *Quant. Struct.-Act. Relat.* **1983,** *1,* 59.
113. Skibo, E.B.; Meyer, R.B., Jr. *J. Med. Chem.* **1981,** *24,* 1155.
114. Hulbert, P.B. *Mol. Pharmacol.* **1974,** *10,* 315.
115. Silipo, C.; Hansch, C.; Grieco, C.; Vittoria, A. *Arch. Biochem. Biophys.* **1979,** *194,* 552.
116. Hansch, C.; Fujita, T. *J. Am. Chem. Soc.* **1964,** *86,* 1616.
117. Hansch, C. *J. Org. Chem.* **1970,** *35,* 620.
118. Cornell, N.W.; Hansch, C.; Kim, K.H.; Henegar, K. *Arch. Biochem. Biophys.* **1983,** *227,* 81.
119. Kubinyi, H. *Drug Res.* **1979,** *23,* 97.
120. Kauzmann, W. *Adv. Protein Chem.* **1959,** *14,* 37.
121. Frank, H.S.; Evans, M.W. *J. Chem. Phys.* **1945,** *13,* 507.
122. Collander, R. *Physiol. Plant* **1954,** *7,* 420.
123. Gaudette, L.E.; Brodie, B.B. *Biochem. Pharmacol.* **1959,** *2,* 89.
124. Brodie, B.B.; Hogben, C.A.M. *J. Pharm. Pharmacol.* **1957,** *9,* 345.
125. Fieser, L.F.; Ettlinger, M.G.; Fawaz, G. *J. Am. Chem. Soc.* **1948,** *70,* 3228.
126. Leo, A.; Hansch, C.; Elkins, D. *Chem. Rev.* **1971,** *71,* 525.
127. The Hydrophobic Interaction. *Faraday Symp. Chem. Soc.* **1982,** *17,* 7.
128. Rekker, R.F. "The Hydrophobic Fragmental Constant". Elsevier: Amsterdam, 1977.
129. Hansch, C.; Leo, A.; Nikaitani, D. *J. Org. Chem.* **1972,** *37,* 3090.
130. Kühne, R.; Bocek, K.; Scharfenberg, P.; Franke, R. *Eur. J. Med. Chem.* **1981,** *16,* 7.
131. Harris, M.J.; Higuchi, T.; Rytting, J.H. *J. Phys. Chem.* **1973,** *77,* 2694.
132. Leo, A.; Hansch, C.; Jow, P.Y.C. *J. Med. Chem.* **1976,** *19,* 611.
133. Hermann, R.B. *Proc. Natl. Acad. Sci. U.S.A.* **1977,** *74,* 4144.
134. Cramer, R.D., III. *J. Am. Chem. Soc.* **1977,** *99,* 5408.
135. Leo, A. *J. Chem. Soc. Perkin Trans. 2* **1983,** 825.
136. Fujita, T. *J. Pharm. Sci.* **1983,** *72,* 285.
137. Smith, R.N.; Hansch, C.; Ames, M.M. *J. Pharm. Sci.* **1975,** *64,* 599.
138. Scholtan, W. *Arzneim.-Forsch.* **1968,** *18,* 505.

139. Lopez de Compadre, R.L.; Campadre, C.M.; Castillo, R.; Dunn, W.J., III. *Eur. J. Med. Chem.* **1983**, *18,* 569.
140. Hansch, C.; Coats, E. *J. Pharm. Sci.* **1970**, *59,* 731.
141. Grieco, C.; Silipo, C.; Vittoria, A.; Hansch, C. *J. Med. Chem.* **1977**, *20,* 586.
142. Kutter, E.; Hansch, C. *Arch. Biochem. Biophys.* **1969**, *135,* 126.
143. Verloop, A.; Hoogenstraaten, W.; Tipker, J. In "Drug Design", Vol. VII, Ariens, E.J., Ed.; Academic Press: New York, 1976, p. 165.
144. Gitler, C.; Ochoa-Solano, A. *J. Am. Chem. Soc.* **1968**, *90,* 5004.
145. Hansch, C. *J. Org. Chem.* **1978**, *43,* 4889.
146. Smith, R.N.; Hansch, C.; Kim, K.H.; Omiya, B.; Fukumura, G.; Selassie, C.D.; Jow, P.Y.C.; Blaney, J.M.; Langridge, R. *Arch. Biochem. Biophys.* **1982**, *215,* 319.
147. Drenth, J.; Kalb, K.H.; Swen, H.M. *Biochemistry* **1970**, *15,* 3731.
148. Hansch, C.; Smith, R.N.; Rockoff, A.; Calef, D.F.; Jow, P.Y.C.; Fukunaga, J.Y. *Arch. Biochem. Biophys.* **1977**, *183,* 383.
149. Hansch, C.; Blaney, J.M. In "Drug Design: Fact or Fantasy?" Wooldridge, K.R.H., Ed.; Academic Press: New York, 1984.
150. Hansch, C.; Hathaway, B.A.; Guo, Z.R.; Selassie, C.D.; Dietrich, S.W.; Blaney, J.M.; Langridge, R.; Volz, K.W.; Kaufman, B.T. *J. Med. Chem.* **1984**, *27,* 129.
151. Roberts, G.C.K. *Pharmacochem. Libr.* **1983**, *6,* 91.
152. Carotti, A.; Hansch, C.; Mueller, M.M.; Blaney, J.M. *J. Med. Chem.* **1984**, *27,* 1401; **1985**, *28,* 261.
153. Carotti, A.; Casini, G.; Hansch, C. *J. Med. Chem.* **1984**, *27,* 1427.
154. Hawkins, H.C.; Williams, A. *J. Chem. Soc. Perkin Trans. 2* **1976**, 723.
155. Carotti, A.; Raguseo, C.; Hansch, C. *Chem.-Biol. Interactions* **1985**, *52,* 279.
156. Hansch, C.; Grieco, C.; Silipo, C.; Vittoria, A. *J. Med. Chem.* **1977**, *20,* 1420.
157. Coats, E.A.; Genther, C.S.; Dietrich, S.W.; Guo, Z.R.; Hansch, C. *J. Med. Chem.* **1981**, *24,* 1422.
158. Selassie, C.D.; Hansch, C.; Khawja, T.S.; Dias, C.D.; Pentecost, S. *J. Med. Chem.* **1984**, *27,* 347.
159. Craig, P.N.; Hansch, C. *J. Med. Chem.* **1973**, *16,* 661.
160. Kim, K.H.; Hansch, C.; Fukunaga, J.Y.; Steller, E.E.; Jow, P.Y.C.; Craig, P.N.; Page, J. *J. Med. Chem.* **1979**, *22,* 366.
161. Hansch, C.; Fukunaga, *J. Chem. Technol.* **1977**, *7,* 120.
162. Hansch, C. *Pharmacochem. Libr.* **1977**, *2,* 47.
163. Denny, W.A.; Cain, B.F.; Atwell, G.J.; Hansch, C.; Panthananickal, A.; Leo, A. *J. Med. Chem.* **1982**, *25,* 276.
164. Topliss, J.G.; Costello, R.J. *J. Med. Chem.* **1972**, *15,* 1066.
165. Topliss, J.G.; Edwards, R.P. *J. Med. Chem.* **1979**, *22,* 1238.
166. Breiman, L.; Freedman, D. *Am. Stat. Assoc.* **1983**, *78,* 131.
167. Wooton, R.; Cranfield, R.; Sheppey, G.C.; Goodford, P.J. *J. Med. Chem.* **1975**, *18,* 607.
168. Unger, S.H. In "Drug Design", Vol IX; Ariëns, E.J. (ed) Academic Press: New York: 1980, p. 48.
169. Dunn, W.J., III; Wold, S.J. Martin, Y. C.; *J. Med. Chem.* **1978**, *21,* 922.
170. Wold, S.; Dunn, W.J., III. *J. Chem. Inf. Comput. Sci.* **1983**, *23,* 6.
171. Lewi, P.J. In "Drug Design", Vol. X, Ariëns, E.J. (ed) Academic Press: New York, 1980, p. 308.
172. Parascandola, *J. Pharm. Hist.* **1974**, *16,* 54.
173. Parascandola, J. *Trends Pharm. Sci.* **1980**, 417.
174. Hansch, C. *J. Chem. Educ.* **1974**, *51,* 360.

# Addendum

## CHAPTER 1

A periodic table for polycyclic aromatic hydrocarbons (PAH) has been developed which allows predictions of PAH reactivity based upon topology and Hückel MO theory.[1] The approach has been extended to enumeration of radical benzenoid hydrocarbons.[2] A very important series of papers has been published by the University of California at Riverside group. In the gas phase, naphthalene, a very volatile PAH, is nitrated in significant yield by $N_2O_5$.[3] This specific reaction is of no particular environmental significance since (a) the lifetime of naphthalene in the presence of ambient $N_2O_5$ concentrations is greater than 10 days while it is only about 12 hours in the presence of OH radical, (b) the primary reactions products are both weak direct and activated mutagens.[3] Concentrations of $N_2O_5$ peak at night since it is in thermal equilibrium with $NO_2$ and $NO_3$, the latter rapidly photolyzing, and a concentration as high as 14 ppb was measured in Riverside, California.[4] $N_2O_5$ exhibits an interesting dichotomy in its reactions with certain PAH. When it is sublimed and passed over fluoranthene adsorbed on a filter, the products are the 1-, 3-, 7-, and 8-isomers of nitrofluoranthene (NF).[5] Similar behavior is observed when $N_2O_5$ is employed in polar solvents such as acetonitrile and nitromethane and the products are similar to those formed by reaction of fluoranthene with nitric acid or $NO_2$. Apparently, the nitrating agent acts as $NO_2^+NO_3^-$ under these polar conditions.[5] In contrast, gas phase reaction or reaction in carbon tetrachloride each produce 2-nitrofluoranthene (2-NF) as the only mononitro derivative. Utilization of the $L_r^+$ values indicate that the 2-position should be least reactive toward protons and Lewis bases. Thus, the nitrating agent under nonpolar conditions appears to be undissociated $N_2O_5$.[5] This explains the presence of 2-NF on ambient airborne particulates, which is formed in the atmosphere, most probably with gas phase fluoranthene.[6] Previously, this compound had been misidentified as 3-NF.[6] The reasons for the confidence in this statement rest upon (a) the previously noted observations that 2-NF is not formed on filters, thus ruling out artifacts, and (b) the nonobservation of this compound in fresh emissions.[6] Similar observations were also reported for 2-nitropyrene.[6] Ambient levels of 2-NF were reported to be in the range of 0.07–0.3

ng/m$^3$, which may be compared with 0.008–0.03 ng/m$^3$ for 1-nitropyrene (1-NP) and 0.003–0.02 ng/m$^3$ for 2-NP.[6] An initial rough determination of the direct mutagenicity of 2-NF indicated a value of 4000 rev/$\mu$g which could be compared with 2000 rev/$\mu$g for 1-NP and 9000 rev/$\mu$g for 2-NP. Our own studies[7] indicated a value of 920 rev/$\mu$g for 2-NF,[8] 1000 rev/$\mu$g for 1-NP and 13,000 rev/$\mu$g for 3-NF. Additional studies of reactions of a variety of tetracyclic and pentacyclic PAH with N$_2$O$_5$ on filters and in solution again indicates the dichotomy in the order of relative reactivities.[9] A mechanism for the formation of 2-NF in ambient air has been proposed.[10]

A very interesting study[11] of photochemical decomposition of fifteen PAH using a rotary photoreactor indicated highly variable, and for some compounds, short lifetimes on alumina and silica gel, two substrates that are light in color and have high surface areas—factors favorable for photochemistry. The lifetimes on carbon black are usually in the 500 to over 1000 h range, while those on fly ash are almost uniformly in the 30–50 h range. These carbon-containing particulates have small surface areas and are dark, possibly introducing an "inner filter" effect. The interesting point suggested by the fly ash study is the possibility that all exposed PAH decompose at essentially the same rate in sunlight. This could explain the similarity between summer and winter PAH profiles. Recent studies[12–14] on wood soot indicate that light is a much more significant determinant of ambient reactivity than O$_3$ or NO$_2$ and relative activation parameters for decomposition have been derived.

Although the potential use of PAH profiles for apportionment of emission sources has been reviewed,[15] convincing evidence of its widespread applicability is not yet apparent. However, some success has been achieved in a lightly urbanized town in New Zealand, in which domestic wood combustion and automobiles are the chief PAH sources.[16] A more extensive discussion of the relative decay index of PAH has appeared along with a series of characteristic source profiles.[17]

# CHAPTER 2

The addendum to Chapter 2 consists of two parts, a brief update involving the exotic species HOC$^+$ and R$_4$B earlier discussed, and a simple analysis combining substituent effect reasoning and data on the protonated forms of most of the reference bases referred to in the main body of the chapter. HOC$^+$, O-protonated CO, remains of continuing high interest to experimentalists[18a–c] and theorists[18d,e] alike. There is some dissension as to the value for the O-proton affinity of CO and the range of suggested values is surprisingly high for such a seemingly simple species. It is, however, generally agreed that HOC$^+$ is ca. 40 kcal/mol less stable than its much more conventional isomer, HCO$^+$, so that the O-proton affinity is ca. 100 kcal/mol. If this value is correct, then the protonation enthalpy on oxygen of CO must be comparable to the protonation enthalpy of $H_2$. Many of these same studies have discussed the competing protonation reactions

$$H_3^+ + CO \rightarrow H_2 + COH^+ \tag{A2-1}$$

$$H_3^+ + CO \rightarrow H_2 + HCO^+ \tag{A2-2}$$

and the intermediacy of a $(H_2 \cdots H \cdots CO)^+$ complex. This complex is an isomer of a singly charged cation containing three hydrogens, one carbon and one oxygen other than the more standard and better characterized C- and O-protonated formaldehyde, $CH_3O^+$ and $CH_2OH^+$ respectively.

The neutral 7 electron tetracoordinate species $R_4B$ were discussed earlier in connection with the proton affinities of ethylene, propene and isobutene. Since writing the chapter, these species have been demonstrated to be useful reactive intermediates in synthetic organic chemistry.[19] More precisely, both methyl and benzyl triphenylboron have been postulated in the photoalkylation of dicyanoarenes by the more conventional corresponding alkyl triphenylborate salts. While thermochemistry is still absent, it is noteworthy that there is selectivity as to which $C-B$ bond is broken.

The second part of the addendum to this chapter compares the stabilization of a variety of simple cations by $\alpha$-hydroxy and $\alpha$-methyl substituents. In the particular, consider the two hydride transfer reactions:

$$HOCH_2^+ + CH_4 \rightarrow HOCH_3 + CH_3^+ \tag{A2-3}$$

$$CH_3CH_2^+ + CH_4 \rightarrow CH_3CH_3 + CH_3^+ \tag{A2-4}$$

The heats of these reactions may be recognized as the stabilization energies of $CH_3^+$ by HO- and $CH_3$-substitution, respectively. These reactions may also be recognized as involving the protonated forms of formaldehyde (O-protonation) and of ethylene (the "classical" ion). Both reactions are endothermic and from the experimental heats of formation we deduce that $CH_3^+$ ion is stabilized by 64 and 39 kcal/mol, respectively, by HO- and $CH_3$-substitution. Analogously, using reactions involving protonated propene and isobutene such as A2-5,

$$(CH_3)_2CH^+ + CH_3CH_3 \rightarrow (CH_3)_2CH_2 + CH_3CH_2^+ \tag{A2-5}$$

we find that $CH_3CH_2^+$ and $(CH_3)_2CH^+$ are stabilized by 46 and 34 kcal/mol respectively by HO-substitution and by 25 and 18 kcal/mol by $CH_3$-substitution. It appears that stabilization by $HO-$ and $CH_3-$ correlate. Indeed a simple linear regression on these sets of stabilizations (x equalling that due to $CH_3-$ and y for $HO-$) where we forced the line through the origin results in a slope of 1.72, and a correlation coefficient, $r^2$, of 0.996. If we simply use this new equation with the energetics of $HOO^+$ and $HO^+$, ie, protonated molecular and atomic oxygen and the accompanying 69 kcal/mol stabilization of $OH^+$ on hydroxyl substitution, the corresponding $CH_3$-stabilization to be 40 kcal mol. This gives us a handle on the C-protonation of formaldehyde and indeed we predict a heat of formation of $CH_3O^+$ of 279 kcal/mol. This corresponds to a difference of the C- and O-proton affinities of formaldehyde of 113 kcal/mol. This value is considerably higher than that found earlier in the chapter. This may occur because only in "real" $CH_3O^+$ and not in any of the above methylated or hydroxylated cations can the substituent interact with more than one unfilled orbital on the cationic center. The three-fold symmetry of the $CH_3O^+$ cation allows there to be two coequal stabilizing interactions, and this has been ignored in our analysis above. While we don't suggest multiplying the previous stabilization by two, nonetheless, we do conclude that the stability of "real" $CH_3O^+$ may have been seriously underestimated.

Another case where there are two such coequal stabilizing interactions is in $CH_3CO^+$, a species recognized as protonated ketene. The directly deduced stabilization of $HCO^+$, protonated carbon monoxide upon methyl substitution is 26 kcal/mol. This value is no doubt too high when used to derive the HO-stabilization energy on transforming $HCO^+$ into $HOCO^+$, ie, protonated carbon dioxide. The statistically deduced 45 kcal/mol worth of stabilization in $HOCO^+$ is also clearly an overestimate. However, that only 9 kcal/mol is found using experimental numbers documents that $CO_2$ is an anomalously weak base. This new analysis, unlike the earlier one in the text, does not refer to $CO_2$ at all, and so we deduce that the anomalous basicity is due to both the stabilization of the neutral base and destabilization of the protonated ion.

# CHAPTER 3

## Physical Factors in the Hydration of Protonated Oxygen and Nitrogen Bases.

Section 5 of Chapter 3 described a model for the solvation of the partially hydrated clusters $BH^+ \cdot 4H_2O$ by bulk water. This analysis yielded the hydrophobic solvation energies of onium ions. Once this factor is known, the model can be extended to calculate all the solvation factors of the onium ions $BH^+$ themselves.

The solvation of ions involves a complex combination of interactions between the ion and the solvent. The solvent acts both as a collection of molecules, leading to specific structure-dependent interactions, and as a continuum dielectric resulting in interactions which depend only on the ionic size. The total interaction energy may be decomposed to the physical factors as in Equation A3-1:

$$\Delta H^\circ_{g \to aq}(BH^+) = \Delta H^\circ_{cavity}(BH^+) + \Delta H^\circ_{dielectric}(BH^+)$$
$$+ \Delta H^\circ_{hydrophobic} + \Delta H^\circ_{IHB} \quad (A3\text{-}1)$$

These terms are calculated by the following procedure:

1. Partial hydration by four $H_2O$ molecules, in the cluster $BH^+4H_2O$, contains the ionic hydrogen bonding (IHB) energies. Therefore, the further hydration energy of the clusters by bulk water, $\Delta H^\circ_{g \to aq}(BH^+ \cdot 4H_2O)$ does

not involve further *ionic* hydrogen bonding. The solvation energy of the cluster is calculated using clustering data and Born-Haber cycles, and its cavity, dielectric, and residual hydrogen bonding (RHB, see below) factors are calculated.

2. The difference between $\Delta H^\circ_{g\rightarrow aq}(BH^+ \cdot 4H_2O)$ and its first three components yields $\Delta H^\circ_{hydrophobic}$ (ie., the total solvation energies of the alkyl groups due to $CH^{\delta+} \cdots OH_2$ interactions).

3. Next, the experimental solvation enthalpy of the naked ion $BH^+$ itself is decomposed to cavity, dielectric, hydrophobic and IHB factors (Equation A3-1). The hydrophobic energy has been found in the preceding step. The cavity and dielectric energies are calculated from Equations 3-19–3-21 in the chapter, using the equivalent radii of the $BH^+$ ions.

4. Finally, the ionic hydrogen bonding contribution to the solvation of the ion $BH^+$, ie., $\Delta H^\circ_{IHB}$, is obtained from Equation A3-1, using the experimental heats of solvation.

The results in Table A3-1 show a reasonable variation of the solvation factors with ion structure as follows. The sum of the cavity and dielectric terms reflects the interaction of the ion with the solvent as a continuum. The cavity term becomes more positive and the dielectric term less negative (less stabilizing) with increasing size. Therefore, for example, $\Delta H^\circ_{continuum}$ decreases from $-40.6$ to $-22.8$ kcal/mol in going from $Me_2COH^+$ to $(i\text{-}Pr)_2COH^+$. At the same time, the hydrophobic solvation becomes more stabilizing. As a consequence, solvation is dominated by continuum factors for, for example, $Me_2COH^+$, but molecular (hydrophobic plus ionic hydrogen bonding) interactions become dominant in the solvation of $(i\text{-}Pr)_2COH^+$.

The ionic hydrogen bonding (IHB) factor changes regularly, increasing by about 10 kcal/mol for each available protic hydrogen, eg, from $R_2OH^+$ to $ROH_2^+$ and $R_3O^+$, and from $R_3NH^+$ to $R_2NH_2^+$ to $H_3O^+$. The ionic hydrogen bonding of water to oxonium ions is stronger than to ammonium ions, as is expected on the basis of charge densities. It is encouraging that the model gives such reasonable trends, since $\Delta H^\circ_{IHB}$ is calculated as a residue term and involves the errors in all the other terms.

Finally, the hydrophobic energies also increase regularly with alkyl size as expected. The contribution of hydrophobic solvation is similar to or larger than IHB already in ions of modest alkyl size, such as $(Me_2)COH^+$ and $Me_2NH_2^+$, and becomes the dominant intermolecular force in larger ions. However, the hydrophobic $CH^{\delta+} \cdots O$ interaction energy per alkyl hydrogen atom ($\Delta H^\circ_{hydrophobic}/n_{CH}$) decreases somewhat, from about $-3$ kcal/mol in smaller ions to a roughly constant value of about $-2.4 \pm 0.2$ kcal/mol in larger ions. The latter value is probably a good approximation for hydrophobic solvation per alkyl hydrogen in aliphatic biological ions (and, as we shall see presently, also of the hydrophobic hydration of neutral bases). We also note that the hydrophobic solvation of aromatic groups is less efficient than of aliphatic groups, due to the smaller number of CH hydrogens available for $CH^{\delta+} \cdots OH_2$ interactions.

TABLE A3-1. Physical Factors in the Hydration of Protonated Oxygen and Nitrogen Bases BH$^+$ (Enthalpies in kcal/mol, Entropies in cal/mol K)

| | $r(\text{Å})$ | $-\Delta H_{diel}$ | $\Delta H_{cav}$ | $-\Delta H_{hydph}$ | $-\Delta H^\circ_{hydph(B)}$ | $\Delta H_{IHB}$ | $-\Delta H^\circ_{hydph}/n_{CH}$ | $\Delta S^\circ_{cav}$ | $-\Delta S^\circ_{hydph}$ |
|---|---|---|---|---|---|---|---|---|---|
| **Water, Alcohols and Ethers** | | | | | | | | | |
| H$_3$O$^+$ | 1.74 | 72.8 | 6.5 | 0 | 8.2 | 44.1 | 0 | 8.5 | 0 |
| MeOH$_2^+$ | 2.08 | 62.6 | 9.2 | 9.1 | 15.2 | 34.8 | 3.0 | 12.2 | 14.2 |
| EtOH$_2^+$ | 2.35 | 56.5 | 11.8 | 16.3 | 19.8 | 34.4 | 3.2 | 15.5 | 32.1 |
| Me$_2$OH$^+$ | 2.36 | 56.2 | 11.9 | 18.1 | 20.8 | 26.2 | 3.0 | 15.7 | 43.5 |
| **Aldehydes, Ketones, Esters** | | | | | | | | | |
| Me$_2$COH$^+$ | 2.48 | 53.8 | 13.2 | 20.2 | 22.8 | 24.2 | 3.4 | 17.4 | 47.4 |
| Me(OMe)COH$^+$ | 2.58 | 52.1 | 14.3 | 21.2 | 23.6 | 24.4 | 3.5 | 18.8 | 50.2 |
| Me(t$-$Bu)COH$^+$ | 3.02 | 45.6 | 19.5 | 28.2 | 31.2 | 24.6 | 2.3 | 25.6 | 54.8 |
| Me(c$-$C$_6$H$_{11}$)COH$^+$ | 3.20 | 43.4 | 21.8 | 32.4 | 34.3 | 25.9 | 2.3 | 28.8 | 57.2 |
| (i$-$Pr)$_2$COH$^+$ | 3.15 | 44.0 | 21.2 | 31.5 | 33.7 | 23.4 | 2.2 | 27.9 | 64.5 |
| (c$-$Pr)$_2$COH$^+$ | 2.97 | 46.3 | 18.8 | 30.2 | 33.1 | 17.3 | 3.0 | 24.8 | 59.1 |
| **Aromatic Aldehyde, Ketone** | | | | | | | | | |
| C$_6$H$_5$CHOH$^+$ | 2.92 | 46.8 | 18.3 | 25.5 | 29.4 | 25.7 | 4.2 | 24.1 | 52.7 |
| Me(C$_6$H$_5$)COH$^+$ | 3.07 | 45.0 | 20.1 | 25.9 | 32.8 | 22.4 | 3.2 | 26.4 | 47.1 |
| **Amide** | | | | | | | | | |
| Me(NMe$_2$)COH$^+$ | 2.81 | 48.4 | 16.9 | 25.2 | 32.6 | 10.8 | 2.8 | 22.2 | |
| **Ammonia, Primary Amines** | | | | | | | | | |
| NH$_4^+$ | 1.81 | 70.4 | 7.0 | 0 | 2.8 | 23.4 | 0 | 9.2 | 0 |
| MeNH$_3^+$ | 2.16 | 60.6 | 10.0 | 9.7 | 12.4 | 20.2 | 3.2 | 13.2 | 16.6 |
| EtNH$_3^+$ | 2.41 | 55.2 | 12.4 | 14.8 | 16.9 | 22.3 | 2.9 | 16.4 | 24.2 |
| i$-$PrNH$_3^+$ | 2.62 | 51.4 | 14.7 | 17.2 | 19.4 | 23.9 | 2.5 | 19.4 | 29.4 |
| n$-$BuNH$_3^+$ | 2.81 | 48.6 | 16.8 | 21.8 | 22.4 | 25.9 | 2.4 | 22.2 | 34.7 |
| **Secondary Amines** | | | | | | | | | |
| Me$_2$NH$_2^+$ | 2.42 | 55.0 | 12.6 | 16.2 | 21.1 | 15.6 | 2.7 | 16.5 | 25.8 |
| Et$_2$NH$_2^+$ | 2.82 | 48.3 | 17.1 | 24.2 | 27.9 | 16.6 | 2.4 | 22.5 | 42.4 |
| n$-$Pr$_2$NH$_2^+$ | 3.20 | 43.3 | 21.9 | 32.8 | 34.6 | 18.1 | 2.3 | 28.8 | 52.7 |
| **Tertiary Amines** | | | | | | | | | |
| Me$_3$NH$^+$ | 2.64 | 51.2 | 14.8 | 23.8 | 27.5 | 6.4 | 2.6 | 19.5 | 36.5 |
| Et$_3$NH$^+$ | 3.12 | 44.4 | 20.8 | 34.5 | 37.0 | 6.1 | 2.3 | 27.4 | 57.1 |
| **Aniline, Pyridine** | | | | | | | | | |
| C$_6$H$_5$NH$_3^+$ | 2.87 | 47.7 | 17.6 | 16.5 | 22.0 | 32.8 | 3.3 | 23.1 | 32.8 |
| PyridineH$^+$ | 2.72 | 49.9 | 15.8 | 21.3 | 27.1 | 10.1 | 4.2 | 20.8 | 36.2 |
| 4-MePyridineH$^+$ | 2.88 | 47.4 | 17.8 | 25.8 | 30.5 | 8.8 | 3.7 | 23.4 | 53.7 |

Interestingly, the hydrophobic solvation of ammonium and oxonium ions of similar size is very similar, though the charge densities of the protonated function, and the hydrogen bonding solvation are quite different. This suggests that the nature of the ionic groups does not affect significantly the solvation of the alkyl groups. Even more surprisingly, the very presence of the charge also does not affect the hydrophobic solvation. This is seen as $\Delta H^\circ_{hydrophobic}(BH^+) = \Delta H^\circ_{hydrophobic}(B)$ (Table A3-1), (the hydrophobic solvation of the neutrals is calculated as shown in Section 5.C. of Chapter 3). Overall, these findings show that, surprisingly, alkyl groups of ions in solution behave as alkyl groups of neutrals molecules. This is explained by the fact that the ionic charge is delocalized from the ion onto the solvent by the hydrogen bonding of the protonated functional groups, leaving the partial charge on the alkyl substituents of ions similar to those in the corresponding neutrals.

It should be pointed out that Taft and co-workers presented a qualitative analysis of solvation factors of ions previously, and identified ionic hydrogen bonding as an important contribution.[20] However, the quantitative analysis summarized here became possible only after the measurement of clustering energies which contain the IHB contributions.

In summary, the bulk hydration of onium ions involves complex forces. Gas-phase data allow a quantitative evaluation of these forces since: (1) gas-phase ionic heats of formation allow the calculation of total solvation energies; and (2) clustering energies allow accounting for the ionic hydrogen bonding contributions, and this in turn allows calculation of the other solvation factors. The solvation factors obtained in this manner show a physically reasonable variation with ion structure. In particular, for most ions (and the corresponding neutral bases as well) each alkyl hydrogen contributes $2.8 \pm 0.1$ kcal/mol to the hydrophobic solvation. Also, for ammonium ions, each protic $NH^+$ hydrogen contributes $7 \pm 1$ kcal/mol to the ionic hydrogen bonding factor.

In conclusion, therefore, the analysis based on gas-phase data breaks down the complex phenomenon of the solvation of organic ions into simple and predictable physical factors.

# CHAPTER 4

A number of papers on hydrogen bonding have appeared in the literature in the last few months. Among them are three articles by Scheiner and co-workers[21-23] in which they examined proton transfers in several additional systems and a review article on their work.[24] In the first manuscript[21] they compared proton transfers in $(H_2S)_2H^+$ and $(H_2O)_2H^+$; in the second[22] they looked at the effect of an external ion on the proton transfers in $(H_2O)_2H^+$ and in $(H_2O)(NH_3)H^+$; in the third[23] they compared proton transfers in $(H_2O)(H_2CO)H^+$ and in $(H_2O)_2H^+$, ie, proton transfers between two hydroxyl oxygens and between a carbonyl and a hydroxyl oxygen. From this work the authors found:

1. In contrast to the single minimum observed for $(H_2O)_2H^+$, when the potential energy surface of $(H_2S)_2H^+$ is calculated as a function of S $\cdots$ S distance and position of the proton, the surface contains two equivalent minima with a small energy barrier between them.[21]
2. Angular distortions of the hydrogen bond and lengthening of the hydrogen bond lead to increased energy barriers for proton transfer for both the sulfur and oxygen systems.[21]
3. There is a larger charge transfer between the two $H_2S$ subunits than between the two $H_2O$ subunits due to the larger polarizability of $H_2S$. Otherwise, the electronic charge distributions are similar.
4. The effects of external ions on the proton transfer barrier are large and extend up to distances of approximately 10 Å. The effects are quite sensitive to the distance and angular placement of the ion with respect to the hydrogen bond but are relatively insensitive to the specific nature of the ion. The latter is indicative of the dominating influence of electrostatics.[22]
5. Approach of an anion or cation toward a symmetric system such as $(H_2O)_2H^+$ introduces a large amount of asymmetry into the transfer potential, while approach of an ion toward an asymmetric system such as $(H_2O)(NH_3)H^+$ reduces the amount of asymmetry.[22]
6. The configuration with the proton bound to the $H_2CO$ moiety is more stable than that with the proton bound to the $H_2O$ moiety when the hydroxyl is located along the direction of a carbonyl lone pair. The stabilities are reversed when the hydroxyl is located along the C=O bond axis.
7. The barrier to proton transfer is consistently higher between two hydroxyl oxygens than between a carbonyl and a hydroxyl oxygen for every O $\cdots$ O distance considered.

Cao and associates[25] have investigated the effect of basis set on proton transfer energies in several ionic symmetric and asymmetric systems. Among the complexes studied are $(H_2O)_2H^+$, $(H_2O)OH^-$, and $(NH_3)_2H^+$. They concluded that the 4-31 + G basis is insufficient to obtain quantitatively accurate dissociation energies for asymmetric anionic clusters. They also applied a modified Marcus theory to aid them in analyzing double-well proton-transfer systems.

Scheiner and co-workers have probed the influence of the basis set on the calculated properties of several neutral hydrogen-bonded systems. They found that obtaining quantitatively accurate geometries and energies for the $(H_2O)HF$ and $(H_2O)HCl$ complexes requires either a triple valence (6-311G**) basis set with electron correlation included via Møller-Plesset theory to the third order (MP3) or a double $\zeta$ basis set augmented by two sets of polarization functions with electron correlation included via MP2 theory.[26] For $(H_3N)HCl$ the 4-31G* basis yields results in fairly good agreement with those from larger bases.[27] However, minimal basis sets (STO-3G, STO-6G, and MINI-1) and the 3-21G basis set lead to exaggerated dissociation energies and incorrect equilibrium structures.[27]

Other investigations of the properties of cluster ions include the work of Sannigrahi and Peyerimhoff on $(HCl)F^-$,[28] Sapse and Jain[29] on $(H_2O)_nF^-$ and $(H_2O)_nCl^-$, n = 1, 2, Yamabe and co-workers[30] on $(H_2S)_nH^+$, n = 2-6, Nguyen[31] on $(N_2)_nNO^+$, n = 1, 2, and Welti and co-workers[32] and Pullman and co-workers[33] on $(H_2O)NH_4^+$. These researchers examined the effect of basis set, geometry optimization, and electron correlation on the structures and stabilities of these complexes. Their results reaffirmed the importance of diffuse functions, polarization functions, and the dispersion contribution in reproducing experimentally measured $\Delta H°$ values.

Schwencke and Truhlar[34] have carried out a systematic study of basis set superposition errors in the computed dimerization energy of (HF)HF. They used the counterpoise method to determine the BSSE for 34 basis sets. They found that utilizing a basis sufficiently large to minimize the counterpoise correction does not guarantee accurate results. In addition, correcting for BSSE for small basis sets does not consistently improve the accuracy of the calculated interaction energies.

Gussoni and associates[35] have employed monomer charge distributions derived from infrared intensities to evaluate which monomers will form weak, neutral CH $\cdots$ B hydrogen bonds. In this work, the authors calculated the hydrogen bond energies of $(NH_3)HCN$, $(HCN)HCN$, and $(NH_3)HCCH$ with a simple electrostatic model. Ray and associates[36] have utilized potential derived point charges to obtain electrostatic interaction energies for a series of neutral hydrogen-bonded dimers of $NH_3$, $H_2O$, $HF$, $CH_3OH$, $CH_3CN$, $HCONH_2$, and $H_2CO$. Their results showed that the potential derived point charges gave a better estimate of electrostatic energies than Mulliken point charges.

# CHAPTER 6

After Chapter 6 was forwarded for publication, a paper by Schaad and Hess[37] came to our attention in which homodesmotic reactions of the type 6-31, 6-32 and 6-33 are called "hyperhomodesmotic" to signify the additional matching of the bonding adjacent to $=CH-$, and a mean $E_T$ increment for pairs of polyenes differing by a $C_2$ unit was employed to evaluate resonance energies for benzene and cyclobutadiene. More recently, reactions analogous to 6-31 have also been used in the case of hexasilabenzene[38] and monosilabenzene[39]. In this latter paper we have employed the alternative methodology developed in this chapter, assessing matching of the bonding in terms of NN and NNN atoms, to show how the presence of the hetroatom in monosilabenzene reactions affects the degree of matching compared to that in benzene reactions. In addition, we have enlarged the scope of the methodology to include the atoms at more distant levels, ie, NNNN and NNNNN, so as to establish the extent to which matching can be achieved in the limit as longer and longer polyene chains are used in the benzene reaction.

## CHAPTER 8

Since the preparation of Chapter 8, a series of directional polarizability constants, $\sigma_\alpha$, have been defined[40]. These have been obtained by calculating polarization potentials of various HX and $CH_3X$ molecules at a point some 3Å away. The values have been successfully tested[40,41] against a variety of experimental data, although most usually polarizability occurs in combination with other electronic effects.

## CHAPTER 9

Substitution of Equation 9-47 into the LD equation (Equation 9-4) gives the relationship

$$Q_X = L\sigma_{lX} + D(\xi\eta + \xi^\circ) + h \qquad (A9\text{-}1)$$

or

$$Q_X = L\sigma_{lX} + D\xi\eta + D\xi^\circ + h \qquad (A9\text{-}2)$$

As $\eta$ is constant throughout a data set we may write

$$R = D\eta \qquad (A9\text{-}3)$$

Equation A9-3 is a triparametric relationship that describes the electrical effect in terms of three different parameters. We have rewritten it as the LDR equation in the form

$$Q_X = L\sigma_{lX} + D\sigma_{dX} + R\sigma_{eX} + h \qquad (A9\text{-}4)$$

where

$$\sigma_l = \sigma_l, \sigma_d = \xi^\circ, \sigma_e = \xi \qquad (A9\text{-}5)$$

Thus, $\sigma_l$ is the localized electrical effect parameter, $\sigma_d$ represents the intrinsic delocalized electrical effect observed when the electronic demand of the active site Y is minimal, and $\sigma_e$ accounts for the variation of the delocalized electrical effect with the electronic demand of Y. The LDR equation should be capable of describing the complete range of electrical substituent effects.

Results obtained for more than 300 data sets suggest that this is indeed the case, and that it is applicable to *all electrical substituent effects encoun-*

*tered in data sets with substituents bonded to carbon.* It follows from the separation of electrical effects into three constituents that the $\sigma_D$ constants are composite parameters that are a function of $\sigma_d$ and $\sigma_e$. The LDR equation has the advantage that the magnitude and sign of the electronic demand are easily determined, and that there is no question about the choice of substituent parameters. As a triparametric equation it requires large data sets for successful application than does the diparametric LD equation, which is a special case of the LDR equation and of course remains valid. When the nature of the electronic demand of the data set is known and the number of available data points is not large, the LD equation is the preferred model of substituent effects; while when the data set is sufficiently large and the nature of the electronic demand is unknown, the LDR equation is preferred.

## CHAPTER 10

The following information updates Tables 10-6 and 10-7, respectively, (see the notes at tops of these tables):

TABLE **10-6.** Some Small-Molecule Modeling Systems

**INSIGHT**

| | |
|---|---|
| Author: | H. Dayringer and A. Tramontano, Monsanto Laboratories/University of California San Francisco |
| Distributor: | BIOSYM Technologies Inc., San Diago, California 92121 |
| Capabilities: | Molecular, macromolecular structure creation, modification, display and manipulation. Interfaces to DISCOVER energetics and simulation package. |
| Hardware: | VAX mainframe, E and S PS 300 graphics processor; others. |

TABLE **10-7.** Some Macromolecular Modeling Systems

**MACROMODEL**

| | |
|---|---|
| Author: | Clark Still, Columbia University, New York, N.Y. |
| Distributor: | Columbia University, New York, N.Y. |
| Capabilities: | Molecular, macromolecular structure creation, modification, display and manipulation. Conformation generation and analysis; energetics. |
| Hardware: | VAX mainframe, E and S PS 300 graphics processor; VT 100 retrographics, Tektronix 4105 and Lundy/Sigma. |

# ADDENDUM REFERENCES

1. Dias, J.R. *Acc. Chem. Res.* **1985,** *18,* 241.
2. Dias, J.R. *J. Mol. Struct. (THEOCHEM)* **1986,** *137,* 9.
3. Pitts, J.N., Jr.; Atkinson, R.; Sweetman, J.A.; Zielinska, B. *Atmos. Environ.* **1985,** *19,* 701.
4. Atkinson, R.; Winer, A.M.; Pitts, J.N., Jr. *Atmos. Environ.* **1986,** *20,* 331.
5. Zielinska, B.; Arey, J.; Atkinson, R.; Ramdahl, T.; Winer, A.M.; Pitts, J.N., Jr. *J. Am. Chem. Soc.* In press.
6. Pitts, J.N., Jr.; Sweetman, J.A.; Zielinska, B.; Winer, A.M.; Atkinson, R. *Atmos. Environ.* **1985,** *19,* 1601.
7. Greenberg, A.; Darack, F.; Wang, Y.; Hawthorne, D.; Natsiashvili, D.; Harkov, R.; Louis, J.; Atherholt, T. Presentation at 1986 EPA/APCA Symposium on Measurement of Toxic Air Pollutants, April 27–30, 1986, Raleigh, NC.
8. 2-Nitrofluoranthene was a gift from Drs. Janet Sweetman and Barbara Zielinska.
9. Pitts, J.N., Jr.; Sweetman, J.A.; Zielinska, B.; Atkinson, R.; Winer, A.M.; Harger, W.P. *Environ. Sci. Technol.* **1985,** *19,* 1115.
10. Sweetman, J.A.; Zielinska, B.; Atkinson, R.; Ramdahl, T.; Winer, A.M.; Pitts, J.N., Jr. *Atmos. Environ.* **1986,** *20,* 235.
11. Behymer, T.D.; Hites, R.A. *Environ. Sci. Technol.* **1985,** *19,* 1004.
12. Kamens, R.M.; Perry, J.M.; Saucy, D.A.; Bell, D.A.; Newton, D.L.; Brand, B. *Environ. Int.* **1985,** *11,* 131.
13. Kamens, R.M.; Fulcher, J.N.; Guo, Z. *Atmos. Environ.* In press.
14. Kamens, R.; Guo, Z.; Fulcher, J.; Bell, D. Presentation at Tenth International Symposium on Polynuclear Aromatic Hydrocarbons, Battelle Laboratory, Columbus, Ohio, October, 1985.
15. Daisey, J.M.; Cheney, J.L.; Lioy, P.J. *J. Air Poll. Control Assoc.* **1986,** *36,* 17.
16. Cretney, J.R.; Lee, H.K.; Wright, G.J.; Swallow, W.H.; Taylor, M.C. *Environ. Sci. Technol.* **1985,** *19,* 397.
17. Masclet, P.; Mouvier, G.; Nikolaou, K. *Atmos. Environ.* **1986,** *20,* 439.
18. Some of the more recent relevant experimental studies include: a) Burgers, P.C., Holmes, J.L., Mommers, A.A. *J. Amer. Chem. Soc.* **1985,** *107,* 1099; b) Wagner-Redeker, W., Kemper, P.R., Jarrold, M.F., Bowers, M.T. *J. Chem. Phys.* **1985,** *83,* 1121; c) McMahon, T.B., Kebarle, P. *J. Chem. Phys.* **1985,** *83,* 3919. Some of the recent relevant theoretical studies include: d) Dixon, D.A., Komornicki, A., Kraemer, W.P. *J. Chem. Phys.* **1984,** *81,* 3606; e) DeFrees, D.J., McLean, A.D., Herbst, R. *Astrophys. J.* **1984,** *279,* 222.
19. Lan, J.Y., Schuster, G.B. *J. Amer. Chem. Soc.* **1985,** *107,* 6710.
20. Taft, R.W. *Prog. Phys. Org. Chem.* **1983,** 14, 247.
21. Scheiner, S.; Bigham, L.D. *J. Chem. Phys.* **1985,** *82,* 3316.
22. Scheiner, S.; Redfern, P.; Szczesniak, M.M. *J. Phys. Chem.* **1985,** *89,* 262.
23. Scheiner, S.; Hillenbrand, E.A. *J. Phys. Chem.* **1985,** *89,* 3053.
24. Scheiner, S. *Acc. Chem. Res.* **1985,** *18,* 174.
25. Cao, H.Z.; Allavena, M.; Tapia, O.; Evleth, E.M. *J. Phys. Chem.* **1985,** *89,* 1581.
26. Szczesniak, M.M.; Scheiner, S.; Bouteiller, Y. *J. Chem. Phys.* **1984,** *81,* 5024.
27. Latajka, Z.; Scheiner, S. *J. Chem. Phys.* **1985,** *82,* 4131.
28. Sannigrahi, A.B.; Peyerimhoff, S.D. *Chem. Phys. Lett.* **1984,** *112,* 267.
29. Sapse, A.M.; Jain, D.C. *Int. J. Quantum Chem.* **1985,** *27,* 281.
30. Yamabe, S.; Minato, T.; Sakamota, M.; Hirao, K. *Can. J. Chem.* **1985,** *63,* 2571.
31. Nguyen, M.T. *Chem. Phys. Lett.* **1985,** *117,* 571.
32. Welti, M.; Ha, T.-K.; Pretsch, E. *J. Chem. Phys.* **1985,** *83,* 2959.
33. Pullman, A.; Claverie, P.; Cluzan, M.-C. *Chem. Phys. Lett.* **1985,** *117,* 419.
34. Schwenke, D.W.; Truhlar, D.G. *J. Chem. Phys.* **1985,** *82,* 2418.
35. Gussoni, M.; Castiglioni, C.; Zerbi, G. *Chem. Phys. Lett.* **1985,** *117,* 263.
36. Ray, N.K.; Shibata, M.; Bolis, G.; Rein, R. *Int. J. Quantum Chem.* **1985,** *27,* 427.
37. Hess, B.A., Jr; Schaad, L.J. *J. Am. Chem. Soc.* **1983,** *105,* 7500.
38. Nagase, S.; Kudo, J.; Aoki, M. *J. Chem. Soc. Chem. Commun.* **1985,** 1121.
39. George, P.; Bock, C.W.; Trachtman, M. In press.
40. Hehre, W.J.; Pau, C.-F.; Headley, A.D.; Taft, R.W.; Topsom, R.D. *J. Amer. Chem. Soc.* **1986,** *108,* 1711.
41. Taft, R.W.; and Topsom, R.D. *Prog. Phys. Org. Chem.* In press.

# CUMULATIVE CONTENTS: VOLUMES 1 To 4

(Article title followed by volume number with chapter in parentheses)

# GENERAL INDEX

## A

## C

# S

Saturation Kinetics, in, 145, 153
    Enzymatic Reactions, 159
Segment, 194
Semi-Empirical Calculations, 236
Serine Protease,
    Active Site, 144–145
    Model, 145, 146, 152ff
Side-Chain Interaction, 196
$\sigma_a$, 261, 262
$\sigma_F$, 261–265
$\sigma_F$, Conformational Effects on, 263, 264
$\sigma_I$, 264
$\sigma_R$, 261
$\sigma_R^0$, 262, 264–267
$\sigma_R^+$, 262
$\sigma_R^-$, 262
$\sigma_X$, 261, 262, 264, 265
Sigma-Inductive Effects, 249
Single-Cis (Cis'), 191, 192, 205
Single-Trans (Trans'), 191, 192, 205, 207
Singlet Oxygen, 5, 25–26, 41
Singlet-Triplet Energy Difference, 69
Solvent Shells, 87
South Form (see Conformation)
Specific Base Catalysis, 156
Standard Molecular Fragments, 326
Stanton's Theorems,
    Application, 52 et seq (Chapter 2)
    Statement, 52–53
Stereoalphabet, 193, 194
Stereoscopic Display, 334
Steric Effect, 281, 298
Steric Hindrance, 85, 102
Steric Strain, 199, 205, 207, 221
Structure-Activity Relationships (see QSAR)
Structure Display, 332, 334
Substituent Constants,
    Definition of, 282
    Estimation of, 309
Substituent Effects, 236
Substituent Effect Parameters, 261–267
Substituents, Classification of, 277
Substituent-Substituent Interactions, 256–258
Structure Count (SC), 9–10, 15–18, 21, 24, 25, 27–28
Sulfur Dioxide, 31–32, 36
SYBYL Computer Program, 336

# T

Tetramethyl Ammonium Ion, 126–128
Theoretical Bracketing,
    Application, 52 et seq (Chapter 2)
    Definition, 52

Theoretical Calculations, 235, 327
    Semiempirical Quantum Mechanical Methods, 327
    Ab Initio Computations, 329
    Classical Mechanical Calculations, 329
    Dynamics Calculations, 330
    Reaction Pathways, Transition State Calculations, 330
Thiols, 144, 146
2-(2-Thiophenyl)-2-Hydroxybutyric Acid, Structure Representation, 321
Through-Conjugation, 258
Torsion Angle,
    Bends, 195, 199, 200
    Definition, 227
    Pro-Middle, 194, 200
    Pro-Single, 191
Transaminase, Model, 144
Transesterification, Kinetics of, by, Enzyme Models, 153ff
Trans Peptide Bond, 221, 222, 224
TRIBBLE Computer Program, 337
Turn, 190, 192, 194, 195, 197, 199

# U

Ultimate Carcinogen, 2
Upper Bounds (see Inequalities)
Urea, Cyclic, in Ammonium Binding, 150–152

# W

Water,
    Comparison with Ethers, 64
    Ionization Potential, 64
    Proton Affinity, 51, 64
Water Complexes,
    $(H_2O)_nH^+$, n = 2–6, 120–123
    $(H_2O)_nOH^-$, n = 1–4, 120–123
    $(H_2O)_nCH_3OH_2^+$, n = 1, 2, 128–135
    $CH_3^+$, 128–135
    $C_2H^-$, 117, 118
    $C_2H_5^+$, 128–135
    i-$C_3H_7^+$, 128–135
    t-$C_4H_9^+$, 128–135
    $CH_3-C_5H_5NH^+$, 113–115
    $C_5H_5NH^+$, 113–115
    $(CH_3)_nNH_{4-n}^+$, n = 1–3, 113–115, 117
    4-CN-$C_5H_5NH^+$, 113–115
    $F^-$, 106
    $HCOO^-$, 115, 116
    HF, 106
    $H_2O$, 106
    $H_3O^+$, 106, 112, 113, 117–123
    $NH_4^+$, 106, 112–115, 117–120